A Dictionary of
SCIENCE

J.L. Sharma M.Sc., Ph.D.
Department of Chemistry
Kirorimal College
University of Delhi
Delhi, India

P.L. Buldinì C.Chem., M.R.S.C.
Centro Radioelettrico Sperimentale
Guglielmo Marconi
Laboratorio Analesi Chimica Materiali
Bologna, Italy

CBS

CBS PUBLISHERS & DISTRIBUTORS PVT. LTD.

New Delhi • Bengaluru • Chennai • Kochi • Kolkata • Mumbai • Pune

ISBN: 81-239-0292-1

First Edition: 1987
Reprint: 1990, 1992
Second Edition: 1994
Reprint: 1998, 2000, 2002, 2003, 2005, 2006, 2007,
        2008, 2009, 2010, 2011, 2012, 2013, 2015

Published by:
Satish Kumar Jain for CBS Publishers & Distributors Pvt. Ltd.,
4819/XI Prahlad Street, 24 Ansari Road, Daryaganj, New Delhi - 110002
delhi@cbspd.com, cbspubs@airtelmail.in • www.cbspd.com
Ph.: 23289259, 23266861, 23266867 • Fax: 011-23243014

Corporate Office: 204 FIE, Industrial Area, Patparganj, Delhi - 110 092
Ph: 49344934 • Fax: 011-49344935
E-mail: publishing@cbspd.com • publicity@cbspd.com

Branches:
• *Bengaluru:* 2975, 17th Cross, K.R. Road, Bansankari 2nd Stage,
  Bengaluru - 70 • Ph: +91-80-26771678/79 • Fax: +91-80-26771680
  E-mail: cbsbng@gmail.com, bangalore@cbspd.com
• *Chennai:* No. 7, Subbaraya Street, Shenoy Nagar, Chennai - 600030
  Ph: +91-44-26681266, 26680620 • Fax: +91-44-42032115
  E-mail: chennai@cbspd.com
• *Kochi:* Ashana House, 39/1904, A.M. Thomas Road, Valanjambalam,
  Ernakulum, Kochi • Ph: +91-484-4059061-65
  Fax: +91-484-4059065 • E-mail: cochin@cbspd:com
• *Kolkata:* 6-B, Ground Floor, Rameshwar Shaw Road, Kolkata - 700014
  Ph: +91-33-22891126/7/8 • E-mail: kolkata@cbspd.com
• *Mumbai:* 83-C, Dr. E. Moses Road, Worli, Mumbai - 400018
  Ph: +91-9833017933, 022-24902340/41 • E-mail: mumbai@cbspd.com
• *Pune:* Bhuruk Prestige, Sr. No. 52/12/2+1+3/2,
  Narhe, Haveli (Near Katraj-Dehu Road Bypass), Pune - 411041
  Ph: +91-20-64704058/59, 32342277 • E-mail: pune@cbspd.com

Representatives:

• Hyderabad: 0-9885175004   • Nagpur: 0-9021734563
• Patna: 0-9334159340   • Vijayawada: 0-9000660880

Printed at:
J.S. Offset Printers, Delhi

Preface

This dictionary of science has been compiled as a clear, convenient, and immediately accessible reference book. Every aspect of its conception and layout has been planned to help the reader to find the scientific information he needs with minimum difficulty. Most of the terms come from a wide range of disciplines-particularly chemistry, physics, computers, life sciences, earth sciences, and numerous branches of engineering and technology. While compiling this dictionary, emphasis has been placed on basic scientific principles, phenomena and processes, all major classifications, materials and their utilities, and important definitions. Restraint must prevail, particularly in a dictionary such as this of modest proportions; the terms included have been selected in a somewhat pragmatic manner, partly on the basis of those most frequently encountered by the authors within the context of everyday science. Comments by the users which might be useful in considering the later revisions of, and additions to, this dictionary will be welcomed and appreciated.

J.L. SHARMA
P.L. BULDINI

ACKNOWLEDGMENTS

Donatella Ferri
Presidio Multizonale Prevenzione
Bologna, Italy

Dr. P. Saxena
Senior Scientific Officer
Ministry of Non-Conventional Energy
Government of India, India

Prof. P.C. Mathur
Dean, Interdisciplinary Applied Science
University of Delhi, Delhi

Dr. Rakesh Parashar
Department of Chemistry
Kirorimal College
University of Delhi, Delhi

Dr. V.L. Pandit
Matrayi College
University of Delhi, Delhi

Dr. Neena Shireesh
Shyam Lal College
University of Delhi, Delhi

Dr. A. Dhar
Senior Scientist
National Physical Laboratory
New Delhi

Shikha Sharma
S.S.S. School
Nizamuddin East, Delhi

Silvano Cavali
Dionex Corpn. Srl,
Milan, Italy

Prof. Ishwar Singh
Department of Chemistry
M. Dayanand University
Rohtak, India

Dr. O.P. Sharma
Shaheed Bhagat Singh College
University of Delhi, Delhi

Dr. Kshama Bhardwaj
Gargi College
University of Delhi, Delhi

Dr. Nitin Kohli
Department of Chemistry
Kirorimal College
University of Delhi, Delhi

Dr. V. Kumar
Department of Chemistry
Kirorimal College
University of Delhi, Delhi

Dr. Vibha Saxena
Department of Chemistry
Sri Venkateshwara College
University of Delhi, Delhi

A

A Abbreviation for ampere(s).

A1-time [Astronomy] An atomic time scale, established by the U.S. Naval Observatory, with the unit (second) equal to 9, 192, 631, 770 cycles of caesium at zero field.

aa-channel [Geology] A narrow, sinuous channel in which a lava river moves down and away from a central vent to feed an aa flow.

aa-lava [Geology] Type of lava with a rough, fragmental surface.

ab- [Electricity] A prefix attached to the names of practical electric units (*e.g.* ampere, volt) indicating the corresponding unit in the electromagnetic system (*e.g.*, abampere, abvolt).

1 abampere=10 amperes

1 abvolt=10^{-8} volt

1 abohm=10^{-9} ohm

abacus [Mathematics] An instrument for performing arithmetical calculations manually by sliding markers on rods or in grooves.

abampre [Electronics] The unit of electric current in electromagnetic centimetre-gram-second system; 1 abampere=10 amperes.

abandon [Engineering] to stop drilling and remove the drill rig from the site of a borehole before the intended depth or target is reached.

abandoned well [Engineering] An oil or gas well whose yield has fallen below that necessary for profit.

abasia [Medicine] Lack of muscular coordination in walking.

A battery [Electricity] American term for low tension battery, the battery, that supplies power for filaments or heaters of electron tubes in battery operated equipment.

abb [Material] A coarse wool from the fleece areas of lesser quality.

Abbe condenser [Optics] An optical condenser used in microscopes. It consists of a variable large-aperture lens system arranged substage to image a light source into the focal plane of a microscope objective.

Abbe refractometer [Optics] An optical instrument for the measurement of the refractive index of liquids.

ABC [Optics] Abbreviation for automatic brightness control.

aberration [Physics] Distortion of an image by a lens or mirror. [Astronomy] A variation in the apparent position of a star or other heavenly body, due to the motion of the observer with the earth.

aberration, chromatic [Optics] The formation, by a lens, of an image with coloured fringes, due to the refractive index of glass being dfferent for light of different colours. The light is thus dispersed into a coloured band. The effect is corrected by the use of achromatic lenses.

aberration, spherical [Optics] The distortion of the image produced by a lens or mirror due to different rays from any one point of the object making different angles with the line joining that point to the optical centre of the lens or mirror and coming to a focus in slightly different positions, so that a point is not imaged as a point, a stright line as straight, or an angle as an equal angle.

abfard [Electricity] A unit of capacitance in the electromagnetic centimetre-gram-second system of units equal to 10^9 farads. Abbreviated as aF.

abiogenesis [Biology] The hypothetical process by which living organisms are created from non-living organic matter. Also known as autogenesis; spontaneous generation.

abiotic [Biology] Referring to the absence of living organisms.

abiotic environment [Biology] All physical and non-living chemical factors such as soil, atmosphere and water, which influence living organisms.

ablation [Chemistry] The rapid removal of heat from a substance by pyrolysis of a material of low thermal conductivity. [Geology] The wearing away of rocks, as by erosion or weathering. [Hydrology] The reduction in volume of

a glacier due to melting and evaporation. [Medicine] The removal of tissue or a part of the body by surgery.

abort [Engineering] To cut short or break off an action, operation, or procedure with an aircraft, space vehicle because of equipment failure.

abrasion [Chemistry] Gradual erosion of the surface of a material both by physical forces and by chemical degradation.

abrasive [Engineering] A substance used for grinding, rubbing, honing, lapping or superfinishing surfaces.

abscess [Biology] A localized collection of pus surrounded by inflamed tissue.

abscissa [Mathematics] of a point P.in analytical geometry, the portion of the x-axis lying between the origin and a point where the line through P parallel to the y-axis cuts the x-axis.

absolute [Physics] Not relative; independent. e.g., absolute zero of temperature, as distinct from zero on an arbitrary scale such as the Celsius temperature scale.

absolute alcohol [Chemistry] Ethanol containing more than 99% pure ethanol by weight.

absolute code {Computers] Program code in a form suitable for direct execution by the central processor, i.e., code containing no symbolic references.

absolute expansion [Physics] The true expansion, not relative to the containing vessel. The coefficient of absolute expansion is equal to the sum of the coefficient of relative or apparent expansion of the liquid and the coefficient of volume expansion of the containing vessel.

absolute humidity [Physics] The amount of water vapour present in the atmosphere, defined in terms of the numbers of kilograms (or grams) of water in one cubic metre of air. See also relative humidity.

absolute motion [Physics] Motion relative to a point fixed on the earth or to an apparently fixed celestial point.

absolute specific gravity [Physics] The ratio of the weight of a given volume of a substance in a vacuum at a given temperature to the weight of an equal volume of water in a vacuum at a given temperature.

absolute temperature [Physiology] The temperature in Celsius degrees relative to the absolute zero at 273.16. (the Kelvin scale) or in Fahrenheit degrees relative to the absolute zero at -459.69°F (the Rankine scale).

absolute temperatrue scale [Physics] A scale with which temperatures are measured relative to absolute zero.

absolute units [Physics] 1. A set of electrical units based on the C.G.S. system; *e.g.*, the abvolt which is 10^{-9} practical volts. 2. Any system of units using the least possible number of fundamental units. *See* SI units, coherent units.

absolute zero The lowest temperature theoretically possible; the zero of thermodynamic temperature. 0K=-273.16°C. =-459.69°F.

absorbed dose [Nucleonics] The amount of energy imparted by ionizing particles to a unit mass of irradiated material at a point of interest.

absorptance [Physics] The ratio of the flux absorbed by a body to the flux falling on it. The absorptance of a black body is taken as 1.

absorption coefficient [Physics] a. 1. The ratio of the sound energy absorbed at a boundry to the sound energy falling on it. 2. See linear absorption coefficient; linear attenuation coefficient.

absorption edge [Physics] The *x*-ray wavelength at which a discontinuity appears in the intensity of an X-ray absorption spectrum. also known as absorption limit.

absorption of gases [Chemistry] The solution of gases in liquids. Sometimes also applied to the absorption of gases by solids when the gas permeates the whole body of the solid rather than its surface. *Compare* adsorption.

absorption oil [Material] A petroleum or coal-tar oil that is contacted with a vapour or gas mixture to remove heavy components, as in the recovery of natural gasoline from wet natural gas.

absorption of radiation [Physics] Radiant energy is partly reflected, partly transmitted and partly absorbed by the surface upon which it falls, the absorption being accompanied by a rise in temperature of the absorbing body. Dull black surfaces absorb the greatest proportion of the incident energy, and brightly polished (reflecting) surfaces the least. Surfaces which are the best absorbers are also the best radiators. See absorptance; absorption coefficient.

absorption spectrum [Physics] A spectrum consisting of dark lines or bands obtained when the light from a source, itself giving a continous spectrum, is passed through a gas into a spectroscope. The dark lines or bands will occur in the same position as the coloured lines in that substance's emission spectrum and will be characteristic of the substance. When the absorbing medium is in the solid or liquid state the spectrum of the transmitted light shows broad dark regions, which are not resolvable into sharp lines. Characteristic X-ray and ultra-violet absorption spectra are also formed.

absorptivity [Physics] The fraction of the radiant energy incident on the surface that is absorbed. This has now been replaced by absorptance.

abstraction [Computers] The principle of ignoring those aspects of a subject that are not relevant to the current purpose in order to concentrate solely on those that are.

abundance [Geology] 1. The ratio of the total mass of a specfied element in the earth's crust to the total mass of the earth's crust, often expressed as a percentage. For example, the abundance of aluminium in the earth's crust is about 8%. [Nucleonics] The ratio of the number of atoms of a particular isotope of an element to the total number of atoms of all the isotopes present, often expressed as a percentage. For example, the abundance of uranium-235 in natural uranium is 0.71%. This is the natural abundance, i.e, the abundance as found in nature before any enrichment has taken place.

a.c. [Physics] see alternating current.

acaricide [Materials] Chemical that kills acarids, i.e., ticks and mites.

acceleration [Physics] The rate of change of velocity with respect to time.

acceleration of gravity [Physics] The acceleration of a body falling freely in a vacuum; varies slightly from place to place as a result of variations in the distance from the centre of mass of the earth.

Standard accepted value= 9.80665 metres per second per second. Also known as apparent gravity.

acceleration time [Computers] The time taken by a device to reach its operating speed from a quiescent state.

accelerator [Chemistry] A substance that increases the rate of a chemical reaction (*i.e.*,a catalyst), particularly in the manufacture of heavy chemicals rubber. [Physics] A machine for increasing the kinetic energy of charged particles (*e.g.*, protons, electrons, alpha particles) by accelerating them in electric fields. In electrostatic generators the acceleration is achieved directly by using a very high potential differene. In multiple accelerators a lower potential difference is used repeatedly to give the particle successive increments of energy. Multiple accelerators are classified as linear accelerators or cyclic accelerators. *See* cyclotron; synchroton; synchrocyclotron; betatron; bevatron; storage ring.

acclerometer [Physics] An instrument for measuring acceleration, especially the acceleration of an aircraft or rocket.

acceptor [Chemistry] 1. A species that accepts electrons, protons, or molecules. [Electronics] An imperfection in a semiconductor that causes hole conduction.

access [Computers] The reading of

data from storage or the writing of data into storage.

access arm [Computers] The mechanical device which positions the read/write head on a magnetic storage unit.

access-control register [Computers] A storage device which controls the word-by-word transmission over a given channel.

access time [Computers] The time period required by a computers store to provide information to the C.P.U. The access time for high-speed store is of the order of 1 microsecond; for backing storage it may be from a millisecond to some minutes.

accessory plate [Optics] Thin plate of quartz, mica, or gypsum used with a petrological microscope to modify the effect of polarized light and intensify qualities in translucent minerals.

acclimatization [Biology] Adaptation of a species or population to a changed environment over several generations. Also known as acclimation.

accounting machine [Computers] A machine that produces tabulations or accounting records of a specified unchanged format.

acceration [Geology] Gradual build-up of land on a shore due to wave action, tides, currents or alluvial deposits.

accumulator [Electricity] A type of voltaic cell or battery that can be recharged by passing a current through it from an external d.c. supply. The charging current, which is passed in the opposite direction to that in which the cell supplies current, reverses the chemical reactions in the cell. The common types are the led-acid accumulator and the nickel-iron accumulator.

acetals [Chemistry] Organic compounds formed by addition of alcohol molecules to aldehyde molecules. If one molecule of aldehyde (RCHO) reacts with one molecule of alcohol (R'OH) a hemiacetal is formed (RCH(OH)OR'). The rings of aldose sugars are hemiacetals. Further reaction with a second alcohol molecule produces a full acetal ($RCH(OR')_2$). The formation of acetals is reversible; acetals can by hydrolysed back to aldehydes in acidic solutions. In synthetic organic chemistry , aldehyde groups are often converted into acetal groups to protect them before performing other reactions on different groups in the molecule. *See alo* ketals.

acetic fermentation [Biochemistry] Oxidation of ethyl alcohol to produce acetic acid by the action of bacteria of the *genus Acetobacter.*

acetolysis [Chemistry] The conversion of a group of atoms in an organic compound to an acetyl group by reacting the compound with glacial acetic acid.

acetone pyrolysis [Chemistry]

Thermal decomposition of acetone into ketene.

acetone sugar [Chemistry] Any reducing sugar that contains acetone, *e.g.,* 1, 2-monoacetone-D-glucofuranose.

acetyl [Chemistry] The univalent organic radical, CH_3CO-

acetylase. [Biochemistry] Any enzyme that catalyzes the formation of acetyl esters.

acetylation [Chemistry] The introduction of an acetyl group into an organic compound.

acetyl number [Chemistry] A measure of free hydroxyl groups in fats or oils; determined by the amount of potassium hydroxide required to neutralize the acetic acid formed by saponification of acetylated fat or oil.

achromatic lens [Optics] A lens free from chromatic aberration thus forming an image free from coloured frings; consists of a pair of lenes, one of crown glass, the other of flint glass, the latter correcting the dispersion caused by the former.

acid [Chemistry] 1. A class of compounds that contain hydrogen and dissociate in water to produce positive hydrogen ions. The reaction, for an acid HX, is commonly written :

$$HX = H^+ + X^-$$

In fact, the hydrogen ion (the proton) is solvated, and the complete reaction is:

$$HX + H_2O \rightleftharpoons H_3O^+ + X^-$$

The ion H_3O^+ is the oxonium ion (or hydroxonium ion or hydronium ion). This definition of acids comes from th Arrhenius theory. Such acids tend to be corrosive substances with a sharp taste, which turn litmus red and give colour changes with other indicators. They are referred to as protonic acids and are classified into strong acids, which are almost completely dissociated in water (e.g. sulphuric acid and hydrochloric acid), and weak acids, which are only partially dissociated (e.g. ethanoic acid and hydrogen sulphide). The strength of an acid depends on the extent to which it dissociates, and is measured by its dissociation constant. See also base.

2. In the Lowry-Brö nsted theory) of acids and bases (1923), the definition was extended to one in which an acid is a proton donor, and a base is a proton acceptor. for example, in the $H_2O + HCN \rightleftharpoons H_3O^+ + CN^-$ the HCN is an acid, in that it donates a proton to H_2O. The H_2O is acting as a base in accepting a proton. similarly, in the reverse reaction H_3O^+ is an acid and CN^- a base. In such reactions, two species related by loss or gain of a proton are said to be conjugate. Thus, in the reaction above HCN is the conjugate acid of the base CN^- and CN^- is the conjugate base of the acid HCN. Similarly,

H_3O^+ is the conjugate acid of the base H_2O. An equilibrium, such as that above, is a competition for protons between an acid and its conjugate base. A strong acid has a weak conjugate base, and vice versa. Under this definition water can act as both acid and base. Thus in $NH_3 + H_2O \rightleftharpoons NH_4^+ + OH^-$ the H_2O is the conjugate acid of OH^-. The definition also extends the idea of acid-base reaction to solvents other than water. For instance, liquid ammonia, like water, has a high dielectric constant and is a good ionizing solvent. Equilibria of the type $NH_3 + Na^+Cl^- = Na^+NH_2 + HCl$ can be studied, in which NH_3 and HCl are acids and NH_2^- and Cl^- are their conjugate bases.

3. A further extension of the idea of acids and bases was made in the Lewis theory (G. N. Lewis, 1923). In this, a Lewis acid is a compound , ion or atom that can accept a pair of electrons and a Lewis base is one that can donate an electron pair. This definition encompasses 'traditional' acid-base reactions. In

$$HCl + NaOH \rightarrow NaCl + H_2O$$

the reaction is essentially

$$H^+ + :OH^- \rightarrow H:OH$$

i.e. donation of an electron pair by OH^-. But it also includes reactions that do not involve ions, e.g. $H_3N: + BCl_3 \rightarrow H_3NBCl_3$ in which NH_3 is the base (donor) and BCl_3 the acid (acceptor). The Lewis theory establishes a relationship between acid-base reactions and oxidation-reduction reactions.

acid anhydrides [Chemistry] Compounds that react with water to form an acid. For example, carbon dioxide reacts with water to give carbonic acid:
$$CO_2(g)+H_2O(aq) \rightleftharpoons H_2CO_3(aq)$$
A particular group of acid anhydrides are anhydrides of carboxylic acids. They have a general formula of the type $R.CO.O.CO.R'$, where R and R' are alkyl or aryl groups.

acid bronze [Material] A copper-tin alloy containing varying amounts of lead and nickel; used in making pumping equipments.

acid dyes [Chemistry] Compounds in which the chromophore is part of a negative ion (usually an organic sulphonate RSO_2O^-). They can be used for protein fibres (*e.g.* wool and silk) and for polyamide and acrylic fibres. Originally, they were applied from an acidic bath. Metallized dyes are forms of acid dyes in which the negative ion contains a chelated metal atom. Basic dyes have chromophores that are part of a positive ion (usually an amine salt or ionized imino group). They are used for acrylic fibres and also for wool and silk, although they have only moderate fastness with these matrials.

acid halide [Chemistry] Also known as. An organic compound with the general formula RCOX, where R is a hydrocarbon group and X is a halogen atom. They are obtained from carboxylic acids by replacing the hydroxyl group with a halogen atom. They are used in halogenation.

acidic hydrogen [Chemistry] That portion of the hydrogen in an acid that is replaceable by metals to form salts.

acidimetry [Chemistry] Determination of the amount of acid present in a solution by titration. *See* volumetric analysis.

acidolysis [Chemistry] Hydrolysis by means of an acid.

acid salt [Chemistry] An acid in which only a part of the acid hydrogen has been replaced by a metal, *e.g.*, sodium hydrogen carbonate, $NaHCO_3$, sodium dihydrogen phosphate, NaH_2PO_4.

acid value [Chemistry] A measure of the free fatty acid present in a sample; the number of milligrams of potassium hydroxide required to neutralize the free fatty acids in one gram of the substance.

aclinic line *See* magnetic equator.

acoustic amplifier [Electronics] A device that amplifies mechanical vibrations directly at audio and ultrasonic frequencies.

acoustic delay [Physics] A delay which is deliberately introduced in sound reproduction by having the sound travel a certain distance before conversion into electric signals.

acoustic detector [Electronics] The stage in a receiver at which demodulation of a modulated radio wave into its audio component takes place.

acoustic energy [Physics] *See* sound energy

acoustic insulator [Materials] A material used to diminish sound energy that passes through it or strikes its surface.

acoustic lens [Materials] Certain materials shaped to refract sound waves in accordance with the principles of geometrical optics, as is done for light.

acoustics [Physics] The scientific study of the production, transmission, and effect of sound.

acoustic transformer [Physics] A device used to increase the efficiency of sound radiation, *e.g.*, megaphone.

acrylic resins [Chemistry] Class of plastics obtained by the polymerization of derivatives of acrylic acid. They are transparent, colourless, and thermoplastic.

ACTH [Biology] Adreno-corticotropic hormone. A polypeptide hormone secreted by the pituitary gland which controls the adrenal glands.

actinic radiation [Physics] Electromagnetic radiation that can cause photochemical reactions, especially radiation that can be used as a source of illumination in photography. It includes X-rays, infrared and ultraviolet radiation, as well as light.

actinides [Chemistry] A series of elements in the periodic table, generally considered to range in atomic number from thorium (90) to lawrencium (103) inclusive. The actinides all have two outer s-electrons (a $7s^2$ configuration), follow actinium, and are classified together by the fact that increasing proton number corresponds to filling of the $5f$ level. In fact, because the $5f$ and $6d$ levels are close in energy the filling of the $5f$ orbitals is not smooth. The outer electronic configurations are as follows:

89 actinium (Ac) $6d^17s^2$

90 thorium (Th) $6d^27s^2$

91 protactinium (Pa) $5f^26d^27s^2$

92 uranium (U) $5f^36d7s^2$

93 neptunium (Np) $5f^57s^2$ (or $5f^46d^17s^2$)

94 plutonium (Pu) $5f^67s^2$

95 americium (Am) $5f^77s^2$

96 curium (Cm) $5f^76d^1s^2$

97 berkelium (Bk) $5f^86d7s^2$ (or $5f^97s^2$)

98 californium (Cf) $5f^{10}7s^2$

99 einsteinium (Es) $5f^{11}7s^2$

100 fermium (Fm) $5f^{12}7s^2$

101 mendelevium (Md) $5f^{13}7s^2$

102 nobelium (Nb) $5f^{14}7s^2$

103 lawrencium (Lw) $5f^{14}6d^1s^2$

The first four members (Ac to U) occur naturally. All are radioactive and this makes investigation difficult because of self-heating, short lifetimes, safety precautions, etc. Like the lanthanoids, the actinides show a smooth decrease in atomic and ionic radius with increasing proton number. The lighter members of the series (up to americium) have f-electrons that can participate in bonding, unlike the lanthanoids. Consequently, these elements resemble the transition metals in forming coordination complexes and displaying variable valency. As a result of increased nuclear charge, the heavier members (curium to lawrencium) tend not to use their inner f-electrons in forming bonds and resemble the lanthanoids in forming compounds containing the M^{3+} ions. The reason for this is pulling of these inner electrons towards the centre of the atom by the increased nuclear charge. Note that actinium itself does not have a $5f$ electron, but it is usually classified with the actinides because of its chemical similarities. Also known as actinoids.

actinium [Chemistry] Ac. A silvery white radioactive metallic element belonging to group IIIB of the periodic table; A.No.89; mass number of most stable isotope 227 (half-life 21.7 years), m.p. 1323-40 K; b.p. 3573K (estimated). Actinium-227 occurs in natural uranium to an extent of about 0.715%. actinium-228 (half-life 6.14 hours) also occurs in nature. There are 22 other artificial isotopes, all radioactive and all with very short half-lives. Its chemistry is similar to that of lanthanum.

actinometer [Chemistry] Any instrument that measures the intensity of electromagnetic radiation, especially one that is based on fluorescence or a photographic process.

actinon [Nucleonics] Actinium emanation. a gaseous radioisotope of radon, $^{219}_{86}Rn$ produced by the disintegration of actinium. It is now known as radon-219.

action [Physics] The product of work and time. Planck's constant of action is measured in SI units of joule seconds.

activated alumina [Chemistry] Aluminium oxide which has been dehydrated in such a way that a porous structure of high surface area is obtained. Activated alumina has the power of absorbing water vapour and certain gaseous molecules, used for drying air and other gases.

activated carbon [Chemistry] Active charcoal. Carbon, especially charcoal characterized by very large surface area per unit volume, which has been treated to remove hydrocarbons and to increase its power of absor-ption, used in many industrial processes for recovering valuable materials out of gaseous mixtures, as a deordorant, and in gas masks.

activated cathode [Electronics]. A

thermionic cathode consisting of a tungsten filament to which thorium has been added and then brought to the surface by heating in absence of an electric field in order to increase thermionic emission.

activation [Physics] The process of inducing radioactivity. [Chemistry] Treatment of a substance by heat, radiation or any activating agent to produce a more rapid or complete chemical or physical change.

activation analysis [Chemistry] A sensitive analytical technique that can be used to detect the presence of several elemens in a sample by first activating it, usually by neutron bombardment in a nuclear reactor, and then examining the gamma-ray spectrum of decay products to detect characteristic emission lines.

activation energy [Chemistry] The minimum energy required for a chemical reaction to take place. In a reaction, the reactant molecules come together and chemical bonds are stretched, broken, and formed in forming the products. During this process the energy of the system increases to a maximum, then decreases to the energy of the products. The activation energy is the difference between the maximum energy and the energy of the reactants; *i.e.,* it is the energy barrier that has to be overcome for the reaction to proceed. The activation energy determines the

way in which the rate of the reaction varies with temperature (*see* Arrhenius equation). It is usually to expressed activation energies in joules per mole of reactants.

active region [Astronomy] A localized, transient, non-uniform region on the surface of the sun, penetrating well down into the lower chromosphere.

active satellite [Engineering] A satellite capable of transmitting signal.

active voltage [Electricity] Component of the voltage in an alternating circuit which is in phase with the current flowing in the circuit.

activity [Chemistry] *a.* A thermodynamic function used in place of concentration in equilibrium constants for reactions involving nonideal gases and solutions. For example, in a reaction

$$A \rightarrow B + C$$

the true equilibrium constant is given by

$$K = a_B . a_C / a_A$$

where a_A a_B, and a_C are the activities of the components, which function as concentrations (or pressures) corrected for nonideal behaviour. Activity coefficients (symbol γ) are defined for gases by $\gamma = a/p$ (where p is pressure) and for solutions by $\gamma = aX$ (where X is the mole fraction). Thus, the equilibrim constant of a gas reaction has the form

$$K_p = \gamma_B P_B \gamma_c Pc/\gamma_A P_A$$

The equilibrium constant of a reaction in solution is

$$K_c = \gamma_B X_B \gamma_c X_c/\gamma_A X_A$$

The activity coefficients thus act as correction factors for the pressures or concentrations. [Nucleonics] A. The number of atoms of a radioactive substance that disintegrate per unit time. the specific activity (a) is the activity per unit mass of a pure radioisotope.

acyclic [Chemistry] Describing a compound that does not have a ring in its molecules having an open-chain structure.

acylation [Chemistry] The introduction of an acyl group, –RCO, into a compound.

adapter [Computers] A device which converts bits of information received serially into parallel bit form for use in the inquiry buffer unit. [Electronics] Device by whose agency a plug of one size or type may be inserted into a socket of a different size or type. [Optics] An attachment to a camera that permits its use in a manner for which it was not designed.

adaptive system [Science Technology] A system that can change itself to meet new requirements.

addition reaction [Chemistry] A chemical reaction in which one molecule adds to another. Addition reactions occur with unsatu-

rated compounds containing double or triple bonds, and may be electrophilic or nucleophilic. An example of electrophilic addition is the reaction of hydrogen chloride with an alkene, e.g.

$$HCl + H_2C = CH_2 \rightarrow CH_3CH_2Cl$$

An example of nucleophilic addition is the addition of hydrogen cyanide across the carbonyl bond in aldehydes to form cyanohydrins. *Addition-elimination* reactions are ones in which the addition is followed by elimination of another molecule.

addition polymer [Chemistry] a polymer formed by the chain addition of unsaturated monomer molecules, such as olefins, with one another without the formation of a by product; e.g., polyethylene, polypropylene and polystyrene. also known as addition resin.

addition resin *See* addition polymer.

additive process [Optics] The process of forming any colour by a mixture of red, green, and blue lights. The colours add together to form a new colour, the colour obtained depending on the proportions of each additive primary color. Equal proportions give white light. *Compare* subtractive process.

address [Computers] The number or name that uniquely identifies a register, memory location, or storage device in a computer.

address format [Computers] A description of the number of addresses included in a computer instruction.

adduct [Chemistry] A compound formed by an addition reaction. The term is used particularly for compounds formed by coordination between a Lewis acid (acceptor) and a Lewis base (donor). *See* acid.

adhesion [Physics] Sticking to a surface. The effect is produce by forces between molecules.

adhesive [Materials] Any substance used for sticking surfaces together; *e.g.*, glues, cement, etc.

adiabatic [Physics] Referring to any change in which there is no loss or gain of heat.

adiabatic demagnetization [Physics] A technique for cooling a paramagnetic salt, such as potassium chrome alum, to a temperature near absolute zero. The salt is placed between the poles of an electromagnet and the heat produced during magnetization is removed by liquid helium. The salt is then isolated thermally from the surroundings and the field is switch off the salt is demagnetized adiabatically and its temperature falls. This is because the demagnetized state, being less ordered, involves more energy than the magnetized state. The extra energy can come only from the internal, or thermal, energy of the substance.

adiabatic process [Physics] **Any** process that occurs without heat entering or leaving a system. In general, an adiabatic change involves a fall or rise in temperature of the system. For example, if a gas expands under adiabatic conditions, its temperature falls (work is done against the retreating walls of the container). The adiabatic equation describes the relationship between the pressure (p) of an ideal gas and its volume (V), *i.e.* pV^γ, where γ is t he ratio of the principal specific heat capacities of the gas and K is a constant.

admittance [Electronics] Quantity which is the reciprocal of impedance, and is thus a measure of the facility with which a current can flow in a circuit under the pressure of a given electromotive force.

adolescence [Physiology] The period of life from puberty to maturity.

adsorbate [Chemistry] The substance that absorbs: Sillca gel , anhydrous alumina and many porous or powdered materials are effective absorbents by virtue of their large specific surface in conjunction with their ability to form bonds with adsorbates.

adsorption [Chemistry] The concentration of a substance on a surface; *e.g.*, molecules of a gas or dissolved or suspended substance on the surface of a solid. In chemisorption a single **layer** of atoms or molecules **of** the adsorbed substance **is held to** the solid surface by covalent bonds. In physisorption, several layers of atoms or molecules are held by van der Waals forces.

advection [Physics] The process in which either matter or energy is transferred from one place to another by a horizontal stream of gas.

aeon [Astronomy] A unit of time equal to billion (10^9) years.

aeration cell [Chemistry] An electrolytic cell whose electromotive force is due to electrodes of the same material located in different concentration of dissolved air. Also known as oxygen **cell**.

aerial [Electronics] The part of a radio system from which energy is transmitted into, or received from, space (or the atmosphere).

aerial sound ranging [Engineering] The process of locating an aircraft by means of the sound emitted by it.

aerobe [Biology] An organism that requires air or free oxygen to maintain its life processes.

aerodynamics [Physics] The study of the motion and control of solid bodies like aircraft, rockets, missiles, etc. in air. Also , the study of air or other gases in motion.

aerolites [Astronomy] Meteroites, especially those consisting of stony material other than iron.

aero metal [Material] A casting alloy consisting chiefly of aluminium, zinc, and copper.

aerosol [Chemistry] A dispersion of solid or liquid particles in a gas; e.g., smoke.

aerospace [Engineering] The earth's atmosphere and the space beyond.

affiniy [Chemistry] Chemical attraction; the force binding atoms together.

afterburning [Engineering] 1. The combustion that results from the addition of fuel to the exhaust of a jet engine in order to increase thrust and reduce fuel consumption. 2. The irregular burning of residual propallant in a rocket motor when the main combustion has finished.

after-damp [Materials] A poisonous mixture of gases, containing carbon monoxide, formed by the explosion of fire-damp (methane CH_4) in coal mines.

after-glow [Astronomy] A glow sometimes observed high in the western sky after sunset; caused by fine dust particles in the upper atmosphere scattering the light coming from the sun.

after-heat [Nucleonics] Heat derived from residual radioactivity after a reactor has been shut down.

agamogony [Biology] A sexual reproduction.

agar [Material] A gelatinous material obtained from certain seaweeds; it is chemically related to the carbohydrates. a solution in hot water sets to a firm jelly, which is used as a base for culture media for growing bacteria.

agate [Materials] a very hard natural form of silica, used for knife-edges of balances, for mortars, for grinding hard materials, and in ornaments.

aggressive carbon dioxide [Chemisty] The carbon dioxide dissolved in water in excess of the amount required to precipitate a certain concentration of calcium ions as calcium carbonate; used as a measure of the corrosivity.

aggressive water [Geology] Any of the water which force their way into place.

agitator [Engineering] A mixing, stirring or shaking device that keeps liquids, and solids in liquids, in motion.

aglycone [Chemistry] A non-sugar component of a glycoside.

agonic line [Geography] A line of zero magnetic declination.

agricultural sciene [Agriculture] The scientific study dealing with the selection, breeding, and management of crops and domestic animals for more economical production.

agronomy [Agriculture] The principles and procedures of soil management and of field crop and special-purpose plant improvement, management, and production.

air conditioning [Engineering] The maintenance of certain aspects of environment within a limited space to facilitate the function of that space; the aspects controlled include air temperature and motion, radiant heat level, moisture, and concentration of pollutants.

aircraft [Engineering] A vehicle designed to be supported by the air, either by dynamic action of the air upon the surfaces of the structure or by its own buoyancy. Also kown as air vehicle.

aircraft rocket [Engineering] A rocket missile designed to be launched from an aircraft.

air engine [Engineering] An engine in which compressed air is the actuating fluid.

airflow orifice [Engineering] An opening through which air moves out of an enclosed space.

albedo [Optics] The ratio of the radiant flux falling on a surface to that reflected by it. [Nucleonics] The probability that a neutron entering a material will be reflected by that material through the surface by which it entered.

albumens, [Biochemistry] A group of soluble proteins occurring in many animals tissues and fluids e.g., egg-white (egg albumen), milk (lactalbumen), and blood (serum albumen).

alchemy [Chemistry] The predecessor of scientific chemistry. An art by which its devotees sought, with the aid of a mixture of mysticism, astrology, practical chemistry, quackery, to transmute base metals into gold, prolong human life, etc. it flourished from about A.D. 500 until the middle ages, when it gradually fell into disrepute.

alcohol [Chemistry] Any of a class of organic compounds derived from the hydrocarbons, one or more hydrogen atoms in molecules of the latter being replaced by hydroxyl groups, OH– e.g., ethanol (ordinary 'alcohol') is C_2H_5OH; theoretically derived from ethane, C_2H_6. Alcohols that contain more than one hydroxyl group are called polyhydric alcohols. *See also* diols and triols.

alcoholometry [Chemistry] The determination of the proportion of ethanol in spirits and other solutions; usually performed by measuring the relative density of the liquid at a standard temperature by a specially graduated hydrometer.

alcosol [Chemistry] The mixture of an alcohol and a colloid.

aldehyde [Chemistry] Any of a class of organic compounds of the type R.CO.H where R is an alkyl or aryl group.

aldose [Chemistry] A monosaccharide containing an aldehyde (formyl) group in the molecule.

algebraic sum [Mathematics] The total of a number of quantities of the same kind, with due consideration of the sign. thus the algebraic sum of 4,-5, and -6 is -7.

algology [Botany] The scientific study of algae.

algorithm [Mathematics] A systematic mathematical procedure that enables a problem to be solved in a fnite number of steps. Problems for which no algorithms exist require heuristic solutions. Also known as algorism.

alicyclic compound [Chemistry] A type of organic compound that is essentially aliphatic, although it contains a ,saturated ring of carbon atoms.

alidade [Engineering] An instrument used for measuring vertical heights and distances.

aliphatic compounds [Chemistry] Organic compounds containing open chains of carbon atoms, in contradistinction to the closed rings of carbon atoms of the organic compounds; comprise the paraffins, the olefins and the acetylenes as well as all their derivatives and substitution products.

aliquot part [Mathematics] A divisor of a number or quantity that will give an integer. Thus 4 is an aliquot part of 8, but 7 is not.

alkalemia [Biology] An increase in the blood pH above normal levels.

alkali [Chemistry] A soluble hydroxide of a metal, particularly of one of the alkali metals; term is often applied to any substance that has alkaline reaction in solution.

alkali metals [Chemistry] The univalent metals lithium, sodium, potassium, rubidium, and caesium, belonging to Group 1A of the periodic table.

alkalimetry [Chemistry] The determination of the amount of alkali present in a solution by titration. *See* volumetric analysis.

alkaline earth metals [Chemistry] The bivalent group of metals comprising beryllium, magnesium, calcium, strontium, barium, and radium, belonging to Group IIA of the periodic table.

alkaloids [Chemistry] A group of basic organic substances of plant origin, containing at least one nitrogen atom in the ring structure of the molecule. Many have important physiological actions and are used in medicine, *e.g.*, codeine, cocaine, nicotine, quinine, morphine.

alkanes [Chemistry] A homologous series of saturated hydrocarbons having the general formula C_nH_{2n-2}. They are chemically inert, stable, and inflammable.

alkanization [Chemistry] The process of converting an unsaturated hydrocarbon into an alkane.

alkenes [Chemistry] A homologous series of unsaturated hydrocarbons having the general formula C_nH_{2n}. Also known as olefins.

alkoxy [Chemistry] A general name for univalent organic radicals having the formula RO, where R is an alkyl group.

alkyl [Chemistry] A general name for univalent saturated hydrocar-

bon radicals having the general formula C_nH_{2n+1}, derived from alkanes; *e.g.*, methyl CH_3^-; ethyl, $C_2H_5^-$.

alkylarne [Chemistry] An arene (*e.g.*, benzene) with one or more hydrogen atoms in the molecule replaced by alkyl group; *e.g.*, methylbenzene, $CH_3C_6H_5$.

alkylation [Chemistry] The introduction of an alkyl group into a molecule; e.g., the addition of alkenes to form alkanes.

alkynes [Chemistry] A homologous series of unsaturated hydrocarbons having the general formula C_nH_{2n-2} and containing a triple bond between two of the carbon atoms in the molecule; *e.g.*, acetylene, $HC \equiv CH$.

allo- [Chemistry] Prefix meaning 'other' used in chemistry to denote a variation from the standard or normal form.

allobar [Chemistry] a mixture of the isotopes of an element that does not occur naturally.

allocation [Computers] The process of specifying computer memory locations during the execution of a program.

allochromatic crystal [Crystallography] A crystal having photoconductive properties due to the presecne of small particles within it.

allomerism [Crystallography] A similarity in the crystalline structure of substances of differemt chemical composition.

allomorphism [Crystallography] A variability in the crystalline structure of certain substances. Allomorphs are different crystalline forms of the same compound.

allopathy [Medicine] A system of medicine that employs remedies whose effects are unlike those of the disease, in contrast to homeopathy.

allotropes [Chemistry] The existance of a chemical element in two or more forms differing in physical properties but giving rise to identical chemical compounds, e.g., sulphur exists in a number of different allotropic forms.

allowed bands [Chemistry] *See* energy bands.

alloy [Material] A composition of two or more metals; an alloy may be a compound of the metals, a solid solution of them, a heterogeneousmixture, or any combination of these. The term is sometimes extended to include non-metallic components, *e.g.,* iron-carbon alloys.

alloy plating [Metallurgy] The process of co-deposition of two or more metals on any surface by electrolysis.

alloy steel [Material] A steel whose distinctive properties are due to the presence of one or more elements other than carbon.

alluvial [Geology] Deposited by rivers.

allyl group [Chemistry] The univalent radical, $CH_2=CHCH_2^-$ derived from propylene.

allyl resin [Chemistry] Synthetic resin formed by the polymerization of chemical compounds containing the allyl group. Also known as allyl plastic.

alpha decay [Nucleonics] A form of radioactive decay in which a nucleus spontaneously emits an alpha particle.

alpha iron [Chemistry] Allotropic form of pure iron that exists upto 1183 K.

alpha particle [Physics] A positively charged particle consisting of two neutrons and two protons, identical with the nucleus of helium atom; emitted by several radioactive elements.

alpha rays [Physics] Streams of fast-moving alpha particles. Alpha rays produce intense ionization in gases through which they pass; are easily absorbed by matter; and produce fluorescene on a fluorescent screen.

altazimuth [Engineering] An instrument for the measurement of the altitude and azimuth of heavenly bodies. Also known as universal instrument.

alternating current [Electricity] *a.c.* A flow of electricity that, after reaching a maximum in one direction, decreases, finally reversing and reaching a maximum in the opposite direction, the cycle being repeated continuously.

alternator [Electricity] A mechanical, electrical, or electromechanical device which supplies alternating current.

altimeter [Engineering] An instrument used to measure height above sea-level. It usually consists of an aneroid barometer calibrated to read zero at sea-level and the height above sea-level in metres .

altitude [Astronomy] 1. Height, 2. The altitude of a heavenly body is its angular distance from the horizon on the vertical circle passing through the body, the zenith, and the nadir.

aluminate [Chemistry] A salt formed when aluminium hydroxide or γ · alumina is dissolved in solutions of strong bases, such as sodium hydroxide. Aluminates exist in solutions containing the aluminate ion, commonly written $[Al(OH)_4]$. In fact the ion probably is a complex hydrated ion and can be regarded as formed from a hydrated Al^{3+} ion by removal of four hydrogen ions. $[Al(H_2O)_6]^{3+} + 4OH^- \rightleftharpoons 4H_2O + [Al(OH)_4(HO)_2]^-$

Other aluminates and polyaluminates, such as $[Al(OH)_6]^{3+}$ and $(HO)_3AlOAl(OH)_3]^{2-}$ are also present.

aluminium [Chemistry] Al. A silvery-white lustrous metallic element belonging to group III of the periodic table; A No. 13; r.d.

2.7; m.p. 933 K; b.p. 2740 K. The metal itself is highly reactive but is protected by a thin transparent layer of the oxide, which forms quickly in air. Aluminium and its oxide are amphoteric. The metal is extracted from purified bauxite (Al_2O_3) by electrolysis; the main process uses a Hall-Heroult cell but other electrolytic methods are under development, including conversion of bauxite with chlorine and electrolysis of the molten chloride. Pure aluminium is soft and ductile but its strength can be increased by work-hardening. A large number of alloys are manufactured; alloying elements include copper, manganese, silicon, zinc, and magnesium. Its lightness, strength (when alloyed), corrosion resistance, and electrical conductivity (62% of that of copper) make it suitable for a variety of uses, including vehicle and aircraft construction, building (window and door frames), and overhead power cables.

alums [Chemistry] A group of double salts with the formula $A_2SO_4.B_2(SO_4)_3.24H_2O$ where A is a monovalent metal and B a trivalent metal. The original example contains potassium and aluminium (called *potash alum* or simply *alum*); its formula is often writen $AlK(SO_4)_2. 12H_2O$ (aluminium potassium sulphate-12-water). Ammonium alum is

$AlNH_4(SO_4)_2 . 12H_2O$, chrome alum is $KCr(SO_4)_2 . 12H_2O$. The alums are isomorphous and can be made by dissolving equivalent amounts of the two salts in water and recrystallizing.

aluminium brass [Metallurgy] Brass to which small amounts of aluminium has been added as a flux to improve the casting qualities, and with the addition of lead, the machining qualities.

aluminium bronze [Metallurgy] An alloy of copper-aluminium which may also contain manganese, iron, nickel, or zinc.

aluminium paint [Material] A mixture of aluminium pigment and oil varnish; reflects the sun's radiation well and retains well the heat in hot-air or hot water pipes or tanks.

aluminium paste [Material] Aluminium powder finely ground in oil; used in aluminium paints.

aluminium soap [Chemistry] Any salt of higher carboxylic acid and aluminium which is insoluble in water and soluble in oils; used in paints, varnishes, lubricating grease and waterproofing substances.

aluminium solder [Metallurgy] A solder containing upto 15% aluminium, having melting point above that of the tin-lead solders.

aluminize [Engineering] 1. To apply a film of aluminium to a material such as glass. 2. To form a protective surface alloy on a metal by treatment at elevated temperature with aluminium or an aluminium compound.

aluminosilicates [Chemistry] A large class of minerals, both natural and synthetic, containing aluminium and silicon combined with oxygen in their structure. It includes clays, zeolites, micas, and many other important mineral materials.

aluminothermic reduction [Chemistry] High-temperature reduction of metal oxides to the corresponding metals by the thermite method.

amalgam [Chemistry] An alloy of mercury with one or more other metals. Most metals form amalgams (iron and platinum are exceptions), which may be liquid or solid. Some contain definite intermetallic compounds, such as $NaHg_x$.

amatol [Materials] An explosive mixture of 80% ammonium nitrate and 20% Trinitrotoluene [T.N.T.]

amaurosis [Biology] Total or partial blindness.

amber [Materials] 1. A fossil resin, derived from an extinct ; species of pine obtained from mines in East Prussia, and found on seashores. 2. A yellow to brown solid, which contains succinic acid, used for ornamental purposes. Also known as succinite.

ambergris [Material] A grey or

waxy material that occurs (probably as the result of disease) in the intestines of the sperm whale, used in the manufacture of perfumes.

amber oil [Material] A yellowish to brown essential oil prepared by destructive distillation of amber or rosin.

ambient light [Optics] The surrounding light, such as that reaching a television picture tube screen from light sources in a room.

ambeba [Biology] The coomon name for a number of species of unicellular protozoans of the order of Amoebida.

amebiasis [Medicine] A parasitic disease of man caused by the ameba Entamoeba histolytica. Alos known as amebic dysentery.

amebicide [Material] A chemical used to kill parasitic amebas.

americium [Chemistry] Am. Transuranic element. A.No. 95. Radioactive. A member of the actinide series. Most stable isotope, Am, has half life of about 8.0×10^3 years; m.p. 1267 K, b.p. 2880 K.

amethyst [Minerals] A violet variety of quartz; impure crystalline silica, SiO_2.

amidases [Biochemistry] Enzymes that control hydrolysis of amides.

amide [Chemistry] A class of organic compound formed by replacing hydrogen atoms of ammonia, NH_3, by acyl radicals, *e.g.*, acetamide, CH_3CONH_2.

The general formula is $RCONH_2$, where $CONH_2$ is the amide group.

aminases [Biochemistry] Enzymes that catalyze the hydrolysis of amines.

amination [Chemistry] The conversion of an aldehyde or ketone into an amine, by reacting them with hydrogen and ammonia in the presence of a catalyst.

amines [Chemistry] Compounds formed by replacing hydrogen atoms of ammonia, NH_3 by organic radicals; classified into primary amines of the type NH_2R; secondary; NHR_2 and tertiary; NR_3.

amino acid [Biochemistry] A carboxylic acid that contains the amino group $-NH_2$. These acids are the units that link together into polypeptide chains to form proteins; they are therefore of fundamental importance to life. Several important amino acids occur in nature, all of which have the general formula: $R-CH-NH_2-COOH$. Only 20 of the more than 80 amino acids found in nature serve as building blocks for proteins; examples are lysine and tyrosine. 'Essential' amino acids are those that an organism is unable to synthesize and therefore has to obtain from its environment. There are eight 'essential' amino acids for man.

amino group [Chemistry] The univalent group $-NH_2$.

aminoplastic resins [Materials] Synthetic resins derived from the reaction of urea, melamine or allied amino compounds with aldehydes. They form the basis of thermosetting moulding materials.

amitosis [Biology] Cell division by simple fission of the nucleus and cytoplasm without chromosome differentiation.

ameter [Physics] An instrument for the measurement of electric current. In moving iron ammeters, a strip of soft iron is caused to move in the magnetic field set up by the current flowing through a coil; for the measurement of direct current, the more accurate moving coil instruments contain a permanent magnet between the poles of which is pivoted a coil carrying the current to be measured. In each type of instrument a pointer attatched to the moving portion moves over a scale graduated in amperes.

ammine [Chemistry] Any coordination compound containing ammonia molecules as ligands; complex compound formed by ammonia with salt or base.

ammonal [Material] A high-explosive mixture of ammonium nitrate, $NH_4 NO_3$, powdered aluminium, and trinitrotoluene (TNT).

ammonification [Chemistry] Addition of ammonia or its compounds to the soil.

ammonolysis [Chemistry] A chemical reaction in which one group of an organic compound is converted to an amine group, by reacting the compound with ammonia.

ammunition [Engineering] All kinds of missiles to be thrown against an energy.

amorphous [Physics] Non-crystalline; having neither definite form nor shape.

amorphous sky [Physics] A sky characterized by an abundance of fractus clouds, usually accompanied by rain falling from a higher, over city cloud layer.

ampere [Electric] A. The unit of electric current. The constant current that, maintained in two straight parallel infinite conductors of negligible cross section placed one metre apart in a vacuum, would produce a force between the conductors of 2×10^{-7} Nm^{-1}. This definition replaced the earlier international ampere defined as the current required to deposit .0011800 gram of silver from a solution of silver nitrate in one second.

ampere-hour [Electric] A practical unit of electric charge equal to the charge flowing in one hour through a conductor passing on ampere. It is equal to 3600 coulombs.

Ampere's law [Electricity] 1. The strength of the magnetic field induced by current flowing through a conductor is, at any point, directly proportional to the product of the current and the length of the conductor and inversely proportional to the square of the distance between the point and the conductor. The direction of the field is perpendicular to the plane joining the point and the conductor. The force between two parallel current carrying conductors in free space is given by:

$$dF = \mu_0 I_1 ds_1 I_2 ds_2 \sin\theta / 4\pi r^2$$

where I_1 and I_2 are currents, ds_1 and ds_2 the incremental lengths, r the distance between the incrementals lengths θ the angle, and μ_0 permeability of free space. also known as Ampere-Laplace law.

ampere-turn [Electricity] A measure of magnetomotive force. The product of the number of turns in a coil and the current in amperes which flows through it.

amperometry [Chemistry] Chemical analysis by techniques which involve the measurement of electric currents.

amphiboles [Minerals] A large group of rock-forming metasilicate minerals. They have a structure of silicate tetrahedra linked to form double endless chains, in contrast to the single chains of the pyroxenes, to which they are closely related. They are present in many igneous and metamorphic rocks. The amphiboles show a wide range of compositional variation but con-

form to the general formula X_2._$_3Y_5Z_8O_{22}(OH)_2$, where X = Ca, Na, K, Mg, or Fe^{2+}, Y = Mg, Fe^{2+}, Fe^{2+}, Al, Ti, or Mn; and Z = Si or Al.

The hydroxyl ions may be replaced by F, Cl, or O. Most amphiboles are monoclinic, including cummingtonite, $(Mg, Fe^{2+})_7 (Si_8O_{22}) (OH)_2$; tremolite, $Ca_2Mg_5 (Si_8O_{22}) (OH,F)_2$; actinolite, $Ca_2(Mg,Fe^{2+})_5 (Si_8O_{22}) (OH,F)_2$; hornblende, $NaCa_2 (Mg, Fe^{2+}, Fe^{3+}, Al)_5 (Si,Al)_8O_{22})(OH,F)_2$; edenite, $NaCa_2 (Mg, Fe^{2+})_5 (Si_7AlO_{22}) (OH,F)_2$; and riebeckite, $(Si_8O_{22})(OH,F)_2$. Anthophyllite, $(Mg, Fe^{2+})_7(Si_8O_{22})(OH,F)_2$, and gedrite, $(Mg,Fe^{2+})_6 Al(Si,Al)_8O_{22}) (OH,F)_2$, are orthorhombic amphiboles.

amphichoric [Chemistry] Giving one colour on reaction with an acid and another colour on reaction with a base. Also known as amphichromatic.

amphiprotic [Chemistry] Capable both of accepting and of yielding protons in solution; amphoteric.

ampholyte [Chemistry] A substance that can act as either an acid in the presence of a strong base, or a base, in the presence of a strong acid.

amphoteric [Chemistry] Chemically reacting as acidic to strong bases and as basic towards strong acids, e.g., the amphoteric oxide, zinc oxide, gives rise to zinc salts with strong acids and zincates with the alkali metals.

amplifier [Electronics] An electronic device for producing an electrical output which is a function of the corresponding input parameter and increases the magnitude of the input by means of energy drawn from an external source.

amplitude [Physics] If any quantity is varying in an oscillatory manner about an equilibrium value, the maximum departure from that equilibrium value is called the amplitude; *e.g.*, in the case of a pendulum the amplitude is half the length of the swing; for a wave motion, *e.g.*, electromagnetic waves or sound waves, the amplitude of the wave determines the amount of energy carried by the wave.

amplitude distortion [Electronics] Variation in the ratio of output to input of a system as the input amplitude is varied.

amplitude filter [Electronics] A device embodying a tetrode or pentode to maintain an output of constant voltage even though the voltage of its input varies.

amplitude modulation [Electronics] One of the principal methods of transmitting informaion by radio waves. The amplitude of a carrier wave is modulated (*see* modulation) in accordance withe the frequency of the signal to be transmitted.

AMU [Physics] See atomic mass units.

amylase [Biochemistry] Any of a group of closely related enzymes that degrade starch, glycogen, and other polysaccharides. Plants contain both α and β-amylases; animals possess only α amylases, found in pancreatic juice and also (in humans and some other species) in saliva. Amylases cleave the long polysaccharide chains, producing a mixture of glucose and maltose.

anabolism [Biochemistry] The metabolic synthesis of proteins, fats, and other constituents of living organisms from molecules or simple precursors. This process requires energy in the form of ATP. *See* metabolism. *Compare* catabolism.

anaerobic [Biology] In the absence of free oxygen.

anaesthetic [Medicine] A substance used in medicine to produce insensibility or loss of feeling.

analgesic [Medicine] A substance used in medicine to relieve pain, *e.g.*, salicylates, morphine.

analog [Electronics] A physical variable that remains similar to another variable in so far as the proportional relationships are the same over some specified range; for example, a temperature may be represented by a voltage which is its analog.

analog computer [Computers] A computer in which numerical magnitudes are represented by physical quantities such as electric current, voltage, or resistance.

See also digital computer.

analysis [Chemistry] The determination of the components in a chemical sample. Qualitative analysis involves determining the nature of a unknown compound or the compounds present in a mixture. Various chemical tests exist for different elements or types of compound , and systematic analytical procedures can be used for mixtures. Quantitative analysis involves measuring the proportions of known components in a mixture. Chemical techniques for this fall into two main classes: volumetric analysis and gravimetric analysis. In addition, there are numerous physical methods of qualitative and quantitative analysis, including spectroscopic techniques, mass spectrometry, polarography, chromatography, activation analysis, etc.

analytic curve [Mathematics] A curve whose parametric equations are real analytic functions of the same real variable.

analytical geometry [Mathematics] A form of geometry based upon the use of coordinates to define positions in space. Also known as coordinate geometry.

anastigmatic lens [Optics] A lens designed to correct astigmatism and curvature of field.

anatomy [Biology] A branch of science dealing with the structure of the body by means of dissection.

androgen [Biochemistry] A class of steriod hormones produced in testis and adrenal cortex which act to regulate sexual characteristics.

anmo- [Science Technology] Prefix denoting the wind.

anemometer [Engineering] A device for measuring the speed of wind or any other moving gas.

aneroid [Engineering] without liquid.

aneroid barometer [Engineeing] An instrument for measuring atmospheric pressure; it consists of an exhausted metal box with a thin corrugated metal lid. Variations in atmospheric pressure cause changes in the displacement of the lid; this displacement is magnified and made to actuate a pointer moving over a scale by means of a system of delicate levers.

angiology [Medicine] The branch of medicine concerned with the study of blood vessels or blood.

angstrom [Science Technology] A unit of length equal to 10^{-10} metre. It was formerly used to measure wavelengths and intermolecular distances but has now been replaced by the nanometre. 1A=0.1 nanometre.

angular acceleration [Physics] The rate of change of angular velocity.

angular displacement [Physics] The angle through which a point, line, or body has been rotated in a specified direction, about specified axis.

angular distance [Physics] The distance between two bodies, measured in terms of the angle subtended by them at the point of observation; used in astronomy.

angular frequency [Physics] The frequency of a periodic process expressed in radians per second equal to 2π times the number of cycles per second. Also known as radian frequency.

angular length [Physics] A length expressed in the unit of the length per radian or degree of a specified wave.

angular momentum [Physics] The product of moment of inertia and angular velocity.

angular velocity [Physics] Rate of motion through an angle about an axis; measured in degrees, radians, or revolutions per unit time.

anhydride [Chemistry] A chemical compound derived from an acid or a base, by elimination of one or more molecules of water. Thus sulphur trioxide, SO_3; is an anhydride of sulphuric acid, H_2SO_4. The term should not be confused with anhydrous.

anhydrous [Chemistry] Descriptive of a compound (generally inorganic) that does not contain water either adsorbed on its surface or combined as water of crystallization.

anilide [Chemistry] An organic compound analogous to an amide but derived from an aromatic amine, especially aniline, *e.g.*,

$$C_6H_5NHCOC_6H_5.$$

aniline ink [Materials] A fast drying printing ink which is a solution of a coal-tar dye in an organic solvent.

animal black [Chemistry] Finely divided carbon obtained by calcination of animal bones or ivory; used in pigments, decolorizers, and purifying agents; varieties include ivory black and bone black.

animal charcoal [Chemistry] Material containing 10% carbon and 90% inorganic matter, chiefly calcium phosphate, $Ca_3 (PO_4)_2$; obtained by destructive distillation of animal matter at high temperatures. It is used as a decolorizing agent. Also known as bone black; bone char.

animal starch *See* glycogen.

anion [Chemistry] A negatively charged ion; an ion that is attracted towards the anode during electrolysis.

anisometric [Crystallography] Not isometric; applied to crystals that have axes of different lengths.

anisotrophic [Crystallography] Denoting a medium in which certain physical properties are different in different directions. Wood, for instance, is an anisotropic material; its strength along the grain difers from that perpendicular to the grain. Single crystals that are not cubic are anisotropic with respect to some physical properties, such as the transmission of electromagnetic radiation.

annealing [Engineering] Very slow regulated cooling, of metals, alloys, or glass to relieve strains set up during heating or other treatment, and to make the material less brittle.

annihilation radiation [Physics] The electromagnetic radiation arising from the collision, and subsequent annihilation, of a particle and its corresponding anti-particle. In the collision between an electron and a positron the annihilation radiation usually consists of two photons of γ-radiation emitted in opposite directions. The energy of the annihilation radiation is derived from the mass of the annihilated particles according to the mass-energy equation.

annual variation [Physics] A very small regular variation that the magnetic declination undergoes during the course of a year.

annular [Engineering] Ringed; an annular space is the space between a inner and outer ring.

annular eclipse [Astronomy] An eclipse of the sun in which a ring of its surface is visible surrounding the darkened moon.

anode [Electricity] The positive electrode of a primary cell or a storage battery. [Electric] The collector of electrons in an electron tube.

anode corrosion [Metallurgy] The disintegration of a metal acting as an anode.

anode current [Electronics] The electron current flowing through an electron tube from the cathode to the anode. Also known as plate current.

anode mud [Chemistry] The insoluble substance(s) that collect at the anode during an electrolytic refining or plating process. Also known as anode slime.

anode rays [Electronics] Positive ions coming from the anode of an electron tube.

anode slime *See* anode mud.

anodic cleaning [Metallurgy] The removal of foreign substances from a metallic surface by electrolysis with the metal as the anode. Also known as anodic pickling; reverse-current cleaning.

anodize [Metallurgy] The formation of a decorative or protective inert film on a metallic surface by making it he anode of a cell and applying electric current.

anomaly [Astronomy] A term used to describe the position of a planet in its orbit. The 'true anomaly' is the angle between the perihelion, the sun, and the planet, in the direction of the planet's motion. The 'mean anomaly' is the angle between perihelion, the sun and a fictitious planet having the same period as the real planet, but assumed to be moving with a constant velocity.

anotron [Electronics] A cold cathode glow-discharge diode having a copper anode and a large cathode of sodium or any other material.

antacid [Medicine] A pharmaceutical term for a substance that counteracts stomach acidity.

antenna *See* aerial.

anthracite [Minerals] A hard form of coal, containing more carbon and far less hydrocarbons than other forms; probably the oldest form of coal.

anti- Prefix denoting opposite, against e.g., antichlor.

antibiotics [Biochemistry] Chemical substances produced by microorganisms such as moulds and bacteria, which are capable of destroying bacteria or preventing their growth. Numerous antibiotics have been discovered, the first of which was penicillin. Some of the common antibiotics are: Aureomycin; Streptomycin; Chloromycetin; Erythromycin; Terramycin; Nystatin.

antibody [Biochemistry] A protein produced by animal plasma cells (of the reticuloendothelial system) as a result of the presence of an antigen. Specific antigens stimulate the formation of specific antibodies. The function of the antibodies is to combine chemically with antigens and thereby to render them harmless to the organism that they are invading.

anticentre [Geology] The point on the surface of the earth that is diametrically opposite the epicentre of an earthquake.

antichlor [Chemistry] A chemical used to remove excess of chlorine from materials after bleaching.

antidepressant [Medicine] Name given to drugs that reduce depression.

antidote [Medicine] A remedy for a particular poison, which generally acts chemically upon the poison and rendering it harmless.

antiferromagnetism [Physics] A type of magnetism that occurs in some metals, alloys, and transition-element compounds. This occurs below a certain temperature, called the Néel temperature, when an ordered array of atomic magnetic moments spontaneously forms in which alternate moments have opposite directions. There is therefore no net resultant magnetic moment in the absence of an applied field. In manganese fluoride, for example, this antiparallel arrangement occurs below a Néel temperature of 72 K. Below this temperature the spontaneous ordering opposes the normal tendency of the magnetic moments to align with the applied filed. Above the Néel temperature the substance is paramagnetic.

anti-freeze [materials] A substance added to water in radiators of motor-car engines in order to lower the freezing point of the water e.g., Ethanediol (ethylene glycol).

antigen [Biochemistry] A protein or carbohydrate that is foreign to an organism and capable of stimulating the formation of antibodies.

antihistamines [Medicine] A group of chemicals that counteract the effect of histamine in the body and are therefore used in the treatment of allergic diseases.

antiknock rating [Engineering] The measurement of ability of an automotive gasoline to resist detonation or pinging in spark ignited engines.

anti-matter [Physics] Hypothetical matter consisting of atoms which are composed of positrons, antiprotons, and antineutrons.

antimony [Chemistry] Sb. An element belonging to group VA of the periodic table; A. No. 51; r.d. 6.73; m.p. 903.6 K; b.p. 1653 K. Antimony has several allotropes. The stable form is a bluish-white metal. Yellow antimony and black antimony are unstable nonmetallic allotropes obtained at low temperatures. the main source is stibnite (Sb_2S_3), from which antimony is extracted by reduction with iron metal or by roasting (to give the oxide) followed by reduction with carbon and sodium carbonate. The main use of the metal is as an alloying agent in lead-accummulator plates, type metals, bearing alloys, solders, Britannia metal, and pewter. It is also an agent for producing pearlitic cast iron.

antinode [Physics] A point of maximum displacement in a series of standing waves. Two similar and equal wave motions travelling with equal velocities in opposite directions along a straight line give rise to antinodes an nodes alternately along the line. The antinodes are separated from their adjacent nodes by a distance corresponding to a quarter of the wavelength of the wave motions.

antioxidant [Chemistry] Any substance added to certain materials, such as rubber, plastics, paints, and oils, to prevent oxidation.

anti-particle [Physics] Every elementary particle has a corresponding real or hypothetical anti-particle of equal mass but opposite electric charge, with which annihilation can take place. The anti-particle of the electron is the positron. Anti-neutrons, anti-neutrinos, and anti-protons, amongst others, have been detected. The anti-neutron has the same mass and spin as a neutron, but opposite magnetic moment.

antipruritic [Medicine] An agent, such as camphor that relieves itching.

antipyretic [Medicine] A substance used medically to lower the body temperature, e.g., aspirin. Also known as febrifuge.

antiseptic [Medicine] Substances which destroy or inhibit the growth of micro-organisms. They can be applied to living tissues.

anti squawk agents [Materials] Substances added to lubricating oils to suppress noise in the operation of automatic clutches, etc.

apareon [Astronomy] The point on a Mars-centered orbit where a satellite is at its greatest distance from the Mars.

apastron [Astronomy] The point of the orbit of one member of a binary star system at which the stars are farthest apart.

aperture [Optics] Opening; in optical instruments, the size of the opening admitting light to the instrument. In spherical mirrors or lenses, the diameter of the reflecting surface.

aperture synthesis [Electronics] The use of two small aerials in a radio telescope to synthesize a large aperture. This principle can be used both with parabolic reflectors and radio interferometers, but it is usually best employed in conjunction with an unfilled aperture.

aphelion [Astronomy] The point, in a planet's orbit when it is farthest from the sun; the opposite of perihelion.

aplanatic [Optics] If any refracting surface produces a point image at X of a point at P irrespective of the angle at which the rays fall on the surface from P, then that surface is said to be aplanatic with respect to P and X.

apocynthion [Astronomy] The time or point, of greatest distance of a satellite in lunar orbit from the moon's surface, the opposite of pericynthion.

apogee [Astronomy] The moon or any other earth satellite is said to be in apogee when it is at its greatest distance from the earth, the opposite of perigee.

Apollo program [Engineering] The scientific program of United States which involved placing men on the moon and returning them safely to earth.

apothem [Mathematics] A perpendicular from the centre of a regular polygon to one of its sides.

apozymase [Biochemistry] The protein part of a zymase.

apparent depth [Physics] The depth of a liquid viewed from above appears to be less than the true depth, owing to the refraction of light. The ratio of the true depth to the apparent depth is equal of the refractive index of the liquid.

apparent expansion [Physics] Relative expansion of a liquid. **apparent volume** [Physics] The difference between the volume of a binary solution and the volume of the pure solvent at the same temperature.

appleton layer *See* ionosphere.

appliance [Engineering] An equipment that draws electric or other energy and produces a desired work, *e.g.*, electric heater, radio.

aprotic solvent [Chemistry] A solvent which does not accept or yield a proton.

apsis [Astronomy] One of the extremities of the major axis of the orbit of a planet or comet. The 'line of apsides' joins one apsis to the other.

aqua fortos [Chemistry] Trivial name for concentrated nitric acid, HNO_3.

aqua regia [Chemistry] Mixture of concentrated nitric and hydrochloric acids (1 to 3 by volume). Highly corrosive liquid which dissolves gold and attacks many substances that are unaffected by other reagents; turns orange-yellow owing to the formation of nitrosyl chloride, $NOCl$, and free chlorine.

aqueous [Chemistry] Watery. usually applied to solutions, indicating that water is the solvent.

arbor [Engineering] The axle of a wheel in a clock or watch.

arc *See* electric arc.

arc discharge [Electronics] A d.c. electrical current between electrodes placed in a gas or vapour, having high current density and relatively low voltage drop.

Archimedes principle [Physics] The apparent loss in weight of a body totally or partially immersed in a liquid is equal to the weight of the liquid displaced. *See* **buoyancy.**

arc lamp [Electronics] An electric lamp in which the light i produced by an arc made whe. current flows through ionized gas between two electrodes. Also known as electric-arc lamp, e.g., mercury-arc lamp.

area [Mathematics] A measure of the size of a two-dimensional surface or of a region on such a surface or in the plane; measured in 'square' units of length, e.g., square metres.

arenes [Chemistry] Aromtic hydrocarbons such as benzene and toluene.

argentometer [Engineering] A hydrometer used to find the amount of silver salt present in a solution.

argentometry [Chemistry] The volumetric analysis of silver which employs precipitation of insoluble silver salt like chloride or chromate.

argol [Chemistry] A reddish-brown crystalline deposit consisting mainly of potassium hydrogen tartrate, which separates in winevats. also known as tartar.

argon [Chemistry] Ar. A monoatomic noble gas present in air (0.93%); A. No. 18; d. 0.00178g gcm^{-3}; m.p. 84 K;b.p.88K. Argon is separated from liquid air by fractional distillation. It is slightly soluble in water, colourless, and has no smell. Its uses include cinert atmospheres in welding and special-metal manufacture (ti and Zr), and (when mixed with 20% nitrogen) in gas-filled electric-light bulbs. The element is inert and forms no true compounds.

arithmetical progression [Mathematics] Series of quantities in which each term differs from the preceding by a constant common difference. for an Arithmetical Progression in which the first term is x, the common difference d, the number of terms n, the last term L, and the sum of n terms

$$S, \quad S = n[2x + (n-1)d]/2$$

$$S = n(x+L)/2$$

$$L = x + (n-1)d$$

armature [Electricity] The coil or coils, usually rotating, of a dynamo or electric motor. Also more widely used as any part of an electric apparatus or machine in which a voltage is induced by a magnetic field, $e.g.$, in gramophone pick-ups, electro-magnetic loudspeakers, relays, etc.

armor [Electricity] Metal sheath enclosing a cable, primarily for mechanical protection.

aromatic [Chemistry] Compounds chracterized by the presence of atleast one benzene ring.

aromaticity [Chemistry] The degree to which a cyclic organic compound or ion with double bonds in ring exhibits the high stability and specific reactivity ($i.e.$ tendency to undergo substitution rather than addition reactions) characteristic of benzene and its derivatives. It is exhibited to a high degree by such compounds as pyridine, tropolones , and thiophene.

Arrhenius equation [Chemistry] An equation of the form

$$k = A \ exp \ (-E_A/RT)$$

where k is the rate constant of a given reaction and E_A the activation energy. A is a constant for a given reaction, called the pre-exponential factor. Often the equation is written in logarithmic form

$$ln \ k = O \ lnA - E_A/RT$$

A graph of ln k against $1/T$ is a straight line with a gradient $-E_A/R$ and an intercept on the ln$k$ axis of ln$A$

arsenic [Chemistry] As. A metalloid element of group of the periodic table; A. No. 33; r.d. 5.7; sublimes at 886 K. It has three allotropes -yellow, black and grey. They grey metallic form is the stable and most common one. Over 150 minerals contain arsenic but the main sources are as impuritls in sulphide ores and in the minerals orpiment (As_2S_3) and realgar (As_4S_4). Ores are roasted in air to form arsenic oxide and then reduced by hydrogen or carbon to metallic arsenic. Arsenic compounds are used in insecticides and as doping agents in semiconductors.

arthroscope [Medicine] An instrument used for the visualization of the interior of a joint cavity.

articulation [Physics] Percentage of speech-sounds correctly received over a radio-communication or reproducing system.

artificial gold [Chemistry] SnS_2. A yellowish-brown powder; insoluble in water but soluble in alkaline sulphides; used as a pigment and for imitation gilding. Also known as mosaic gold.

artificial radioactivity *See* induced radioactivity.

artificial satellite [Engineering] Any man-made object placed in a near-periodic orbit in which it moves under the gravitational influence of one celestial body, such as the earth, or another planet.

aryl [Chemistry] An organic univalent radical derived from an arene; *e.g.*, phenyl, C_6H_5-derived from benzene.

asbestos [Minerals] A variety of fibrous silicate minerals, mainly calcium magnesium silicate; used as a heat-insulating material and for fire-proof fabrics.

aseptic [Biochemistry] Free from bacteria.

ash [Material] Incombustible residue left after the complete combustion of any substance. It consists of the non-volatile, inorganic constituents of the substance.

asphalt [Material] A black semi-solid sticky substance composed of bitumen with mineral matter. It consists mainly of complex hydrocarbons; occurs naturally in asphalt lakes or in deposits mixed with sandstone and lime stone; made artificially by adding mineral matter to bitumen; used in road-making and building and in paints and varnishes.

aspirator [Engineering] Apparatus for drawing a current of air or other gas through a liquid.

assay [Chemistry] Analysing for one constituent of a mixture, particularly the estimation of metals in ores.

assembly language [Computers] A programming language that allows a computer user to write a program using mnemonics instead of numeric instructions.

assign [Computers] A control statement in FORTRAN which assigns a computed value i to a variable k, the latter representing the number of statements to which control is then transferred.

association [Chemistry] Under certain conditions, *e.g.*, in solution, the molecules of some substances associate into groups of several molecules, thus causing the substance to have an abnormally high molecular weight.

astatic [Physics] Having no tendency to change position; without orientation or directional characteristics.

astatic coils [Electricity] An arrangement used in sensitive electrical instuments; the coils are arranged to give zero resultant external magnetic field when an electric current passes through them, and to have zero electromotive force induced in them by an external magnetic field.

astatic galvanometer [Physics] A highly sensitive type of moving magnet galvonometer, in which two equal small magnets are arranged parallel but in opposition at the centres of two oppositely wound coils, the system being suspended by a fine torsion fibre. Since the resulting magnetic moment is zero, the earth's magnetic field exerts no controlling torque on the moving system. Instead, the restoring torque is supplied by the suspending fibre and is made very small by using a fine quartz fibre.

astatic pair [Physics] Arrangement of magnets used in astatic galvanometers.

astatine [Chemistry] At. Element. A. No. 85; placed in group VIIA of periodic the table. The most stable isotope, At-210 has a half-life of only 8.3 hours.

asterism [Astronomy] A constellation of small group of stars. [Optics] A star like optical phenomenon seen in gemstones due to reflection of light by lusturous inclusions reduced to sharp lines of light by a domed cabochon style of cutting.

asteroids [Astronomy] A belt of small bodies rotating round the sun in orbits between those mars and jupiter. the largest, Ceres, has a diameter of about 685 km, but most are much smaller. It is thought that there are many thousands of these bodies. Also

known as planetoids; minor planets.

astigmatism [Optics] A defect of lenses (including the eye) caused by the curvature being different in two mutually perpendicular planes; thus rays in one plane may be in focus while those in the other are out of focus producing distortion. Astigmatism of eye is corrected by the use of cylindrical lenses. [Electronics] In an electron-beam tube, a focus defect in which electrons in different axial planes come to focus at different points.

astringent [Medicine] A substance that by contracting body tissues, veins etc., reduces the discharge of mucus or blood.

astro-compass [Engineering] An instrument for determining direction relative to the stars; unaffected by the errors to which magnetic or gyro-compasses are subject; it is used to determine the errors of such instruments.

astrolable [Engineering] An instrument used to observe position and measure the altitude of heavenly bodies. The simplest form consists of a graduated circular ring with a movable sighting arm

astrometry [Astronomy] The branch of astronomy concerned with measurements of the positions of celestial bodies on the celestial sphere.

astron [Nucleonics] A proposed thermo-nuclear device in which a deuterium plasma is confined by an axial magnetic field produced by a shell of relativistic electrons.

astronautics [Astronomy] The scientific study of travel outside the earth's atmosphere. The art skill or activity of operating spacecraft.

astronomical equator [Astronomy] An imaginary line on the surface of the earth connecting points having 0° astronomial latitude. Also known as terrestrial equator.

astronomical latitude [Astronomy] Angular distance between the direction of gravity (plumb line) and the plane of the celestial equator.

astronomical longitude [Astronomy] Angle between the plane of the reference meridian and the plane of the local celestial meridian.

astronomical tide [Astronomy] An equilibrium tide due to attraction of the sun and the moon.

astronomical unit [Astronomy] The mean distance from the centre of the earth to the centre of the sun; equal to 1.495×10^{11} metres, approximately 92.9×10^{11} miles.

astronomy [Science Technology] The science concerned with the study of the heavenly bodies, their motions, relative positions, and nature. Its main branches are astrometry, celestial mechanics, and astrophysics. *See also* radioastronomy and cosmology.

astrophysics [Astronomy] The branch of astronomy concerned with the physical properties of celestial bodies, and the interaction between matter and energy within them.

asymmetric Not symmetric.

asymmetric carbon atom [Chemistry] A carbon atom in a molecule of an organic compound with four different atoms or groups attached to its four valences. Such a grouping permits of two different arrangements in space, leading to the existence of optical isomers. *See* streoisomerism.

asymtote [Mathematics] A line approaching a curve, but never reaching it within a finite distance.

atactic polymer [Chemistry] A polymer in which the groups attached to the main chain are not arranged regularly. In isotatic polymers the same irregularity is repeated along the chain, where as in syndiotactic polymers there are asymmetric carbon atoms in the chain and successive groups lie on alternate sides of the chain. *Compare* tactic polymer.

athermancy [Physics] The property of being opaque to radiant heat; i.e., of absorbing heat radiations.

atmolysis [Mechanics] The separation of a mixture of gases through the walls of a porous vessel by taking advantage of the different rates of diffusion of the constituents.

atmometer [Engineering] The general name for an instrument which measure the rate of evaporation of water. Also known as evaporometer.

atmosphere 1. [Chemistry] The gaseous envelope surrounding the earth (or other heavenly body). The composition of the earth's atmosphere varies very slightly in different localities and according to altitude. Volume composition of dry air sea-level (average values): nitrogen, 78.03%; oxygen, 20.98%; argon, 0.94%; carbon dioxide, 0.04%; neon, 0.0018%; helium, 0.0005%; krypton, 0.0001%; xenon, 0.00001%. Air generally contains, in addition to the above, water vapour, hydrocarbons, hydrogen peroxide, sulphur compounds and dust particles in small and very variable amounts. [Mechanics] A unit of pressure. The pressure that will support a column of mercury 760 mm high (29.92 inches) at 273 K., sea-level and latitude 45°. 1. normal atmosphere=101325 newtons per square metre. Atmospheric pressure fluctuates about this value from day to day.

atmospheric attenuation [Physics] A process in which the flux density of parallel beam of energy decreases with increasing distance from the source as a result of absorption or scattering by the atmosphere.

atmospheric noise [Electronics] Noise heard during radio reception due to atmospheric interference.

atmospherics [Electronics] Electrical discharges that take place in the atmosphere, causing crankling sounds in radio receivers.

atom [Chemistry] The smallest part of an element that can exist chemically. Atoms consist of a small dense nucleus of protons and neutrons surrounded by moving electrons. The number of electrons equals the number of protons so the overall charge is zero. The electrons are considered to move in circular or elliptical orbits or, more accurately, in regions of space around the nucleus.

atomic bomb [Nucleonics] A device for suddenly producing a rapid explosion as a result of neutron chain rection in a fissile material such as uranium-235 or plutonium. Also known as fission bomb.

atomic clock [Engineering] A very accurate form of clock in which the basis of the time scale is derived from the vibrations of atoms or molecules. See ammonia clock; caesium clock.

atomic constants [Physics] physical constants which play a fundamental role in atomic physics, including the electronic charge, electronic mass, speed of light, Avogadro number, and Planck's constant.

atomic energy See nuclear energy.

atomic heat [Chemistry] The numerical product of the atomic weight and the specific heat capacity of an element. Dulong and Petit's law states that the atomic heat of all solid elements is approximately 25 joules per mole per degree. The law is obeyed by many elements at ordinary temperatures but at lower temperatures the atomic heat of all elements falls below this value, tending to zero as absolute zero of temperature is aproached.

atomic hydrogen maser [Physics] An active clock oscillator using a hydrogen maser to produce a very stable signal at 1420 megahertz.

atomic mass [Chemistry] The mass of an element measured in atomic mass units.

atomic mass unit [Chemistry] AMU. A unit used for expressing the masses of individual isotopes of elements; approximately equal to 1.66×10^{-27} kg. Formerly defined so that the most 'abundant isotope of oxygen, $^{16}_{8}O$, had a mass of 16 atomic mass units. In 1961 the 'unified atomic atomic mass unit' was defined as 1/12 of the mass of an atom of $^{12}_{6}C$, and was adopted by the International Union of Pure and Applied Physics and the International Union of Pure and Applied Chemistry. Atomic Wei ghts mentioned in this dictionary are based upon this scale.

atomic nucleus See nucleus, atomic.

atomic number [Chemistry] Z. The number of electrons rotating round the nucleus of the neutral atoms of an element, or the number of protons in the nucleus. Also known as proton number.

atomic orbital [Physics] The space-dependent part of a wave function describing an electron in an atom.

atomic particle [Physics] One of the particles of which an atom is constituted, an electron, proton, or neutron.

atomic pile [Nucleonics] The original name for a nuclear reactor.

atomic power plant [Nucleonics] An assembly for converting stored nuclear energy into work, such as a nuclear electric power generating station.

atomic second [Physics] The duration of 9, 192, 631, 770 periods of the radiation corresponding to the two hyperfine levels of the fundamental state of the atom of caesium-133.

atomic theory [Chemistry] Hypothesis that the matter is composed of particles called atoms which cannot be further sub-divided.

atomic volume [Chemistry] The atomic weight of an element divided by its density. The volume occupied by one gram-atom of an element in the solid state.

atomic weight [Chemistry] Also known as relative atomic mass. The ratio of the average mass per atom of a specified isotopic composition of an element to 1/12 of the mass of an atom $^{12}_{6}$of C. The natural isotopic composition is assumed unless otherwise stated.

atomization [Engineering] The mechanical subdivision of a liquid or meltable solid to produce drops which vary in diameter depending upon the process.

atom smasher [Physics] A popular name for a particle accelerator.

attenuated vaccine [Biochemistry] A suspension of bacteria or viruses used to produce active immunity.

attenuation [Physics] Loss of power sufered by radiation as it passes through matter. [Biochemistry] Weakening or reduction of the virulence of a micro-organism.

atto- [Physics] Prefix denoting one million millionth; 10^{-18} metre.

audibility limits [Physics] The limits of frequency of sound waves that are audible as sound to the human ear. The lowest is about 30 hertz, corresponding to a very deep vibrating rumble, and the highest in the region of 20,000 hertz.

audio-frequency [Physics] A frequency between 30 and about 20,000 hertz, which in the case of sound waves would be audible.

audimeter [Physics] An instrument for measuring human hearing both loss due to deafness and the masking produced by noise.

Auer metal [Metallurgy] A pyrophoric alloy of 65% misch metal (a mixture of cerium and other metals) an 35% iron, used as 'flint' in lighters.

Aufbau principle [Chemistry] A principle that gives the order in which orbitals are filled in successive elements in the periodic table. The order of filling is $1s$, $2s$, $2p$, $3s$, $3p$, $4s$, $3d$, $4p$, $5s$, $4d$, $5p$, $6s$, $4f$, $5d$, $6p$, $7s$.

Auger effect [Physics] The ejection of an electron from an atom without the emission of an X-or gamma-ray photon, as a result of the de-excitation of an excited electron within the atom. This type of transition occur in the X-ray region of the emission spectrum. The kinetic energy of the ejected electron, called an Auger electron, is equal to the energy of the corresponding X-ray photon minus the binding energy of the Auger electron.

Auger shower [Astronomy] A very large cosmic-ray shower. Also known as extensive air shower.

auric [Chemistry] Containing trivalent gold, i.e., Au (III).

auriferous [Chemistry] Pertaining to gold-bearing.

aurora borealis [Astronomy] A display of coloured light streaks and glows, mainly red and green, visible in the regions of the north and south poles. Probably caused by streams of electrified particles from the sun; most prominent when large sunspots are observed. In southern latitudes the effect is called the aurora Australis. *See* Solar wind.

aurous [Chemistry] Containing univalent gold, i.e., Au (I).

austenite [Metallurgy] A solid solution of carbon or of iron carbide in the gamma form of iron; normally stable only at high temperatures, but may be preserved at normal temperatures by certain alloying elements or by rapid cooling.

autocapacitive coupling [Electronics] Coupling of two circuits by a capacitor common to both circuits.

autocatalysis [Chemistry] Catalysis in which the catalyst is produced during the course of the reaction that is being catalysed.

autoclave [Engineering] A thick-walled vessel with a tightly fitting lid, in which substances may be heated under pressure to above their boiling points; used in the manufacture of chemicals; for sterilizing medical instruments; etc.; and in cooking.

autocode [Computers] The process of using a computer to convert automatically a symbolic code into a machine code. Also known as automatic code.

autograft [Biology] A tissue transplanted from one part to another part of an individual's body.

autoimmunity [Biology] An immune state in which antibodies are formed against the person's own body tissues.

autolysis [Biology] The self-destruction of biological cells after death, as a result of the action of their own enzymes.

automatic alarm [Electronics] Alarm device such as electric bell, buzzer or lamp, operated by the automatic making or breaking of an electrical contact when an emergency occurs.

automatic computer [Computers] A computer which can carry out a special set of operations without human intervention.

automation [Engineering] The application of mechanical, or more commonly electronic or computerized, techniques to minimize the use of manpower in any process.

auxochrome [Chemistry] A group in a dye molecule that influences the colour due to the chromophore. Auxochromes are groups, such as -OH and -NH$_2$, containing lone pairs of electrons that can be delocalized along with the delocalized electrons of the chromophore. The auxochrome intensifies the colour of the dye. Formerly, the term was also used of such groups as -SO$_2$O$^-$ which make the molecule soluble and affect its application.

autosome [Biology] Any chromosome other than a sex chromosome.

auxins [Biochemistry] A type of plant hormone that promotes the elongation and growth of plant cells and stimulates rooting; e.g., indole-3-acetic acid.

avalanche [Physics] A shower of particles caused by the collision of a high energy particle (*e.g.*, a cosmic ray) with any other form of matter. [Electronics] The cumulative process in which an electron or other charged particle accelerated by a strong magnetic field collides with and ionizes gas molecules, thereby releasing new electrons which in turn have more collisions, so that the discharge is thus self-maintained.

Avogadro constant [Chemistry] Symbol N_A or L. The number of atoms or molecules in one mole of substance. It has the vaue 6.022 52 X 10^{23}. Formerly it was called *Avogadro's number*.

Avogadro's law [Chemistry] Equal volumes of all gases contain equal number of molecules at the same pressure and temperature. The law often called Avogadro's hypothesis, is true only for ideal gases.

axis of symmetry [Crystallography] A line about which a given figure is symmetrical; *e.g.*, the diameter of a circle.

azeotrope [Chemistry] Constant-boiling mixture. A mixture of two or more liquids that distils at a certain constant temperature and has a constant composition at a given pressure. Its boiling point may be a maximum or a minimum relative to the original components.

azide [Chemistry] A derivative of hydrazoic acid. A compound containing the univalent azido group, $-N_3$, *e.g.*, potassium azide, KN_3.

azimuth [Astronomy] The angular distance from the north or south point of the horizon to the foot of the vertical circle through a heavenly body. The azimuth of a horizontal direction is its deviation from the north or south.

azimuthal quantum number *See* quantum numbers.

azines [Chemistry] Organic derivatives of hydrazine, of the general formula RR'C=N-N=CRR', where R and R' are univalent organic radicals. The suffix-azine is also used in systematic naming of six-membered unsaturated heterocyclic compounds containing nitrogen in the ring. Such compounds are sometimes described as azines.

azino [Chemistry] The quadrivalent radical=N-N=.

azo compound [Chemistry] A compound containing an azo group attached to two carbon atoms (-CN=NC-). Aromatic azo compounds are usually prepared by azo coupling.

azo coupling [Chemistry] The formation of an azo compound by the reaction of an aromatic diazo compound with a suitable nucleophilic reagent, such as an amine or a phenol.

azoic [Geology] The portion of the earlier Precambrian time in which there is no trace of life.

azoic dye [Chemistry] Water insoluble azo dye that is formed within the fibre by the azo coupling of a diazo compound with a suitable azo coupling component, often a naphthol derivative. Also known as ice colour.

azonal soil [Geology] Any group of soils without well-developed profile characteristics, owing to their youth, conditions of parent material, or relief that prevents development of normal soil-profile characteristics.

azotemia [Medicine] Presence of excessive amounts of nitrogenous compounds in the blood.

azulite [Mineral] A translucent pale-blue variety of carbonate mineral of zinc, found in Arizona and Greece.

azygote [Biology] An individual produced by haploid parthenogenesis.

B

Babbit metal [Metallurgy] Any of a group of related alloys used for making bearings. They consist of tin containing antimony (about 10%) and copper (1-2%), and often lead.

babble [Physics] Unwanted disturbing sounds in a carrier or other multiple channel system which result from the aggregate crosstalk or mutual interference from other channels.

Babo's law [Chemistry] The vapour pressure of a liquid is decreased when a solute is added, the amount of the decrease being proportional to the amount of solute dissolved.

bacillus [Biology] In general, a rodshaped bacterium. In particular, a genus of spore-producing bacteria.

back bias [Electronics] Voltage applied to the grid of a tube or electrode of another device to reduce a condition which has been upset by some external cause.

back E.M.F. of cell [Electricity] When the poles of a cell become polarized (see polarization, electrolytic) a back E.M.F. is set up opposing the natural E.M.F. of the cell.

back E.M.F. of electric motor [Electricity] E.M.F. set up in the coil of an electric motor, opposing the current flowing through the coil, when the armature rotates.

background [Physics] The counting rate of a counter tube caused by sources other than the one being measured. Due primarily to natural radioactivity in the soil, and cosmic rays.

backing storage [Computers] Computer stores with a capacity to store enormous quantities of information, but with an access time much greater than the main store. The commonest types are magnetic tape decks, fixed magnetic disc stores, and exchangeable magnetic disc stores.

backlash [Electronics] The incomplete rectification of an alternating current in a thermionic valve due to the pressure of positive ions in the residual gas in the valve.

bacteria [Biology] A group of small micro-organisms, also called schizomycetes. They are typically small cells of about 1 micron in transverse diameter. They may be free living, saprophytic or parasitic; some are pathogenic to man, animals and plants.

bactericide [Materials] A substance that kills bacteria.

bacteriology [Biology] The scientific study of bacteria.

bacteriolysis [Biology] Dissolution of bacterial cells.

bacteriophage [Biology] A virus that requires a bacterium in which to replicate. Also known as phage.

bakelite [Materials] Trade name for various synthetic thermosetting plastics of which penolformaldehyde resins are amongst the most widely known. They have high electrical and chemical resistance.

baking powder [Materials] A mixture that produces carbon dioxide gas on wetting or heating, thus causing the formation of bubbles in dough and making it 'rise'. Usually contains sodium hydrogen carbonate, $NaHCO_3$,

and tartaric acid or cream of tartar.

baking soda [Chemistry] $NaHCO_3$. Sodium hydrogen carbonate.

balance [Science Technology] An accurate weighing device. The simple beam balance consists of two pans suspended from a centrally pivoted beam. Known masses are placed on one pan and the substance or body to be weighed is placed in the other. When the beam is exactly horizontal the two masses are equal. An accurate laboratory balance weighs to the nearest hundredth of a milligram. Specially designed balances can be accurate to a millionth of a milligram. More modern substitution balances use the substitution principle. In this calibrated weights are removed from the single lever arm to bring the single pan suspended from it into equilibrium with a fixed counter weight. The substitution balance is more accurate than the two-pan device and enables weighing to be carried out more rapidly. In automatic electronic balances, mass is determined not by mechanical deflection but by electronically controlled compensation of an electric force. A scanner monitors the displacement of the pan support generating a current proportional to the displacement. This current flows through a coil forcing the pan support to return to his original postion by means of a magnetic force. The signal generated enables the mass to be read from a digital display. The mass of the empty container can be stored in the balance's computer memory and automatically deducted from the mass of the container plus its contents.

balanced transmission line [Electronics] Transmission line in which the two conductors are at equal and opposite potentials above and below zero potential when the line is used to transmit intelligence.

balance voltages [Electricity] Voltages that are equal in magnitude and opposite in polarity with respect to ground. Also known as push-pull voltages.

balata [Materials] A natural rubber-like material very similiar to gutta-percha.

ballast lamp [Electricity] A light producing electrical resistance device which maintains nearly constant current by increasing resistance as the current increases.

ballistic galvanometer [Electronics] An instrument for measuring the total quantity of electricity passing through a circuit due to a momentary current. Any galvanometer may be used ballistically provided that its period of oscillation is long compared with the time during which the current flows.

ballistic missile [Engineering] A ground-to-ground missile with a parabolic flight path. A missible

that is propelled and guided only during the initial phase of its flight.

ballistics [Mechanics] A branch of mechanics which deals with the motion, behaviour and characteristics of projectiles, bombs, rockets, and guided missiles.

Balmer series [Physics] The visible portion of the spectrum of hydrogen. It consists of a series of sharp distinct lines, the wavelentgths, λ of which may be represented by the formula :

$$1/\lambda = R(1/2^2 - 1/n^2) :$$

$n=3, 4, 5$, etc., R is a constant known as Rydberg's constant, which has the value 1.09677×10^7 m^{-1}.

band [Electronics] A specific range of frequencies used in communications for a definite purpose, such as the long-wave band in radio; certain frequencies within the band are assigned to different transmitting stations and the receiver may be turned to any desired frequency within the band. [Physics] A closely spaced group of molecular energy levels that appear as fluted bands separated by dark spaces in the spectra of compounds. [ADP] A set of circular or cyclic recording tracks on a storage device such a magnetic drum, tape, or disc.

band spectrum [Physics] An emission or absorption spectrum consisting of groups or bands each having one sharp edge. Each band is composed of **a large**

number of closely spaced lines. Band spectra is characteristic of chemical compounds.

bands theory of solids [Physics] A quantum mechanical theory of motion of electrons in solids that predicts certain restricted ranges or bands for the energies of these electrons.

bandwidth [Electronics] The range of frequencies within which the performance of a circuit, receiver, or amplifier does not differ from its maximum value by a specified amount. The bandwidth of a radio emission is the width of the frequecny band that carries a specified proportion (usually 99%) of the total power radiated.

bar [Physics] A C.G.S. unit of pressure equal to 10^6 dynes per square centimetre or 10^5 pascals (approximately 750 mmHg or 0.987 atmosphere). The millibar (100 Pa) is commonly used in meteorology.

barbiturates [Chemistry] Class of organic compounds derived from barbituric acid. Many of these compounds have a powerful soporific effect. They were formerly used extensively in sleeping tablets, but as an overdose could be fatal, they have been largely replaced by safer substances.

Barff process [Chemistry] Prevention of rusting of iron by the action of steam upon the surface of the red-hot metal, resulting in a surface coating of black oxide of iron, Fe_3O_4.

barium [Chemistry] Ba Element. At. No. 56. A.W. 137.34. A silvery-white soft metal, which tarnishes readily in air. m.p. 798K b.p. 1913K; occurs as barytes, $BaSO_4$ and as barium carbonate, $BaCO_3$. A radioactive isotope of barium has atomic mass 140, half life 12.8 days, and decays by negative beta-particle emission.

barium enema [Medicine] A suspension of barium sulphate administered as enema into the lower bowl to render it radiopaque.

barium meal [Medicine] A suspension of barium sulphate taken orally to render the upper gastrointestinal tract radiopaque.

Bárkhausen effect [Physics] The effect observed when a ferromagnetic substance is magnetized by a slowly increasing magnetic field; the magnetization does not take place continuously, but in a series of small steps. The effect is due to orientation of magnetic domains present in the substance.

barn [Physics] Unit of area for measuring the cross-section of nuclei 1 barn = 10^{-24} cm^2.

Barnet effect [Physics] The production of a small magnetization in an initially unmagnetized iron rod when it is rotated at very high speed around its axis.

barodynamics [Mechanics] The mechanics of very heavy structures which may collapse under their own weight.

barogram [Engineering] An absolute pressure gauze specifically designed to measure atmospheric pressure.

barometer [Engineering] An instrument for measuring atmospheric pressure. Mercury barometer consists of a long tube closed at the upper end, filled with mercury and inverted in a vessel containing mercury.

baroscope [Engineering] An apparatus for demonstrating the equality of the weight of air displaced by an object and its loss of weight in air.

barostat [Engineering] A system which maintains constant pressure inside a chamber.

barretter [Electricity] 1. A bolometer used for making power measurements in microwave devices. 2. A device to obtain a constant voltage over a range of currents and may be used as a voltage regulator.

barrier penetration [Physics] The passage of a particle through a potential barrier, *i.e.*, through a region of finite extent in which the potential energy of the particle is greater than its total energy.

barye [Physics] Unit of pressure in the C.G.S. system, equal to one dyne per sq.cm.

baryon [Physics] A collective name for nucleons and hyperons. They are all hadrons and are believed to consist of three quarks bound together. The number of baryons minus the number of corresponding antibaryons taking part in a process is called the baryon number, a quantity that appears to be conserved in all processes. All baryons have spin 1/2.

basal [Geology]1. A rock of volcanic origin, chemically resembling feldspar. 2. Pertaining to or located at the base.

base [Chemistry] A compound that reacts with a protonic acid to give water (and a salt). The definition comes from the Arrhenius theory of acids and bases. Typically, bases are metal oxides, hydroxides, or compounds (such as ammonia) that give hydroxide ions in aqueous solution. Thus, a base may be either: (1) An insoluble oxide or hydroxide that reacts with an acid, *e.g.*

$$CaO(s) + 2HCl(aq) \rightarrow CaCl_2(aq) + H_2O(l)$$

Here the reaction involves hydrogen ions from the acid,

$$CaO(s) + 2H^+(aq) \rightarrow H_2O(l) + Ca^{2+}(aq)$$

(2) A soluble hydroxide, in which case the solution contains hydroxide ions. The reaction with acids is a reaction between hydrogen ions and hydroxide ions :

$$H^+ + OH^- \rightarrow H_2O$$

(3) A compound that dissolves in water to produce hydroxide ions. For example, ammonia reacts as follows :

$$NH_3(g) + H_2O(l) = NH_4^+(aq) + OH^-$$

Similar reactions occur with organic amines (*see also* nitrogenous base; amine salts). A base that dissolves in water to give hydroxide ions is called an *alkali*. Ammonia and sodium hydroxide are common examples.

The original Arrhenius definition of a base has been extended by the Lowry-Brönsted theory and by the Lewis theory. *See* acid.

base band [Electronics] The band of frequencies occupied by all transmitted signals used to modulate the radio wave that is produced by transmitter in the absence of signals.

base metals [Chemistry] In contradistinction to the noble metals, metals that corrode, tarnish, or oxidize on exposure to air, moisture, or heat. [Metallurgy] The principal metal of an alloy.

base unit [Physics] A unit that is defined in terms of a primary standard, *e.g.*, the unit of mass (kilogram) in SI units.

BASIC [Computers] A procedure level computer language well suited for conversational mode on a terminal usually connected with a remotely operated computer.

basic dyes [Materials] A group of dyes that are organic bases, the cations of which are the colouring agents. Used for dyeing wool and cotton. Also known as cationic dyes.

BASIC-PLUS [Computers] An extension of the BASIC programing language. It includes more powerful capabilities, especially for data manipulation.

basic salt [Chemistry] A compound that can be regarded as being formed by replacing some of the oxide or hydroxide ions in a base by other negative ions. Basic salts are thus mixed salt-oxides (*e.g.* bismuth(III) chloride oxide, BiOCl) or salt-hydroxides (*e.g.* lead (II) chloride hydroxide, Pb(OH)Cl).

basic slag [Materials] A mixture resulting form the steelmaking process containing tetracalcium phosphate, $Ca_4P_2O_9$, calcium silicate, $CaSiO_3$, lime, CaO, and ferric oxide, Fe_2O_3. Its high phosphorus content makes it a valuable fetilizer.

bass [Physics] Sound waves of frequencies at the lower end of the audio range, *i.e.*, below 250 hertz.

bathochromatic shift [Chemistry] The shift of the fluorescence of a compound towards the red part of spectrum due to the presence of a bathochrome radical in the molecule.

bath salts [Chemistry] The main constituent is generally sodium sesquicarbonate, Na_2CO_3. $NaHCO_3.2H_2O$, or some other soluble sodium salt to soften the water.

bathymetry [Engineering] Science of measuring ocean depths.

battery [Electricity] A number of primary or secondary cells arranged in series or parallel. In series, they give a multiple of the E.M.F. of the cell; in parallel, they give the same E.M.F. as the cell, but have a greater capacity, *i.e.*, a given current can be supplied for a longer period. The common 'dry batteries' usually consist of Leclanche' cells.

bauxite [Mineral] Natural hydrated aluminium oxide, Al_2O_3. xH_2O containing silica, iron hydroxides and dry minerals as impurities; the most important ore of aluminium.

bauxite cement [Material] A rapid-hardening cement consisting mainly of calcium aluminate; made by heating bauxite and lime in an electric furnace. Also known as ciment fondu.

bead thermistor [Electronics] A thermistor made by applying the semiconducting material to two wire leads, as a viscous droplet which cements the leads upon firing.

beam [Physics] Radiation travelling in a particular direction.

beam hole [Nucleonics] A hole made in the shield, and usually through the reflector, of a nuclear reactor to permit the escape of a beam of radiation, particularly neutrons, for experimental purposes.

beam riding [Engineering] A method of rocket guidance in which the missile steers itself along the axis of a beam of radiation, usually a conically scanned radar beam.

beam transmission [Electronics] Radio transmission in which the electromagnetic waves are sent in a particular direction in a beam instead of being radiated in all directions.

beam width [Electronics] The angle, measured in horizontal plane, the directions at which the intensity of an eletromagnetic beam, such as a radio or radar beam, is one half its maximum value. Also known as beam angle.

beat frequency [Electricity] The different frequency resulting from the interaction between radio frequency signals of different wavelengths.

beats [Physics] A periodic increase and decrease in loudness heard when two notes of nearly the same frequency are sounded simultaneously; caused by interference of sound waves, the number of beats produced per second is equal to the difference in frequencies of the two notes.

Beaufort scale [Mechanics] A numerical scale for the estimation of wind force, based on its effect on common objects. It is based on a system of code numbers from 0 to 12 classifying wind speeds into groups from 0.1 mile per hour to those over 75 miles per hour.

Beckmann thermometer [Engineering] A sensitive thermometer for measuring small differences or changes in temperature. The quantity of mercury in the bulb can be varied by causing it to overflow into a reservoir at the top, thus enabling the thermometer to be used over various ranges of temperature.

becquerel [Nucleonics] Bq. The derived SI unit of nuclear activity (radio-active). The number of atoms of a radioactive substance that disintegrate in one second.

Becquerel effect [Electricity] The phenomenon of current flowing between two unequally illuminated electrodes of a certain type when they are immersed in an electrolyte.

Becquerel rays [Nucleonics] Radiations emitted by radioactive substances; later renamed as alpha, beta and gamma rays.

beeswax [Material] A yellow to greyish-brown wax consisting of a mixture of compounds, secreted by bees for the purpose of building their honeycombs; used in polishes, cosmetics, textile finishes and in pharmacy.

bell metal [Metallurgy] An alloy of copper and tin containing 15-25% tin and 60-85% copper.

Benedict's solution [Chemistry] A solution of sodium and potassium tartrates, copper sulphate and sodium carbonate; used to detect reducing sugars.

beneficiation [Metallurgy] The separation of ores into valuable components and wastes. It can be achieved in various ways *e.g.*, by froth flotation, magnetic separation, leaching. Also known as ore dressing.

Benham top [Optics] A disc with many black and white portions which, when rotated at different speeds and subjected to certain lighting, produces sensations of colour.

benzene ring [Chemistry] The six-carbon ring structure found in benzene, C_6H_6 and in organic compounds formed from benzene by replacement of one or more hydrogen atoms by other atoms or group of atoms.

benzine [Chemistry] A mixture of hydrocarbons (mainly alkanes) obtained from petroleum; it boils between 310 and 350K, and is used as a solvent. Because of possible confusion with benzene, the word 'benzine' should be avoided in scientific writing. Also known as petroleum benzin; petroleum ether; solvent naphtha.

benzoate [Chemistry] A salt or an ester of benzoic acid.

benzoyl [Chemistry] The univalent radical C_6H_5CO-.

benzyl [Chemistry] The univalent radical $C_6H_5CH_2$-.

benzylidene [Chemistry] The bivalent radical $C_6H_5CH=$.

benzylidyne [Chemistry] The trivalent radical $C_6H_5C=$.

Bergius process [Chemistry] A process for the manufacture of oil from coal. Coal, made into a paste with heavy oil, is heated with hydrogen under a pressure of 250 atmosphere to a temperature of 720-740K, in the presence of a catalyst. The carbon of the coal reacts with the hydrogen to give a mixture of various hydrocarbons.

berkelium [Chemistry] Bk. Transuranic element. At. No. 97. A radioactive element, member of the actinide series; properties resemble those of the rare-earth cerium. Most stable isotope, $^{247}_{97}$Bk, has a half-life of about 1400 years.

Bernoulli's theorem [Physics] At any point in a tube through which a liquid is flowing, the sum of pressure energy, potential energy, and kinetic energy is constant.

Berthollide compounds [Chemistry] Chemical compounds the composition of which does not conform to a simple ratio of atoms in the molecule.

beryllium [Chemistry] Symbol Be. A grey metallic element of group II of the periodic, table: At. No. 4; r.d. 1.85; m.p. 1558K; b.p. 3243K. Beryllium occurs as beryl ($3BeO.Al_2O_36SiO_2$)and chrysoberyle ($BeO.Al_2O_3$). The metal is extracted from a fused mixture of BeF_2/NaF by electrolysis or by magnesium reduction of BeF_2. It is used to manufacture Be-Cu alloys, which are used in nuclear reactors as reflectors and moderators because of their low absorption cross section.

Bessemer process [Metallurgy] A process for making steel from cast iron. Molten iron from the blast furnace is run into the Bessemer converter, a large egg-shaped vessel with holes below. Through these, air is blown into the molten metal, and the carbon is oxidized to its oxide. The requisite amount of spiegel is then added to introduce the correct amount of carbon to obtain the desired variety of steel. In some modern converters, instead of air a mixture of oxygen and steam is blown into the molten metal to avoid the absorption of nitrogen by the steel. This is known as the VLN (very low nitrogen) process.

beta decay [Nucleonics] A radioactive disintegration of an unstable nucleus in which a neutron changes to a proton with the emission of an electron and an antineutrino or in which a proton changes to a neutron with the emission of a positron and a neutrino. Thus a beta decay involves unit change of atomic number but no change of mass number. It is a form of weak interaction.

beta emitter [Nucleonics] A radioactive nucleus that disintegrates by the emmission of a negative or positive electron.

beta factor [Physics] In plasma physics, the ratio of plasma kinetic pressure to the magnetic pressure.

beta-iron [Metallurgy] An allotropic (see allotropy) form of pure iron, stable between 1041K to 1183K, similar to alpha-iron except that it is non-magnetic.

beta particle [Nucleonics] An electron or positron emitted by a radio-active nucleus. *See* beta decay.

beta rays [Nucleonics] A stream of beta particles; they possess greater penetrating power than alpha rays and are emitted with velocities in some cases exceeding 98% of the velocity of light.

betatron [Nucleonics] A cyclic accelerator for accelerating a continuous beam of electrons to high speeds by means of the electric field produced by a changing magnetic flux. The electrons move in stable circular orbits in an evacuated torus-shaped chamber. By allowing the fast electrons to strike a metal target a continuous source of gamma rays with energies up to 300 MeV can be produced. Also known as induction accelerator; rheotron.

BeV [Physics] A billion (10^9) electron volts.

bevatron [Nucleonics] A cyclic accelerator for accelerating protons and other particles to very high energies (up to 6 gigaelectron volts).

bi- [Science Technology] Prefix denoting two; formerly used in chemical nomenclature to indicate an acid salt of a dibasic acid. *See* bicarbonate.

bias [Electronics] A d.c. bakcground signal; in magneto-optic disk drives, a magnetic bias field is used in recording and erasing.

bicarbonate [Chemistry] Acid salt of carbonic acid, H_2CO_3; in which one of the acidic hydrogen has been replaced by a metal, *e.g.*, sodium bicarbonate, $NaHCO_3$.

bi-concave [Optics] A term used to describe a lens that is concave on both sides.

bi-convex [Optics] A term used to describe a lens that is convex on both sides.

big-bang theory *See* superdense theory.

bilateral amplifier [Electronics] An amplifier capable of receiving as well as transmitting signals; mainly used in transceivers.

bile [Biochemistry] An alkaline fluid secreted by the liver of vertebrates; important in the digestion of fats. It consists of cholesterol, bile salts (salts of cholic acid), and bile pigment (degradation products of haemoglobin).

billion [Science Technology] Million million, 10^{12} (British); thousand million, 10^9 (American).

bimetallic strip [Engineering] A strip composed of two different metals welded together in such a way that a rise of temperature will cause it to buckle as a result of unequal expansion; used in thermostats.

bimorph cell [Physics] Two plates

of piezoelectric material joined together so that they bend in proportion to an applied voltage.

binary [Science Technology] Characterized by or composed of two parts. [ADP] Possessing a property for which there exist two choices or conditions excluding the other.

binary cell [Computers] An element in a computer that can store information by virtue of its ability to remain stable in one of two possible states.

binary code [Computers] A code in which each allowable position has one of two possible states, commonly 0 and 1.

binary compound [Chemistry] A chemical compound of two elements only. Denoted by the suffix-ide; *e.g.*, calcium carbide, CaC_2.

binary notation [Mathematics] Binary number sytem. A system of numbers that has only two different digits, usually 0 and 1. There are several ways of representing numbers in the binary notation; one common method is shown below. Because it has only two digits, which can be represented by an electric current switched on or switched off, this notation is used in computers.

Decimal system	Decimal system
1	0001
2	0010
3	0011
4	0100

Binary system	Binary system
5	0101
6	0110
7	0111
8	1000
9	1001
10	1010

binary stars [Astronomy] Two stars gravitationally attracted to each other, so that they revolve around their common centre of gravity, thus forming a double star.

binary word [Computers] A group of bits which occupies one storage address and is treated by the computer as a unit.

binding energy [Physics] The energy that must be supplied to a nucleus in order to cause it to decompose into its constituent neutrons and protons. The binding energy of a neutron or a proton is the energy required to remove a neutron or a proton from a nucleus.

binocular [Optics] Any optical instrument designed for the simultaneous use of both eyes; *e.g.*, binocular field glasses, binocular microscope.

binomial [Mathematics] A mathematical expression consisting of the sum or difference of two terms; , $x^2 - 3y$.

binomial nomenclature [Biology] The method of naming plants and animals introduced by Linnaeus in the mid-eighteenth century. Every plant or animal has two Latin names; a generic name designating its genus, and a specific name indicating the species; , *Felis tigris,* the tiger.

binomial theorem [Mathematics] The expansion of $(x + y)^n$.

biochemical fuel cell [Electricity] A fuel cell in which small amounts of electric power can be produced continuously for a long time by some form of biological system.

biochemical oxygen demand [Biology] BOD. The amount of oxygen dissolved in water sufficient to meet the metabolic needs of anaerobic micro-organisms in water rich in organic matter, such as sewage. Also known as biological oxygen demand.

biochemistry [Chemistry] The chemical study of substances present in living organisms along with the reactions and methods for the identification of these substances.

biochrome [Biochemistry] Any naturally occurring plant or animal pigment.

biocide *See* pesticide.

biodegradation [Biochemistry] Chemical degradation by biological influences; especially the break down of substances potentially detrimental to the environment in waste products *e.g.,* detergents in waste water.

biology [Biology] The branch of science concerned with the study of life and living organisms. The main branches of which are botany and zoology. Other branches include cytology, histology, morphology, physiology, embryology, ecology, genetics, and microbiology. Related subjects are biochemistry, biophysics, and biometry.

bioluminescence [Biology] A form of luminescence found in living creatures, such as fire flies, glow worms, etc. The light is emitted when the substance luciferin is oxidized in the presence of the enzyme luciferase.

biomass [Biology] The mass of living matter in a population of particular organisms in a particular area.

biometry [Biology] The application of mathematical and statistical methods to study the biology.

biosphere *See* ecosphere.

biosynthesis [Biochemistry] Synthesis of chemical compounds by living organism.

biotype [Biology] A group of individual organisms having the same genetic characteristic.

biprism [Optics] An optical device used to obtain interference fringes. It consists of two acute-angled prisms placed base to base.

bismuth [Chemistry] Symbol Bi. A white crystalline metal with a pinkish tinge belonging to group V of the periodic table; At. No. 83; r.a.m. 208.98; r.d. 9.78; m.p. 544.5K; b.p. 1833K. The most important ores are bismuthinite (Bi_2S_3) and bismite (Bi_2O_3). Peru, Japan, Mexico, Bolivia, and Canada are major producers. The metal is extracted by carbon reduction of its oxide. Bismuth is the most diamagnetic of all metals and its thermal conductivity is lower than any metal except mercury. The metal has a high electrical resistance and a high Hall effect when placed in magnetic fields. It is used to make low-melting-point casting alloys with tin and cadmium. These alloys expand on solidification to give clear replication of intricate features. It is also used to make thermally activated safety devices for fire-detection and sprinkler systems. More recent applications include its used as a catalyst for making acrylic fibres, as a constituent of malleable iron, as a carrier of uranium-235 fuel in nuclear reactors, and as a specialized thermocouple material. Bismuth compounds (when lead-free) are used for cosmetics and medicinal preparations. It is attacked by oxidizing acids, steam (at high temperatures), and by moist halogens. It burns in air with a blue flame to produce yellow oxide fumes.

bit [Computers] A unit of information in information theory. The amount of information required to specify one of two alternatives, *e.g.*, to distinguish between 1 and 0 in the binary notation as used in computers. Also used as a unit of capacity in a store. *See also* byte; character; word.

bits and bytes [Computers] Data transfer rates are typically expressed in megabits per second, date density is expressed in bits per square inch or per square centimeter; memory capacity is typically specified in megabytes. Converting data density to capacity requires dividing by 8 (Because there are 8 bits per byte) and multiplying by an efficiency factor that depends on the way the data is encoded on the disk.

bit density [Computers] Number of bits which can be placed, per unit length, area or volume, on a storage medium. Also known as record density.

bit location [Computers] Storage position on a record capable of storing one bit.

bit pattern [Computers] A combination of binary digits arranged in a sequence.

bit-slice architec [Computers] An architecure that allows the customer to modify and integrate logic chips by microprogramming elementary circuit operations.

bittern [Chemistry] The mother-liquor remaining after the crystallization of common salt, NaCl, from sea-water; source of compounds of magnesium, bromine, and iodine.

bitumen [Materials] A term covering numerous mixtures of hydrocarbons, more particularly solid or tarry mixtures, soluble in carbon disulphide.

biuret [Chemistry] H_2CONH_-

$CONH_2.H_2O$. An insoluble, colourless crystalline substance formed from urea. Also known as carbamoylurea *See* biuret reaction.

biuret reaction [Chemistry] A chemical reaction in which an alkaline solution of biuret gives a purple colour on the addition of cupric sulphate; used as biochemical test for protein and urea.

bivalent [Chemistry] Divalent. Having a valence of two.

black ash [Materials] Impure sodium carbonate obtained in the Léblanc process.

black body [Physics] An ideal body which would absorb all incident radiation and reflect none.

black body radiation [Physics] The emission of radiant energy which would take place from a black body at a fixed temperature.

blackdamp [Engineering] Carbon dioxide in coal mines.

black hole [Astronomy] A hypothetical region of space possessing a gravitational field so intense that no matter or radiation can escape from it. Such regions are believed to form when a star collapses, having used up all its nuclear fuel. Smaller stars create supernova explosion when they die, leaving neutron stars; it is the more massive stars that are believed to create black holes.

The boundary of the black hole is thought to be a sphere (called the event horizon) with a radius (called the Schwartzchild radius) $2GM/c^2$, where M is the mass of the region, G is the gravitational constant, and c is the velocity of light. The problem of detecting black holes is that, being unable to emit or reflect radiation, they are invisible. However, it is thought that some X-ray binary star exist in which one member of the pair is a black hole.

blacklead [Materials] Natural crystalline form of carbon. A soft grey-black solid; used for making vessels to resist high temperatures, in pencils, and as a lubricant. Also known as Plumbago; graphite.

black light [Optics] Invisible light, such as ultraviolet rays which fall on fluorescent materials, and cause them to emit visible light.

blanc fixe [Chemistry] $BaSO_4$. Artificial barium sulphate; used as an extender in the paint industry.

blanket [Nucleonics] A layer of fertile material surrounding the core of a nuclear reactor to act as a reflector, or for the purpose of breeding new fuel. *See* breeder reactor.

blast furnace [Engineering] A tall, cylindrical smelting furnace for reducing iron ore to pig iron. It is constructed of refractory bricks covered with steel plates and charged from above with a mixture of the ore, limestone ($CaCO_3$), and coke. The coke is ignited at the bottom of the furnace by a blast of hot air; the carbon monoxide so produced reduces the iron oxide to iron, while the heat of the reaction decomposes the limestone into carbon d'oxide and lime. The lime combines with the sand and other impurities in the ore to form a molten slag. The molten iron and the slag are tapped off at the bottom of the furnace.

blasting gelatin [Material] Jelly-like mixture of gun-cotton with nitroglycerin. A very powerful explosive.

bleaching [Chemistry] Removing the colour from coloured materials by chemically changing the dyestuffs into colourless substances.

bleaching agent [Chemistry] An oxidising or reducing chemical such as chlorine, sodium hypochlorite, sulphur dioxide, or hydrogen peroxide used in bleaching.

bleaching powder [Material] Chloride of lime chlorinated lime. A whitish powder, consisting mainly of calcium oxychloride, $CaOCl_2$, prepared by the action of chlorine on calcium hydroxide, $Ca(OH)_2$. The action of dilute acids liberates chlorine, which acts as an oxidizing agent and so bleaches the material.

blende [Mineral] ZnS. Natural zinc sulphide.

blink comparator [Optics] An instrument used to alternately view two pictures in the same visual field in rapid succession; used to detect small differences in similar images.

blink microscope [Optics] A blink comparator which magnifies the compound pictures.

block [ADP] The set of locations or tape positions in which a group of words is stored or recorded as a unit.

block brazing [Metallurgy] The process of joining metals by applying hot blocks to the joint and using a nonferrous filler metal with a melting point higher than 700 K.

blocking [ADP] Combining two or more computer records into one block. [Electronics] Appling a high negative bias to the grid of an electron tube to reduce its anode current to zero.

block polymer [Chemistry] A copolymer having alternating sequences of identical monomer units.

blood [Biology] The red viscid fluid filling heart and blood vessels. It consists of a colourless fluid, plasma in which are suspended the red blood corpuscles (erythrocytes), the white blood corpuscles (leucocytes) and the platelets (thrombocytes). An average adult human male has about 6.2 litres of blood in his body.

blood bank [Engineering] A special refrigerator in which blood is kept after withdrawal from donors, until required for transfusion.

blood cells [Biology] There are

three types of blood cells: red corpuscles (erythrocytes), white corpuscles (leucocytes), and blood plateletes (thrombocytes). The function of the red corpuscles is to transport oxygen throughout the body, by way of the haemoglobin that they contain. The function of the white cells is to combat infection. Also known as blood corpuscles; haemocytes.

blood clotting [Medicine] Constriction of damaged vessel and adhesion of platelets to the side of injury and to each other. In the secondary phase, it involves coagulation over and through the platelet mass.

blood count [Medicine] Determination of the number of red or white cells per cubic millimetre of the blood.

blood crisis [Medicine] The sudden appearance of large numbers of nucleated erythrocytes (red cells) in the circulating blood.

blood gas [Materials] War gas which when enters the body, primarily by breathing, affects body functions by interfering with transfer of oxygen from the lungs via the blood to tissues.

blood groups [Biology] Four different types of blood possess different clotting agents. These clotting agents can be labelled *a* and *b* and the blood can contain one agent or both or neither. Blood group A contains agent *a* and this type of blood is immediately clotted by agent *b*. Blood group

B contains agent *b*, and is immediately clotted by agent *a*. Blood group AB contains agent *a* as well as *b* and is not clotted by adding either agent. Blood group O contains neither clotting agent *a* nor *b* and is clotted by both *a* and *b*. Hence blood group AB is the universal acceptor and blood group O is the universal donor.

blood plasma [Medicine] Blood from which all blood cells have been removed. Plasma is 90% water, in which the principal solutes are proteins, salts, sugar and urea.

blood platelets [Biology] Small membrane-bounded coinishaped particles that circulate in the blood. If a blood vessel should break, the platelets clump together to form a plug to stop the bleeding. Platelets contain substantial quantities of ATP, and it is the diphosphate that causes the agglutination. Human blood contains about 250000 platelets per cubic millimetre. Also known as thrombocytes; platelets.

blood pressure [Medicine] Pressure exerted by blood on the walls of blood vessels.

blood sugar [Biology] The amount of glucose present in the circulating blood. This level is controlled by various enzymes and hormones.

blown oil [Materials] A thickened oil made by blowing air through a natural vegetable or animal oil.

blowpipe [Engineering] A device

for producing a jet of flame by forcing an inflammable gas mixed with air or oxygen through a nozzle at high pressure.

bluestone *See* blue vitriol.

blue vitriol [Chemistry] Crystalline cupric sulphate, $CuSO_4.5H_2O$; used for copper plating and for spraying on plants. Also known as blue stone.

board of trade unit [Electricity] B.O.T. unit. A British unit of electrical energy, the kilowatt-hour. The energy obtained when a power of 1 kilowatt is maintained for 1 hour.

boart *See* bort.

BOD *See* Biochemical oxygen demand.

body-centred [Crystallography] A crystal is said to be body-centred when there is a lattice point at the centre of the body of the crystal as well as at the corners. It is said to be 'face-centred' when there is a lattice point at the centre of the each face.

bog iron ore [Mineral] $Fe_2O_3.xH_2O$. Impure form of hydrated ferric oxide, found in bogs and marshes.

Bohemian glass *See* hard glass.

Bohr atom [Physics] An atomic model having the structure postulated in Bohr theory.

Bohr magneton [Physics] The amount of magnetic moment. *i.e., (he/ 4p mc)* where *h* is Planck's constant, *e* and *m* are the charge and mass of the elctron and *c* is the speed of light.

Bohr orbit [Physics] One of the electron paths around the nucleus in Bohr's model for hydrogen atom.

Bohr radius [Physics] The radius of ground state orbit of hydrogen atom in Bohr theory.

Bohr Sommerfeld theory [Physics] A modification of Bohr theory in which circular as well as elliptical orbits are allowed.

Bohr theory [Physics] The theory published in 1913 by the Danish physicist Niels Bohr (1885-1962) to explain the line spectrum of hydrogen. He assumed that a single electron of mass *m* travelled in a circular orbit of radius *r*, at a velocity *v*, around a positively charged nucleus. The angular momentum of the electron would then be *mvr*. Bohr proposed that electrons could only occupy orbits in which this angular momentum had certain fixed values, $h/2\pi$, $2h/2\pi$, $3h/2\pi$,.... $nh/2\pi$, where *h* is the Planck constant. This means that the angular momentum is quantized, *i.e.* can only have certain values, each of which is a multiple of *n*. Each permitted value of *n* is associated with an orbit of different radius and Bohr assumed that when the atom emitted or absorbed radiation of freqeucny *v*, the electron jumped from one orbit to another; the energy emitted or absorbed by each jump is equal to *hv*. This theory gave good results in predicting the series of lines observed in the hydrogen spectrum but not with any other spectrum. The idea of quantized values of angular momentum was later explained by the wave nature of the electron. Each orbit has to have a whole number of wavelengths around it; *i.e.* $n\lambda = 2\pi r$, where π is the wavelength and n a whole number. The wavelength of a particle is given by h/mv, so $nh/mv = 2\pi r$, which leads to $mvr = nh/2\pi$. Modern atomic theory does not allow subatomic particles to be treated in the same way as large objects, and Bohr's reasoning is somewhat discredited. However, the idea of quantized angular momentum has been retained.

boiled oil [Materials] Linseed oil boiled with, or containing, metalic drying agent, such as lead monoxide, PbO. Used in paints.

boiler scale [Chemistry] Deposits from the contaminants present in the boiler water which form on the internal surfaces of heat absorbing components. It increases the metal temperatures and results in eventual failure of the pressure parts due to overheating.

boiling [Physics] The state of a liquid at its boiling point when the maximum vapour pressure of the liquid is equal to the external pressure to which the liquid is subject, and the liquid is freely converted into vapour. Also known as ebullition.

boiling point [Physics] The temperature at which the maximum vapour pressure of a liquid is equal to the external pressure; the temperature at which the liquid boils freely under that pressure. Boiling points are normally quoted at standard atmospheric pressure, *i.e.* 760 mm of mercury.

boiling water reactor [Nucleonics] BWR. A nuclear reactor in which water, is used as coolant and moderator. Steam is thus produced in the reactor under pressure, and can be used to drive a turbine.

bolide [Astronomy] A large bright meteor; some of these objects explode on entering the earth's atmosphere.

bolometer [Engineering] An extremely sensitive instrument for measuring heat radiations. Consists essentially of two very thin, blackened platinum gratings, forming two arms of a Wheatstone bridge circuit. Radiant heat falling upon one of the gratings raises its electrical resistance, thus causing a deflection of the needle of a galvanometer in the circuit.

Boltzmann constant [Physics] Symbol k. The ratio of the universal gas constant (R) to the Avogadro constant (N_A). It may be thought of therefore as the gas constant per molecule :

$$k = R/N_A = 1.380\ 622 \times 10^{-23}\ JK^{-1}$$

bomb calorimeter [Engineering] A strong metal vessel used for measuring heats of reaction, especially heats of combustion e.g., for determining the calorific value of a fuel. To do this, a known weight of the substance under test is burnt in the vessel, and by measuring the quantity of heat produced, the calorific value is calculated.

bond [Chemistry] Valence bond, linkage. A representation of a valence link by which one atom is attached to another in a chemical compound.

bond energy [Chemistry] An amount of energy associated with a bond in a chemical compound. It is obtained from the heat of atomization. For instance, in methane the bond energy of the C-H bond is one quarter of the enthalpy of the process

$$CH_4(g) \rightarrow C(g) + 4H(g)$$

Bond energies can be calculated from the standard enthalpy of formation of the compound and from the enthalpies of atomization of the elements. Energies calculated in this way are called average bond energies or bond-energy terms. They depend to some extent on the molecule chosen; the C-H bond energy in methane will differ slightly from that in ethane. The bond dissociation energy is a different measurement, being the energy required to break a particular bond; e.g. the energy for the process:

$$CH_4(g) \rightarrow CH_3.(g) + H.(g)$$

bond ash [Materials] Ash obtained

by heating bones in air. Consists mainly of calcium phosphate, $Ca_3(PO_4)_2$; used in cleaning jewellery and in some pottery.

bone black [Materials] *See* animal charcoal.

bone char [Materials] *See* animal charcoal.

bone oil [Materials] Product obtained by the destructive distillation of bones. Dark-oily, evil-smelling liquid used as a source of pyridine. Also known as Dippel's oil.

Boolean algebra [Mathematics] A branch of symbolic logic used in computers. Logical operations are performed by operators such as 'and', 'or', 'not-and', in a way analogous to mathematical signs.

booster [Electricity] A generator or transformer inserted in a circuit in order to increase or decrease the magnitude or to change the phase of the voltage acting on the circuit. [Electronics] A separate radio-frequency amplifier connected between an antenna and a television receiver to amplify weak signals.

booster dose [Medicine] A part of the immunizing agent given at a later period to stimulate effects of previous dose of the same agent.

boranes [Chemistry] Hydrides of boron, having the general formula BnH_2n+_2; the boron analogues of alkanes.

borate [Chemistry] A salt or ester of boric acid.

borax [Minerals] $Na_2B_4O_7$. $10H_2O$. A white, yellow, blue or grey mineral and is the chief ore of boron. It is used as an antiseptic, flux and a cleaning agent.

borax bead test [Chemistry] A chemical test for the presence of certain metals. A bead of borax fused in a wire loop will react chemically with the salts of a number of metals, often producing colours which help to identify the metal; *e.g.*, manganese compounds give a violet bead, cobalt a deep blue.

Bordeaux mixture [Materials] A mixture of cupric sulphate, $CuSO_4$, calcium oxide, CaO, and water; used for spraying plants as a fungicide for plant diseases.

Born-Haber cycle A cycle of reactions used for calculating the lattice energies of ionic crystalline solids. For a compound MX, the lattice energy is the enthalpy of the reaction

$$M^+(g) + X^-(g) \rightarrow M^+X^-(s)$$
$$\Delta H_L$$

The standard enthalpy of formation of the ionic solid is the enthalpy of the reaction

$$M(s) + 1/2X_2(g) \rightarrow M^+X^-(s)$$
$$\Delta H_f$$

The cycle involves equating this enthalpy (which can be measured) to the sum of the enthalpies of a number of steps proceeding from the elements to the ionic solid. The steps are :

(1) Atomization of the metal :

$$M(s) \rightarrow M(g) \ \Delta H_1$$

(2) Atomization of the nonmetal :

$$1/2X_2(g) \rightarrow X(g) \ \Delta H_2$$

(3) Ionizatin of the metal :

$$M(g) \rightarrow M^+(g) + \bar{e} \ \Delta H_3$$

This is obtained from the ionization potential.

(4) Ionization of the nonmetal :

$$X(g) + \bar{e} \rightarrow X^-(g) \ \Delta H_4$$

This is the electron affinity.

(5) Formation of the ionic solids :

$$M^+(g) + X^-(g) \rightarrow M^+X^-(s)$$
$$\Delta H_L$$

Equating the enthalpies given above , ΔH_L can be found.

boron [Chemistry] B. Element. At. No. 5. A.Wt. 10.811. A brown amorphous powder or yellow crystals, m.p. 2573K. It occurs as borax and boric acid; used for hardening steel and for producing enamels and glasses. It is used in semi-conductors and in filaments for specialized aerospace applications. Amorphous boron is used in flares, giving a green coloration. The isotope boron-10 is used in nuclear reactor control rods and shields.

boron counter tube [Engineering] A counter tube containing a boron chamber used for counting neutrons. The counting pulse results from particles emitted when neutrons interact with the $^{10}_{5}B$ isotope.

boron thermopile [Nucleonics] A thermopile in which alternate thermocouple junctions are coated with boron; exposure to a flux of slow neutrons generates heat in these junctions, producing an output voltage proportional to neutron flux.

bort [Minerals] Impure or discoloured diamond; useless as a gem, it is as hard as pure diamond and is used for drills, cutting tools, etc. Also known as boart.

Bosch process [Chemistry] An industrial process for the manufacture of hydrogen. Water gas, a mixture of carbon monoxide and hydrogen, is mixed with steam and passed over a heated catalyst. The steam reacts chemically with the carbon monoxide to give carbon dioxide, CO_2 and hydrogen. The carbon dioxide is then removed by dissolving it in water under pressure.

Bose-Einstein statistics [Physics] The branch of statistical mechanics used with systems of identical particles having the property that the wave function remains unchanged if any two particles are interchanged.

bosons [Physics] Particles that conform to Bose-Einstein statistics, such as photons and mesons, whose numbers are not conserved in particle interactions. Bosons have integral spin (0, 1, 2).

botany [Biology] A branch of biological sciences dealing with the study of plants and their life.

Bourdon gauge [Engineering] A pressure gauge for steam boilers, etc. It depends on the tendency of a partly flattened curved tube to straighten out when under internal pressure.

Boyle's law [Physics] At a constant temperature, the volume of a given quantity of any gas is inversely proportional to the pressure upon the gas; *i.e.,* V α $1/P$, *or PV* = constant. It is only true for a perfect gas.

Bragg's law [Physics] When a beam of X-ray, of wavelength λ stikes a crystal surface, the maximum intensity of the reflected ray occurs when $\sin\theta = n\lambda/2d$. Where d is the distance separating the layers of the atoms or ions in the crystal, θ is the complement of the angle of incidence, and n is an integer.

Bragg's scattering [Physics] Scattering of X-rays or neutrons by regularly spaced atoms in a crystal, for which constructive interference occurs only at definite angles called Bragg's angle. Also known as Bragg's diffraction.

brake horsepower [Mechanics] The horsepower of an engine measured by the degree of resistance offered by a brake; it presents the useful horsepower that the engine can develop.

branched chain [Chemistry] A chain of carbon atoms in an organic molecule, in which the main chain has one or more branches.

branching [Nucleonics] The occurrence of more than one radioactive disintegration scheme for a particular nuclide.

brandy [Materials] A potable alcholic beverage distilled from wine or fermented fruit juice, usually after the ageing of the wine in wooden casks.

brass [Materials] A large class of alloys, consisting principally of copper and zinc.

braze [Metallurgy] To solder metals by melting a nonferrous filler metal such as brass or hard solder, with a melting point lower than that of base metals, at the point of contact.

breeder reactor [Nucleonics] A nuclear reactor that produces the same kind of fissile material as it burns, *e.g.*, a reactor using plutonium as a fuel can produce more plutonium than it uses by conversion of uranium-238.

Bremsstrahlung [Nucleonics] (German, meaning 'brake radiation'). X-rays emitted when an electron strikes a positively charged nucleus; it results from the direct conversion of kinetic energy into electromagnetic radiation.

brewing [Engineering] The making of beer. Malt is ground and mixed with water. In the resulting 'mash', chemical changes take place, the chief of which is the conversion of starch into maltose, forming a sweetish liquid known as wort. This is boiled with the addition of hops. After cooling and removal of solids, yeast is added and fermentation occurs.

Brewster's law [Optics] The tangent of the angle of polarization is numerically equal to the refractive index of the reflecting medium when the polarization is a maximum.

brightner [Metallurgy] Any substance which when employed in small concentrations in the electrolytic bath for electroplating metal, yields smoother and brighter coatings.

brimestone [Materials] Sulphur fused into blocks or rolls. A common name for native sulphur.

Brinell test [Engineering] A test for the hardness of metals. A ball of chrome steel, or other hard material, of standard size, is pressed by a heavy load into the surface of the metal, and the diameter of the depression is measured. The Brinell Number is the ratio of the load in kilograms to the area of the depression in square millimetres.

Britannia metal [Metallurgy] A silver white tin alloy of variable composition, containing 85% - 90% tin, with some antimony and copper, and sometimes also zinc, lead and bismuth.

British thermal unit [Physics] The quantity of heat required to raise the temperature of 1 lb of water through 1°Fahrenheit; equal to 251.997 calories or 1055.06 joules.

broad-spectrum antibiotic [Medicine] An antibiotic which is effective against gram-positive as well as gram-negative bacterial species.

bromate [Chemistry] A salt of bromic acid, $e.g.$, $NaBrO_3$.

bromide [Chemistry] Salt of hydrobromic acid, HBr; binary compound with bromine, 'Bromide' of pharmacy is potassium bromide, KBr.

bromide paper [Materials] Photographic paper coated with silver bromide, AgBr.

bromination [Chemistry] A reaction in which one or more bromine atoms are substituted for hydrogen atoms in an organic molecule.

bromine [Chemistry] Symbol Br. A halogen element; At.No. 35; r.d. 3.13; m.p. 266 K b.p. 331.8 K. It is a red volatile liquid at room temperature, having a red brown vapour. Bromine is obtained from brines in the USA (displacement with chlorine); a small amount is obtained from sea water in Anglesey. Large quantities are used to make 1,2-dibromoethane as a petrol additive. It is also used in the manufacture of many other compounds. Chemically, it is intermediate in reactivity between chlorine and iodine. It forms compounds in which it has oxidation states of 1, 3, 5, or 7. The liquid is harmful to human tissue and the vapour irritates the eyes and throat.

bromine number [Chemistry] The amount of bromine absorbed by a fatty oil; indicates the degree of unsaturation and purity of the oil.

bronchitis [Medicine] An inflammation of the bronchial tube.

Brönsted acid [Chemistry] A chemical species which can act as a source of protons.

bronze [Metallurgy] 1. A class of alloys of copper and tin. 2. A copper alloy containing no tin, $e.g.$, aluminium bronze is an alloy of copper and aluminium.

brookite [Minerals] TiO_2, A brown, reddish, or black mineral containing titanium.

Brownian movement [Physics] Erratic random movements performed by microscopic particles in a disperse phase; $e.g.$, particles in suspension in a liquid, or smoke particles in air. It is caused by the continuous irregular bombardment of the particles by the molecules of the surrounding medium, $i.e.$, statistical pressure fluctuations over the particle.

brush discharge [Electricity] The discharge of electricty from sharp points on a conductor. The surface density ($i.e.$, quantity of electricity per unit area) is greatest at sharp points; the high charge at such points causes a displacement of the charge on the air particles near the points, and hence an attraction to the points. On reaching the change, the particles acquire some of the points and are repelled. This causes a stream of charged air particles to leave the vicinity of the points.

brush station [Computers] A location in the data processing device which the holes in a punched card are sensed by brushes sweeping electrical contacts.

bubble chamber [Nucleonics] An instrument for making the tracks of ionizing particles visible as a row of bubbles in a liquid. The liquid is heated to slightly above its boiling point and maintained under pressure to prevent boiling. Immediately before the passage of the particles the pressure is reduced, the ionized particles than act as centres for the formation of small vapour bubbles, which can be photographed to give a record of the tracks of the particles.

bucking transformer [Electricity] A transformer whose voltage opposes that of a second transformer.

bucking voltage [Electricity] A voltage having the polarity opposite to that of another voltage against which it acts.

bubble memory [Computers] A method by which information is stored as magnetized dots (bubbles) which rests on a thin film of semiconductor material.

bubble sort [Computers] A sort achieved by exchanging pairs of keys, which begins with the first pair and exchanges successive pairs until the list is ordered.

buffer [Chemistry] A solution the hydrogen ion concentration of which, and hence the acidity or alkalinity, is practically unchanged by dilution or as a result of a chemical reaction. [Computers] A storage device used to compensate for a difference in rate of flow of information or time of occurrence of events when information is transmitted from one computer device to another.

buffing [Engineering] The smoothing and brightening of a surface by an abrasive compound pressed against it by a soft wheel or belt.

bug [Computers] A defect in program code or in designing a routine or computer.

bulk density [Engineering] The density of a powder or of a porous of granular substance, calculated for unit, volume of the substance including the pores or spaces between the grains; it is generally less than the true density of the material.

bulk modulus [Mechanics] Elastic modulus applied to a body having uniform stress distributed over the whole of its surface. Its value is given by the expression pv/v where p = intensity of stress, V = original volume of the body, and v=change in volume.

bulk strength [Mechanics] The strength per unit volume of a solid.

Bunsen burner [Engineering] A gas burner used in laboratories.

It consists of a metal tube with an adjustable air-value for burning a mixture of gas and air.

Bunsen cell [Electricity] A primary cell in which the anode consists of zinc and is immersed in dilute sulphuric acid, and the cathode consists of carbon immersed in concentrated nitric acid.

buoyancy [Physics] The upward thrust on a body immersed in a fluid. This force is equal to the weight of the fluid displaced; *see* Archimedes' principle.

burette [Chemistry] A graduated glass tube with a tap, for measuring the volume of liquid run out from it. It is used in volumetric analysis.

burning *See* combustion.

burnt alum [Materials] A white porous mass of anhydrous potassium aluminium sulphate, $K_2SO_4 \cdot Al_2(SO_4)_3$, obtained by heating alum.

bus [Computers] A channel or path for transferring data or electrical signals.

business applications [Computers] Computer systems involving normal day-to-day accounting procedures.

butter of antimony [Chemistry] Antimony trichloride. $SbCl_3$. A white crystalline substance, m.p. 346K.

butyl [Chemistry] The univalent alkyl radical C_4H_9-.

butylaminobenzoate. [Medicine] Local anaesthetic used as ointment.

butyl rubber [Materials] A synthetic rubber; copolymer (*see* polymerization) of iso-butylene and sufficient isoprene (2%-3%) to enable vulcanization to be effected. Owing to its low permeability to gases, butyl rubber is used in the manufacture of tyre inner tubes.

butyryl [Chemistry] The univalent radical $CH_3(CH_2)_2CO-$.

buzzer [Electricity] An electromagnetic device having an armature which vibrates rapidly, producing a buzzing sound.

BWR *See* boiling water reactor.

Bx-cable [Electricity] Insulated wires in flexible metal tubing used for bringing electric power to electronic equipments.

byerit [Geology] Bituminous coal that does not crack in fire and melts and enlarges upon strong heating.

by-pass capacitor [Electricity] capacitor having a reactance which is small compared with a resistence connected in parallel with it.

by-product [Chemistry] A substance obtained during the manufacture of some other substance but is not considered as the principal material. Often as important as the manufactured substance itself, *e.g.*, the by-products of coal-gas manufacture include ammonia, coal-tar, and coke.

byte [Computers] A single unit of information handled by a computer; a sequence of adjacent binary digits operated upon as a unit in a computer and usually shorter than a word.

C

cable [Electricity] Strands of insulated electrical conductors laid together and wrapped in a heavy insulation.

cacodyl [Chemistry] The dimethylarsino group, $(CH_3)_2As^-$, derived from arsine.

cadaver [Medicine] a dead animal or human body to be studied by dissection.

cadmium [Chemistry] Cd. Element At. No. 48. At.Wt. 112.40. A soft silvery-white metal, placed in group IIB of periodic table, m.p. 594K; b.p. 1038K. The element's name is derived from the ancient name for calamine, zinc carbonate $ZnCO_3$, and it is usually found associated with zinc ores, such as sphalerite (ZnS), but does occur as the mineral greenockite (CdS). Cadmium is usually produced as an associated product when zinc, copper, and lead ores are reduced. Cadmium is used in low-melting-point alloys to make solders, in Ni-Cd batteries, in bearing alloys, and in electroplating (over 50%). Cadmium compounds are used as phospho-rescent coatings in TV tubes. Cadmium and its compounds are extremely toxic at low concentrations; great care is essential where solders are used or where fumes are emitted. It has similar chemical properties to zinc but shows a greater tendency towards complex formation.

cadmium cell [Electricity] Standard primary cell. *See* Weston cell.

cadmium-silver oxide cell [Electricity] An alkaline electrolyte cell which may be used without recharging in primary batteries or which may be recharged for secondary battery use.

cadmium sulphide cell [Electricity] A photoconductive cell in which a small wafer of cadmium sulphide provides an extremely high dark-light resistance ratio.

caesium [Chemistry] Cs. A soft silvery-white metallic element belonging to group I of the periodic table; At.No. 55; r.d. 1.88; m.p. 301.5K; b.p. 951K. It occurs in small amounts in a number of minerals, the main source being carnallite $(KCl.MgCl_2.6H_2O)$. It is obtained by electrolysis of molten caesium cyanide. The natural isotope is caesium-133. There are 15 other radioactive isotopes. Caesium-137 (half-life 33 years) is used as a gamma source. As the heaviest alkali metal, caesium has the lowest ionization potential of all elements, hence its use in photoelectric cells, etc.

caesium clock [Engineering] A device used in the SI unit definition of the second. It is based on the energy difference between two states of the caesium nucleus in a magnetic field. This energy difference corresponds to a frequency of 9192631770 hertz. A beam of caesium atoms is split into the two components by a non-uniform magnetic field. Nuclei in the lower state are irradiated in a cavity by radio-frequency radiation at the difference frequency. Some are excited to the higher frequency by absorbing this radiation. By reanalyzing the mixture of atoms and using a feedback system, the *r-f* oscillator can be locked to the difference frequency with an accuracy of one part in 10^{13}. It thus constitutes an extremely accurate clock which can be used as a standard for time measurements.

CAL [Computers] A higher-level computer language, developed especially for time sharing purposes.

calcareous [Scientific Technique] Resembling, containing or composed of calcium carbonate.

calcic [Scientific Technique] Derived from or containing calcium.

calcination [Chemistry] Strong heating; conversion of hydrates, carbonates or other compounds into corresponding oxides by heating at high temperature in air.

calcite [Minerals] Natural crystalline calcium carbonate, $CaCO_3$. Also known as calcspar.

calcium [Chemistry] Ca. A soft grey metallic element belonging to group II of the periodic table; At.No. 20; r.d. 1.55; m.p. 1118K b.p. 1757K. Calcium compounds are common in the earth's crust; e.g. limestone and marble $(CaCO_3)$, gypsum $(CaSO_4.2H_2O)$, and fluorite (CaF_2). The element is extracted by electrolysis of fused calcium chloride and is used as a getter in vacuum systems and a deoxidizer in producing nonferrous alloys. It is also used as a reducing agent in the extraction of such metals as thorium, zirconium, and uranium.

Calcium is an essential element for living organisms, being required for normal growth and development.

calefaction [Engineering] Warming.

calender [Engineering] To pass a material between rollers or plates to form its thin sheets or to make it smooth and glossy.

calibration [Scientific Technique] The graduation of an instrument to enable measurements in definite units to be made with it; thus the arbitary scale of a galvanometer may be calibrated in amperes, thereby converting the instrument into an ammeter for measuring electric current.

caliche [Minerals] Impure natural sodium nitrate $NaNO_3$, found in chile.

californium [Chemistry] Cf. A radioactive metallic transuranic element belonging to the actinides; A.No. 98; mass number of the most stable isotope 251 (half-life about 700 years). Nine isotopes are known; californium-252 is an intense neutron source, which makes it useful in neutron activation analysis and potentially useful as a radiation source in medicine.

callipers [Engineering] Also spelled calipers. An instrument for measuring the distanc between two points, especially on a curved surface; e.g., for measuring the internal and external diameters of tubes.

calite [Metallurgy] A practically non-corrodible alloy of iron, nickel and aluminium below 1477K.

calling device [Electronics] A device which generates the pulses required for establishing connections in an automatic telephone switching system.

calomel [Chemistry] Hg_2Cl_2. mercurous chloride.

calomel electrode [Chemistry] A frequently used half cell in which the electrode is mercury coated with calomel (HgCl) and the electrolyte is a solution of potassium chloride and saturated calomel. The standard electrode potential is -0.2415 volt (298K)

In the calomel half cell the reactions are

$$HgCl(s) \rightleftharpoons (aq) + Cl^-(aq)$$

$$Hg + (aq) + \bar{e} \rightleftharpoons Hg(s)$$

The overall reaction is

$$HgCl(s) + e^- \rightleftharpoons Hg(s) + Cl^-(aq)$$

equivalent to a $Cl_2(g) \mid Cl^-(aq)$ half cell in which the pressure is the dissociation pressure of HgCl.

calorescence [Physics] Phenomenon of absorption of infrared light radiations by a surface, their conversion into heat, and the consequent emission of visible radiation. The production of visible light by infrared radiation.

calorie [Physics] A unit for measuring the quantity of heat. The amount of heat required to raise temperature of 1 g. of water through 1°C. The 15° calorie is defined as the amount of heat required to raise the temperature of 1 g. of water from 14.5°C. to 15.5°C. This calorie is equal to 4.1855 joules. The International Table Calorie is defined as 4.1868 joules. The joule is the SI unit of heat.

calorific value (fuel) [Chemistry] The quantity of heat produced by a given mass of the fuel on complete combustion. Expressed in joules per kilogram (SI units), calories per gram (C.G.S. units).

calorize [Metallurgy] A common process by which a coating of aluminium and aluminium-iron alloys is produced on iron and steel.

calutron [Nucleonics] An electromagnetic apparatus for the separation of uranium from other elements.

calx [Materials] 1. The powdery oxide of a metal formed when an ore or a mineral is roasted. 2. Quicklime (calcium oxide).

camera, photographic [Optics] A device for obtaining photographs or exposing cinematic film either coloured or black and white. A camera consists essentially of a light-proof box with a lens at one end and a light-sensitive film or plate at the other. An 'exposure' is made by opening a 'shutter' over the lens for a predetermined period during which as image of the object to be photographed is thrown upon the light-sensitive film.

camera, television [Electronics] The part of a television system that converts optical image into electrical signals. It consists of an optical lens system similar to that used in a photographic camera, the image from which is projected into a 'camera tube'. The camera tube comprises a photosensitive mosaic that is scanned by an electron beam housed in an evacuated glass tube. The output signals of the camera tube are usually preamplified within the body of the camera.

Canada balsam [Materials] A yellowish liquid derived from fir trees with a refractive index similar to that of glass. Used for mounting microscopic slides and as an adhesive for optical instruments.

canal rays [Physics] Positively charged ions produced during the discharge of electricity in gases, driven to the cathode by the applied potential difference and allowed to pass through canals bored in the cathode.

candela [Optics] Cd. The SI unit of luminous intensity. Defined as the luminous intensity, in the perpendicular direction, of a surface of 1/600000 square metre of a black body at the temperature of freezing platinum under a pressure of 101325 Nm^{-2}. The candela now replaces the international candle. Also known as new candle.

candle power [Optics] Luminous intensity as formerly expressed in terms of the international candle but now expressed in candela.

candle wax [Materials] Usually either paraffin wax or stearine.

canonical form [Chemistry] One of the possible structures of a molecule that together form a resonance hybrid.

canton's phosphorus [Materials] Impure calcium sulphide, CaS, having the property of phosphorescence after exposure to light; used in luminous paints.

caoutchouc [Materials] Crude rubber which has been cured over fire into a solid dark mass.

capacitance [Electricity] Electrical capacity. C. The property of a system of electrical conductors and insulators that enables it to store electric charge when a potential difference exists between the conductors. It is measured by the charge that must be communicated to such a system to raise its potential by one unit. The SI unit of capacitance is the farad.

capacitor [Electricity] A deivce used to store electric charge in an electrical circuit. It consists of two parallel metal plates separated by an insulating material. When the elecrical circuit is switched off the condenser discharges, giving rise the electrical energy in the curcuit. If Q is the charge stored by the condenser at a potential difference V, then the capacitor is said to have a capacity C equal to Q/V.

capillary action [Physics] A general term for phenomena observed in liquids due to unbalanced inter-molecular attraction at the liquid boundry; e.g., the rise or depression of liquids in narrow tubes, the formation of films drops, bubbles, etc. Also known as capillarity.

capillary tube [Engineering] A tube of very small internal diameter.

capture [Nucleonics] A process by which an atomic or nuclear system acquires an additional particle, e.g., the capture of electrons by ions or of neutrons by nuclei. 'Radiative capture' is a nuclear capture process that results in the emission of gamma rays only.

caramel [Chemistry] a brown substance of complex composition, formed by the action of heat on sugar.

carat [Physics] Measure of weight of diamonds and other gems; formerly 3.17 grains (0.2053g), now standardized as the international carat, 0.200g. [Metallurgy] A measure of fineness of gold, expressed as parts of gold in 24 parts of the alloy. Thus 24 carat gold is pure gold, 18 carat gold contains 18 parts in 24.

carbanion [Chemistry] An organic ion with a negative charge on a carbon atom; i.e. an ion of the type R_3C^-. Carbanions are intermediates in certain types of organic reaction (e.g. the aldol reaction).

carbazole [Chemistry] A group of organic heterocyclic compounds containing dibenzopyrole system.

carbide [Chemistry] Binary compound of carbon; loose term for calcium carbide.

carbmol [Chemistry] Former name for methanol, CH_3OH.

carbocyclic compounds [Chemistry] A class of organic compounds containing closed rings of carbon atoms in their molecules. It includes alicyclic (e.g., cyclohexane) and aromatic (e.g., benzene) compounds.

carbohydrases [Biochemistry] A class of enzymes that hydrolyze (*see* hydrolysis) carbohydrates; *e.g.*, amylase, lactase, and maltase.

carbohydrates [Biochemistry] A large group of organic compounds composed of carbon, hydrogen, and oxygen only, with the general formula $C_x(H_2O)_y$. Comprises monosaccharides, disccharides (both sugars), and polysaccharides (starch and cellulose). Carbohydrates play an essential part in the metabolism of all living organisms, starch being the principal form in which energy is stored and cellulose being the principal structural material of plants.

carbolic acid [Chemistry] Former name for phenol, C_6H_5OH.

carbon [Chemistry] C. A nonmetallic element belonging to group IV of the periodic table; At.No. 6; m.p.3823K b.p. 4562K. Carbon has two main allotropic forms (*see* allotropy). Diamond (r.d. 3.52) occurs naturally and small amounts can be produced synthetically. It is extremely hard and has highly refractive crystals. The hardness of diamond results from the covalent crystal structure, in which each carbon atom is linked by covalent bonds to four others situated at the corners of tetrahedron. The C-C bond length is 0.154 nm and the bond angle is 109.5°.

Graphite (r.d. 2.25), the other allotrope, is a soft black slippery substance (sometimes called *black lead* or *plumbago*). It occurs naturally and can also be made by the Acheson process. In graphite the carbon atoms are arranged in layers, in which each carbon atom is surrounded by three others to which it is bound by single or double bonds. The layers are held together by weaker van der Waals' forces. The carbon-carbon bond length in the layers is 0.142 nm and the layers are 0.34 nm apart. Graphite is a good conductor of heat and electricity. It has a variety of uses including electrical contacts, high-temperature equipment, and as a solid lubricant. Graphite mixed with clay is the 'lead' in pencils (hence its alternative name). There are also several amorphous forms of carbon, such as carbon black and charcoal.

carbon-12 [Nucleonics] A stable isotope of carbon with mass number 12. forming about 98.9% of natural carbon; basis of the new scale of atomic masses.

carbon-14 [Nucleonics] A naturally occurring radioisotope of carbon having a mass number 14 and half life of 5780 years; used in radiocarbon dating and in the elucidation of the metabolic path of carbon in photosynthesis.

carbonado [Mineral] A black, dark-coloured, impure variety of diamond, useless as a gem but very hard and is used for drills, etc. Also known as black diamond.

carbonate [Chemistry] A salt of carbonic acid, H_2CO_3.

carbonation [Chemistry] Treatment with carbon dioxide, usually for the formation of carbonates.

carbon black [Material] A finely divied soot-like form of carbon produced by pyrolysis or by incomplete combustion from carbon-rich materials, such as mineral oils, acetylene, or natural gas. Used mainly as a rein forcing pigment in rubber, and as a black pigment in inks, plastics, etc.

carbon cycle [Biology] The circulation of carbon (as carbon dioxide) between living organisms and the atmosphere. Carbon dioxide is built in complex carbon compounds by plants during photo-synthesis; animals obtain their carbon atoms by feeding on plants or other animals; during respiration, and by decay after death, some of this carbon is returned to the atmosphere in the form of carbon dioxide [Nucleonics] A series of nuclear reactions in which four hydrogen nuclei combine to form a helium nucleus with the liberation of energy and two positrons. The process is believed to be the source of energy in many stars and to take place in six stages. In this series carbon-12 acts as if it were a catalyst, being reformed at the end of the series:

$$^{12}_{6}C + {}^{1}_{1}H \longrightarrow {}^{13}_{7}N + \gamma$$
$$^{13}_{7}C \longrightarrow {}^{13}_{6}C + e^{+}$$
$$^{13}_{6}C + {}^{1}_{1}H \longrightarrow {}^{14}_{7}N + \gamma$$
$$^{14}_{7}N + {}^{1}_{1}H \longrightarrow {}^{15}_{8}O + \gamma$$

$$^{15}_{8}O \longrightarrow {}^{15}_{7}N + e^{+}$$
$$^{15}_{7}N + {}^{1}_{1}H \longrightarrow {}^{12}_{6}C + {}^{4}_{2}He.$$

carbon dating [Nucleonics] A method of estimating the ages of archaeological specimens of biological origin. As a result of cosmic radiation a small number of atmospheric nitrogen nuclei are continuously being transformed by neutron bombardment into radioactive nuclei of carbon-14:

$$^{14}_{7}N + {}^{1}_{0}n \longrightarrow {}^{14}_{6}C + {}^{1}_{1}p$$

Some of these radiocarbon atoms find their way into living trees and other plants in the form of carbon dioxide, as a result of photosynthesis. When the tree is cut down photosynthesis stops and the ratio of radiocarbon atoms to stable carbon atoms begins to fall as the radiocarbon decays. The ratio $^{14}C/^{12}C$ in the specimen can be measured and enables the time that has elapsed since the tree was cut down to be calculated. The method has been shown to give consistent results for specimens up to some 40,000 years old, though its accuracy depends upon assumptions concering the past intensity of the cosmic radiation. Also known as radio carbon dating.

carbon fibre [Materials] 1. A material consisting of black silky threads of pure carbon that can be made stronger and stiffer than any other material of the same weight 2. Commercial material made by pyrolyzing and spun, felted, or woven raw material to char at temperatures from 950K to 2100K.

carbon film resistor [Electricity] A resistor made by depositing a thin film of carbon on a ceramic base.

carboniumion [Chemistry] An organic ion with a positive charge on a carbon atom; *i.e.* an ion of the type R_3C^+. Carbonium ions are intermediates in certain types of organic reaction (*e.g.* Williamson's synthesis).

carbonization *See* destructive distillation.

carbon pile [Electricity] A variable resistor consisting of a stack of carbon discs mounted between a fixed metal plate and a movable one which serve as the terminals of the resistor; the resistance value is reduced by applying pressure to the movable part.

carbon steel [Metallurgy] Steel containing about 2% carbon.

carbonyl [Chemistry] The divalent group-CO; characteristic of aldehydes and ketones.

carbonyls [Chemistry] Complex compounds of metals with carbon monoxide as ligand; *e.g.*, nickel tetra carbonyl, $Ni(CO)_4$.

carborundum [Chemistry] Silicon carbide, SiC. A dark crystalline solid, nearly as hard as diamond, used as an abrasive and as a refractory material. Made by heating silica, SiO_2, with carbon in an electric furnace.

carboxyl group [Chemistry] The univalent group, --COOH, char-

acteristic of the organic carboxylic acids.

carboxylic acids [Chemistry] Organic acids containing one or more carboxyl groups in the molecule; *e.g.*, acetic acid, CH_3COOH; phthalic acid, $C_6H_4(COOH)_2$. They form salts with bases and esters with alcohols. *See also* fatty acid.

carburettor [Engineering] A device in the internal-combustion petrol engine for mixing air with petrol vapour prior to explosion.

carcinogen [Medicine] A substance capable of producing cancer (carcinoma).

card [Computers] An information carrying medium for the introduction of data and instructions into computer.

card code [Computers] The representation of characters on a punched card by means of punching one or more holes per column.

card feed [Computers] A device which inserts cards into the machine one at a time.

card verifier [Computers] An electromechanical device which allows the operator ot check that a card has been properly keypunched.

carnallite [Minerals] Natural potassium magnesium chloride, $KCl.MgCl_2.6H_2O$, found in the Stassfurt deposits. An important source of potassium salts.

carnotite [Minerals] $K(UO_2)_2$ $(VO_4)_2.xH_2O$. Uranium potassium vanadate of variable composition.

Carnot's cycle [Physics] An ideal reversible cycle of operations for the working substance of a heat engine. The four steps in the cycle are : (*a*) isothermal expansion, the substance taking in heat and doing work; (*b*) adiabatic expansion, without heat change, external work done; (*c*) isothermal compression, heat given out, work done on the substance by external forces; (*d*) adiabatic compression, no heat change, work done on the substance.

Carnot's principle [Physics] The efficiency of any reversible heat engine depends only on the temperature range through which it works and not upon the properties of any material substance. If all the heat is taken up at absolute temperature T_1 and all given out at absolute temperture T_2 (as in Carnot's cycle), the efficieny is $(T_1-T_2)/T_1$.

Caro's acid [Chemistry] H_2SO_5 peroxymonosulphuric acid.

carotenoid [Biochemistry] Any of a group of yellow, orange, red, or brown plant pigments chemically related to terpenes. Carotenoids are responsible for the characteristic colour of many plant organs, such as ripe tomatoes, carrots, and autumn leaves.They also function in the light reactions of photosynthesis.

carpology [Botany] The study of the morphology of fruits and seeds.

carrier [Chemistry] A substance assisting a chemical rection by combining with part or all of the molecule of one of the reacting substance to form a compound that is then easily decomposed again by the other reacting substance; the carrier is thus left unchanged. [Nucleonics] An inactive substance used to transport a radiosotope in radio-active tracing. A radioisotope is said to ba 'carrier-free' if it can be used without a carrier. [Physics] The mobile electrons or holes that carry charges in a semiconductor.

carrier channel [Electronics] The equipment and lines that make up a complete carrier-current circuit between two or more points.

carrier current [Electronics] A high frequency alternating current superimposed on ordinary telephone, telegraph, and power-line frequencies for communication and control purposes.

carier density [Electronics] The density of electrons and holes in a semiconductor.

carrier gas [Chemistry] The gas that carries the sample in gas chromatography.

carrier line [Electronics] An transmission line used for multiple channel carrier communication.

carrier noise [Electronics] Noise produced by undesired variation of a radio-frequency signal in absence of any intended modulation.

carrier transmission [Electronics] Transmission in which the transmitted electric wave is resulted from the modulation of a single frequency wave by a modulating wave.

carrier wave [Physics] A continuous electromagnetic radiation, of constant amplitude and frequency, amplitude and frequency, emitted by a radio transmitter. By modulation of the carrier wave, oscillating electric currents caused by sounds at the transmitting end are conveyed by it to the receiver. [Electronics] The radio wave produced by a transmitter when there is no modulating signal, or any other wave, recurring series of pulses, or direct current capable of being modulated.

cascade [Engineering] An arrangement of separation devices, such as isotope separators, connected in series so that they multiply the effect of each individual device.

cascade amplifier [Electronics] A vacuum tube or transisterised amplifier containing two or more stages arranged in conventional series manner.

cascade liquefier [Engineering] An apparatus used for liquefying air, oxygen, etc. A gas cannot be liquefied until it is brought to a temperature below its critical temperature. In the cascade liquefier the critical temperature of the gas is reached step by step, using a series of gases having successively lower boiling points.

cascade process [Engineering] A process used in the separation of isotopes. It consists of a series of stages connected so that the separation produced by one stage is multiplied in subsequent stages. In a 'sample cascade' the enriched fraction is fed to the succeeding stages and the depleted fraction to the preceeding stage.

cascade shower See shower.

cased glass [Materials] Glass composed of two or more layers of different colours.

Cassegranian telescope [Optics] A reflecting telescope in which a hole in the centre of the primary mirror allows the light to pass through it to the eye-piece or the photographic plate.

cassiopeium See lutetium.

cassiterite [Mineral] SnO_2. Natural stannic oxide; a principle ore of tin.

casting alloy [Metallurgy] An alloy which cannot be forged or rolled but can be shaped only as casting.

cast iron [Metallurgy] Impure, brittle form of iron, such as produced in the blast furnace. It contains from 2%-4.5% carbon in the form of cementite and usually also some manganese, phosphorus, silicon, and sulphur. Generally not used direct, but converted in to steel or wrought iron. Also known as pig-iron.

catabolism [Biochemistry] The

metabolic breakdown of large molecules in living organisms to smaller ones, with the release of energy. Respiration is an example of a catabolic series of reactions. *See* metabolism. *Compare* anabolism.

catalase [Biochemistry] An enzyme that decomposes hydrogen peroxide.

catalysis [Chemistry] The alteration of the rate at which a chemical reaction proceeds, by the introduction of a substance (catalyst) that remains unchanged at the end of the reaction. Small quantities of the catalyst are usually sufficient to bring the action about or to increase its rate substantially.

catalyst [Chemistry] A substance that increases the rate of a chemical reaction without itself undergoing any permanent chemical change (*see also* inhibition). Catalysts that have the same phase as the reactants are homogeneous catalysts (*e.g.* enzymes in biochemical are reactions). Those that have a different phase are hetereogeneous catalysts (*e.g.* metals or oxides used in many industrial gas reactions). The catalyst provides an alternative pathway by which the reaction can proceed, in which the activation energy is lower. It thus increases the rate at which the reaction comes to equilibrium, although it does not alter the position of the equilibrium. The catalyst itself takes part in the reaction and consequently may undergo physical change (*e.g.* conversion into powder). In certain circumstances, very small quantities of catalyst can speed up very large reactions. Some catalysts are also highly specific in the type of reaction they catalyse, particularly in biochemical reactions.

catalytic cracking [Engineering] The use of a catalyst to bring about the cracking of high boiling mineral oils to get low boiling hydrocarbons.

cataphoresis *See* electrophoresis.

catenary [Mathematics] A curve formed by a chain or string having two fixed points. Equation, $y = cosh$ x/k, where k is the distance between the vertex of the curve and the origin.

catenation [Chemistry] The property of an element to link to itself to form long chain as seen in the case of carbon and sulphur.

cathetometer [Optics] A telescope or microscope fitted with cross-wires in the eyepiece and mounted so that it can slide along a graduated scale. Cathetometers are used for accurate measurment of lengths without mechanical contact. The microscope type is often called a travelling microscope.

cathode [Electricity] A negative electrode. In electrolysis cations are attracted to the cathode. In vacuum electronic devices electrons are emitted by the cathode and flow to the anode. It is therefore from the cathode that electrons flow into these devices. However, in a primary or secondary cell the cathode is the electrode that spontaneously becomes negative during dischargbe, and from which therefore electrons emerge.

cathode ray [Physics] A stream of electrons emitted from the negatively charged electrode or cathode when an electric discharge takes place in a vacuum tube.

cathode-ray oscilloscope [Electronics] CRO. An instrument based upon a cathode-ray tube, which provides a visible image of one or more rapidly varying electrical quantities. Also used as an indicator in a radar system.

cathode-ray tube [Electronics] CRT. A vacuum tube that allows the direct observation of the behaviour of cathode rays. It consists essentially of an electron gun producing a beam of electrons that after passing between horizontal and vertical deflection plates, falls upon a luminescence screen; the position of the beam can be observed by the luminescence produced upon the screen. Electric potentials applied to the deflection plates are used to control the position of the beam, and its movement across the screen, any desired manner. It is used as the picture tube in television receivers and in cathode-ray oscilloscopes.

cathodic polarization [Electricity] Portion of electric cell; polarization occurring at the cathode.

cathodic protection [Metallurgy] The protection of iron or steel against corrosion (*see* rusting) by using a more reactive metal. A common form is galvanizing (*see* galvanized iron), in which the iron surface is coated with a layer or zinc. Even if the zinc layer is scratched, the iron does not rust because zinc ions are formed in solution in preference to iron ions. Pieces of magnesium alloy are similarly used in protecting pipelines etc.

cathophosphorescence [Electronics] The phosphorescence produced when high velocity electrons bombard a metal in a vacuum.

catholyte [Chemistry] Electrolyte near the cathode during electrolysis.

cation [Chemistry] Positively charged ion; that, during electrolysis, is attracted towards the negatively charged cathode.

cationic dyes *See* basic dyes.

causality [Physics] The relating of causes of the effects that they produce. Many contemporary physicists believe that no coherent causal description can be given of events that occur on the sub-atomic scale.

caustic [Chemistry] Corrosive towards organic matter (but not applied to acids). *e.g.*, caustic soda. [Physics] Parallel rays of light falling on a concave sphercial mirror do not form a point image at the focus. Instead, there is a region of maximum concentration of the rays forming a curve or surface of revolution, called a caustic, the apex or cusp of which is at the focus of the mirror. A similar caustic is formed in the image formed by a convex lens receiving parallel light. Such a curve may be seen on the surface of a liquid in a container, formed by the reflection of light upon the curved wall of the container.

caustic alkali [Chemistry] Sodium or potassium hydroxide.

caustic potash [Chemistry] Potassium hydroixde, KOH.

caustic soda [Chemistry] Sodium hydroxide, NaOH.

cavitation [Physics] The formation of cavities in fluids when the pressure drops as a result of high velocity, in accordance with Bernouilli's theorem. These vapour-filled cavities collapse when they are carried to regions of higher pressure and the resulting impact pressure can cause pitting of such parts as propellers.

cavity magnetron [Electronics] A magnetron having a number of resonant cavities forming the anode; used as microwave oscillator.

cavity radiator [Physics] A hot enclosure with a small opening which allows some radiation to escape or enter; the escaping radiation approximates that of a blackbody.

celestial equator [Astronomy] The circle in which the plane of the earth's equator meets the celestial sphere. Also known as equinoctial.

celestial latitude [Astronomy] Angular distance north or south of the ecliptic; the arc of a circle of latitude between the ecliptic and a point on the celestial sphere measured northward or southward from the ecliptic through 90°.

celestial longitude [Astronomy] Angular distance east of the vernal equinox, along the ecliptic; the arc of the ecliptic or the angle at the ecliptic pole between the circle of latitude of the vernal equinox and the circle of latitude of a point of the celestial sphere, measured eastward from the circle of latitude of the vernal equinox through 360°.

celestial mechanics [Astronomy] The branch of astronomy concerned with the motions of celestial bodies or systems under the influence of gravitational fields.

celestial meridian [Astronomy] A great circle on the celestial sphere, passing through the two celestial poles and the observer's zenith.

celestial sphere [Astronomy] The imaginary sphere to the inner surface of which the heavenly bodies appear to be attached; the observer is situated at the centre of the sphere.

celestine *See* celestite.

celestite [Minerals] Natural, crystalline strontium sulphate, $SrSO_4$ ore of sky-blue colour mined as a source of strontium. Also known as celestine.

cell [Computers] A storage or memory localion capable of containing one character, one byte, or one word, the capacity being defined by the natural storage capacity of the computer. [Biology] The unit of life. All living organisms are composed of discrete, membrane-bounded units, which usually comprise two distinct forms of protoplasm: the nucleus and the cytoplasm. The former contains the nucleic acids responsible for organizing the synthesis of the cell's enzymes and for controlling the characteristics of its progeny, while the latter contains the enzyme systems that control the cell's metabolism and manufacture its constituents. Many micro-organisms (*e.g.*, bacteria, protozoa, etc.) consist of only one cell, whereas a man consists of billions of cells. [Electricity] A system in which two electrodes are in contact with an electrolyte. The electrodes are metal or carbon plates or rods or, in some cases, liquid metals (*e.g.* mercury). In an electrolytic cell a current from an outside source is passed through the electrolyte to produce chemical change (*see* electrolysis). In a voltaic cell, spontaneous reactions between the electrodes and electrolyte(s) produce a potential difference between the two electrodes.

Voltaic cells can be regarded as made up of two half cells, each composed of an electrode in contact with an electrolyte. For example, a zinc rod dipped in zinc sulphate solution is a $Zn \mid Zn^{2+}$ half cell. In such a system zinc atoms dissolve as zinc ions, leaving a negative charge on the electrode

$$Zn(s) \longrightarrow Zn^{2+} (aq) + 2e-$$

The solution of zinc continues until the charge build-up is sufficient to prevent further ionization. There is then a potential difference between the zinc rod and its solution. This cannot be measured directly, since measurement would involve making contact with the electrolyte, thereby introducing another half cell. A rod of copper in copper sulphate solution comprises another half cell. In this case the spontaneous reaction is one in which copper ions in solution take electrons from the electrode and are deposited on the electrode as copper atoms. In this case, the copper acquires a positive charge.

The two half cells can be connected by using a porous pot for the liquid junction (as in the Daniell cell) or by using a salt bridge. The resulting cell can then supply current if the electrodes are connected through an external circuit. The cell is written:

$$Zn(s) \mid Zn^{2+}(aq) \mid \mid Cu^{2+}(aq) \mid Cu, E=1.10V$$

where, E e.m.f. of the cell equal

to the potential of the right-hand electrode minus that of the left-hand electrode for zero current. Note that 'right' and 'left' refer to the cell as written. Thus, the cell could be written

Cu(s) I Cu^{2+}(aq) I I Zn^{2+}(aq) I Zn(s)$_3$1.10V

The overall reaction for the cell is:

Zn(s)+Cu^{2+}(aq) → Cu(s) + Zn^{2+} (aq)

This is the direction in which the cell reaction occurs for a positive e.m.f.

The cell above is a simple example of a chemical cell; *i.e.* one in which the e.m.f. is produced by a chemical difference. Concentration cells are cells in which the e.m.f. is caused by a difference of concentration. This may be a difference in concentration of the electrolyte in the two half cells. Alternatively, it may be an electrode concentration difference (*e.g.* different concentrations of metal in an amalgam, or different pressures of gas in two gas electrodes). Cells are also classified into cells without transport (having a single electrolyte) and with transport (having a liquid junction across which ions are transferred). Various types of voltaic cell exist, used as sources of current, standards of potential, and experimental set-ups for studying electrochemical reactions.

cell division [Biology] The process by which living cell multiply; may be mitotic or amitotic.

cell membrane [Biology] A thin layer of protoplasm, consisting mainly lipids and proteins, which is present on the surface of all cell. Also known as plasma membrane.

cell wall [Biology] A semi-rigid, permeable structure which is composed of cellulose, lignin, or other substances and which envelops most plant cells.

Celsius temperature [Physics] Temperature measured on a scale in which the melting point of ice is 0° and the boiling point of the water is 100°. This definition has been superseded by the International Practical Temperature Scale of 1968, which is expressed in both Kelvins and degrees Celsius. The unit for both means of expressing temperature is the Kelvin, and temperature differences may be expresed in Kelvins even when using Celsius temperatures. The relation between the Kelvin temperature (T) and the Celsius temperature (t) is given by; $T=t+273.15$. Also known as centigrade temperature.

celtium *See* hafnium.

cement [Chemistry] Any bonding material. [Materials] Portland cement and allied cements are made from materials containing lime, alumina, and silica (*e.g.*, limestone and clay), which are heated strongly in a kiln to form clinker (consisting mainly of calcium silicates and aluminates). The finely ground clinker undergoes complex hydration process when mixed with water setting and hardening to a stone like material.

cementation [Metallurgy] An early process for steel manufacture. Bars of wrought iron were heated to 1000°C for 7-10 days in charcoal at red heat. [Chemistry] setting of a plastic material. [Engineering] Plugging a cavity or drill hole with cement.

Cementite [Metallurgy] Iron carbide. Fe_3C. A hard, brittle compound that is responsible for the brittleness of cast iron and is present in steel. Also known as iron carbide.

centi- Prefix denoting one hundredth of, i.e., 10^{-2} in metric units.

central processing unit Central processor. See C.P.U.

centre of curvature [Optics] Of a spherical mirror. The centre of the sphere of which the mirror is a part.

centre of gravity [Mechanics] The fixed point through which the resultant force of gravity always passes, irrespective of the position of the body. This is identical to the centre of mass in a uniform gravitational field.

centre of mass [Mechanics] The point at which the mass of a body may be considered to be concentrated. The point from which the sum of the moments of inertia of all the component particles of a body is zero.

centre of symmetry [Scientific Technique] A point in an object through which any straight line encounters exactly similar points on opposite sides.

centrifugal [Mechanics] Acting or moving in a direction away from the axis of rotation or the centre of a circle along which the body is moving.

centrifugal force [Mechanics] The outward force acting on a body rotating in a circle round a central point. The centripetal force is the radial force imposed by the constraining system, necessary to keep the body moving in its circular path. The centrifugal and centripetal forces are equal and opposite. The centrifugal force acting on a body of mass m moving in a circle radius r, with a velocity v is mv^2/r.

centrifuge [Engineering] An apparatus for separating particles from a suspension. Balanced tubes containing the suspension are attached to the opposite ends of arms rotating rapidly about a central point by centrifugal force the suspended particles are forced outwards, and collect at the bottoms of the tubes. See also ultracentrifuge.

centigrade temperature See Celsius temperature.

centripetal force See centrifugal force.

ceramic [Materials] Pertaining to products or industries involving the use of clay or other silicates.

ceramic amplifier [Electronics] An amplifier which utilizes the piezoelectric properties of semiconductors such as silicon.

ceramic capacitor [Electricity] A capacitor whose dielectric is a ceramic material.

ceramic glaze [Engineering] A glazy finish on a clay surface, obtained by sparying metallic oxides and other chemicals and firing at high temperature.

cerargyrite *See* horn silver.

cerate [Chemistry] A metallic salt or soap made from lard.

Cerenkov radiation [Electronics] Light emitted when charged particles pass through a transparent medium at a velocity greater than the velocity of light in that medium. Also spelled as (Cherenkov)

Ceres [Astronomy] The largest of the asteroids.

ceresin [Materials] Hard, brittle paraffin wax with a melting point in the range of 343-373K; Used in the manufacture of candles, shoe polishes, electrical insulation and floor waxes.

ceric [Chemistry] Containing tetravalent cerium.

cerium [Chemistry] Ce. Element At. No. 58. At.Wt. 140.12. A steelgrey soft, rare-earth metal, m.p. 1068K, b.p. 3706K. It occurs in several rare minerals, *e.g.* monazite sand, and is used in pyrophoric alloys for lighter 'flints'; compounds are used in the manufacture of gas mantles and as an opacifier and polisher in the glass industry.

cermet [Materials] Ceramet. Abbreviation of CER(A)mic and METal. A very hard mixture of a ceramic substance and sintered metal, used where resistance to high temperature, corrosion, and abrasion is required.

cerography [Chemistry] Painting in which wax is used as a binder for pigments.

cerous [Chemistry] Containing trivalent cerium.

cesium [Chemistry] *See* caesium.

cetane number [Engineering] A measure of the ignition characteristics of a diesel fuel by comparison with a range of mixtures, in which cetane is given a value of 100 and α-methylnaphthalene is assigned zero value.

c.g.s. units [Sci Tech] A system of units based on the centimetre, gram, and second. Derived from the metric system, it was badly adapted to use with thermal quantities (based on the inconsistently defined calorie) and with electrical quantities (in which two systems, based respectively on unit permittivity and unit permeability of free space, were used). For scientific purposes c.g.s. units have now been replaced by SI units.

chabazite [Minerals] $CaAl_2Si_4O_{12} \cdot 6H_2O$. A natural zeolite, calcium aluminium silicate.

chain reaction [Nucleonics] In general, any self-sustaining molecular or nuclear reaction, the products of which contribute to the propagation of the reaction. In particular a fission chain reaction is a process in which one nuclear transformation is capable of initiating a chain of similar transformation. When nuclear fission occurs in a uranium-235 nucleus, between 2 and 3 neutrons are emitted, each of which is capable of causing the fission of further uranium-235 nuclei. The chain reaction so created is the basis of the atomic bomb and the nuclear reactor. [Chemistry] Chemical chain reactions usually involve free radicals as intermediates. An example is the reaction of chlorine with hydrogen initiated by ultraviolet radiation. A chlorine molecule is first split into atoms :

$$Cl_2 \longrightarrow Cl\cdot + Cl\cdot$$

These react with hydrogen as follows

$$Cl\cdot + H_2 \longrightarrow HCl + H\cdot$$
$$H\cdot + Cl_2 \longrightarrow HCl + Cl\cdot \text{ etc.}$$

chalcedony [Minerals] Cryptocrystalline. A variety of natural impure silica, SiO_2, that has a fibrous structure and a waxy lustre. Used for ornaments.

chalcocite [Minerals] Cu_2S. Copper glance. Natural copper sulphide.

chalcogens [Chemistry] The elements of group VI A of the periodic table : oxgen, sulphur,

selenium, tellurium, and polonium.

chalcopyrite [Minerals] Copper pyrites. A natural sulphide of copper and iron, $CuFeS_2$; the most abundant ore of copper.

chalones [Biology] Physiologically active substances produced within tissues that appear to control the mitosis of the cells of the specific tissues that produce them.

change of phase [Physics] A change of matter in one physical phase (solid, liquid, or gas) into another. The change is invariably accompanied by the evolution or absorption of energy, even if it takes place at constant temperature.

channel [Computers] A path along which digital or other information may flow in a computer. [Electronics] 1. A path for a signal, as an audio amplifier may have several input channels. 2. The main current path between the source and drain electrodes in a field-effect transistor or other semiconductor device.

channel capacity [Electronics] The number of signals per second that can be transmitted through a channel. Also, in information theory, the hypothetical limiting rate at which information could be communicated by a given channel, with the frequency of errors tending to zero.

channel effect [Electronics] A leakage current flowing over a surface path between the collector and the emitter in some types of transistors.

character [Computers] A unit of information as handled by computers, usually six bits.

characteristic [Mathematics] The integral or whole-number part of a logarithm.

charcoal [Materials] A general name for numerous varieties of carbon, usually impure; generally made by heating vegetable or animal substances with exclusion of air. Many forms are very porous and adsorb various materials readily.

charge, electric See electric charge.

charge carrier [Physics] A mobile conduction electron or mobile hole in a semiconductor.

Charles' law [Physics] The volume of a fixed mass of gas at constant pressure expands by a constant fraction of its volume at 0°C for each Celsius degree or Kelvin its temperature is raised. For any ideal gas the fraction is approximately 1/273. This can be expressed by the equation $V = V_o(1 + t/273)$, where V_o is the volume at 0°C and V is its volume at $t°C$. This is equivalent to the statement that the volume of a fixed mass of gas at constant pressure is proportional to its thermodynamic temperature, $V = kT$, where k is a constant. The law resulted from experiments begun around 1787 by the French scientist J.A.C. Charles but was properly established only by the more accurate results published in 1802 by the French scientist Joseph Gay-Lussac. Thus the law is also known as Gay-Lussac's law. An equation similar to that given above applies to pressures for ideal gases : $p = p_o(1 + t/273)$, a relationship known as Charles' law of pressures.

check [Computers] A test necessary to detect a mistake in computer programming or a computer malfunction.

chelate laser [Optics] A laser in which coherent pulses of light are produced by chelate molecules, such as chelates of europium.

cheddite [Materials] Class of explosives containing sodium or potassium chlorate with dinitrotoluene and other organic substances.

chelating agent [Chemistry] A compound the atoms of which form more than one coordinate bonds with metal ions in solution.

chelation [Chemistry] The formation of a closed ring of atoms by the attachment of compounds or radicals to a central polyvalent metal ion (occasionally nonmatallic); usually due to the sharing of a lone pair of electrons, from donor atoms in the compounds or radicals, with the central ion, e.g., two molecules of ethylenediamine NH_2CH_2-CH_2NH_2 form a 'chelate ring' with a cupric ion as shown in the diagram.

$$\begin{array}{ccc} CH_2\text{-}NH_2 & & NH_2\text{-}CH_2 \\ | \quad \searrow & \overset{++}{Cu} & \swarrow \quad | \\ CH_2\text{-}NH_2 \quad \nearrow & & \nwarrow \quad NH_2\text{-}CH_2 \end{array}$$

chelometry [Chemistry] A technique used in chemical analysis which involves the formation of 1:1 metal-chelates with aminopolycarboxylate and polyamine reagents, which are soluble in suitable medium; a form of complexometric titration.

chemical [Chemistry] A substance which takes parts in a chemical reaction.

chemical affinity [Chemistry] The tendency of an atom or compound to react or combine with atoms or compounds of different chemical nature. The free energy decrease is a quantitative measure of chemical affinity.

chemical bonds [Chemistry] A strong force of attraction holding atoms together in a molecule or crystal. Typically chemical bonds have energies of about 1000kJ mol^{-1} and are distinguished from the much weaker forces between molecules (*see* van der Waals' forces). There are various types. *See* also covalent bond; ionic bond; coordinate bond; homopolar bond; double bond; triple bond.

chemical burn [Medicine] Destruction of living tissues by caustic agents, irritant gases or other chemicals.

chemical change [Chemistry] A change in a substance involving an alteration in its chemical composition, due to an increase, decrease, or rearrangement of atoms within its molecules.

chemical combination [Chemistry] The combination of elements to give compounds. There are three laws of chemical combination.

(1) The law of constant composition states that the proportions of the elements in a compound are always the same, no matter how the compound is made. It is also called the law of constant proportions or definite proportions.

(2) The law of multiple proportions states that when two elements A and B combine to form more than one compound, then the masses of B that combine with a fixed mass of A are in simple ratio to one another. For example, carbon forms two oxides. In one, 12 grams of carbon is combined with 16 grams of oxygen (CO); in the other 12g of carbon is combined with 32 grams of oxygen (CO_2). The oxygen masses combining with a fixed mass of carbon are in the ratio 16:32, *i.e.* 1:2.

(3) The law of equivalent proportions states that if two elements A and B each form a compound with a third element C, then a compound of A and B will contain A and B in the relative proportions in which they react with C. For example, sulphur and carbon both form com-

pounds with hydrogen. In methane 12 g of carbon react with 4 g of hydrogen. In hydrogen sulphide, 32 g of sulphur react with 2 g of hydrogen (*i.e.* 64 g of S for 4 g of hydrogen). Sulphur and carbon form a compound in which the CS ratio is 12:64 (*i.e.* CS_2). The law is sometimes called the law of reciprocal proportions.

chemical dating [Chemistry] The determination of age of minerals and ancient materials by measuring and determining their chemical compositions.

chemical energy [Chemistry] That part of the energy stored within an atom or molecule that can be released by a chemical reaction as a result of rearrangement of the atoms in reacting compounds.

chemical engineering [Engineering] The design, operation, and manufacture of plant or machinery used in industrial chemical processes.

chemical equation [Chemistry] The short form representation of a chemical reaction with the help of symbols and formulae. While writing a chemical equation, the formulae or symbols of the reactants are written on the left hand side while the formulae or symbols of the products are written on the right hand side. The reactants and the products are separated from each other by the sign($\rightarrow$).

chemical equilibrium [Chemistry]

A reversible chemical reaction in which the concentrations of reactants and products are not changing with time because the system is in thermodynamic equilibrium. For example, the reversible reaction:

$$3H_2 + N_2 \rightleftharpoons 2NH_3$$

is in chemical equilibrium when the rate of the forward reaction

$$3H_2 + N_2 \rightarrow 2NH_3$$

is equal to the rate of the back reaction

$$2NH_3 \rightarrow 3H_2 + N_2$$

See also equilibrium constant.

chemical equivalent [Chemistry] The mass of an element which will combine with or displace directly or indirectly, 1 g of hydrogen, or 8 g of oxygen. The gram-equivalent, or equivalent weight, is the equivalent expressed in grams. The equivalent weight of an acid is the mass, in grams, of an acid that contains 1 g of replaceable hydrogen. The equivalent weight of an alkali, or a base, is the mass that neutralizes the equivalent weight of an acid. The equivalent weight of an oxidising agent is the mass of the agent which provides 8 g oxygen. The equivalent weight of an element is obtained by dividing its atomic mass by its valence. Equivalent weight of an ion is the formula weight divided by the charge present on it.

chemical etching [Metallurgy] Formation of desired surface characteristics when a polished metal surface is etched by suitable reagents.

chemical formula *See* formula.

Chemical fuel [Materials] The principal fuels used in internal combustion engines (automobiles, diesel, and turbojet) and in the furnaces of stationary power plants are organic fossil fuels. These fuels, and others derived from them by various refining and separation processes, are found in the earth in the solid (coal), liquid (petroleum), and gas (natural gas) phases.

chemical inhibitor [Chemistry] A substance which can retard or stop a chemical reaction, chemical inhibitors are specific in nature.

chemical kinetics [Chemistry] The branch of physical chemistry which deals with the study of the mechanisms and rates of chemical reactions.

chemical reaction [Chemistry] The interaction of two or more substances, resulting in the formation of one or more new substances; there is a minute change, m, in the mass of the system, given by $E=mc^2$, where$=E$ is the energy evolved or absorbed during the course of reaction and c is the speed of light.

chemical shift [Chemistry] Shift in the nuclear magnetic resonance spectrum as a result of diamagnetic shielding of the nuclei by the surrounding molecules.

chemical symbol [Chemistry] A shorthand notation for the chemical name of an element, consisting of letters, *e.g.*, B,C,H, Na represent boron, carbon, hydrogen and sodium respectively.

chemiluminescence [Chemistry] The emission of light accompanied by some heat during a chemical reaction without any apparent change in temperature. Also known as cold flame.

chemisorption [Chemistry] *See* adsorption.

Chemistry [Chemistry] The study of the properties, composition, and structure of matter, the changes in structure and composition which matter undergoes, and the accompanying energy changes. The objective of the chemist is to aid in the interpretation of the universe, Much progress has been made toward meeting this objective because not only has the structure and composition of many of the materials on the Earth been elucidated, but also those of the planets, the satellites, the stars, and the materials of interstellar space.

Chemolithotrophic bacteria [Biochemistry] Those bacteria capable of generating metabolically useful energy, that is, adenosinetriphosphate (ATP), by the oxidation of inorganic compounds Molecular hydrogen, elemental sulfur and salts, of NH_4^+, NO_2, Fe^{2+}, Mn^{2+}, S^{2-} and $S_2O_3^{2-}$ can serve as exclusive energy sources for the growth of specific bacteria.

chemotherapy [Medicine] The treatment of disease by chemical substances that are toxic to the causative micro-organisms or directly attack neoplastic growths.

chemurgy [Engineering] The study of chemical industrial processes based on organic substances of agricultural origin.

chert [Minerals] A hard variety of silica, SiO_2, resembling flint.

china clay [Minerals] A white, pure natural form of hydrated aluminium silicate, $Al_2Si_2O_5(OH)_4$. On heating, it loses water and changes its chemical composition. It is used in the manufacture of ceramics, paper, rubber and inks. Also known as kaolin.

Chinese white [Chemistry] Zinc oxide, ZnO. Also known as zinc white.

chip [Electronics] 1. A well processed semiconductor die mounted on a suitable base to form a diode, transistor or other semiconductor device. 2. An integrated microcircuit performing a significant number of functions.

chirality [Chemistry] The concept of 'handedness' (right-or left-handedness) applied to stereoisomerism. A geometrical figure representing the configuration of a molecule in space is said to have chirality if its image in a plane mirror cannot be made to coincide with it.

chloracne [Medicine] A disfiguring skin disease that is caused by certain chlorinated aromatic hydrocarbons. It can result from contact, ingestion, or inhalation of the chemicals.

chlordization [Chemistry] See chloriaration [Metallurgy] Treatment of mineral ores with chlorine or hydrochloric acid to form the chloride of the main metal present in the ore.

chlorinated lime [Chemistry] Calcium oxychloride, $CaOCl_2$. See bleaching powder.

chlorination [Chemistry] 1. A chemical reaction in which a chlorine atom is introduced into a compound. See halogenation. 2. The treatment of water with chlorine to disinfect it.

chlorine [Chemistry] Symbol Cl. A halogen element; At. No. 17; d. 3.214 g dm^{-3}; m.p. $272k$ b.p. 238.5K. It is a poisonous greenish-yellow gas and occurs widely in nature as sodium chloride in seawater and as halite (NaCl), carnallite ($KCl.MgCl_2.6H_2O$), and sylvite (KCl). It is manufactured by the electrolysis of brine and also obtained in the Downs process for making sodium. It has many applications, including the chlorination of drinking water, bleaching, and the manufacture of a large number of organic chemicals.

It reacts directly with many elements and compounds and is a strong oxidizing agent. Chlorine compounds contain the element in the 1, 3, 5, and 7 oxidation states.

chlorinity [Geology] A measure of the chloride and other halogen content by mass, of sea water.

chlorite [Chemistry] 1. A salt of chlorous acid. 2. A group of mineral silicates of aluminium iron, and magnesium.

chloro [Chemistry] A prefix describing an organic compound which contains chlorine atoms substituted for hydrogen.

chlorocarbon [Chemistry] A compound of chlorine and carbon only, such as carbon tetrachloride, CCl_4.

chlorohydrin [Chemistry] Organic compounds containing a chlorine atom and a hydroxyl group attached to adjacent carbon atoms in a hydrocarbon molecule; they are formed by addition of hypochlorous acid at the double bond to alkenes.

chlorophyll [Biochemistry] The general name for any of several oil-soluble green tetrapyrole pigments found in plants, which absorbs energy from sunlight, enabling them to build up carbohydrates from atmospheric carbon dioxide and water by photosynthesis. Chlorophyll-α ($C_{55}H_{72}O_5N_4Mg$) and chlorophyll-β ($C_{55}H_{70}O_6N_4Mg$) occur in green plants and algae.

Chloroplast [Botany] A type of cell plast occurring in the green parts of plants, containing chlorophyll pigments, and functioning in photosynthesis and protein synthesis.

chlorosis [Medicine] A type of anemia in young females characterized by reduction in hemoglobin and a greenish skin colour. [Botany] A plant disease in which green parts of the plants become yellow.

choke [Elect] A coil of low resistance and high inductance used in electrical circuits to pass direct currents whilst suppressing alternating currents. Also known as choking coil.

choke-damp *See* after-damp.

choking gas [Chemistry] Any gas which can casue irritation and inflammation of the bronchial tubes and lungs, *e.g.*, phosgene.

chondrification [Physiology] Formation or conversion into cartilage (A specialized connective tissue).

chondrite [Astronomy] A type of stony meteorite that contains the small round mases of olivine or pyroxene known as chondrules.

chord [Physics] A combination of two or more tones. [Mathematics] A straight line joining two points on a curve. *See* circle.

chromascope [Optics] An instrument used for the determination of optical effects of colour.

chromate [Chemistry] A salt of chromic acid.

chromatic aberration *See* aberration, chromatic.

chromatics [Optics] The branch of optics concerned with the study of colours.

chromatids [Biochemistry] The two identical strands into which a chromosome splits during cell reproduction.

chromatogram [Chemistry] A record obtained by chromatography. The term is applied to the developed records of paper chromatography and thin-layer chromatography and also to the graphical record produced in gas chromatography.

chromatography [Chemistry] A technique for analysing or separating mixtures of gases, liquids, or dissolved substances. A vertical glass tube is packed with an adsorbing material, such as alumina. The sample is poured into the column and continuously washed through with a solvent (a proces known as elution). Different components of the sample are adsorbed to different extents and move down the column at different rates. In Tswet's original application, plant pigments were used and these separated into coloured bands in passing down the column (hence the name chromatography). The usual method is to collect the liquid (the eluate) as it passes out from the column in fractions.

In general, all types of chromatography involve two distinct phases-the stationary phase (the adsorbent material in the column in the example above) and the moving phase (the solution in the example). The separation depends on competition for molecules of sample between the moving phase and the stationary phase. The form of column chromatography above is an example of adsorption chromatography, in which the sample molecules are adsorbed on the alumina. In partition chromatography, a liquid (*e.g.* water) is first absorbed by the stationary phase and the moving phase is an immiscible liquid. The separation is then by partition between the two liquids. In ion-exchange chromatography (*see* ion exchange), the process involves competition between different ions for ionic sites on the stationary phase. Gel filtration is another chromatographic technique in which the size of the sample molecules is important is known as 'gas-liquid chromatography. This is one of the most powerful methods of analysis. When the stationary phase is an active solid, the process is known as 'gas-solid chromatography'. When the mobile phase is a liquid, it can be applied to a column of the active solid or to a thin layer of the solid on a plate. Filter paper can also be used as the stationary phase.

chromatophore [Biochemistry] Any pigmentary cell or colour producing plastid, such as those of the deep layers of the epidermis; a chlorophyll-containing granule in certain bacteria.

chromatron [Electronics] A type of cathode ray tube that has four screens or a colour picture tube having colour phosphors deposited on the screen in strips instead of dots; used as a colour picture-tube in television. Also known as chromoscope.

chrome cake [Materials] A green form of salt cake (sodium sulphate) containing small amount of chromium compounds as impurity.

chrome dye [Materials] A mordant dye, most frequently one in which sodium dichromate is used as the mordant.

chrome pigment [Materials] An inogranic pigment containing chromium compounds. These are quite stable and resistant to sunlight, weathering, and chemical action than the brighter organic dyes.

chrome red [Materials] Basic lead chromate, $PbO.PbCrO_4$. Used as a pigment in paints.

chrome yellow [Materials] Lead chromate, $PbCrO_4$. Used as a pigment.

chromic [Chemistry] Containing trivalent chromium.

chromite [Chemistry] 1. A salt of bivalent chromium. 2. A natural oxide of ferrous iron and chromium, $FeCr_2O_4$; a commerical source of chromium and its compounds. Also known as chrome iron ore.

chromium [Chemistry] Symbol Cr. A hard silvery transition element; A.No. 24; r.d. 7.19; m.p., 2130K, b.p. 2913K. The main ore is chromite ($FeCr_2O_4$). The metal is extracted by heating chromite with sodium chromate, from which chromium can be obtained by electrolysis. Alternatively, chromite can be heated with carbon in an electric furnace to give ferrochrome, which is used in making alloy steels. The metal is also used as a shiny decorative electroplated coating and in the manufacture of certain chrominum compounds.

At normal temperatures the metal is corrosion-resistant. It reacts with dilute hydrochloric and sulphuric acids to give chromium(II) salts. These readily oxidize to the more stable chromium(III) salts. Chromium also forms compounds with the +6 oxidation state, as in chromates, which contain the CrO_4^{2-} ion.

chromium steel [Engineering] Steel containing varying amounts of chromium; strong and tough, used for tools, etc.

chromizing [Metallurgy] Surface-alloying of metals in which an alloy is formed by diffusing chromium into the base metal.

chromocyte [Biochemistry] A pigmented cell.

chromogen [Biochemistry] A micro-organism capable of producing colour under suitable conditions.

chromomere [Biochemistry] One of a linear series of bedlike structures composing a chromosome.

chromphore [Chemistry] An arrangement of atoms such as --N=N-, as a result of which many organic compounds become coloured.

chromophyll [Biochemistry] Any plant pigment.

chromoplast [Biochemistry] Any coloured cell plastid, excluding chloroplasts.

chromoscope [Optics] An instrument for analyzing colour values and intensities.

chromosome [Biochemistry] The heredity-bearing genes; carrier of the living cell, derived from chromatin, and largely consist of nucleoproteins, the nucleic acid being DNA. The unit of genetic information is the gene (see also cistron and operon) and each chromosome may be regarded as comprising a number of genes. Chromosomes occur in pairs in somatic cells, each species being characterized by the different number of chromosomes that its cells contain (man has 46 chromosomes per cell).

chromosphere [Astronomy] The layer of the sun's atmosphere surrounding the photosphere, which is visible during a total eclipse. The chromosphere is several thousand kilometres thick and has an estimated temperature of 20000 K.

chromous [Chemistry] Containing bivalent chromium.

chromyl [Chemistry] The bivalent radical CrO_2^{--}, containing sexivalent chromium; e.g., in chromyl chloride, CrO_2Cl_2.

chronoamperometry [Chemistry] Electro-analysis by measuring at a working electrode the rate of change of current versus time during a titration at controlled potential.

chronograph [Engineering] An accurate time-recording instrument.

chronometer [Engineering] An accurate clock, especially one used on a ship in navigation.

chronometry [Engineering] The technique of measuring time by chronometer.

chronotron [Engineering] A device that measures the time between two events, by measuring the positions on a transmission line of pulses initiated by the events.

chrysotherapy [Medicine] The use of gold compound in the treatment of diseases.

chu [Physics] Abbrev. for centigrade heat unit. It is the amount of heat required to raise the temperature of one pound of water by one degree centigrade from 15°C to 16°C). It is sometimes called *pcu* (pound centigrade unit).

ciment fondu See bauxite cement.

cinder [Material] Slag obained from a metal furnace.

cinders [Material] Incombustible residue from a burning process; in particular small pieces of clinker from the burning of soft coal.

cingulum [Botany] The part of a plant between stem and root.

cinnabar [Mineral] Natural mercuric sulphide, HgS. A bright red crytalline solid, a principal ore of mercury.

ciphony equipment [Electronics] Any equipment attached to a radio transmitter, radio receiver or telephone for scrambling or unscrambling voice messages.

circuit, electrical [Electricity] The complete path traversed by an electric current.

circuit breaker [Electricity] An electromagnetic device which opens a circuit automatically when the circuit exceeds a predetermined value.

circuit interrupter [Electricity] A device in a circuit breaker to remove energy from an arc in order to extinguish it.

Circuit switching [Electronics] A method of communication in which communicating devices use a dedicated end-to-end path that is held for the duration of the call; most telephone connections are circuit-switched.

circular bifringence [Optics] The phenomenon in which an optically active compound transmits right circularly polarized light with a different velocity from left circularly polarized light.

circularly polarized light [Optics] Light that can be resolved into two vibrations lying in planes at right angles, of equal amplitude and frequency and differing in phase by 90°. The electric vector of the wave describes, at any point in the path of the wave, a circle about the direction of propagation of the light as axis.

circular mil [Mechanics] A unit of area. The area of a circle whose diameter is 0.001 inch, i.e. 0.785 X 10^{-6} sq in. Used in measuring the cross-section of fine wire.

circulating reactor [Nucleonics] A nuclear reactor in which the fissionable material circulates through the core in fluid form, or as small particles suspended in fluid.

cis-trans isomerism [Chemistry] A type of geometrical isomerism associated with compounds containing a double bond and each atom is tetravalent with two different atoms or groups joined to them. The double bond prevents the rotation of two atoms linked to it, so two spatial arrangements of the atoms are possible. Like groups in such compounds may be either on the same side of the plane of the double bond (*cis*-form) or on opposite (*trans*-form), e.g., maleic acid and fumaric acid are respectively cis-and trans-forms.

H--C--COOH
$\|$
H--C--COOH   Maleic acid

H--C--COOH
$\|$
HOOC--C--H   Fumaric acid

cistron [Biology] The functional unit of genetic information, taking account the distribution of abnormal (mutant) genes among pairs of chromosomes, and the way in which an abnormal gene in one chromosome may be compensated for by a normal gene either in the same chromosome (cis-configuration) or its pair (trans-configuration). Also known as structural gene.

citric acid cycle [Biochemistry] A complex of enzyme controlled biochemical reactions, which occur within living cells, as a result of which pyruvic acid is broken down into carbon dioxide and energy. The citric acid cycle is a most important clearing-house of metabolic intermediates, since it deals with the final stages of the oxidation of carbohydrates and fats and is also involved in the synthesis of some amino acids. Also known as Kreb's

civil day [Aston] A mean solar day begining at midnight which may be based on either apparent solar time or mean solar time,

civil time [Astronomy] Solar time in a civil day that begins at midnight.

civil twilight [Astronomy] The time interval of incomplete darkness between sunrise and the time when the centre of sun's disc is 6o below horizon.

cladding [Engineering] Process of covering one material with another and bonding them at high temperature and pressure. Also known as bonding. [Nucleonics] **The covering** of a fuel element in a nuclear reactor by a thin layer of another metal, to prevent corrosion by the coolant and the release of fission products.

clad metal [Metallurgy] A metal covered on one or both sides with a different metal.

Clark cell [Electricity] An early form of standard cell of 1.433 volts at 288K; consists of zinc-amalgam anode and mercury cathode, both immersed in a saturated solution of zinc sulphate. Now replaced by Weston cell.

clarke [Geology] A unit of average abundance of an element in the earth's crust, expressed as percentage.

Clark process [Chemistry] A method of softening water by adding alkaline solutions of calcium hydroxide so that the acid carbonates are converted into normal carbonates.

classical mechanics [Mechnics] Mechanics based on Newton's laws of motion.

clathrate [Chemistry] A solid mixture in which small molecules of compound or element are trapped in holes in the crystal lattice of another substance. Clathrates are sometimes called enclosure compounds, but they are not true compounds (the molecules are not held by chemical bonds). Quinol and ice both from clathrates with substances such as sulphur dioxide and xenon.

Claude process [Engineering] A process for producing liquid air, based on the cooling that results from the adiabatic expansion of a gas that is performing external work. Air under pressure is divided into two separate channels. The first channel leads to a compressor, where the air performs external work by driving the compressor. The cool air so produced is used to reduce the temperature of the compressed air from the second channel in a counter-current heat exchanger.

Clausius Clapeyron equation [Physics] An equation governing phase transitions of a substance, $dp/dt = \Delta H(T \Delta V)$, where p is the pressure, T is the temperature ΔH is the change in heat content and ΔV is the change in volume during the transition.

Clausius equation [Physics] An equation of state for gases which applies a correction to the Van der Waals equation,

$[P + (n^2 a / \{T(V + c^2)\})](V - nb) = nRT;$

where P, V and T are the pressure, volume and temperature of the gas, n the number of moles of the gas, R the gas constant, a depends upon only temperature, b is a constant and c is a function of a and b.

clay [Geology] A natural material with partiles less than 1/256 mm in diameter clays are plastic when wet, and are composed of silica, alumina and are often associated with compounds of iron, alkalies and alkaline earths.

clay wash [Materials] A light oil

such as kerosine used to clean fuller's earth after it has been in a filter.

clean room [Engineering] A room in which special precautions are observed to reduce the level and entry of dust particles and other contaminants present in air.

cleavage [Physics] The manner of breaking of a crystalline substance, so that more or less smooth surfaces are formed.

climatology [Sci] The branch of meteorology concerned with the mean physical state of the atmosphere along with statistical variations in space as well as time as reflected in the weather behaviour over a period of past years.

clinial genetics [Biology] The study of biological inheritance by direct observation of the living patient.

clinical pathology [Biochemistry] The diagnostic study of diseases by means of laboratory tests of blood, urine, stool etc. of the living patient.

clinical pharmacology [Medicine] The scientific study of the effects of drugs on human beings.

clock [Engineering] A device for indicating the passage of time.

clock frequency [Electronics] The master frequency of periodic pulses which schedule the operation of a digital computer of like devices.

clock oscillator [Electronics] An

oscillator which controls an electronic clock.

clock star [Astronomy] Any bright star used to measure time whose right ascension is well known.

clot [Physics] A semisolid coagulum of blood or lymph.

clotting [Chemistry] The formation of solid deposits or clots in liquids, often due to the coagulation of soluble proteins dissolved in the liquid, e.g., the clotting of blood.

cloud chamber [Physics] An apparatus for making the tracks of ionizing particles visible as a row of droplets.

cloud point [Chemistry] The temperature at which a homogeneus liquid becomes cloudy or turbid, owing to separation into two phases, when cooled under specified conditions.

clusius column [Nucl.Sc.] A device for separating gaseous isotopes. It consists of a high column with a central heated wire. As a result of thermal diffusion the lighter isotope collects at the top of the tube.

cluster [Astronomy] An aggregation of stars that move together.

clutch [Engineering] A device for the connection and disconnection of shafts in equipment drives while running.

coagulation [Chemistry] The process in which colloidal particles come together to form larger masses. Coagulation can be brought about by adding ions to neutralize the charges stabilizing the colloid. Ions with a high charge are particularly effective (e.g. alum, containing Al^{3+}, is used in styptics to coagulate blood). Another example of ionic coagulation is in the formation of river deltas, which occurs when colloidal silt particles in rivers are coagulated by ions in sea water. Heating is another way of coagulating certain colloids (e.g. boiling an egg coagulates the albumin)

coal [Geology] The general name given to stratified accumulations of carbon containing matter derived from vegetation. The main varieties are : peat, lignite, ordinary, and anthracite.

coal blasting [Engineering] Breaking coal with explosives.

coal dust [Engineering] A finely divided coal which can pass through 100 mesh screens.

coalesce [Sci Tech] To bring together small masses to form larger mass

coal gas [Materials] Flammable gas derived from coal either naturally or by destructive distillation of coal in iron retorts; average composition by volume is : hydrogen 50 %, methane 30%, carbon monoxide 8%, nitrogen, carbon dioxide and oxygen 8%, and other hydrocarbons 4%.

coal gasification [Engineering] The conversion of coal, coke or char into a gaseous product by reaction with air, oxygen, steam, carbon dioxide, or mixture of these.

coalification [Geology] Conversion of plant remains in coal by the processes of diagenesis and metamorphism. Also known as carbonification; incarbonization; incoalation.

coal liquefaction [Engineering] The process of making a liquid mixture of hydrocarbons by destructive distillation of coal.

coal-tar [Materials] A thick black oily liquid obtained as a byproduct of coal-gas manufacture. Distillation and purification yeilds, amongst other valuable products: benzene, C_6H_6; toluene, $C_6H_5CH_3$; xylene, $C_6H_4(CH_3)_2$; phenol, C_6H_5OH; naphthalene; $C_{10}H_8$; cresol, $CH_3C_6H_4OH$, and anthracene, $C_{14}H_{10}$. Pitch is left as a residue.

coaxial [Mechnics] Sharing a common axis.

coaxial cable [Elect] A transmission line in which one conductor is mounted in side and insulated from the outer metal tube which acts as the second conductor.

cob [Engineering] To chip away waste material from an ore, using hand hammers.

cobalt [Chemistry] Symbol Co. A light-grey transition element; At, No. 27; r.d. 8.9 m.p. 1768 K.b.p. 2913 K. Cobalt ores are usually roasted to the oxide and then reduced with carbon or water gas. Cobalt is usually alloyed for use. Alnico is a well-known magnetic alloy and cobalt is also used to make stainless steels and in high-strength alloys that are resistant to oxidation at high temperatures (for turbine blades and cutting tools).The metal is oxidized by hot air and also reacts with carbon, phosphorus, sulphur, and dilute mineral acids. Cobalt salts, usual oxidation states II and III, are used to give a brilliant blue colour in glass, tiles, and pottery. Anhydrous cobalt(II) chloride paper is used as a qualitative test for water and as a heat-sensitive ink. Small amounts of cobalt salts are essential in a balanced diet for mammals (*see* essential element). Artificaly produced cobalt-60 is an important radioactive tracer and cancer-treatment agent.

cobalt bomb [Nucl Sc] A theoretical atomic or hydrogen bomb encased in cobalt, in which cobalt would be transfomed into deadly radioactive dust upon detonation.

cobaltic [Chemistry] Containing trivalent cobalt, *e.g.,* cobaltic chloride, $CoCl_3$.

COBOL [Computers] A business data processing computer language which can be fed as a series of English statement describing a complete business operation. Derived from Common Business Oriented Language.

coccus [Biology] A globular or spherical-shaped bacterium.

cochineal [Chemistry] A natural red dyestuff obtained from the dried body of the Coccus cacti insect.

Cockcroft Walton generator [Electricity] A high voltage direct current accelerator used for accelerating nuclear particles (particularly protons). The D.C. voltage is obtained by multiplying a low A.C. voltage by an arrangement of rectifiers and capacitors.

codimer [Chemistry] A copolymer formed from the polymerization of two dissimilar olefin molecules, e.g., the product of polymerizaion of isobutylene with one of the two normal butylenes.

codon [Biochemistry] A sequence of three adjacent nucleotides in a nucleic acid that codes for a specific amino acid.

coefficient [Mathematics] A number of other known factor by which a variable quantity is multiplied, e.g., in $ax^2 + bx + c=0$, a is the coefficient of x^2. and b is the coeffcient of x. [Physics] A measure of a specified property of a particular substance under specified conditions, e.g. the coefficient of friction of a substance.

coelostat [Engineering] A device consisting of a clock-driven mirror used in conjunction with an astronomical telescope to follow the path of a celestial body and reflect its light into the telescope.

coenzyme [Biochemistry] The nonprotein portion of an enzyme which functions as an acceptor of electrons or functional groups, and plays an essential part in some reactions catalysed by enzymes, it often acts as a temporary carrier of an intermediate product of the reaction.

coercive force [Physics] the force necessary to reverse the field-mag-netization vector in a magnetic material;the force is exerted by an external magnetic field, and may vary with temperature.

cofactor [Biochemistry] A low-molecular weight, heat-stable inorganic or organic substance required for the action of an enzyme.

Coffey still [Electricity] Apparatus for the fractional distillation of solutions of ethanol as obtained by fermentation on an industrial scale ; the product is known as rectified spirit.

coffin [Nucl.Sc.] A lead box used for transporting radioactive substances.

coherent [Physics] A beam of light or other electromagnetic radiation of the same or almost the same wavelength is said to be coherent if its waves are in phase.

coherent precipitate [Chemistry] A precipitate that is a continuation of the lattice structure of the solvent and has no phase or grain boundary.

coherent scattering [Physics] Scattering in which there is a definite phase relationship between incoming and scattered particles or radiation.

coherent source [Physics] A source in which there is a constant phase difference between waves emitted from different parts of the source.

coherent units [Scientific Techniques] A system or units in which the quotient or product of any two units in the system yields the units of the resultant quantity, e.g., when unit length is divided by unit time, the unit of velocity results. The basic units of a coherent system are arbitrarily defined physical quantities. All other units are obtained from these basic units by defining relations and are called 'derived units. The coherent units now in scientific use are the SI units.

cohesion [Physics] The tendency of parts of a body of like composition to hold together, as result of intermoleculer attractive forces.

coinage metals [Metallurgy] The metals copper, silver, gold.

coke [Materials] A greyish porous brittle solid containing about 80% carbon; obtained as a residue in the manufacture of coal-gas('gas -coke') ; also made specially in coke ovens, in which the coal is treated at lower temperatures than in gas manufacture.

cold cathode [Electronics] A cathode whose operation does not depend on its temperature being above the ambient temperature.

cold emission [Physics] The emission of electrons by a solid without the use of heat (thermal emission), either as a result of field emission or secondary emission.

cold light [Physics] Light emitted in luminescence which is accompanied by very little or no infrared radiation, and therfore has very little heating effect.

collargol [Chemistry] A powder containing protein material and finely divided silver; with water it forms a colloidal solution of silver.

collector [Electronics] A semiconductive region through which a primary flow of charge carriers leaves the base of a transistor. 2.The electrode in a transistor through which a primary flow of carriers leaves the inter-electrode region.

collector modulation [Electronics] Amplitude modulation in which the modulator varies the collector voltage of a transistor.

collector plate [Electronics] One of the several metal inserts which are sometimes embedded in the lining of an electrolyte cell to make the resistance between the cell lining and the current leads as small as possible.

colligative properties [Chemistry] properties that depend on the concentration of particles (molecules, ions, etc.) present in a solution, and not on the nature of the particles. Examples of colligative properties are osmotic pressure (*see* osmosis). Relative lowering of vapour pressure, depression of freezing point, and elevation of boiling point.

collimator [Optics] 1. A tube containing a convex achromatic lens at one end and an adjustable slit at the other, the slit being at the focus of the lens. Light rays entering the slit thus leave the collimator as a parallel beam. 2. An arrangement of absorbers for limiting a beam of radiation to the required dimensions and angular spread in radiology. 3. A small fixed telescope attached to a larger one for the purpose of accurately setting the line of sight of the larger instrument.

collision [Physics] An interaction resulting from the close approach of two or more particles (bodies), or systems of particles, and confined to a relatively short time interval during which the motion of atleast one of the particles (or systems) changes abruptly.

collision frequency [Physics] The average number of collisions undergone by a particle travelling through a medium, in a unit time.

collision theory [Chemistry] Theory of chemical reaction according to which, the rate of chemical reaction is equal to the number of reactant-molecule collisions multiplied by a factor that corrects for low-energy-level collisions.

collodion [Materials] A solution of cellulose nitrate in a mixture of 40% alcohol and 60% ether.

colloid [Chemistry] A system of which one phase is made up of particles having dimensions of 10-10,000 Å and which is dispersed in a different phase.

colloidal metals [Chemistry] Colloidal solutions or suspensions of metals, the metal being distributed in the form of very small electrically charged particles. They are prepared by striking an electric arc between poles made of the metal, under water or by the chemical reduction of solution of a salt of the metal. Used in medicine.

colloidal solution [Chemistry] A solution in which the solute is present in the colloidal state. Common examples include solutions of starch, albumen, colloidal metals, etc. The solvent is termed the dispersion medium and the dissolved substance the disperse phase. Several types of colloidal solution are possible, depending upon whether the dispersion medium and the disperse phase are respectively liquid and solid (suspensoid sols), liquid and liquid (emulsoid sols), gas and solid, etc.

colloidal state [Chemistry] A system of particles in a dispersion medium, with properties distinct from those of a true solution because of the larger size of the particles. The presence of these particles, which are approximately 10^{-4} to 10^{-6}mm across, can often be detected by means of the ultramicroscope. As a result of the grouping of the molecules, a solute in the colloidal state cannot pass through a suitable semipermeable membrance and gives rise to negligible osmotic pressure, depression of freezing point, and elevation of boiling point effects.

colloider [Engineering] A device used to remove colloids from sewage.

cologarithm [Mathematics] The logarithm of the reciprocal of a number, expressed with a positive mantissa. Abbrev. colog.

colon [Biology] The main part of the large intestine, concerned with the absorption of water and mineral salts.

colorimeter [Engineering] Apparatus used in colorimetric analysis for comparing intensities of colour. *See also* tintometer.

colorimetric analysis [Chemistry] A form of quantitative analysis in which the quantity of a substance is estimated by comparing the intensity of colour produced by it with specific reagents, with the intensity of colour produced by a standard amount of the substance.

colour [Optics] The visual sensation resulting from the impact of light of a particular wavelength on the cones of the retina of the eye.

colour index [Astronomy] Of a star, the numerical difference between the apparent photographic magnitude and the apparent photovisual magnitude.

colour temperature [Physics] The temperature of a full radiator that would emit visible radiation of the same spectral distribution as the radiation from the light source under consideration.

colourtron [Electronics] A type of cathode-ray tube , used as a colour picture-tube in television, that has three electron guns, one for each primary colour.

colour vision [Optics] The ability to discriminate light on the basis of wavelength composition. White light, such as daylight, consists of a mixture of electromagnetic radiations of various wavelengths. A surface that reflects all of these will appear white ; some surfaces, however, have the propety of absorbing some of the radiations they receive, and reflecting the rest. Thus, a surface that absorbs all light radiations excepting those correspnding to blue, will apper blue by reflecting only those radiations. In the cases of colour seen by transmitted light, as in coloured glass, the glass absorbs all the radiations except those that are visible and pass through. *See* surface colour; pigment colour.

columbium *See* niobium.

column bleed [Chemistry] The loss of carrier liquid during gas chromatography due to evaporation.

column chromatography [Chemistry] A form of chromatography in which the mobile phase is liquid and the stationary phase is activated alumina, or a similar substance, contained in a vertical glass column. The mixture is introduced at the top of the column and washed through the stationary phase by a solvent. The components of the mixture are selectively adsorbed, forming coloured bands down the length of the column (if the components are coloured).

colza oil [Materials] Rapeseed oil. Yellow oil obtained from the seeds of various Brassica plants; used as an edible oil, lubricant, and in the quenching of steel.

coma [Astronomy] The gaseous envelope which surrounds the nucleus of a comet. [Optics] An error in the optical system, so that a point has an asymmetrical image.

coma cluster [Astronomy] A particular group of several thousand galaxies.

comb growth unit [Biology] A unit for the standardization of male sex hormones.

combination, laws of chemical *See* chemical combiantion, laws of.

combustion [Chemistry] A chemical reaction, in which a substance combines with oxygen producing heat, light, and flame.

comes [Astronomy] The smaller star in a binary-star system.

comet [Astronomy] A heavenly body, moving under the attraction of the sun in an eccentric orbit. It consists of a hazy gaseous cloud containing a brighter nucleus and a fainter tail. The nucleus is thought to consist of ice and dust particles.

command guidance [Engineering] A method of missile or rocket guidance in which computed information is transmitted to the missile and causes it to follow a directed flight path.

comminution [Engineering] Size reduction of materials by any of several means, *e.g.*, cutting, grinding, chopping, etc.

commission ore [Minerals] Uranium-bearing ore of 0.10% or higher U_3O_8 content, for which the U.S. Atomic Energy commission has an established price.

common intermediate [Chemistry] A chemical compound common to two chemical reactions, as a product of one and a reactant in the other.

Common-channel signaling [Computers] A scheme whereby all the signaling information for establishing a connection is carriied on a special channel separate from the one carrying the voice or data; it replaces the scheme of in-band signaling, whch used the same channel for signaliing as for carrying the voice.

communicable disease [Medicine] An infectious disease which can be transmitted directly or indirectly from one person to another.

communication [Electronics] The transmission of intelligence between two or more points over wires or by radio.

communication channel [Electronics] The wire or radio channel which serves to convey intelligence between two or more points.

communication satellite [Engineering] An orbiting artificial satellite of the earth which relays radio, television, and other signals between ground terminal stations thousands of kilometres apart.

commutator [Engineering] A device for altering or reversing the direction of an electric current; used in the dynamo to convert the alternating current into a direct one, if required.

compile [Computers] The act of translating a computer program written in a high-level language (such as LISP) into the machine language that controls the computer's basic operations.

complementary angles [Mathematics] Angles together totalling 90° or one right angle.

complementary colours [Optics] 1. Pairs of colours that, when combined, give the effect of white. 2. Two colours which lie on opposite sides of the white point in the chromaticity diagram, so that an additive mixture of the two in appropriate proportions, can be made to yield an achromatic mixture.

complete radiation *See* black radiation.

complex [Chemistry] A compound in which molecules or ions form coordinate bonds to a metal atom or ion. The complex may be a positive ion *e.g.* $[Cu(H_2O)_4]^{2+}$), a negative ion *e.g.* $Fe[(CN)_6]^{3-}$). or a neutral molecule *e.g.* $[PtCl_2(NH_3)_2]$. The formation of such coordination complexes is typical behaviour of transition metals. The complexes formed are often coloured and have unpaired electrons (i.e. are paramagnetic). *See also* ligand; chelate.

complexing agent [Chemistry] A substance capable of forming a complex compound with metal ions in solution, *e.g.,* ammonia, ethylenediamine. Also known as ligand; complexon.

complex number [Mathematics] A complex number consists of two parts, 'real' and 'imaginary', and can be expressed in the form $a+ib$, where both a and b are real quantities and i is the square root of -1, *i.e.,* $1^2 = -1$. The real part of the complex number is 'a' and the imaginary part 'ib'. Such numbers obey the ordinary laws of algebra except that in equations containing them the real and imaginary parts are equated separately.

complexometric analysis [Chemistry]
A method of chemical analysis
based on titration of metal ions
in solution with chelating agents
(*see* chelation), such as EDTA or
other complexons.

component [Chemistry] 1. A term
in the phase rule. The number of
components in a system is the
least number of substances from
which every phase of the system
may be constituted., *e.g.*, each of
the phases ice, water, and water
vapour in equilibrium is composed of one component, H_2O.
2. The smallest number of chemical substances able to form all
the constituents of a system in
whatever proportion they may be
present.

component force and velocities
[Mechanics] Two or more forces
or velocities that produce the
same effect upon a body as a
single force or velocity, known as
the resultant.

compost [Materials] A mixture of
decaying organic matter used as
fertilizer.

composting [Biochemistry] Aerobic bacterial-decomposition of
solid organic waste. Decomposition is accelerated by the addition of ammonium carbonate.
The product can be used as a soil
conditioner and for landfill.

compound [Chemistry] A substance formed by the combination of elements in fixed proportions. The formation of a compound involves a chemical reaction; *i.e.* there is a change in the
configuration of the valence
electrons of the atoms. Compounds, unlike mixtures, cannot
be separated by physical means.
See also molecule.

compound, interstitial [Chemistry]
A compound of a metal and a
non-metal or certain metalloids
in which the metalloid or non-
metal atoms occupy the intersti-
ces between the atoms of the
metal lattice.

compound lens [Optics] A combination of two or more lenses
cemented together, second surface of one lens has the same
radius as the first surface of the
following lens.

compound mircoscope [Optics] A
microscope having two lenses or
lens systems ; one lens forms the
enlarged image of the object
while the second lens magnifies
the image formed by the first
lens.

compound nucleus [Nucl.Sc.] An
intermediate state in a nuclear
reaction in which the incident
particle combines with the target
nucleus and its energy is shared
among all the nucleons of the
system.

compressibility [Mechnics] The
reciprocal of bulk modulus (*See*
elastic modulus). The compressibility (k) is given by $-dV/Vdp$,
where dV/dp is the rate of change
of volume (V) with pressure.

compressional wave [Physics] A
disturbance travelling in an elastic medium; characterized by

changes in volume and by particle motion parallel with the direction of wave-motion. Also known as pressure wave.

compression wave [Mechanics] A wave in a fluid in which a compression is propagated.

Compton effect [Physics] The reduction in the energy of a photon, as a result of its interaction with a free electron. Part of the photon's energy is transferred to the electron and part is redirected as a photon of reduced energy.

computed radiography [Computers] A diagnostic imaging technique that dcirects X-ray at the patient as cconventional X-ray units do, but develops the image immediately by scanning with a laser beam, a process that also digitizes the image.

computed tomography [Computers] a diagnostic imaging technique (Sometimes called a CAT scan) that directs X-rays axially around a patient and computes a two-dimensional image of the body slice that is displayed on a cathode-ray tube (CRT).

computer [Computers] An electronic device that processes information according to a set of instructions, called the program. The most versatile type of computer is the digital computer, in which the input is in the form of characters, represented within the machine in binary notation. The three basic components of a digital computer are the peripheral input and output devices, the memory, and the central processing unit (CPU).

Computer hardware consists of the actual electronic or mechanical devices used in the system; the software consists of the programs and data.

computer architecture [Computers] the way various computational elements are interconnected to achieve a computational function.

computer logic [Computers] A science designed to make use of computers in logic calculus.

computer system [Computers] 1. A set of related but connected components of a computer or data processing system. 2. A set of hardware parts that are related and connected, and thus form a computer.

concave [Sci Tech] Curving inwards; thus, a conave (or biconcave) lens thinner at the centre than at the edges.

concave convex [Sci Tech] A term used to describe a lens that curves inwards on one side and outwards on the other.

concentration [Chemistry] The quantity of dissolved substance per unit quantity of solvent in a solution. Concentration is measured in various ways. The amount of substance dissolved per unit volume (symbol c) has units of mol dm^{-3} or mol 1^{-1}. It is now called 'concentration' (formerly molarity). The mass concentration (symbol P) is the mass of solute per unit volume of solvent. It has units of kg dm^{-3}, g cm^{-3}, etc. The molal concentration (or molality; symbol m) is the amount of substance per unit mass of solvent, commonly given in units of mol kg^{-1}.

concentration cell [Electricity] A primary cell whose E.M.F. is due to a difference in concentration between different parts of the electrolyte.

concentric [Sci Tech] Having the same centre, *e.g.*, two concentric tubes would appear, in cross-section, as two concentric circles.

conceptual dependency [Computers] an approach to natural-language understanding iin whch sentences are translated into basic concepts that are expressed as a small set off semantic primiitives.

conception [Biology] The act of becoming pregnant as a result of the fertilization of an ovum by the sperm to form a viable zygote.

concrete [Materials] A building material composed of stone, cement, sand and water. Reinforced concrete has steel rods of meshes imbedded in it to increase its tensile strength.

condensation [Physics] Change of state from vapour, or gas to liquid by decrease in temperature. [Electricity] An increase of electrical charge on a capacitor conductor. [Optics] Focusing or collimation of light.

condensation polymer [Chemistry] A high molecular weight compound obtained by condensation polymerization.

condensation polymerization [Chemistry] Polymerization in which monomers combine together to form high-molecular weight polymers.

condensation reaction [Chemistry] A reaction in which two compounds or two different parts of the same compound combine to form one compound with the elimination of small molecules such as water, or hydrogen chloride.

condensation pump [Engineering] Apparatus used to obtain high vacua, *i.e.*, pressures of the order of 10^{-6} mm mercury. Also known as diffusion pump.

condenser [Chemistry] Liebig condenser. Apparatus for converting vapour into liquid during distillation. In its simplest form it consists of a tube along which the vapour passes and is cooled, usually by cold water flowing through an outer jacket surrounding the tube. [Electricity] *see* capacitor. [Optics] A device used in optical instruments to converge rays of light; *e.g.*, in the microscope a condenser lens is used to converge upon the object to be viewed.

conductance [Electricity] The conductance of a direct current circuit is the reciprocal of its resistance. The conductance of an alternating current circuit is its resistance divided by the square of its impedance. The SI unit is the siemens, formerly called the mho or reciprocal ohm. Also known as electrical conductance.

conduction [Electricity] 1. The passage of electrical charge. [Physics] Tranmission of energy by a medium which does not involve movement of medium itself. 2. The transmission of heat from places of higher to places of lower temperature in a substance by the interaction of atoms or molecules possessing greater kinetic energy with those possessing less. In gases the heat energy is transmitted by collision of the gaseous molecules those possessing the greater kinetic energy imparting, on collision, some of their energy to molecules having less. Conduction in liquids is mainly due to the same process. In solid electrical conductors, the chief contribution to thermal conduction arises from a similar process taking place between the free electrons present. The interaction of the molecules responsible for thermal conduction in solid electrical insulators arises from the elastic binding forces between the molecules, which are effectively fixed in space.

conduction band [Physics] The range of energies (*i.e.*, band) in which electrons can move freely in a solid, producing net transport of charge.

conduction current [Physics] A current due to flow of conduction electrons through a body.

conduction electron [Physics] An electron in the conduction band of a solid, where it is free to move under the influence of an electric field. Also known as valence electron.

conductivity [Physics] A measure of the ability of a substance to conduct heat. For a block of material of cross section A, the energy transferred per unit time E/t, between faces a distance, l, apart is given by $E/t=1A(T_2 - T_1)/l$, where 1 is the conductivity and T_2 and T_1 are the temperatures of the faces. This equation assumes that the opposite faces are parallel and that there is no heat loss through the sides of the block. The SI unit is therefore $Js^{-1}m^{-1}K^{-1}$. [Electricity] The reciprocal of the resistivity of a material. It is measure in siemens per metre in SI units. When a fluid is involved the electrolytic is given by the ratio of the current density of the electric field strength.

conductivity cell [Electricity] A glass vessel with two electrodes at a definite distance apart and filled with a solution whose conductivity is to be measured.

conductometric titration [Chemistry] A titration method in which the end point of a neutralization reaction is determined by changes in the electrical conductivity of the titrated solution during the addition of the titrant.

conductor [Electricity] A body capable of carrying an electrical current; a body that, if given an electric charge, will distribute that charge over itself. [Physics] A body that will pemit heat to flow through it by conduction.

conduit [Electricity] A flexible metal or plastic tubing through which insulated wires are run.

Condy's fluid [Materials] A solution of sodium or calcium (or sometimes aluminium) permanganate, $NaMnO_4$, or $Ca(MnO_4)_2$; used as a disinfectant.

cone [Mathematics] A solid figure traced by a straight line passing through a fixed point, the vertex, and moving a fixed circle. For a cone of vertical height h, slant height l, and radius of base r, the volume is given by $V = (\pi r^2 h)/3$, and the area of the curved surface $a = \pi rl$.

conformation [Chemistry] Any of the large number of possible shapes of a molecule resulting from rotation of one part of the molecule about a single bond.

conformational analysis [Chemistry] The determination of the arrangement of atoms of a molecule in space, which may rotate about a single bond.

conformation theory [Chemistry] The principle that three-dimensional structure of a molecule enables its stability and reactivity to be predicted. The theory pays special attention to the conformation of substituted hydrogen atoms in organic compounds; the axial (vertical) or equatorial (horizontal) disposition of substituents has been shown to be of great importance in predicting physical and chemical properties.

congeners [Chemistry] Elements that belong to the same group in the periodic table.

congenital disease [Medicine] Any disease or disorder which is present at birth.

congestin [Biochemistry] Toxin produced by certain sea anemines.

congestion [Medicine] An abnormal accumulation of fluid within the vessels of an organ or part.

congruent figures [Mathematics] Geometrical figures equal in all respects.

conjugate acid-base pair [Chemistry] An acid and a base related by the ability of the acid to generate the base by the loss of a proton, *e.g.*, conjugate base of HCl is Cl⁻.

conjugated [Chemistry] Describing double or triple bonds in a molecule that are separated by one single bond. For example, the organic compound buta-1, 3-diene, $H_2C=CH-CH=CH_2$, has conjugated double bonds. In such molecules, there is some delocalization of electrons in the pi orbitals between the carbon atoms linked by the single bond.

conjugate foci *See* conjugate points.

conjugate particles [Physics] A particle and its antiparticle.

conjugate points [Optics] Points on

either side of the lens, such that an object placed at either will produce an image at the other. Also known as conjugate foci.

conjunction [Astronomy] A planet (or other heavenly body) is said to be in superior conjunction when it is in a straight line with the sun and the earth; a planet with its orbit inside that of the earth is in inferior conjunction when it is between the sun and the earth and in line with them.

conservation law [Physics] Any physcial quantity associated with an isolated system is constant.

conservation of charge [Electricity] The principle that the total electric charge associated with a system remains constant: that electric charge can be neither created nor destroyed.

conservation of energy [Physics] The principle that energy can neither be created nor destroyed, although it can be changed from one form to the other.

conservation of mass [Physics] The notion that mass cannot be created or destroyed; now seen to be an approximation which can be applied to systems not involving nuclear reactions or velocities approaching the velocity of light.

conservation of momentum [Physics] The principle that the total momentum of two colliding bodies before impact is equal to their total momentum after impact.

When velocities comparable to the speed of light are being considered, the variation of mass with velocity (*see* relativity, theory of) must be taken into account, and the expression for the momentum becomes :

$$\text{Momentum} \quad = mV$$
$$= m_o V/\sqrt{1 - V^2/c^2}$$

where m_o =rest mass, V=velocity of the body, and c=velocity of light.

conservative field [Physics] A field of force in which the work done in moving a body from one point to another is independent of the path taken. The force required to move the body between these points in a conservative field is called a conservative force.

console [Computers] The part of a computer which is used to control the machine manually, rectify errors, revise the contents of storage, and provide communication between operator and the central processing unit.

consolute liquids [Chemistry] Liquids which are perfectly miscible in all proportions under certain conditions.

consolute temperature [Chemistry] The temperature at which two partially miscible liquids become completely miscible. Also known as critical solution temperature.

constant [Sci Tech] 1. A component of a relationship between variables that does not change its value, e.g. in $y = ax + b$, b is a constant. 2. A fixed value that has to be added to an indefinite integral. Known as the constant of integration, it depends on the limits between which the integration has been performed.

constantan [Metallurgy] An alloy of copper containing 10% -55% nickel. Its electrical resistance does not vary with temperature. It is used in electrical equipment.

constant boiling mixture *See* azeotropic mixture.

constant composition, law of *See* chemical combination, law of.

constellation [Astronomy] A group of stars, fixed relative to each other and forming configuration in the sky, *e.g.*, Leo, Orion.

constipation [Medicine] The passage of dry and hard stools.

contact lens [Optics] A thin lens fitted over the cornea to correct defects of vision.

containment [Nucl. Sc.] In a controlled thermonuclear reaction, the process of preventing the plasma from coming into contact with the walls of the containing vessel is referred to as containment. Also known as confinement.

contamination [Scientific Techniques] The act of soiling by the introduction of foreign material or bacteria. [Nucl. Sc] The deposit of radioactive materials on objects or in the atmosphere.

continuous function [Mathematics] A function $f(x)$ is continuous at $x = a$ if the limt of $f(x)$ as x approaches a is $f(a)$. A function

that does not satisfy this condition is said to be a discontinuous function.

continuous spectrum [Physics] A radiation spectrum which is continuously distributed over a frequency region without being broken up into lines or bands.

continous wave [Physics] Radio or radar transmissions generated continously and not in short pulses.

continuum [Mathematics] A continuous series of component parts passing into one another; *e.g.*, the three space dimensions and the time dimension are considered to form a four-dimensional continuum.

contraceptive [Medicine] Any chemical agent or mechanical device used to prevent conception, *e.g.*, condom, cervical cap.

control card [Computers] A punched card containing input data or parameters which are necessary to begin or modify a program.

control grid [Electronics] An electrode placed between the cathode and the anode of a thermionic valve for controlling the flow of electrons through the valve.

controller [Computers] Electronic circuitry that enables a host computer to operate or control the drive, and to send and receive data.

controlled thermonuclear reaction CTR. *See* thermonuclear reaction.

control rod [Nucl. Sc.] Part of the control system of a nuclear reactor that directly affects the rate of reaction therein. Usually a rod or tube, which can absorb neutrons and is made of steel or aluminium containing boron, and cadmium.

control structure [Computers] reasoning strategy. The strategy for manipulating the domain knowledge to solve a problem.

control word [Computers] A series of bits whose contents provide directives to control the operation of microprocessor computing elements.

convection [Physics] A process by which heat is transferred from one part of a fluid to another by movement of the fluid itself. In natural convection the movement occurs as a result of gravity; the hot part of the fluid expands, becomes less dense, and is displaced by the colder denser part of the fluid as this drops below it.

conventional current [Electricity] An electric current that is, by convention, regarded as flowing from a point of high potential to one of low potential. In fact, a current consisting of a flow of electrons flows in the opposite direction.

converging lens [Optics] A lens capable of bringing to a point a beam of light passing through it; *i.e.*, a convex lens.

conversion [Nucl. Sc.] The process in a nuclear reactor as a result of which fertile material is transformed into fissile material, *e.g.*, the conversion of thorium-232 into uranium-233.

conversion factor [Nucl. Sc.] The number of fissile atoms produced from the fertile material per fissile atom destroyed in the fuel.

conversion electron [Physics] An orbital electron ejected from an atom as a result of the energy it acquires from a transition of the nuclens from one energy state to another in the absence of gamma-ray emission.

converter [Electricity] An electrical machine for converting alternating current to direct current or vice-versa. [Metallurgy] The retort used in the Bessemer process.

converter reactor [Nucl. Sc.] A nuclear reactor that produces fissile material form fertile material by conversion.

convex [Scientific Techniques] Curving outwards; *e.g.*, a convex lens, that is thicker at the centre than at the edges.

coolant [Materials] Generally a fluid used for cooling, usually extracting heat from one source and transferring it to another. In a nuclear reactor the coolant transfers the heat from the nuclear reaction to the steam-forming plant.

coordinate bond [Chemistry] A kind of covalent bond between two atoms in which shared pair of electrons is suplied by one atom only called donor and the atom which accepts the electron-pair is called the acceptor. Also known as coordinate valence dative bond; dative bond.

coordinate geometry *See* analytical geometry.

coordinates [Mathematics] Magnitudes used to define the position of a point or line within a fixed frame of reference. *See* Cartesian coordinates and polar coordinates.

coordination chemistry [Chemistry] The chemical study of the interaction of metal ions with other molecules or ions capable of forming coordinate bond.

coordination compound [Chemistry] A compound in which the molecule or a component ion of the molecule contains a central atom surrounded by atoms or groups of atoms (called ligands) attached to the central atom by a number of valence bonds in excess of the stoichiometric valence of the central atom. Thus, potassium ferrocynide is a coordination compound in its anion $[Fe(CN)_6]^{4-}$, the central iron atom, which has a valence of two is attached to six CN^- groups.

coordinate induction [Biochemistry] Induction of a set of related enzymes by a single inducer.

coordination number [Chemistry] 1. In a crystal lattice, the number of anions that surround a cation. 2. In the molecule of a coordi-

nation compound, the number of atoms directly linked to the central atom.

coordinate repression [Biochemistry] Repression of a set of related enzymes.

copal [Materials] A hard and resinous substance obtained from certain trees; used in varnishes.

coplanar [Mathematics] In the same plane.

copolymerization *See* polymerization.

copper Cu. [Chemistry] Element. At. No. 29 At.Wt. 63.546. A red malleable and ductile metal, m.p. 1357K; b.p. 2855K; placed in group IB of periodic table; before silver, the best conductor of electricity. It is unaffectted by water or steam. Used for steam boilers, electrical wire and apparatus, in electrotyping, and in numerous alloys, *e.g.*, bronze, brass, monel metal, gun metal, bell metal, Dutch metal, manganin, constantan, nickel silver, etc.

coprecipitation [Chemistry] Simultaneous precipitation of more than one substances.

coral [Minerals] Deposits of impure calcium carbonate, $CaCO_3$, formed of the hard skeletons of various marine organisms.

cordite [Materials] An explosive prepared from cellulose nitrate and nitroglycerine.

core [Electricity] Magnetic material that is used to increase the **inductance** of /a **coil** through which it passes. It may be laminated or made of compressed ferromagnetic particles. 2. [Nucleonics] The central part of a nuclear reactor that contains the fissile material. 3. [Computers] The devices, semiconductors, ferrite rings, etc., that constitute the memory of a computer.

corona [Astronomy] A white irregular holo surrounding the sun, which is visible during a total eclipse.

corona discharge [Electricity] A luminous discharge that appears round the surface of a conductor due to ionization of the air (or other gas surrounding it), caused by the voltage gradient exceeding a critical value, but not being sufficient to cause sparking.

corpuscle *See* blood cell.

corpuscular radiation [Physics] Radiation consisting of subatomic particles, such as electrons, protons, neutrons, and deutrons.

corpuscular theory [Optics] The theory that light consists of minute corpuscles in rapid motion; now abandoned.

corrosion [Chemistry] The slow destruction of a metal by chemical action such as an acid water or atmospheric oxygen.

corrosive [Chemistry] A substance which attacks and destroys the surface of solids and of living things, *e.g.*, mineral acids.

corrosive sublimate [Chemistry]

$HgCl_2$, mercuric chloride; used as an antiseptic.

corrugated lens [Optics] A lens having circular sections cut out from the surface to reduce its weight without lowering its focal power.

Corubin [Materials] Crystalline aluminium oxide, Al_2O_3; obtained as by-product of the aluminothermic reduction.

corundum [Materials] Natural aluminium oxide, Al_2O_3. A crystalline substance nearly as hard as diamond; used as an abrasive.

cosecant *See* trigonometrical ratios.

cosine *See* trigonometrical ratios.

cosmetic [Medicine] A substance used to improve the appearance or prevent disfigurement of the skin.

cosmic [Astronomy] Pertaining to cosmos, the vast extraterrestrial regions of the universe.

cosmic dust [Astronomy] Small particles of matter, probably ranging in size from one hundredth to one ten-thousandth of a millimetre, forming clouds in intersteller space.

cosmic noise [Astronomy] Radio disturbances caused by a phenomenon outside the earth's atmosphere, such as sunspots.

cosmic radio waves [Astronomy] Radio waves reaching the earth from intersteller or intergalactic sources.

cosmic radiation [Astronomy] Very

energetic radiation falling upon the earth from outer space with nearly the speed of light, and consisting chiefly, if not entirely, of charged particles. The majority of these are most probably protons, although electrons and alpha particles are also present. There is also evidence that a small component (about 2%) of the primary radiation consists of heavy atomic nuclei. The primary particles, when incident upon our atmosphere, cause several secondary processes. Proton-neutron collisions in the top lenth of the atmosphere give rise to mesons. High-energy electrons are created in the atmosphere by meson decay, by interaction of high energy protons with nuclei, by knock on collisions of mesons with electrons, etc. These high-energy electrons give rise to cosmic ray showers resulting in the creation of photons, and further electrons.

cosmic ray shower [Physics] The simultaneous appearance of a number of downward-directed ionizing particles with or without photons, caused by a single cosmic ray. Also knówn as cascade shower.

cosmic year [Astronomy] The period of roṭation of the Milky Way Galaxy, about 220 million years.

cosmogony [Physics] The science of the nature of the heavenly bodies, with particular reference to the formation of planets, stars, and galaxies.

cosmology [Astronomy] The science of the nature, origin, and history of the universe. A more general and widely used term than cosmogony when referring to the universe as a whole. *See* steady state theory; superdense theory.

cosmotron [Nucl.Sc.] A proton accelerator containing a very large ring-shaped electromagnet.

cotangent *See* trigonometrical ratios.

Cottrell precipitator [Engineering] A device used for removing the dust and other suspended particles from gases by electrostatic precipitation.

coulomb [Electricity] C. The derived SI unit of electric charge, defined as the quantity of electricity transferred by 1 ampere in one second.$=10^{-1}$ electromagnetic unit; 3×10^9 electrostatic units.

coulomb force [Electricity] The electrostatic force of attraction or repulsion exerted by one charged particle on another in accordance with Coulomb's law.

coulomb friction [Mechanics] Friction occurring between dry surfaces.

coulomb scattering · [Electricity] The scattering of sub-atomic particles caused by the electrostatic (coulomb) field surrounding an atomic nucleus.

coulomb's law [Electricity] The force of attraction or repulsion between two charged bodies (whose charges behave as though they were concentrated at a point) is proportional to the magnitude of the charges and inversely proportional to the square of the distance between them. In SI units, the equation is written: $F=q_1q_2/4\pi\varepsilon_o d^2$, where F = force in newtons, q_1 and q_2 = charges in coulombs, d = distance between them in metres, and ε_o = dielectric constant.

coulometer Coulombmeter. *See* voltameter.

coulometry [Chemistry] A technique for the determination of the amount of an electrolyte released during electrolysis by measuring the amount of electricity used.

count down [Engineering] A process used in leading up to the launch of a large rocket vehicle.

counter tube [Electricity] A device for counting individual ionizing events.

couple [Chemsitry] Joining two molecules. [Electronics] To connect to electrical circuits. [Electronics] Two metal placed in contact, as in a thermocouple. [Physics] Two equal and opposite parallel, but not collinear, forces acting upon a body. The moment of a couple is the product of either force and the perpendicular distance between the line of action of the forces.

coupling reactions [Chemsitry] 1. *See* azo coupling. 2. Two chemical reactions that have a common

intermediate and thus a means of energy transfer from one to another.

covalency [Chemsitry] The number of covalent bonds an atom can form.

covalent bond [Chemsitry] Formed by sharing of valence electrons rather than by transfer. For instance, hydrogen atoms have one outer electron ($1s^1$). In the hydrogen molecule, H_2, each atom contributes 1 electron to the bond. Consequently, each hydrogen atom has control of 2 electrons one of its own and the second from the other atom giving it the configuration of an inert gas [He]. In the water molecule, H_2O, the oxygen atom, with six outer electrons, gains control of an extra two electrons supplied by the two hydrogen atoms. The gives it the configuration [Ne]. Similarly, each hydrogen atom gains control of an extra electron from the oxygen, and has the [He] electron configuration.

covalent crystal [Chemsitry] A crystal in which the atoms are held in the lattice by covalent bonds (*See* vaience, electronic theory of). Typical examples are diamond, silicon, and most organic crystals. *See also* semiconductors.

C.P.U. [Computers] Central processing unit. The central electronic unit in a computer that processes input information, and information from the store, and produces the output information. The C.P.U. and the store form the central part of the computer. The devices connected to them, known as peripherals, include the backing storage and the input and output equipment.

cracking [Chemsitry] The process of breaking down chemical compounds by heat. The term is applied particularly to the cracking of hydrocarbons in the kerosine fraction oobtained from petroleum refining to give smaller molecules, both as a source of branched-chain hydrocarbons suitable for gasoline (for motor fuel) and as a source of ethene and other alkenes. Catalytic cracking is a similar process in which a catalyst is used to lower the temperature required and to modify the products obtained.

cream of tartar [Chemsitry] $C_4O_6H_5K$. potassium hydrogen tartrate.

creep [Metallurgy] A permanent change in the physical dimensions of a metal caused by the application of a continuous strees.

creosote [Material] A distillation product obtained from coal-tar or from the tar obtained by the destructive distillation of wood. An oily, transparent liquid containing phenol and cresol; it is used for preserving timber.

crith [Mechanics] The weight of 1 litre of hydrogen at 273.16K, and a pressure of 760 mm; approximately 0.09g.

critical angle [Optics] The least angle of incidence at which total internal reflection occurs. When a ray of light passing from a denser to a less dense, *i.e.*, rarer medium, *e.g.*, glass to air, meets the surface, a portion of the light does not emerge, but is internally reflected. As the angle of incidence increases, the intensity of the internally reflected beam also increases until an angle is reached when the whole beam is thrown back and total internal reflection taking place.

critical current [Physics] The current in a superconductive substance above which the material is normal and below which the substance is superconducting, in absence of external magnetic fields at a specified temperature.

critical damping [Physics] A measuring instrument is said to be critically damped when it takes up its equilibrium deflection in the shortest possible time, The oscillations of the indicator (needle) about the equilibrium position being quickly damped out.

critical frequency [Physics] The limiting frequency below which a radiowave will be reflected by an ionospheric layer at vertical incidence at a given time.

critical grid current [Electronics] The value of grid current when the anode currents start to flow in a gas filled vacuum tube.

critical mass [Nucl Science] The minimum amount of fissile material required in a nuclear reactor or a nuclear weapon to sustain a chain reaction.

critical potential [Physics] The minimum energy required to raise the energy level of an orbital electron (*See* excitation) or to remove it from the atom. *See* ionization potential.

critical pressure [Physics] The pressure of a fluid in its critical state; *i.e.* when it is at its critical temperature and critical volume.

critical reaction *See* chain reaction.

critical point [Physics] 1. The temperature and pressure at which two phases of a substance in equilibrium with each other become identical, forming one phase. 2. The temperature and pressure at which two partially miscible liquids become completely miscible.

critical state [Physics] The state of a fluid in which the liquid and gas phases both have the same density. The fluid is then at its critical temperature, critical pressure, and critical volume.

critical temperature [Chemsitry] The temperature above which a gas cannot be liquified applying pressure alone.

critical velocity [Mechanics] The velocity at which the flow of a liquid ceases to be streamline and becomes turbulent.

critical volume [Physics] The volume of a fixed mass of a fluid in its critical state; i.e., when it is at its critical temperature and

critical pressure. The critical specific volume is its volume per unit mass in this state: in the past this has often been called the critical volume.

cross linking [Chemsitry] The joining of polymer molecules (*See* polymerization) to each other by valence bonds. A polymar may be imagined, in the simplest case, to consist of very long chain-like molecules; cross-linkage would have the effects of joining adjacent chains by lateral links.

crown glass [Materials] A variety of hard glass which is highly transparent for visible light and is used in making optical instruments.

crucible [Engineer] A refractory vessel of heat-resisting material used for containing high temperature chemical reactions, like melting or calcination.

crust [Geology] The outermost solid layer of the earth. Also known as lithosphere.

cryobiology [Biology] The use of low temperature techniques in the study of living animals and plants.

cryogen *See* freezing mixture.

cryogenics [Physics] The study of materials and phenomena at temperatures close to absolute zero.

cryogenic temperature [Physics] The temperature range within a few degrees of absolute zero.

cryohdrates [Chemsitry] Crystal-line substance containing the solute with a definite molecular proportion of water that crystal-lize out from solutions cooled below the freezing point of pure water.

cryolite [Minerals] Natural sodium aluminium fluoride, Na_2AlF_6; used for the extraction of alumin-ium.

cryometer [Engineering] A ther-mometer especially designed for measuring low temperatures.

cryophorus [Engineering] An ap-paratus used to demonstrate and study the cooling effect of evapo-ration.

cryoscopic method [Chemsitry] The determination of the molecular weight of a dissolved substance by noting the depression of freezing point produced by a known concentration. Also known freezing point method.

cryostate [Engineering] A vessel in which a specified low tempera-ture may be maintained.

cryotron [Electronics] A switch based on superconductivity. The simplest form consists of a coil of wire of one superconducting meterial wound round a length of wire of another superconductor, all immersed in a bath of liquid helium. A control current passed through the coil produces a magnetic field strong enough to destroy the superconductivity of the coil. Thus the current in the coil controls the resistance of the wire, switching if from zero to a finite value.

crystal [Crystallography] A solid with a regular polyhedral shape. All crystals of the same sub-stance grow so that they have the same angles between their faces. However, they may not have the same external appearance be-cause different faces can grow at different rates, depending on the conditions. The external form of the crystal is referred to as the crystal habit. The atoms, ions, or molecules forming the crystal have a regular arrangement and this is the crystal structure.

crystal defect [Crystallography] Any departure from crystal symmetry caused by free surface, disorder, vacancies and interstitials dislo-cations, impurities and lattice vibrations. Also know lattice defect.

crystal detector [Electronics] A fine wire ('cat's whisker') in contact with a crystal of galena (PbS) or other suitable semicon-ductor. This arrangement is a good conductor of electricity in one direction, and suppresses most of the flow of electricity in the other directions.

crystal diffraction [Physics] Dif-fraction of X-rays, electrons, or neutrons by a crystal.

crystal glass [Materials] A water-clear lead glass which has a high index of refraction and can be polished easily.

crystal lattice [Crystallography] The

regular pattern of atoms, ions, or molecules in a crystalline sub-stance. A crystal lattice can be regarded as produced by re-peated transactions of a unit cell of the lattice. *See also* crystal system.

crystalline [Crystallography] Hav-ing the regular internal arrange-ment of atoms, ions, or mole-cules characteristic of crystals. Crystalline materials need not necesarily exist as crystals; all metals, for example, are crystal-line although they are not usually seen as regular geometric crys-tals.

crystallite [Crystallography] A small crystal, *e.g.* one of the small crystals forming part of a mi-crocrystalline substance.

crystalliods [Chemsitry] Substances that, in solution, are able to pass through a semipermeable mem-brane; substances that do not usually form colloidal solutions.

crystal microphone [Electronics] A microphone in which the sound waves to be amplified or trans-mitted vibrate a piezoelectric crystal, which generates a vary-ing E.M.F.

crystal oscillator [Electronics] A source of electrical oscillations of very constant frequency deter-mined by the physical character-istics of a quartz crystal. *See* quartz clock.

crystal pick-up [Electronics] A pickup in a record player in which the varying E.M.F. is

produced by a piezoelectric crystal as a result of the vibrations obtained from the undulations in the groove of the record.

crystal rectifier [Electronics] A semiconductor used as rectifier, usually in a manner similar to a diode valve; also called a semi-conductor diode.

crystal system [Crystallography] A method of classifying crystalline substances on the basis of their unit cell. There are seven crystal systems. If the cell is a paral-lelopiped with sides a, b, and c and if α is the angle between b and c, β the angle between a and c, and γ the angle between a and b, the systems are

cubic $a = b = c$ and $\alpha = \beta = \gamma = 90°$

tetragonal $a = b \neq c$ and $\alpha = \beta = \gamma = 90°$

rhombic $a \neq b \neq c$ and $\alpha = \beta = \gamma = 90°$

hexagonal $a = b \neq c$
and $\alpha = \beta = 90°$, $\gamma = 120°$

trigonal $a = b = c$ and $\alpha = \beta = \gamma < 120°$

monoclinic $a \neq b \neq c$
and $\alpha = \gamma = 90° \neq \beta$

triclinic $a \neq b \neq c$ and $\alpha \neq \beta \neq \gamma \neq 90°$

cube 1. [Mathematic] A regular hexahedron; a regular solid fig-ure with six square faces. 2. The third power of a number, *e.g.*, 64 is the cube of 4, 4^3.

cube root [Mathematics] $\sqrt[3]{}$. The cube root of a number is the quantity that when raised to the third power gives that number. Thus 3 is the cube root of 27.

culture medium [Biology] A preparation used for growing and cultivating microorganisms for experimental purposes.

cupel [Metallurgy] A cup made of bone ash used in the extraction of the noble metals by cupellation.

cupellation [Metallurgy] The separation of silver, gold, and other noble metals from impurities that are oxidized by hot air. The impure metal is placed in a cupel, a flat dish made of porous refractory material and a blast of hot air is directed upon it in a special furnace. The impurities are oxidized by the air and are partly swept away by the blast and partly absorbed by the cupel.

cupric [Chemsitry] Containing bivalent copper, Cu(II).

cuprous [Chemsitry] Containing univalent copper, Cu(I).

curare [Materials] A very poisonous extract from the plants containing certain alkaloids.

curie [Nucl.Science] A measure of the activity of a radioactive substance (*see* radioactivity). Originally defined as the quantity of radon in radioactive equilibrium with 1 g of radium. Now extended to cover all radioactive isotopes and defined as that quantity of a radioactive isotope which decays at the rate of 3.7×10^{10} disintegrations per second.

Curie point [Physics] The temperature of a given ferromagnetic substance above which it becomes merely paramagnetic. Also known as curie temperature.

Curie's law [Physics] The susceptibility *(X)* of a paramagnetic substance is proportional to the thermodynamic temperature *(T)*, i.e. $x = C/T$, where C is the Curie constant. A modification of this law, the *Curie-Weiss* law, is more generally applicable. It states that $X = C/(T - \mu)$, where μ is the Weiss constant, a characteristic of the material.

curium [Chemsitry] Cm. Transuranic element, At. No. 96. A radioactive actinide whose most stable isotope, $^{247}_{96}Cm$, has a half-life of 1.6×10^7 years.

current, electric, *See* electric current.

current amplifier. [Electronics] An amplifier capable of delivering more signal current than is fed in.

current antinode. [Electronics] A point at which current is maximum along a transmission line, antenna, or other circuit element having standing waves.

current density [Electronics] 1. The current flowing through a conductor, plasma, etc. per unit crossectional area. It is usually expressed in amperes per square metre. 2. During electrolysis, the current flowing per unit area of electrode.

current node [Electronics] A point at which current is zero along a transmission line, antenna, or other circuit element having standing waves.

current regulator [Electronics] A device which maintains the output current of a voltage source at a predetermined, essentially constant, value, despite changes in load impedance.

current relay [Electronics] A relay which operates at a particular current value rather than at a particular voltage value.

cursor [Engineering] A transparent slider with a fine hair-line, used in slide-rules.

cyanide process [Metallurgy] A process for the extraction of gold from its ores by dissolving it in a solution of potassium cyanide, KCN, reducing the resulting potassium aurocyanide, $KAu(CN)_2$, with zinc, filtering off, melting down, and cupelling the metal.

cyanite [Minerals] Al_2SiO_5. A blue mineral consisting of aluminium silicate; used as a refractory.

cyanosis [Medicine] A bluish tingue observed most frequently under the nails and lips. It is always due to lack of oxygen.

cybernetics [Engineering] The theory of communication and control mechanisms in living beings and machines.

cybotaxis [Chemistry] The tendency of molecules in liquids to form regularly arranged groups, resembling crystals. See liquid crystals.

cycle [Physics] Any series of changes or operations performed by or on a sytem, which brings it back to its original state, e.g., the frequency of an alternating current is measured in cycles per second. See also hertz.

cyclic [Chemistry] Describing a compound that has a ring of atoms in its molecules. See carbocyclic and heterocyclic compounds.

cyclic figure [Mathematics] A figure through all the vertices or corners of which a circle may be drawn; a figure inscribed in a circle.

cyclization [Chemistry] The formation of cyclic molecule from an open-chain compound.

cyclo [Chemsitry] Prefix designating a cyclic compound, e.g. a cycloalkane or cyclosilicate.

cycloalkanes [Chemsitry] Cyclic saturated hydrocarbons containing a ring of carbon atoms joined by single bonds. They have the general formula C_nH_{2n}, for example cyclohexane, C_6H_{12}, etc. In general they behave like the alkanes but are rather less reactive.

cycloid [Mathematics] A figure traced out in space by a point on the circumference of a circle, which rolls without slipping along a fixed straight line.

cyclotron [Nucl Science] An accelerator for imparting to charged particles of atomic magnitudes energies of several million electron volts. The ions or charged particles are caused to traverse a spiral path between two hollow semicircular electrodes, called dees, by means of a suitable magnetic filed applied perpendicularly to the plane of the dees.At each half-revolution the particles receive an energy increase of some tens of thousands of electron-volts from an oscillating volatge applied between the dees.

cyclotron radiation [Electronics] The electromagnetic radiation emitted by charged particles orbiting in a magnetic field.

cyst [Medicine] A sac with a distinct wall, containing fluid or other material.

cystic disease [Medicine] A disorder of women, generally at or near menopause, characterized by the development of large cysts in the breast.

cystometer [Medicine] An instrument used to determine pressure in the urinary bladder under standard conditions.

cytochemistry [Biochemistry] The chemistry of living cells.

cytochrome [Biochemistry] Any of a group of proteins, each with an iron-containing haem group, that form part of the electron transport chain in mitochondria and chloroplasts. Electrons are transferred by reversible changes in the iron atom between the reduced Fe(II) and oxidized Fe(III) states.

cytodiagnosis [Biology] The determination of the nature of an abnormal liquid by studying the **cells** contained in it.

cytogenetics [Biology] A branch of molecular pathology combining the methods of cytology and genetics.

cytokinesis [Biology] Division of the cytoplasm following nuclear division.

cytokinins [Botany] Plant hormones that promote cell division in plants. They have potential uses in prolonging the freshness of vegetables ánd cut flowers.

cytology [Biology] The study of the structure and function of living cells.

cytolysis [Biology] The dissolution or disintegration of cells, particularly by the destruction of their surface membranes.

cytopathic [Medicine] Pertaining to disease of the living cell.

cytoplasm [Biology] The extra-nuclear component of the living cell, containing mitochondria, plastids, spherosomes, etc. This together with the nucleus, constitutes protoplasm. The chemical constituents are chiefly proteins alongwith a high proportion of water.

cytosol [Biology] The continuous aqueous phase of cytoplasm with its dissolved solutes.

cytotoxic [Medicine] Any substance which is toxic to cells; applied to the drugs used for the treatment of carcinomas and the reticuloses.

cytotropism [Biology] The tendency of individual cells and groups of cells to move toward or away from each other.

dacite glass [Geology] A natural glass formed by rapid cooling of dacite (glassy rock of volcanic origin) lava.

dactylography [Science Technology] The scientific study of fingerprints as a method for indentifying people.

dactylology [Science Technology] The finger sign method of communication with deaf and dumb people.

dadur [Meteorolgy] In India, a wind blowing down the Ganges valley from Sivalik hills at Hardwar.

Dalton's atomic theory [Chemistry] A theory of chemical combination, first stated by John Dalton in 1803. It involves the following postulates :

(1) Elements consist of indivisible small particles (atoms).

(2) All atoms of the same element are identical; different elements have differnt types of atom.

(3) Atoms can neither be created nor destroyed.

(4) 'Compound elements' (i.e. compounds) are formed when atoms of different elements join in simple ratios to form 'compound atoms' (i.e. molecules).

Dalton's law [Chemistry] of partial pressure. The total pressure of a mixture of two or more gases or vapours is equal to the sum of the pressure that each component would exert if it were present alone, and occupied the same volume as the whole mixture.

damp [Engineering] 1. To reduce the fire in a boiler or a furnace by putting damp coals or ash on the fire bed. 2. A poisonous gas in coal mines. [Physics] A decrease in amplitude of a vibration or oscillation with time.

damp air [Meteorolgy] Air with a high relative humidity.

damp oscillation [Physics] Any oscillation in which the amplitude of the oscillating quantity decreases with time. Also known as damp vibration.

damped wave [Physics] A wave whose amplitude drops exponentially with distance because of energy losses which are proportional to the square of amplitude.

damping [Physics] The dissipation of energy in motion of any type, especially oscillatory motion and the consequent reduction of decay of the motion. [Engineering] Reducing or eliminating reverberation in a room by putting sound absorbing materials on the walls and ceiling. Also known as sound proofing.

Daniell cell [Electronics] A type of primary voltaic cell with a copper positive electrode and a negative electrode of a zinc amalgam. The zinc-amalgam electrode is placed in an electrolyte of dilute sulphuric acid or zinc sulphate solutio⸱

in a porous pot, which stands in a solution of copper sulphate in which the copper electrode is immersed. While the reaction takes place ions move through the porous pot, but when it is not in use the cell should be dismantled to prevent the diffusion of one electrolyte into the other. The e.m.f. of the cell is 1.08 volts with sulphuric acid and 1.10 volts with zinc sulphate.

daraf [Electronics] The practical unit of elastance equal to the reciprocal of farad.

dark conduction [Electronics] Residual conduction in a photosensitive substance that is not illuminated.

dark discharge [Electronics] An invisible electrical discharge in a gas.

dark ground illumination [Optics] A device used in microscopy, whereby transparent or unstained objects are made to appear as bright particles on a black background.

dark reaction [Biochemistry] In photosynthetic cells, the light independent enzymatic reactions concerned with the synthesis of glucose from carbon dioxide, ATP and NADPH.

Darwins theory [Biology] The theory that different species arise by the process of natural selection.

dash-pot [Engineering] A mechanical damping device that depends upon the fact that when a body moves through a fluid medium, viscous forces are set up, which damp the motion of the body. It usually consists of a piston, attached to the part whose movements are to be damped, fitting loosely into a cylinder containing either air or oil.

dasymeter [Physics] A thin glass globe for determining the density of a gas.

data base [Computers] An organized collection of data a subject.

data-base management system [Computers] A computer system for the storage and retrieval of information about some domain.

data formating [Computers] Structuring the presentation of data as alphabetic or numerical and specifying the size and type of each item.

data manipulation [Computers] The standard operations of sorting, merging input/output and report generation.

datamation [Computers] A shortened term for automatic data processing.

data origination [Computers] The process of putting data in a form which can be read and understood by a machine.

data processor [Computers] Any device which can perform operations on data, *e.g.*, desk calculator, analog computer, digital computer.

data structure [Computers] The form in which data are stored in a computer.

dating [Science Technology] The use of methods and techniques to determine the age of minerals, fossils, or ancient objects. *See* also potassium argon dating, radiocarbon dating; radioactive age; and rubidium-strontium dating.

dative bond. *See* coordinate bond.

daughter element. [Nucl Science] An element produced by the radioactive decay of a pre-existing element such as uranium. Also known as decay product.

d-block elements [Chemistry] The block of elements in the periodic table consisting of scandium, yttrium, and lanthanum together with the three periods of transition elements: titanium to zinc, zirconium to cadmium, and hafnium to mercury. These elements all have two outer *s*-electrons and have *d*-electrons in their penultimate shell; i.e. and outer electron configuration of the form $(n-1)d^x ns^2$, where x is 1 to 10.

deactivation [Chemistry] 1. Rendering inactive. 2. Loss of radioactivity. 3. Chemical removal of the active constituent of a corrosive substance.

dead sea [Geology] A pool, lake, or other source of water that has undergone precipitation of its rock salts, gypsum, or other evaporites.

deaeration [Engineering] Removal of air or any gas from a substance.

deaminase [Biochemistry] An enzyme that catalyses the removal of an amino group from a compound.

deamination [Chemistry] The removal of amino groups from a compound.

Dean and Stark method [Engineering] A method of estimating the quantity of water in an oil or other liquid substance. The liquid under examination is distilled in to a special reflux condenser so constructed that the water is prevented from running back into the distillation flask. The volume of water so collected is measured and thus the water content of a known weight of initial liquid can be calculated.

debris [Geology] Large fragments arising from disintegration of rocks and strata.

de Broglie wavelength [Physics] The wavelength of the wave assoicated with a moving particle. The wavelength (l) is given by $l = h/mv$, where h is the Planck constant, m is the mass of the particle, and v its velocity. The *de Broglie wave* was first suggested by the French physicist Louis de Broglie (1892) in 1924 on the grounds that electromagnetic waves can be treated as particles (photons) and one could therefore expect particles to behave in some circumstances like waves. The subsequent observation of electron diffraction substantiated this argument and the de Broglie wave became the basis of wave mechanics.

debye [Physics] A unit of electrical dipole moment equal to 1×10^{-18} electrostatic unit of 3.33546×10^{-30} coulomb metere.

Debye and Huckel's theory [Chemistry] A theory of the behaviour of strong electrolytes, according to which each ion is surrounded by an ionic atmosphere of oppositely charged ions which retards the movement of ions towards the respective electrode during electrolysis.

deca- [Science Technology] da Prefix denoting ten times in the metric system.

decanning [Nucl Science] Removal of the outer container of an enriched uranium fuel rod during reprocessing of the fuel.

decantation [Chemistry] The separation of a solid form a liquid by allowing the former to settle and pouring off the latter.

decarburize [Metallurgy] The removal of carbon from the surface of a ferrous alloy by heating in a suitable medium.

decay [Nucl Science] The transformation of a radioactive substance into its decay (or daughter) products (*see* radioactivity). Also used in relation to the transformation of -particles into more stable particles.

decay constant *See* disintegration constant.

decay period of *See* half-life.

deci- [Science Technology] d. Prefix denoting one tenth in the metric system.

decible [Physics] One tenth of a bel. A unit for comparing levels of power. Two power levels, P_1 and P_2 said to differ by n decibles when :

$$n = 10 \log_{10} P_2/P_1$$

This unit is often used to express sound intensities. In this case, P_2 is the intensity of the sound under consideration and P_1 is the intensity of some reference level, often the intensity of the lowest audible note of the same frequency.

declination [Astronomy] The angular distance of a heavenly body from the celestial equator. [Geology] The angle between magnetic needle and the geographical meridian. Also known as magnetic deviation.

decomposition [Chemistry] The breaking up of chemical compounds into simpler molecules under various influences; *e.g.*, by chemical action, by heat (pyrolysis), by an electric current (electrolysis).

decomposition potential [Chemistry] The electrode potential at which the electrolysis current begins to increase appreciably.

decompression [Engineering] Any procedure for relieving the pressure or compression.

decontamination [Engineering] The removal of chemical or biological contaminants.

decrepitation [Chemistry] Bursting or cracking of crystal of certain substances on heating, mainly due to expansion of water within the crystals.

defect [Chemistry] A discontinuity in a crystal lattice. A *point defect* consists either of a missing atom or ion creating a *vacancy* in the lattice (a vacancy is sometimes called a *Schottky defect*) or an extra atom or ion between two normal lattice points creating an *interstitial*. A *Frenkel defect* consists of a vacancy in which the missing atom or ion has moved to an interstitial position. If more than one adjacent point defect occurs in a crystal there may be a slip along a surface causing *a line defect (or dislocation)*. Defects are caused by strain or, in some cases, by irradiation. All crystalline solids contain an equilibrium number of point defects above absolute zero; this number increases with temperature. The existence of defects in crystals is important in the conducting properties of semiconductors.

deferent [Astronomy] An imaginary circle around the earth in which a celestial body or its epicycle is supposed to move.

deficiency disease [Medicine] Disease produced by lack of a particular vitamin, mineral or other essential nutrients, *e.g.*, scurvy, by the deficiency of vitamin C.

definition [Optics] The sharpness of an image formed by a lens, mirror or other optical system.

[Electronics] The accuracy of sound or vision reproduction in a radio or television set.

deformation [Engineering] An alter nation in the size or shape of a body.

deformation potential [Electronics] The electric potential that acts on a free electron in a conductor or semiconductor as a result of deformation of the crystal lattice.

degaussing [Physics] The demagnetization of a magnetized substance with a coil carrying an alternating current of everdecreasing magnitude.

degenerate gas [Nucl Science] A state of matter in which electrons and atomic nuclei are packed too closely together for the evolution of nuclear energy; occurs in stars of the white dwarf class.

degradation [Chemistry] In general the breakdown of molecules into simpler fragments; the step wise decrease of the length of polymer macromolecules. *See* also depolymerization.

degree [Science Technology] A subdivision of an interval in a scale of measurement; *e.g.*, the Celsius degree. [Mathematics] A measure of angle. One three hundred and sixtieth of the angle traced by the complete revolution of a line about the point O, until it returns to its original position. 2. The sum of the exponents of the variables in a mathematical expression; the exponent of the derivative of the highest order in a differential equation.

degree of latitude and longitude See latitude; longitude.

degrees of freedom [Chemistry] 1. A term used in the phase rule; the number of independent variables defining the state of a system (*e.g.* the temperature and pressure in the case of a gas) that must be given definite values before this is completely determined. 2. The number of independent ways in which a molecule may possess translational, vibrational, or rotational energies.

degritting [Engineering] Removal of very small solid particles (grit) from a liquid carrier by gravity separation or centrifugation.

dehydration [Chemistry] The elimination or removal of water; usually the removal of chemically combined water, *e.g.*, concentrated sulphuric acid, H_2SO_4; acts as a dehydrating agent on substances that contain hydrogen and oxygen in the proportions in which they occur in water.

deionization [Electronics] The return of an ionized gas to its neutral state after all sources of ionization have been removed.

dehydrogenase [Biochemistry] An enzyme that catalyses oxidation reactions by the removal of hydrogen from the substrate.

dekatron [Electronics] A gas-filled emission tube with a central anode usually surrounded by ten cathodes and associated transfer electrodes. Incoming pulses cause a glow discharge to be transferred from one cathode to the next so that the tube may be used for counting or switching.

delayed neutrons [Nucl Science] Neutrons resulting from nuclear fission that are emitted with a measurable time delay. Only a small proportion of neutrons are delayed, but the average delay period must be taken into account in the control of nuclear reactors. *Also see* prompt neutrons.

delay line [Electronics] A component or circuit designed to introduce a calculated delay in the transmission of a signal.

deletion [Biology] Loss of chromosome segment of any size, down to a part of a single gene.

deliquescent [Chemsitry] The property of picking up moisture from the air to such an extent as to dissolve in it; becoming liquid on exposure to air.

delocalization [Chemistry] In certain chemical compounds the valence electrons cannot be regarded as restricted to definite bonds between the atoms but are 'spread' over several atoms in the molecule. Such electrons are said to be *delocalized*. Delocalization occurs particularly when the compound contains alternating (conjugated) double or triple bonds, the delocalized electrons being those in the pi orbitals.

The molecule is then more stable than it would be if the electrons were localized, an effect accounting for the properties of benzene and other aromatic compounds.

delta-iron [Metallurgy] An allotropic (*see* allotropy) form of pure iron which is nonmagnetic and exists between 1673K and the melting point.

delta metal [Metallurgy] An alloy of copper (55%) and zinc (43%) with small amounts of iron and other transition metals.

delta ray [Physics] A beam of electrons or protons.

demagnetization [Physics] The process of depriving a body of its magnetic properties. The 'demagnetization energy' is the energy that would be released when a body is completely demagnetized.

demodulation [Physics] The process, in a radio, television, or radar receiver, of separating information from a modulated carrier wave. The equipment used is called a demodulator or a detector.

denature [Chemistry] To denature ethanol is to add some poisonous substance to it to make it unfit for human consumption, *e.g.*, methylated spirits. [Nucl Science] To denature a fissile material is to add another isotope to it to render it unsuitable for use in a nuclear weapon. [Biochemistry] To denature a soluble or globular protein is to produce a structural change in it, either chemically or by heating, so that it loses most of its solubility; usually involves an unfolding of the polypeptide chain.

dendrite [Chemistry] A many-branched crystal. [Biology] The branching processes of a neurone that carry impulses into the cell body and form synapses with the axons of other neurones.

dendrology [Botany] The branch of botany concerned with the classification, identificaction, and distribution of trees and other woody plants.

denitrification [Chemistry] Removal of nitrates or nitrogen. [Biochemistry] The reduction of nitrate or nitrite to gaseous products such as nitrogen, nitrous oxide; brought about by denitrifying bacteria.

dentrifying bacteria [Biochemistry] Bacteria present in the soil that reduce nitrates to nitrites or nitrogen.

denominator [Mathesmatic] The number below the line in a vulgar fraction, *e.g.*, 7 in 5/7.

densimeter [Engineering] An instrument which measures the density or specific gravity of a solid, liquid, or gas. Also known as gravitometer.

densitometer [Engineering] An instrument for the measurement of the density of an image produced by light, x-rays, gamma rays, etc., on a photographic plate.

density [Physics] The mass of unit volume of a substance. In SI units density is expresed in kilograms per cubic metre, in *c.g.s.*, units in grams per cubic cetimetre, and in *f.p.s.*, units in pounds per cubic foot. 1 kg $m^{-3} = 10^{-3}$ g $cm^{-3} = 0.062428$ 1b ft^{-3}. *See also* relative density; vapour density. [Optics] 1. If one medium has a greater refractive index than another for light of a given wavelength, then it has the greater optical density for that wavelength. **density modulation** [Electronics] Modulation of an electron beam by varying the density of electrons in beam with time.

dental gold [Metallurgy] An alloy of gold containing 5 to 12% silver and 4 to 10% copper.

deodorant [Materials] Any substance which destroys or masks an unpleasent odour.

deoxyribonucleic (desoxyribosenucleic) acid DNA. [Biochemistry] Long thread-like molecules found in chromosomes and some viruses, consisting of two interwound helical chains of polynucleotides. The sugar of all the nucleotides is 2-deoxy-D-ribose, but each nucleotide in characterized by one of the four nitrogenous bases, which are : adenine, cytosine, guanine, and thymine. DNA molecules are responsible for storing the genetic code by the order of the arrangement of their nitrogenous bases, three bases coding for one amino acid.

depilatory [Materials] A compound used for removing hair temporarily from the skin.

depleted material [Nucl Science] In general, a material that contains less of a particular isotope than it normally possesses. In particular, applied to nuclear fuel, a material that contains less fissile isotopes than natural uranium, *e.g.*, the residue from an isotope separation plant for a nuclear reactor.

depletion layer [Electronices] The region of semiconductor in which the density of mobile carriers is too low to neutralize the fixed charge density of donors and acceptors.

depletion-mode transistor [Electronics] a field-effect transistor(FET) that conducts currents when its gate and source are at the same potential. To modulate current conduction between the drain and source, the gate is negatively biased with respect to the source.

depolarization [Electronics] The prevention of electrical polarization in a cell. For example, in Leclanche cell polarization is reduced by surrounding the positive carbon pole with manganese dioxide, MnO_2. This oxidizes the hydrgoen liberated at the pole, which is mainly responsible for polarization.

depolymerization [Chemistry] The breaking down of polymers into original monomers; the reverse of polymerization.

depth of field [Optics] Depth of focus. The range over which a camera, or other optical instrument, will produce a distinct image of an object.

derivative [Chemistry] A compound derived from (but not necessarily prepared from) some other compound, usually retaining the general structure of the parent compound; *e.g.*, nitrobenzene, $C_6H_5NO_2$, is a derivative of benzene, C_6H_6, one hydrogen atom in the molecule of the latter being replaced by a nitro group. [Mathematics] The result of differentiation of a mathematical function. Also known as derived function.

derived unit *See* base unit.

dermatology [Medicine] The science which deals with the study of skin, its structure, functions, diseases and their treatment.

desalination [Chemistry] The process of removing salt from seawater or soil to make it suitable for drinking or agricultural purposes.

desiccator [Chemistry] An apparatus used in laboratories for drying substances and for preventing hygroscopic substances from picking up moisture. It consists of a glass vessel, with a close-fitting ground lid, that contains some hygroscopic substance, *e.g.*, phosphorus pentoxide, or anhydrous calcium chloride.

desorption [Chemistry] The removal of molecules, ions, etc., from the surface of a solid so that they become gaseous; the reverse of adsorption.

destructive distillation [Chemistry] Heating a complex substance to produce chemical changes in it, and distilling off the volatile substances so formed, *e.g.*, the destructive distillation of coal produces coal-gas and many other valuable products. Also known as carbonization.

detector [Electronics] That part of a radio receiver in which the information is separated from the modulated carrier wave. Now more usually called a demodulator. See demodulation. Also known as crystal detector.

detergent [Materials] A synthetic cleaning agent resembling soap in the ability to emulsify oil and hold dirt, containing surfactants, may also contain whitening agent. These are generally compounds such as alkylarene sulphonates, sulphated aliphatic alcohols, etc. Unlike soaps, they are effective in hard water. Various synthetic ('soapless') detergents have been developed from petrochemicals. The commonest, used in washing powders, is sodium dodecylbenzenesulphonate, which contains $CH_3(CH_2)_{11}C_6H_4SO_2O^-$ ions. This, like soap, is an example of an anionic detergent, i.e. one in which the active part is a negative ion. Cationic detergents have a long hydrocarbon chain connected to a positive ion.

Usually they are amine salts, as in $CH_3(CH_2)_{15}N(CH_3)_3{}^+Br^-$, in which the polar part is the $N(CH_3)_3{}^+$ groups *Nonionic detergents* have nonionic polar of the type $C_2H_4-O-C_2H_4-OX$, which form hydrogen bonds with the water.

Synthetic detergents are also used as wetting agents, emulsifiers, and stabilizers for foam.

detonating gas [Materials] A mixture of hydrogen and oxygen in a volume ratio of 2:1; *i.e.* the volume ratio required to form water. It is extremely explosive when ignited.

detonation [Chemistry] An exothermic chemical reaction that takes place within a high velocity shock wave. Also used loosely to describe the combustion reactions that occur during knocking or 'pinking' in an internal combustion engine.

detoxication [Medicine] The process of removing poisonous property of a substance.

deuterium [Chemistry] D. ^2_1H The isotope of hydrogen with mass number 2, and atomic mass 2.0147. The abundance of deuterium in natural hydrogen is 0.0156%.

deuterium oxide D_2O. *See* heavy water.

deuteron [Chemistry] the nucleus of the deuterium atom, consisting of a proton and a neutron.

developing, photographic [Optics]

The action of certain chemicals, usually organic reducing photographic plate or film in order to bring out the latent image. The developer reduces those areas of the silver salt that have been exposed to light to metallic silver. This remains as a black deposit *See* photography.

deviation [Mathematics] The difference between one of a set of values and the mean of the set. The 'mean deviation' is the mean of all the individual deviations of the set.

device [Engineering] A mechanism, tool, or other piece of equipment designed for specific uses.

devitrification [Chemsitry] The process by which the glassy texture of a material is converted into crystalline texture.

dew [Meterolgy] Liquid water produced by condensation of water vapour in the air when the temperature falls sufficiently for the vapour to reach saturation.

dew point [Physics] The temperature at which the water vapour present in the air saturates the air and begins to condense, *i.e.*, dew begins to form.

dew point hygrometer [Engineering] An instrument for determining the dew point by measuring the temperature at which vapour being cooled in a silver vessel begins to condense.

dextro rotatory [Optics] Rotating

or deviating the plane of vibration of polarized light to the right (observer looking against the oncoming light). *See* polarization of light.

diagnosis [Computers] The processing of locating and explaining detectable errors in a computer routine or hardware components. [Medicine] The act of distinguishing one disease from the other on the basis of their signs and symptoms.

diagonal [Mathematics] A line joining the intersections of two pairs of sides of a rectilinear figure.

dialysis [Chemistry] The separation of colloids in solution from other dissolved substances by selective diffusion through a semipermeable membrane. Such a membrane is slightly permeable to the molecules of the dissolved substances, but not to the larger molecules or groups of molecules in the colloidal state.

dialyzed iron [Chemistry] A colloidal solution of ferric hydroxide, $Fe(OH)_3$. A deep red liquid, used in medicine.

dialyzate [Chemistry] The material which does not diffuse through the membrane during dialysis.

diamagnetic [Physics] Having magnetic permeability less than one; such substances are repelled by a magnet and tend to position themsleves at right angles to the magnetic lines of force.

diamagnetism [Physics] The property of a substance that has a small negative magnetic susceptibility and is repelled by magnets. This type of magnetism is due to a change in the orbital motion of the electrons in the atoms of the substance consequent on the application of an external magnetic field. The phenomenon occurs in all substances, although the resulting diamagnetism is often marked by the much greater effects due to paramagnetism or ferromagnetism.

diameter [Mathematics] A line segment which passes through the centre of a circle, and whose end points lie on the circle.

diamond [Minerals] A natural crystalline allotropic form (*see* allotropy) of carbon. It is colourless when pure, but is sometimes coloured due to the presence of traces of impurities; has a very high refractive index and dispersive power; one of the hardest substances known hardness to 10 Moh's scale owing to the covalent bonds between the atoms in its crystals and is transparent to x-rays (imitations are not); Used for cutting tools and drills, and as a gem.

diaphragm [Optics] Any opening in an optical system which controls the cross-section of a light beam passing through it, to control light intensity, or to increase the depth of focus. [Physics] A separating wall or membrane, especially one which transmits some substances and forces but not others.

diastereo-isomer [Chemistry] A pair of optical isomers which are not mirror image of each other. Also known as diastereomer.

diathermancy [Physics] The property of being able to transmit heat radiation; it is similar to transparency with respect to light.

diathermy [Medicine] A method of medical treatment by heating the body-tissues by the passage of a high frequency electric discharge.

diatomaceous earth *See* kieselguhr.

diatomic [Chemsitry] Consisting of two atoms in a molecule.

diazo compound [Chemsitry] An organic compound containing N=N- grouping.

diazole [Chemsitry] A five membered cyclic hydrocarbon out of which two are nitrogen atoms and three are carbon atoms.

diazonium compounds [Chemsitry] Organic compounds of the general formula $RN_2^+X^-$ where R is an aryl radical, RN_2^+ is a cation, and X- is an anion, *e.g.*, benzenediazonium chloride, $C_6H_5N_2+$ Cl^-, is a typical diazonium salt. Diazonium salts are prepared by diazotization (*see* diazo compound) of amines, an important stage in the production of azo dyes.

diazotization [Chemsitry] Reaction between a primary aromatic amine and nitrous acid to give a diazo compound.

dibasic acid [Cheistry] An acid containing two atoms of acidic hydrogen in a molecule; an acid giving rise to two series of salts, normal and acid salts; *e.g.* sulphuric acid, H_2SO_4, which gives rise to normal sulphates and hydrogen sulphates or bisulphates. Also known diprotic acid.

dicarboxylic acid [Chemistry] An organic compound containing two carboxyl groups.

dicentric [Biology] Having two centromeres.

dichogamy [Biology] Producing mature male and female reproductive structures at different times.

dichotomy [Astonomy] A configuration of three bodies so that they form a right triangle. [Biology] Forked branching produced by division of the growing point into two equal parts.

dichroic mirror [Optics] A mirror obtained by coating with special metal film which reflects certain colours of light while allowing the others to pass through.

dichroism [Optics] The property of some crystals, such as tourmaline, of selectively absorbing light vibrations in one plane while allowing light vibrations at right angles to this plane to pass through.

dichromate cell [Electronics] A primary cell of E.M.F., 2.03V having a positive pole of carbon and a negative pole of zinc in a liquid consisting of a solution of sulphuric acid, H_2SO_4, and potassium dichromate, $K_2Cr_2O_7$ the latter acting as depolarizing agent by its oxidizing action. Also known as bichromate cell.

dichromatism [Medicine] A form of partial colour blindness in which vision is apparently based on two primary colours rather than normal three.

die [Engineering] A mould or tool used to impart shapes or to form impression on materials such as metals and ceramics.

die forging [Metallurgy] Shaping metal by plastic deformation in a die.

dielectric [Materials] Non-conductor of electricity. A substance in which an electric field gives rise to no net flow of electric charge but only to a displacement of charge. Also known as insulator.

dielectric absorption [Electronics] The persistence of electric polarization in certain dielectric material after removal of electric field.

dielectric circuit [Electronics] Any electrical circuit having capacitors.

dielectric constant [Electronices] For an isotropic medium, the ratio of the capacitance of a capacitor filled with a given dielectric to that of the same capacitor having only a vacuum as dielectric. *Also see* permittivity.

dielectric current [Electronics] The current flowing at any instant through a surface of a dielectric that is located in a changing magnetic field.

dielectric heating [Electronics] A form of heating in which electrically insulating material is heated by being subjected to an alternating electric field. It results from energy being lost by the field to electrons within the atoms and molecules of the material. In industrial dielectric heating the material to be heated is placed between the plates of a capacitor connected to a high frequency power source.

dielectric strength [Electronics] The maximum voltage that can be applied to a dielectric material without causing it to break down; usually expressed in volts per mm.

dielectrophoresis [Physics] The motion of electrically polarized (*See* electric polarization) particles in a non-uniform electric field.

Diels-Alder reaction [Chemistry] A method of preparing a benzene ring from a diene and a compound containing a single double bond (*e.g.*, maleic acid) or triple bond.

diene [Chemistry] An alkene containing two double bonds, *e.g.*, butadiene, $H_2C=CHCH=CH_2$.

diesel fuel [Materials] Fuel used in internal combustion engines; usually that fraction of crude oil which distils after kerosine.

diesel rig [Engineering] Any diesel engine apparatus or machinery.

differential calculus [Mathematics] A branch of mathematics that deals with continously varying quantities. It is based upon the differential coefficient of one quantity with respect to another of which it is a function. Used for solving problems involving the rates at which processes occur and for obtaining maximum and minimum values for continously varying quantities.

differential coefficient [Mathematics] Derived function, derivative.

differential equation [Mathematics] An equation that involves differential coefficients. An ordinary differential equation is one in which only one independent variable is involved. The order of a differential equation is the same as that of the derivative of the highest order appearing in it; the degree is given by the largest exponent.

differential thermal analysis [Chemistry] A method of determining the temperature at which thermal reactions occur in a material undergoing continuous heating to elevated temperatures; also involves the determination of nature and intensity of such reaction.

differentiation [Biology] 1. The development of cells so that they are capable of performing specialized functions in the organs and tissues of the organisms to which they belong. 2. In microscopic specimen, the removal of the excess stain from certain parts to show up the structure as a whole. [Mathematics] The operation, used in the calculus, of obtaining the differential coefficient; if $y=xn$, the differential coefficient, $dy/dx=nx^{n-1}$.

diffraction [Physics] A redistribution in space of the intensity of waves which results from the presence of an object causing variations of either the amplitude or phase of the wave. When a beam of light passes through an aperture or past the edge of an opaque obstacle and is allowed to fall upon a screen, patterns of light and dark bands (with monochromatic light) or coloured bands (with white light) are observed near the edges of the beam, and extend into the geometrical shadow. This phenomenon is a particular case of interference and is due to the wave nature of light.

diffraction grating [Optics] An optical device used to disperse a beam of light, X-rays, or other electromagnetic radiation into its constituent wavelengths, *i.e.*, for producing its spectrum. It may consist of any device that acts upon an incident wave front in a manner similar to that of a regular array of parallel slits where the slit is of the same order as the wavelength of the incident radiation.

diffraction zone [Physics] The portion of a radio propagation path which lies outside a line-of-sight path.

diffractometer [Physics] An instrument used for studying the structure of matter by means of diffraction of X-rays, electrons, neutrons, or other waves.

diffuse-cutting filter [Optics] A colour filter whose absorption changes gradually with the change in wavelength of radiation.

diffused junction [Electronics] A semiconductor junction that has been formed by the diffusion of an impurity within a semiconductor crystal.

diffuse transmission [Physics] Transmission of acoustic or electromagnetic radiation in all directions by a transmitted body.

diffusion [Physics] The spreading of molecules of a material through a solid, a liquid, or a gas, caused by the random movement of molecules. 2. The movement of carriers in a semiconductor. [Optics] 1. The distribution of incident light by reflection. 2. Transmission of light through a translucent material.

diffusion of particles [Nucleonics] In nuclear physics, the passage of elementary particles through matter in such a way that the probability of scattering is large compared to that of capture.

diffusion of solutions [Chemsitry] Molecules or ions of a dissolved substance move freely through the solvent, the solution becoming uniform in concentration; the phenomenon is similar to diffusion of gases.

diffusion plant [Nucl Science] A plant for separating isotopes, based on their different rates of diffusion in the gaseous state through a membrane.

diffusion pump *See* condensation pump

digest [Chemistry] To dissolve a solid in a liquid by adding the solid to the hot liquid. When a solid is digested a chemical action takes place between the solid and the liquid. For example, cupric oxide is digested in sulphuric acid to make cupric sulphate.

digestion [Engineering] The process of sewage treatment by the anaerobic decomposition of **organic matter.** [Physiology] The process of converting food into absorable form by the breaking it into simpler chemical compounds. This is brought about by digestive juices.

digestive system [Biology] A system of structures in which food substances are digested.

digit [Astronomy] One twelfth of the diameter of the sun or moon; used to denote the **extent of an eclipse.** [Mathematics] A single figure or numeral; *e.g.*, 67142 is a number of 5 digits.

digital [Computers] Pertaining to data in the form of digits.

digital computer [Computers] A computer which operates on discrete data (in the form of digits), rather than physical quantities, by performing arithmetic and logic processes on the data.

digital display [Computers] A method of indicating the reading of a measuring instrument (*e.g.*, a voltmeter), clock, etc., in which number appear on a screen, as opposed to a pointer moving round a scale. It is often based on a digitron or a light-emitting diode. *See also* liquid-crystal display.

digital subtraction angiography [Computers] a technique for imaging blood vessels that takes X-ray images both with and without a contrast medium injected in the area of interest and subtracts the two to suppress surrounding details.

digitron [Computers] A vacuum tube used to give a digital display of a numerical value. It has cathodes shaped to form the digits 0-9.

dihedral [Mathematics] Formed by two intersecting planes.

dihydric [Chemistry] Containing two hydroxyls groups in a molecule; *e.g.*, ethylene glycol, $OH-CH_2-CH_2-OH$.

dilatancy [Chemistry] The tendency of some viscous or colloidal materials to solidify or become more rigid under the influence of pressure. *Compare* thixotropy.

dilatant [Chemistry] A substance which has ability to increase its volume when its shape is changed.

dilation [Physics] A change in volume. The act of stretching or expanding. Also known as dilatation.

dilatometer [Physics] An apparatus used for measuring volume changes of substances. It generally consists of a bulb with a graduated capillary stem.

dilute [Chemistry] Containing a large amount of solvent, generally water. 'Dilute' laboratory solutions of reagents are generally of twice normal strength, containing upto 2 gram-equivalents per litre.

dilution [Chemistry] 1. The further addition of water or any other solvent to a solution. 2. The reciprocal of concentration; the volume of solvent in which unit quantity of solute is dissolved.

dimensions [Physics] The proudct or quotient of the basic physical quantities, raised to the appropriate powers, in a derived phsycial quantity. The basic physical quantities of a mechanical system are usually taken to be mass (M), length (L), and time (T). Using these dimensions, the derived physical quantity velocity will have the dimensions L/T and acceleration will have the dimensions L/T^2. As force is the product of a mass and an acceleration force has the dimensions MLT^{-2}.

dimer [Chemistry] A chemical species composed of two molecules of a monomer, *e.g.*, Al_2Cl_6 is the dimer of $AlCl_3$.

dimmer [Electronices] An electrical or electronic control for changing the intensity of a light source.

dimorphism [Chemistry] The existence of a substance in two different crystalline forms.

dimorphous [Chemistry] Existing in two different crystalline forms.

di-neutron [Nucl Science] An unstable system comprising two neutrons.

diode [Electronics] A two-electrode electron tube containing an anode and a cathode. The cathode acts as a heated filament, which gives off electrons, which anode is a plate, which attracts electrons. It is used for demodulation and rectification.

diol [Chemistry] Any dihydric alcohol derived from aliphatic hydrocarbons by the substitution of hydroxyl groups for two of the hydrogen atoms in the molecule; general formula $C_nH_{2n}(OH)_2$. Also known as glycol; dihydric alcohol.

dione [Chemistry] Suffix indicating the presence of two keto groups in a molecule

dioptre [Optics] Unit of power of a lens; the power of a lens in dioptres is the reciprocal of its focal length in metres. The power of a converging lens is usually taken to be positive, that of a diverging lens negative.

dioptric [Optics] Produced by means of refraction.

dioptrics [Optics] The branch of optics that treats the refraction of light, especially by the transparent medium of the eye, and by lenses.

dip brazing [Metallurgy] Soldering by dipping into a hot salt or metal bath by using a nonferrous metal with a melting point above 700K.

dip circle [Engineering] An instrument for measuring the angle of magnetic dip. It consists of a magnetized needle mounted to rotate in a vertical plane, the angle being measured on a circular scale, marked in degrees. Also known as inclinometer.

dipeptide [Chemistry] A peptide consisting of two amino acids.

diploid cell [Biochemistry] A cell in which the nucleus contains chromosomes in pairs. Nearly all animal cells are diploid, except gametes. *See* haploid.

dipole [Physics] Two equal point electric charges (elecric dipole) or magnetic poles (magnetic dipole) of opposite sign, separated by a small distance. The dipole moment is the product of either charge (or pole) and the distance between the two. It may also be expressed as the couple that would be required to maintain the dipole at right angles to a field electric or magnetic of unit intensity, Molecules in which the centres of positive and negative charge are separated constitute dipoles, the dipole moments of which are measured by Debye units or coulomb metres. [Electronics] A radio aerial consisting of two rods.

dipole moment *See* dipole.

dippel's oil *See* bone oil.

diprotic acid *See* dibasic acid.

di-proton [Nucl Science] An unstable system comprising two protons.

dipsa [Materials] Foods that cause thirst.

dipsesis [Physiology] Extreme thirst, craving the abnormal kinds of drinks. Also known as dipsosis.

dipsogen [Materials] Thirst provoking agent.

Dirac quantization [Physics] The condition arising from the conversion of angular momentum, that for any electric charge q and magnetic monopole with magnetic charge m; $2qm = nhc$, where n is an integer, h is Planck's constant and c is the speed of light.

Dirac theory [Physics] Theory of electron based on Dirac's equation, which accounts for its spin angular momentum and gives its magnetic moment and its behaviour in an electromagnetic field.

direct current [Electronics] D.C.; d.c. An electric current in which flow of charge is in one direction only.

direct cycle reactor [Nucl Science] A nuclear power plant in which heat transfer fluid circulates through the reactor and then directly pass to the turbine in a continuous cycle.

direct dyes [Materials] A group of dyes that dye cotton, viscose, rayon, and other cellulose direct, without the use of mordants. Generally used with 'assistants' such as common salt or sodium sulphate, which assist absorption by the fibre. Also known as cotton dyes; substantive dyes.

direct motion [Astronomy] The motion of a planet or other celestial body round the sun in the same direction as the earth. All the planets have direct motion, but some comets and satellites do not, and they are said to have 'retrograde motion'. 2. Motion across the sky from west to east.

direct viewfinder [Optics] A viewfinder in which the user views the subject directly through the lens or sight.

disaccharides [Biochemistry] A group of sugars the molecules of which are derived by the condensation of two monosaccharide molecules with the elimination of a molecule of water. On hydrolysis disaccharides yield the corresponding monosaccharides. Some important disaccharides are lactose, maltose, and sucrose.

discharge, electrical [Electronics] The release of the elctric charge stored in a capacitor through an external circuit. [Cheistry] The conversion of the chemical energy stored in an electric cell into electrical energy. [Electronics] The passage of electricity through a gas, usually accompanied by a glow arc, spark or corona.

discharge lamp [Electronics] A

lamp in which light is produced by an electric discharge between the electrodes in a gas at low pressure.

discharge liquor [Chemistry] Liquid that has passed through a processing operation. Also known as effluent.

discharge tube [Electronics] A glass vessel in which electric current passes from anode to cathode in a high vacuum.

discriminator [Electronics] An electronic circuit that converts frequency or phase modulation into amplitude modulation.

disinfectants [Materials] Chemical agents which destroy microorganisms but not bacterial spores. These are too corrosive or toxic to be applied to living tissues but are suitable for application to inanimate objects.

disintegration [Nucl Science] Any process in which the nucleus of an atom emits one or more particles or photons, either due to spontaneous radioactivity or as the result of a collision.

disintegration constant [Nucl Science] The probability of the decay of an atomic nucleus per unit time that characterizes a radio active isotope. It determinès the exponential decrease with time, t of the activity X, given by:

$$X = X_0 e^{-\lambda t}$$

where X_0 is the activity when $t = 0$, and λ is the disintegration constant. Also known as decay constant; transformation constant.

dislocation [Crystallography] A line defect in a crystal, the result of slip along a surface of one or more **lattice constants**. [Biology] A displacement of organs, or articular surfaces, or bones.

disordering [Crystallography] The displacement of atoms from their position in a crystal lattice (*e.g.*, as a result of the effect of ionizing radiation) to positions that are not part of the lattice.

disperse dyes [Materials] Dyes of all chemical types that are applied in the form of fine suspensions in water to man-made fibres, such as cellulose acetate, nylon, and polyester fibres. They are insoluble in water, but are usually soluble in organic solvents, such as esters.

disperse phase [Chemistry] The dissolved or suspended substance in a colloidal solution or suspension.

dispersion [Chemistry] A disperse phase suspended in a disperse medium; a system of particles dispersed and suspended in a solid, liquid, or gas.

dispersion medium [Chemistry] A medium in which a substance in the colloidal state is dispersed; the solvent in a colloidal solution.

dispersion of light [Optics] The splitting of light of mixed wavelengths into a spectrum. A beam of ordinary white light, *e.g.*, sunlight, on passing through an optical prism or a **diffraction**

grating, is divided up or dispersed into light of the different wavelengths of which it is composed if the beam that emerges after dispersion is allowed to fall upon a screen, a coloured band or spectrum is observed. Dispersion by a prism is due to the fact that lightwaves of different wavelenths are refracted or bent through different angles on passing through the prism, and are thus separated.

dispersive power [Optics] A measure of the dispersion of light of two specified wavelengths ('I' and 'II'); given by the ratio

$$(n_1 - n_2)/(n - 1)$$

where n_1, is the refractive index of the medium for wavelength I, n_2 that for wavelength II, and n is the average of n_1 and n_2. When considering the dispersive power of media for ordinary white light, the dispersive power is often defined as n_b-n_r/n_y-1 where n_b, n_r and n_y are the refractive indices for blue, red, and yellow lights respectively.

displacement [Chemistry] A chemical reaction in which an, atom radical, or molecule takes the place of another element which is in a compound. [Physics] The distance of an oscillating particle from its mean position.

display tube [Electronics] A cathode ray tube used to provide a visual display.

disproportionation [Chemistry] A type of chemical reaction in which the same compound is simultaneously reduced and oxidized. For example, copper(I) chloride disproportionates thus :

$$Cu_2Cl_2 \dashrightarrow Cu + CuCl_2$$

dissect [Biology] To cut, divide and separate into different parts.

dissipiation [Physics] Any loss of energy, generally by conversion into heat.

dissociation [Chemistry] A reversible separation of a molecule into two or more fragments, which may be brought about either by collision or by the absorption of electromagnetic radiation.

dissociation constant [Chemistry] A constant whose numerical value depends upon the equilibrium between the dissociated and undissociated forms of a molecule, a higher value indicates greater dissociation.

dissolve [Chemistry] To make a solid or a gas disappear in a liquid.

distillation [Chemistry] The process of converting a liquid into vapour, condensing the vapour, and collecting liquid or distillate.

distribution law [Chemistry] If a substance is dissolved in two immiscible liquids, the ratio of its concentration in each liquid is constant.

diurnal [Astronomy] Daily; performed or completed once every 24 hours.

div (divergence) [Mathmatic] Δ. The expression

$$\delta u_1/\delta x + \delta u_2/\delta y + \delta u_3/\delta z$$

where u_1, u_2, and u_3 are the x, y, and z components of the vector u. In physics, ΔF is used to describe an excess flux leaving a volume in space, where F is a three dimensional vector function. *See* also Laplace equation.

divalent [Chemistry] Having a valence of two. Also known as bivalent.

divergent [Science Technology] Going away in different directions from a common path or point.

diverging lens [Optics] A lens that causes a parallel beam of light passing through it to diverge or spread out; concave lens.

diver's liquid [Chemistry] A solution of ammonium nitrate in liquid ammonia; used as a solvent for some metals and their oxides and hydroxides.

division [Mathematics] An arithmetic operation in which a dividend is divided by a divisor to give a quotient and a remainder.

DNA *See* deoxyribonucleic acid.

DNA chimera [Biochemistry] A DNA containing genes from two different species.

DNA replicase system [Biochemistry] The entire complex of enzymes and specialized proteins required in biological replication.

Doctor solution [Materials] A solution of sodium plumbite used to remove bad smelling compounds from petroleum.

dolomite [Minerals] $MgCO_3 \cdot Ca \cdot CO_3$. Natural double carbonate of magnesium and calcium. A whitish solid that occurs naturally in vast amounts, comprising whole mountain ranges. Also known as pearl spar.

do loop [Computers] A FORTRAN technique which enables any number of instructions to be executed repeatedly.

domain [Electronics] the smallest area of a magnetic material in which all field-magnetization vectors point in the same direction; a recorded bit, typically 1 micrometer across, consists of many domains.

donor [Electronics] An imperfection in a semiconductor that causes electron conduction.

doping [Electronics] The addition of a small quantity to a semiconductor to achieve a particular characteristic.

Doppler broadening [Physics] The broadening of spectral emission or absorption lines (*see* spectrum) due to random motion of the emitting or absorbing molecules, atoms, or nuclei. *See* Doppler effect.

Doppler effect [Physics] The apparent change in the frequency of sound or electromagnetic radiation due to relative motion between the source and the observer. The pitch (frequency) of the sound emitted by a moving object (*e.g.*, the whistle of a moving train) appears to a stationary observer to increase as the object approaches him and to decrease as it recedes from him. The light emitted by a moving object appears more red (red light being of lower frequency than the other colours) when it is receding from the observer (or the observer receding from it). The Doppler effect is used in radar, to distinguish between stationary and moving targets and to provide information concerning their velocity, by measuring the frequency shift between the emitted and the reflected radiation. Also known as Doppler shift; Doppler's principle.

Doppler shift *See* Doppler effect.

dose [Nucl Science] The 'absorbed dose' is the energy imparted by ionizing radiation to unit mass of irradiated matter. Measured in rads (*i.e.*, 100 ergs per gram or 0.01 joule per kilogram). The 'maximum permissible dose (or level)' is the recommended upper limit for the absorbed dose that a person should receive during a specified period.

dosimeter [Nucl Science] A device for measuring a dose of ionizing radiation. Several methods are used, including ionization chambers, the blackening of photographic film, and the extent to which a chemical reaction in solution proceeds. Also known as dosemeter.

double bond [Chemistry] Two co-valent bonds linking two atoms in a chemical compound, *e.g.*, R-HC=CH-R; characteristic of an unsaturated compound.

double decomposition [Chemistry] A chemical reaction between two compounds in which each of the original compounds is decomposed and two new compounds are formed. Also known as metathesis.

double salt [Chemistry] A crystalline salt in which there are two different anions and/or cations. An example is the mineral dolomite, $CaCO_3.MgCO_3$, which contains a regular arrangement of Ca^{2+} and Mg^{2+} ions in its crystal lattice. Alums are double sulphates. Double salts only exist in the solid; when dissolved they act as a mixture of the two separate salts. Double oxides are similar.

double star [Astronomy] Two stars held very close to each other as a result of their mutual gravitational attraction, which move through space together giving the appearance of single point of light to the naked eye.

doublet [Physics] Two stationary states which have the same orbital and spin angular momentum but have different total angular momenta, and therefore have slightly different energies due to spin-orbit coupling.

drain [Engineering] A pipe or channel which carries off surface water or liquid sewage. [Electronics] One of the electrodes in a thin-film transistor through which charge carrier leave the inter-electrode region.

drift [Electronics] The movement of current carriers in a semiconductor under the influence of an applied voltage.

drift mobility [Electronics] The average drift velocity of carriers per unit electric field in a homogeneous semiconductor.

drift velocity [Electronics] The average velocity of a carrier that is moving under the influence of an electric field in a semiconductor, conductor, or electron tube.

drug [Medicine] Any chemical substance used in medicine to cure or prevent disease. 2. A habit-forming narcotic; any substance that causes physiological or emotional dependence.

dry cell [Electronics] A type of small Leclanche cell containing no free liquid. The electrolyte of ammonium chloride is in the form of a paste, and the negative zinc pole forms the outer container of the cell; used for torch batteries, radio batteries, etc. Also known as dry battery.

dry corrosion [Metallurgy] Destruction of a metal or alloy by gases in atmosphere above the dew point.

dry ice [Chemistry] Solid carbon dioxide, CO_2, has a temperature of 195K; used to refrigerate foodstuffs in transit, for carbonation of liquids and for cold traps in laboratories; sublimes from

solid state to a gas without liquefying.

drying oil [Materials] An animal or vegetable oil that will harden to a tough film when a thin layer is exposed to the air. The hardening is due to oxidation or polymerization of the unsaturated fatty acids of which these oils partially consist; used in paints and varnishes, e.g., linseed oil, dehydrated castor oil, and certain fish oils.

ductility [Materials] A property, especially of metals, of being capable of being drawn out into a wire.

ductless glands [Anatomy] Glands or organs producing hormones in the body. Also known as endocrine glands.

Dulong and Petit's law [Chemistry] For a solid element, the product of The atomic weight and the specific heat capacity, i.e., the atomic heat, is a constant, approximately equal to 25 joules per mole. For validity of this law, *see* atomic heat.

duplet [Chemistry] A pair of electrons shared between two atoms forming a single covalent bond.

duplex [Engineering] Consisting of two parts working together or in a similar fashion.

duralumin [Metallurgy] A light and hard aluminium alloy containing about 4% copper, and small amounts of magnesium, manganese, and silicon.

dust core [Electronics] A core for

magnetic devices made of pow-dered metal (often molybdenum) held together with a suitable binder-Particularly suitable for high frequency equipments.

Dutch metal [Metallurgy] A ductile alloy of copper and zinc. It takes a high polish, and is used for making low-priced jewellery.

dwarf star [Astronomy] A star of small volume, high density, and usually low luminosity. *See also* white dwarf star.

dyes [Chemistry] Substances used to impart colour to textiles, leather, paper, etc. Compounds used for dyeing (dyestuffs) are generally organic compounds containing conjugated double bonds. The group producing the colour is the chromophore; other noncoloured groups that influ-ence or intensify the colour are called auxochromes. Dyes can be classified according to the chemi-cal structure of the dye molecule. For example, *azo* dyes contain the -N=N- group (*see* azo com-pounds). In practice, they are classified according to the way in which the dye is applied or is held on the substrate. *See also* acid dyes; direct dyes; disperse dyes; reactive dyes and vat dyes.

dynamic allotropy [Chemistry] A kind of allotropy in which the allotropes are in dynamic equi-librium with each other, *e.g.*, liquid sulphur.

dynamic equilibrium [Chemistry] A state of equilibrium in which the rate of change of two

processes is equal and opposite *e.g.*, a liquid in equilibrium with its saturated **vapour**. [Biology] Biochemical processes involving living tissues which are continu-ally being broken down and resynthesised so that their struc-ture remains constant.

dynamic range [Electronics] the range of a signal's amplitude compared to the maximum noire level;in the case of a CRT display, determined by the number of gray levels at each pixel.

dynamics [Mathematics] A branch of mechanics; the mathematical and physical study of the behavi-our of bodies under the action of forces that produce changes of motion in them.

dynamite [Materials] An explosive consisting of nitroglycerin ab-sorbed in kieselguhr.

dynamo [Electronics] A device for converting mechanical energy into electrical energy, depending on the fact that if an electrical conductor moves across a mag-netic field, an electric current flows in the conductor.

dynamo effect [Physics] A process in the ionosphere in which winds and the resultant movement of ionization in the geomagnetic field give rise to induced current.

dynamometer [Engineering] Any instrument designed for the measurement of power.

dynatron oscillator [Electronics] An oscillator, using a tetrode (screen grid valve) in such a way that the anode current increases as the anode voltage is **reduced.**

dyne [Physics] In C.G.S. system the unit of force; the force that, acting upon a mass of 1 gm, will impart to it an acceleration of 1 cm per second. 1dyne=10^{-5} newton.

dynode [Electronics] An electrode whose primary function is the secondary emission of electrons. Also known as electron mirror.

dyox [Materials] Trade name for chlorine dioxide, ClO_2; used to treat flour.

dysfunction [Medicine] Abnormal functioning of any organ or part.

dysopia [Medicine] Painful or defective vision.

dyspepsia [Physiology] Any pain or discomfort associated with eating.

dysprosium [Chemistry] Dy. A soft silvery metallic element belonging to the lanthanoids; At.No. 66; r.d. 8.551 (293 K); m.p. 1685 K, b.p. 2850 K. It occurs in apatite, gadolinite, and xenotime, from which it is extracted by an ion-exchange process. There are seven natural isotopes and twelve artificial isotopes have been identified.

dystetic mixture [Chemistry] A mixture that has a constant maximum melting point.

dystophic [Biology] Pertaining to an environment which does not supply adequate nutrition.

e [Mathematics] The irrational number defined as the limit as n tends to infinity of $(1+1/n)^n$. It has the value 2.71828...It is used as the base of natural logarithms and occurs in the exponential function, e^x

earth [Astronomy] The third planet of the solar system, lying between venus and mars, at an average distance of 150,000,000 km from the sun. It is almost spherical in shape, being slightly flattened at the poles. Mass=5.976×10^{24} kg; equitorial radius 6378.388 km.

earth core [Geology] Centre of the earth, beginning at a depth of 2900 kilometres. Also known as core.

earthing [Electronics] A conductor making electrical connection between the conductor and the earth; is assumed to have zero potential.

earthlight [Astronomy] A faint illumination of the dark side of the moon during a crescent phase, due to sun light reflected from the earth's surface. Also known as earth shine.

earthmover [Engineering] A machine used to excavate, transport, or push earth.

earthquake [Geology] Shaking of the earth's crust caused mostly by displacement along a fault, or by volcanic action.

earth's crust Lithosphere. *See* crust.

east-west asymmetry of cosmic rays [Physics] The observed intensity of cosmic ray particles coming from the west is greater than that coming from the east at any given lattitude. This asymmetry is due to the deflection of the primary charged cosmic ray particles by the magnetic field of the earth, and indicates a preponderance of positively charged particles in the incoming radiation.

ebonite [Materials] A hard black insulating material made by vulcanizing rubber with high proportions of sulphur; contains about 30% combined sulphur. Also known as vulcanite.

ebulliometer [Physics] An instrument used to measure precisely the absolute or differential boiling points of solutions. Also known as ebullioscope.

ebullition *See* boiling.

eccentricity [Astronomy] A measure of the extent to which an ellipse is elongated, equal to the distance between the foci divided by the length of the major axis. This value is used to express the eccentricity of a planet's orbit round the sun : *e.g.*, the eccentricity of the earth's orbit is 0.0167.

echelom [Physics] A form of interferometer consisting of a stack of glass plates arranged stepwise

with a constant offset. It gives a high resolution and is used in spectroscopy to study hyperfine line structure.

echo [Physics] The effect produced when sound or other radiation is reflected or otherwise returned on meeting a solid obstacle or a reflecting medium with sufficient delay.

echo chamber [Acoustics] A reverbernt room used in a studio to add echo effects to sound at the time of recording.

echo sounder [Engineering] A device for estimating the sea depth beneath a ship by measuring the time taken for a sound pulse to reach the sea bed and for its echo to return.

eclipse [Astronomy] The darkening of a heavenly body when it moves into the shadow of another heavenly body. An eclipse of the moon is seen when the shadow of the earth. falls on the moon, solar eclipse is seen when shadow of the falls moon on the earth.

ecliptic [Astronomy] The sun's apparent path in the sky relative to the star; the intersection of the plane of the earth's orbit with the celestial sphere.

ecliptic pole [Astronomy] Either of the two points 90^0 from the ecliptic on the celestial sphere.

ecology [Biology] The study of the relation of plants and animals to their environment and to each other. Also known as environmental biology.

ecosphere [Biology] 1. The part of the earth's atmosphere in which life can exist; also called the 'biosphere' or 'physiological atmosphere'. 2. The part of the atmosphere surrounding any planet on which life could exist. 3. The part of space surrounding any star in which life could be possible.

ectoderm [Biology] The outer of the two germ layers of gastrula, gives rise to the skin and nervous system.

ectoplasm [Biology] The outer layer of the cytoplasm of a living cell. Usually a semi-solid gel containing relatively few granules.

ectozoa [Biology] Animals which live externally on other organisms.

eddy current [Electronics] Induced (*see* induction) electric currents set up in the iron cores of elctromagnets and other electrical apparatus. These currents cause considerable waste of energy in the cores of armatures of dynamos and in transformers. Also known as Foucault current.

edible oil [Materials] Any fatty oil obtained from the flesh or seeds of plants which is primarily used in foodstuffs. These oils vary in degree of unsaturation ranging from about 78% to 10%.

Edison cell. *See* Nickel-iron accumulator.

eductor [Engineering ejector-like] An device for mixing two fluids.

effective capacitance [Electronics] Total capacitance existing between any two given points of an electric circuit.

effective current [Electronics] The value of alternating current which will give the same heating effect as the corresponding value of direct current. Also known as root-mean-square current.

effective resistance [Electronics] The resistance of a conductor of electricity to alternating currents; in addition to the direct current resistance it includes the effect of any losses caused by the current (*e.g.*, eddy currents); measured by the ratio of the total loss of the square of the root mean square of the current.

effective value *See* root mean square of an alternating quantity.

effervescence [Chemistry] The formation of gas bubbles in a liquid by chemical reaction.

efficiency of a machine [Physics] The ratio of the output energy to the input energy. The efficiency of a machine can never be greater than unity. Often expressed as a percentage.

efflorescence [Chemistry] The process in which a crystalline hydrate loses water, forming a powdery deposit on the crystals.

effluent *See* discharge liquor.

effusion [Chemistry] The flow of a gas through a small aperture. The relative rates at which gases effuse, under the same conditions, is approximately inversely proportional to the square roots of their densities.

eigenfunction [Mathematics] An allowed wave function of a system in quantum mechanics. The associated energies are eigenvalues.

einstein [Physics] A unit of light energy used in photo chemistry, equal to Avogadro's number times the energy of one photon of light of the frequency in question.

einsteinium [Chemistry] Es. Transuranic element, At. No. 99 Most stable isotope, 25499Es, has a half-life of 276 days.

Einstein number [Physics] A dimensionless number used in magnetofluid dynamics, equal to the ratio of the velocity of a fluid to the speed of light.

Einstein equation [Physics] 1. The mass-energy relationship announced by Einstein in the form $E=mc^2$, where E quantity of energy, m mass, and c the speed of light.It presents the concept that energy possesses mass. 2. The relationship $E_{max} = hf - W$, where E_{max} the maximum kinetic energy of electrons emitted in the photoemissive effect, h Planck constant, f frequency of the incident radiation, and W work function of the emitter. This is also written $E_{max}=hf-\phi e$, where e the electronic charge and ϕ a potential difference, also called the work function. (Sometimes W and ϕ are distinguished as *work function energy* and *work function potential*). The equation can also be applied to photoemission

from gases, when it has the form: $E= hf -- I$, where I the ionization potential of the gas.

Einstein shift [Physics] A slight displacement towards the red of the lines of the solar spectrum due to the sun's gravitation field; predicted by Einstein's general theory of relativity and subsequently verified experimentally.

Einstein's principle of relativity [Physics] The principle that all the laws of physics must assume the same mathematical form in any inertial frame of reference, thus it is impossible to determine the absolute motion of a system by any means.

Einstein universe [Physics] A model of the universe which is a four-dimensional cylindrical surface in a five-dimensional space.

elastance [Electronics] The reciprocal of capacitance; measured in reciprocal farads or darafs.

elastic [Mechanics] Capable of sustaining deformation without permanent loss of shape or size.

elastic collision [Mechanics] A collision between bodies under ideal conditions, such that their total kinetic energy before collision equals their total kinetic energy after collision. Referred to nuclear physics, an elastic collision is one in which an incoming particle is scattered without causing the excitation or breaking up of the struck nucleus.

elasticity [Mechanics] The property of a body or material of resuming its original form and dimensions when the forces acting upon it are removed. If the forces are sufficiently large for the deformation to cause a break in the molecular structure of the body or material, it loses its elasticity and the elastic limit is said to have been reached. Hooke's law applies only within the elastic limit.

elastic limit [Mechanics] The limit of stress within which the strain in a material completely disappears when the stress is removed.

elastic modulus *See* modulus of elasticity.

elastic wave [Physics] A wave propagated by a medium having inertia and elasticity, in which displaced particles transfer momentum to neighbouring particles, and are themselves restored to original position.

elastin [Biochemistry] An elastic fibrous protein which occurs in connective tissues.

elastomer [Materials] A natural or synthetic rubber or rubberoid material, which has the ability to undergo deformation under the influence of a force and region its original shape once the force has been removed.

electret [Electronics] A solid dielectric possessing a permanent electric moment.

electric [Electronics] Arising from, producing, containing, or actuated by electricity, *e.g.*, electric cell.

electrical [Electronics] Related to or associated with electricity, but not containing it or having its properties or characteristics ; often used interchangeable with electric.

electrical capacity *See* capacitance.

electrical condenser *See* capacitor.

electrical energy. *See* electric energy.

electrical image [Electronics] A set of point charges on one side of a conducting surface that would produce the same electric field on the other side of the surface (in its absence) as the actual electrification of that surface.

electrical induction *See* induction.

electrical line of force [Electronices] A line in an electric field whose direction is everywhere that of the field.

electrical potential *See* electric potential.

electric arc [Electronics] A highly luminous discharge of electricity through a gas accompanied by high temperature; occurs when an electric current flows through a gap between two electrodes. Also known as arc.

electric-arc furnace [Metallurgy] A steel-making furnace in which an electric arc provides the source of heat. In direct-arc furnaces, the arc is formed between an electrode and the metal being heated.

electric bell *See* bell, electric.

electric charge [Electronics] A basic property of elementary particles of matter; may be positive, negative, or neutral. The elementary particle called an electron is said to be negatively charged with electricity, and the proton is said to be positively charged to an equal but opposite extent. These entities represent the basic units of electrically charged matter. Therefore, matter containing an equal number of protons and electrons is electrically neutral but matter containing an excess of electrons possesses an overall negative charge; similarly matter that has a deficiency of electrons (*i.e.*, an excess of protons) possesses an overall positive charge.

electric constant [Electronics] Permittivity of free space. E_0. per metre. It arises as the constant of proportionality in Coulomb's law, its value depending on the choice of units. *See also* electric field; permittivity.

electric current [Electronics] 1. An electric current is said to flow through a conductor when there is an overall movement of electrons through it. The SI unit of current is the ampere. 2. The net transfer of electric charge per unit time. Symbol I.

electric current, heating effect of [Electronics] When an electric current flows through a conductor of finite resistance, heat energy is continuously generated at the expense of electrical energy. The quantity of heat produced is proportional to the resistance of the conductor, and is equal to VI or I^2R watts (joules per second), V being the potential difference in volts, I the current in amperes, and R the resistance in ohms.

electric displacement [Electronics] The electric field intensity multiplied by permittivity. Symbol D. Also known as dielectric displacement; dielectric flux density; electric induction.

electric double layer [Chemistry] A diffused aggregation of positive and negative charges surrounding a suspended colloidal particle, which helps in maintaining its stability. *See also* zeta potential.

electric energy [Electronics] A form of energy related to the position of an electric charge in an electric field. For a body with charge Q and an electric potential V, its electrical energy is QV. If V is a potential difference, the same expression gives the energy transformed when the charge moves through the p.d.

electric field [Electronics] The region surrounding an electric charge, in which a force is exerted on a charged particle completely defined in magnitude and direction at any point by the force upon unit positive charge situated at that point. The field strength E, or force exerted upon a unit charge at a distance r from a charge Q, is given by : $E = Q/4 \pi r^2 \varepsilon$ where ε is the permittivity. For free space (*i.e.*, a vaccum) ε becomes ε_0, the electric constant, and has the value 8.854185×10^{-12} Fm^{-1}.

electric flux [Electronics] ψ 1. The quantity of electricity displaced across unit area of a dielectric. It is the scalar product of the electric displacement and the area. 2. The electric lines of force in a region.

electric induction. *See* electric displacement.

electricity [Physics] Physical phenomenon involving electrical charges and their effects when at rest and when in motion.

electricity, frictional [Physics] A separation of electric charge that results from the rubbing together of different materials; *e.g.*, on rubbing celluloid with rabbit's fur, is found to possess a positive charge, and the celluloid receives an equal negative charge. The rubbing motion strips some of the electrons from the atoms or molecules of the fur, which collect on the surface of the celluloid. Also known as triboeselectricity.

electricity, static *See* static electricity.

electric light [Electronics] Illumination produced by the use of electricity; it may be produced by virtue of the heating effect of an electric current on a wire or filament (*see* electric-light bulb), by an electric arc (*see* arc lamp), or by the passage of electricity through a vapour, as in the mercury vapour lamp of fluorescent lamps.

electric meter [Engineering] An electricity measuring device which totalizes with time, such as watthour meter or ampere-hour meter.

electric motor [Engineering] A device for converting electrical energy into mechanical energy, depending on the fact that when an electric current flows through a conductor placed in a magnetic field possessing a component at right angles to the conductor, a mechanical force acts upon the conductor. In its simplest form, it consists of a coil or armature through which the current flows, placed between the poles of a powerful electromagnet, the field magnet; the mechanical force upon the conductor causes the armature to rotate.

electric polarizability [Electronics] Induced dipole moment of an atom or molecule in a unit electric field.

electric polarization [Electronics] P. When an electric-field is applied to an electrically-neutral atom, a displacement of the electrons with respect to the positive nucleus occurs. This give rise to a small electric dipole possessing an electric moment in the direction of the field. This effect occurs when a dielectric is placed in an electric field, the electric field acting upon each individual atom of dielectric; the electric polarization is given by

$$P = D - E\,\varepsilon_0$$

where D is the electric displacement,

E is the electric field strength, and *o is the electric constant.

electric potential [Electronics] V. The energy required to bring unit electric charge from infinity to the point in an electric field at which the potential is being specified. The unit of electric potential is the volt. The *potential difference (p.d.)* between two points in an electric field or circuit is the difference in the values of the electric potentials at the the two points, *i.e.*, it is the work done in moving unit charge from one point to the other.

electric power [Electronics] The rate at which electricity is converted to other forms of energy; measured in watts. *See also* watt.

electric-power meter [Engineering] A device which measures electric power consumed, either at an instant, as in a wattmeter, or averaged over a time interval.

electric power substation [Electronics] An assembly of equipment in an electric power system through which electricity is passed for transmission, transformation, distribution, or switching.

electric power transmission [Electronics] Process of transferring electric energy from one point to another in an electric power system.

electric precipitation [Engineering] A process which utilizes an electric field to improve the separation of hydrocarbon reagent dispersions.

electric probe [Physics] A device used to measure density of electrons and ions, space and wall potentials, electron temperatures, and random electron currents in a plasma.

electric spark [Electronics] A discharge of electricity, accompanied by light and sound, through a dielectric or insulator.

electric susceptibility [Electronics] Xe. The ratio of the electric polarization *(P)* produced in a substance to the product of the electric field strength *(E)* to which it is subjected and the electric constant (ε_O), *i.e.*,

$$Xe = P/E\ \varepsilon_O$$

The susceptibility is related to the relative permittivity $((\varepsilon_r)$ by

$$Xe = 1 - \varepsilon_r.$$

electric wave [Electronics] An electromagnetic wave, especially one whose wavelength is at least a few centimetres. Also known as Hertzian wave.

electrification [Electronics] 1. The process of estabilishing a charge in an object. 2. Generation, distribution and utilization of electricity.

electroacoustics [Engineering] The conversion of sound energy into electric energy, or vice-versa.

electrocardiography [Engineering] ECG. An instrument for recording the current and voltage waveforms assoiciated with the contraction of the heart muscle.

electrochemical cell [Chemistry] A

combination of two electrodes for producing an electromotive force; includes dry cells, wet cells, standard cells, fuel cells solid electrolyte cells, and reserve cells.

electrochemical effect [Chemistry] Conversion of chemical energy to electric energy, as in electrochemical cells; or the reverse process used to produce chemicals with the help of electric energy.

electrochemical EMF [Chemistry] Electrical force generated by means of chemical action.

electrochemical equivalent [Chemistry] Z The mass of a given element liberated from a solution of its ions in electrolysis by one coulomb of charge.

electrochemical series [Chemistry] A series of chemical elements arranged in order of their electrode potentials. The hydrogen electrode ($H^+ + e \rightarrow \frac{1}{2}H_2$) is taken as having zero electrode potential. Elements that have a greater tendency than hydrogen to lose electrons to their solution are taken as *electropositive*; those that gain electrons from their solution are below hydrogen in the series and are called *electronegative*. The series shows the order in which metals replace one another from their salts; electropositive metals will replace hydrogen from acids. The chief metals and hydrogen, placed in order in the series, are :

potassium, calcium, sodium, magnesium, aluminium, zinc, cadmium, iron, nickel, tin, lead, hydrogen, copper, mercury, silver, platinum, gold. Also known as electromotive series.

electrochemistry [Chemistry] The study of the processes involved in the interconversion of electrical energy and chemical energy.

electrode [Electronics] 1. A conductor by which an electric current enters or leaves an electrolyte in electrolysis, an electric arc, or a vacuum tube (*See* discharge in gases and thermionic valve); the positive electrode is the anode, the negative one the cathode. 2. [Electronics] In a semiconductor device, an element that emits or collects electrons or holes, or controls their movement by an electric field.

electrodeposition [Metallurgy] The process of depositing by electrolysis, especially the deposition of one metal on another as in electroplating.

electrode potential [Chemsitry] The potential difference produced between the electrode and the solution in a half cell. It is not possible to measure this directly since any measurement involves completing the circuit with the electrolyte, thereby introducing another half cell. *Standard electrode potentials E* are defined by measuring the potential relative to a standard hydrogen half cell using 1.0 molar solution at 25°C. The convention is to designate the cell so that the oxidized form is written first. For example.

$$Pt(s) \mid H_2(g)H^+ \ (aq) \mid Zn^{2+}$$
$$(aq) \mid Zn(s)$$

The e.m.f. of this cell is --0.76 volt (*i.e.*, the zinc electrode is negative). Thus the standard electrode potential of the $Zn^{2+} \mid Zn$ half cell is --0.76V. Electrode potentials are also called *reduction potentials*.

electrodynamics [Electronics] The study of the relationship between electric and magnetic forces and their mechanical causes and effects.

electrodynamometer [Engineering] An instrument for measuring current, voltage, or power, in both direct current and alternating current circuits. It depends upon the interaction of the magnetic fields of fixed and movable coils.

electroforming [Metallurgy] The production, of metal articles by the deposition of a metal upon an electrode during electrolysis.

electrokinetic potential *See* zeta potential.

electroluminescence [Electronics] Fluorescence resulting from bombardment of a substance with electrons.

electrolysis [Chemistry] The occurence of a chemical reaction by passing an electric current through an electrolyte. In electrolysis, positive ions migrate to the cathode and negative ions to

the anode. The reactions occurring depend on electron transfer at the electrodes and are therefore redox reactions. At the anode, negative ions in solution may lose electrons to form neutral species. Alternatively, atoms of the electrode can lose electrons and go into solution as positive ions. In either case the reaction is an oxidation. At the cathode, positive ions in solution can gain electrons to form neutral species. Thus cathode reactions are reductions.

electrolysis, Faraday's laws of *See* Faraday's laws of electrolysis.

electrolyte [Chemistry] A compound which when dissolved or in the molten state, conducts an electric current and is simultaneously decomposed by it. The current is carried not by electrons but by ions.

electrolytic capacitor [Electronics] A capacitor consisting of two electrodes separated by an electrolyte; a dielectric film, usually a thin layer of gas, is formed on the surface of one electrode. Also known as electrolytic condenser.

electrolytic cell [Chemistry] A cell in which electrolysis occurs; *i.e.*, one in which current is passed through the electrolyte from an external source.

electrolytic corrosion [Chemistry] Corrosion that occurs through an electrochemical reaction. *See* rusting.

electrolytic dissociation [Chemistry] Ionization of a compound in a solution.

electrolytic gas *See* detonating gas.

electrolytic potential [Chemistry] Difference in potential between an electrode and the surrounding electrolyte; expressed in terms of some standard electrode difference.

electrolytic rectifier [Electronics] A rectifier consisting of two electrodes immersed in an electrolyte, which is used to convert an alternating current into a direct current. It depends on the properties of certain metals and solutions to allow current to flow in one direction only.

electrolytic refining [Chemistry] The purification of metals by electrolysis. When applied to copper, a large piece of impure copper is used as the anode with a thin strip of pure copper as the cathode. Copper (II) sulphate solution is the electrolyte. Copper dissolves at the anode: Cu →
> Cu²+ + 2e, and is deposited at the cathode.

electrolytic separation [Nuclers] A method of separating deuterium from hydrogen. When water is electrolyzed the hydrogen ions are discharged at the cathode slightly faster than the heavier isotope, deuterium. Thus, over a period, the water becomes enriched with deuterium.

electromagnet [Physics] A coil of wire with a soft iron core ; when an electric current flows through the coil, the iron core behaves like a magent. When the current stops flowing, it loses its magnetism.

electromagnetic [Physics] Pertaining to phenomena in which electricity and magnetism are related.

electromagnetic environment [Physics] Radio-frequency fields existing in a given area or space.

electromagnetic induction [Physics] The production of an electromotive force either by motion of a conductor through a magnetic field so as to cut across the magnetic flux or by a change in the magnetic flux which threads a conductor. *See also* Faradays law of elctromagnetic induction; and Lenz law.

electromagnetic interaction [Physics] The form of interaction that occurs between electrically charged elementary particles. It can be explained by the exchange of virtual photons (*see* virtual state) between the interacting particles.

electromagnetic lens [Electronics] An electron lens in which electron beams are focused by an electromagnetic field.

electromagnetic pump [Engineering] A device used for pumping liquid metals. A curent is passed through the liquid metal, which is contained in a flattened pipe placed between the poles of an electromagnet. The liquid metal is thus subjected to a force which acts along the axis of the pipe.

electromagnetic radiation *See* electromagnetic wave.

electromagnetic spectrum [Physics] The range of wavelengths over which electromagnetic radiation extends. The longest waves (10^5-10^3 metres) are radio waves, the next longest (10^{-3}-10^{-6}m) are infrared waves, then comes the narrow band (4-7 X 10^{-7}m) of visible radiation, followed by ultraviolet waves (10^{-7} -10^{-9}m), X-rays (10^{-9}-10^{-11}m), and gamma rays (10^{-11}--10^{-14}m).

electromagnetic units [Physics] E.M.U. A system of electrical units, within the C.G.S. system, based on the unit magnetic pole, which repels a similar pole, placed 1 cm away, with a force of 1 dyne. The E.M.U. of electric current is that current that flowing in an arc of a circle of unit length and radius (*i.e*, 1 cm), exerts a force of 1 dyne on a unit magnetic pole placed at the centre. The E.M.U. of resistance is that resistance in which energy is dissipated at the rate of 1 erg per second by the flow of 1 E.M.U. of current. The E.M.U. of electromotive force or potential is the potential that applied across the ends of a conductor of 1 E.M.U. resistance, causes 1 E.M.U. of current to flow.

electromagnetic wave [Physics] A disturbance which propagates outward from any electric charge which oscillates, or is accelerated; far from the charge it consists of vibrating electric and magnetic fields which move at the speed of light and arc at right angles to each other and to the direction of propagation.

electrometallurgy [Metallurgy] The study of electrical processes used in separating a metal from its ore; or refining or shaping metals by electrical and electrolytic procedures.

electrometer [Engineering] An instrument for measuring voltage differences, which draws almost no current from the source.

electromotive force [Chemistry] E.M.F. The source of electrical energy required to produce an electric current in a circuit. It is defined as the rate at which electrical energy is drawn from the source and dissipated in a circuit when unit current is flowing in the circuit. The SI unit is the volt. *See* potential difference.

electromotive series *See* electrochemical series.

electron [Physics] An elementary particle with a rest mass of 9.109558×10^{-31} kg and a negative charge of $1.602.192 \times 10^{-19}$ coulomb. Electrons are present in all atoms in groupings called shells around the nucleus; when they are detached from the atom they are called free electrons. The antiparticle of the electron is the positron.

electron affinity [Cheistry] A. The energy change occurring when an atom or molecule gains an electron to form a negative ion. For an atom or molecule X, it is the energy released for the electron-attachment reaction

$$X(g) + e^- \; ---> \; X^-(g)$$

Often this is measured in electronvolts. Alternatively, the molar enthalpy change, ΔH, can be used.

electron capture [Chemistry] The formation of a negative ion when a free electron is captured by an atom or molecule [Nucl Science] A radioactive transformation as a result of which a nucleus captures one of its orbital electrons.

electron conduction [Electronics] Conduction of electricity as a result of movement of electrons, rather than from ions in a gas or solution, or holes in solid.

electron-deficient compound [Chemistry] A compound in which there are fewer electrons forming the chemical bonds than required in normal electron-pair bonds. *See* borane.

electron density [Physics] The density of electronic charge at a given point in a molecule; alternatively defined as the probability of finding an electron at the particular point.

electron diffraction [Physics] A diffraction effect resulting from the passage, of electrons through matter, analogous to the diffraction of visible light or *X*-rays. The phenomenon of electron diffraction is the principal evidence for the existence of waves associated with electrons (*see* de Broglie wavelength). The diffraction of electrons when passed through crystals or thin metal foils is used as a method of investigating crystal structure.

electronegativity [Chemistry] The tendency of an atom to attract shared pair of electrons towards itself in a molecule.

electron exchanger *See* redox exchanger.

electron gun [Electronics] The source of electrons in a cathode ray tube or electron microscope. It consists of a cathode emitter of electrons, an anode with an aperture through which the beam of electrons can pass, and one or more focusing and control electrodes.

electronic [Electronics] Pertaining to electron devices or to circuits or systems utilizing electron devices.

electronic band spectrum *See* band spectrum.

electronic heating [Engineering] Heating by radio frequency current produced by an electron-tube oscillator or any other radio frequency source. Also known as radio-frequency heating; high-frequency heating.

electronics [Electronics] The study and design of control, communication, and computing devices that rely on the movement of electrons in circuits containing semiconductors, thermionic valves, resistors, capacitors and inductors.

electron lens *See* electromagnetic lens.

electron microscope [Electronics] An instrument similar in purpose to the ordinary light microscope, but with a much greater resolving power. Instead of a beam of light to illuminate the object, a parallel beam of electrons from an electron gun is used. In the transmission electron microscope, the object which must be in the form of a very thin film of the material, allows the electron beam to pass through it; but, owing to differential scattering in the film, an image of the object is carried forward in the electron beam. The latter then passes through a magnetic or electrostatic focusing system (*see* electromagnetic lens) which is equivalent to the optical lens system in an ordinary microscope, *i.e.*, it produces a much magnified image. This is received on a fluorescent screen and recorded by a camera. Magnifications up to 250000 can be achieved. In the scanning electron microscope a thick sample can be used and the sample is scanned by the electron beam. Secondary electrons emitted from the surface of the sample are focused onto a screen. The magnification is less with this type of instrument but a three-dimensional image can be formed.

electron multiplier *See* photomultipliers.

electron paramagnetic resonance [Chemistry] EPR. A method of spectroscopic analysis similar to nuclear magnetic resonance except that microwave radiation is employed instead of radio frequencies. It is used for studying free radicals, crystal centres, transition elements and structures involving unpaired electrons. Also known as electron spin resonance.

electron probe microanalysis [Chemistry] A method of analysing a very small quantity of a substance by directing a finely focused electron beam on to it so than an *X*-ray emission is produced characteristic of the elements present in the sample. The diameter of the beam is usually about 1 μm and quantities as small as 10^{-14} gm can be detected by this means. The method may be used quantitatively for elements whose atomic numbers exceed 11.

electron spin resonance *See* electron paramagnetic resonance.

electronvolt [Physics] eV. A unit of energy equal to the work done on an electron in moving it through a potential difference of one volt. It is used as a measure of particle energies although it is not an SI unit. 1 eV = 1.60219 X 10^{-19} joule.

electron wave [Physics] The de Broglie wave or probability amplitude wave of an electron.

electrophilic [Chemistry] Any chemical process in which electrons are acquired from or shared with other molecules or ions.

electrophilic reagent [Chemistry] A substance that reacts at centres of high electron density. Essentially electron acceptors (*e.g.*, halogens) that gain or share electrons from an outside atom or ion.

electrophoresis [Chemistry] The migration of the electrically charged solute particles present in a colloidal solution towards the oppositely charged electrode, when two electrodes are placed in the solution and connected externally to a source of E.M.F. Also known as cataphoresis.

electrophorus [Electronics] A device for showing electrostatic charging by induction.

electroplating [Metallurgy] Depositing a layer of metal by electrolysis, the object to be plated forming the cathode in an electrolytic tank or bath containing a solution of a salt of the metal that is to be deposited.

electroscope [Engineering] An instrument for detecting the presence of an electric charge by means of mechanical forces exerted between electrically charged bodies.

electrostatic energy [Electronics] The potential energy possessed by a collection of electric charges by virtue of their position relative to each other.

electrostatic field [Electronics] A region in which a stationary electrically charged particle would be subjected to a force of attraction or repulsion as a result of the presence of another stationary electric charge. *See* electric field.

electrostatic force [Electronics] Force exerted on a charged particle due to an electrostatic field.

electrostatic generator [Engineering] A machine designed for the continuous separation of electric charge, *e.g.*, Van de Graaf Generator. It consists of a belt of insulating material (*e.g.*, silk) passing over two pulleys. The second electrode is very nearly a sphere; there are corona points for transferring charge to and from the belt. This machine is capable of producing steady D.C. voltage, enabling precision measurements of very high voltages to be made.

electrostatic induction [Electronics] The process of charging a body electrically by bringing it near another charged body, then touching it to ground.

electrostatic precipitation [Electronics] A widely used method of controlling the pollution of air (or other gases). The gas, containing solid or liquid particles suspended in it, is subjected to a uni-directional electrostatic field, so that the particles are attacted to, and deposited upon, the positive electrode. *See* Cottrell precipitator.

electrostatics [Electronics] The study of electric charges at rest, their electric field, and potentials.

electrostatic units [Electronics] ESU. A system of electrical units based upon the electrostatic unit of electric charge. The electrostatic unit of charge (called the staticoulomb) is that quantity of electricity that will repel an equal quantity, 1 cm distant from it in a vacuum, with a force of 1 dyne.

electrostriction [Electronics] The change in the dimensions of a dielectric when placed in an electric field. An example is the contraction of a solvent due to the electrostatic field of a dissolved electrolyte.

electrotyping [Engineering] The production of copies of plates of type, etc., by the electrolytic deposition of a layer of metal on a previously prepared mould. This is a cast of the subject to be copied, made of plastic material and coated with a layer of graphite, which acts as a conductor of electricity. It is then suspended to act as a cathode in an electrolytic bath (*see* electroplating) containing a solution of a salt of the metal required, usually copper. The passage of an electric current will deposit a layer of any required thickness of metal upon the cathode, the layer, being a replica of the original type.

electrovalence [Chemistry] The valence of an atom which has formed ionic bond.

electrovalent bond *See* ionic bond.

electrovalent crystal *See* ionic crystal.

electrum [Metallurgy] A natural alloy of gold (55%-85%) and silver.

element [Chemistry] A substance consisting entirely of atoms of the same atomic number. [Electronics] A part of an electron tube, semiconductor device, or antenna array that contributes directly to the electrical performance.

elementary particles [Physics] The basic units of which all matter is composed. The stable particles protons, electrons, and neutrons combine with neutrons to form stable atoms. But many other short-lived particles and resonances have been detected that play an essential part in the structure of matter. For every particle that exists, there is a corresponding anti-particle, which has the same mass and spin but opposite electric charge. Some electrically netural particles have anti-particles in which some other property is reversed (*e.g.*, strangeness) and some netural particles are regarded as their own anti-particles.

elevation of boiling point [Chemistry] The rise in the boiling point of solution produced by a non-volatile substance dissolved in a solvent. For a dilute solution the elevation is proportional to the number of molecules or ions present (*see* colligative properties), and the elevation produced by the same molecular concentration (or ionic concentration in the case of an electrolyte is a constant for a particular solvent.

eleven-year period [Astronomy] A periodic change in occurrence of sunspots, the cycle being complete in approximately eleven years; associated with this is a cyclic variation in the magnitude of the daily variation.

eliminator [Electronics] A device that takes the place of batteries, generally consists of a rectifier operating from alternating current.

ellipse [Mathematics] A closed plane figure formed by cutting a right circular cone by a plane obliquely through is axis (*see* conic sections). The sum of the distances from any point on the perimeter of an ellipse to its two foci is constant.

ellipsoid [Mathematics] A solid figure traced out by an ellipse rotating about one of its axes.

ellipsometry [Optics] Techniques for the measurement of the degree of ellipticity of polarized light.

elliptically polarized light [Optics] Light can be resolved into two vibrations lying in planes at right angles, and of equal frequency. The electric vector at any point in the path of the wave describes an ellipse about the direction of propagation of the light. The form of this ellipse is determined

by the amplitudes of these two vibration and by the difference of phase between them.

eluant [Chemistry] A liquid used to extract one material from another in chromatography.

eluate [Chemistry] The solution that results from the elution process in chromatography.

elution [Chemistry] The process of removing an adsorbed material (adsorbate) from an adsorbant by washing it in a liquid (eluent). The solution consisting of the adsorbate dissolved in the eluent is the eluate. Elution is the process used to wash components of a mixture through a chromatographic column.

elutriation [Chemistry] The washing, separation, or sizing of fine particles of different weight by suspending them in a current of air or water.

emanation *See* radioactive emanation.

emanometer [Engineering] An instrument used to measure the radon content of the atmosphere.

embolus [Biology] A thrombus or any other particle carried by blood-stream which blocks a blood vessel.

embrittlement [Metallurgy] Loss or reduction in ductility in a metal or plastic by hardening it.

embryology [Biology] The branch of biology concerned with the study of the growth and development of embryos from fertilized eggs.

emersion [Astronomy] The reappearance of a celestial body after an eclipse or occulation.

emission [Physics] Any radiation of energy by means of electromagnetic waves.

emission lines [Physics] Spectral lines resulting from emission of electromagnetic waves by atoms, ions, or molecules during transitions from excited states to the lower energy states.

emission spectroscopy [Physics] Study of the composition of materials and identification of elements by observation of wavelengths of radiation they emit when they return to normal state after excitation by an external energy source.

emissive power See emittance, thermal.

emissivity [Physics] 8. The ratio of the radiation emitted by a surface to the radiation emitted by a perfect black body radiator under similar physical conditions.

emittance, thermal [Physics] Power radiated per unit area of a radiating surface. Also known as emissive power; radiating power.

emitter [Computers] A time pulse generator found in a equipment, such as card-punch. [Electronics] One of the three electrodes in a transistor.

empirical [Scientific Technology] Based upon the results of actual experiment and observation rather than theory.

empirical formula See formula.

emulsification [Chemistry] The process of dispersing one liquid in another immiscible liquid.

emulsifying agent [Chemistry] Material added in small quantities to stablize an emulsion. Also known as emulsifier; dispersers.

emulsion [Chemistry] A colloid in which small particles of one liquid are dispersed in another liquid. Usually emulsions involve a dispersion of water in an oil or a dispersion of oil in water, and are stabilized by an *emulsifier*. Commonly *emulsifiers* are substances, such as detergents, that have lyophobic and lyophilic parts in their molecules.

emulsion paint [Materials] A paint in which the vehicle is an emulsion of a binder such as oil, resin, or latex, in water.

emulsion, photographic [Materials] The lightsensitive coating on a film or plate (*see* photography). A 'nuclear emulsion' is a photographic emulsion specially prepared to record the tracks of elementary particles and nuclear fragments that pass through it.

emulsoid sol *See* colloidal solution.

enamel [Chemistry] 1. A class of substances (vitreous enamels) having similar composition to glass with the addition of stanic oxide, SnO_2, or other infusible substances to render the enamel opaque. 2. A finely ground oil paint that dries relatively harder smoother, and glossier than ordinary paint. 3. The external layer of teeth consisting mainly of calcium phosphate and carbonate salts.

enantiomorphism [Chemistry] The occurrence of substances in two crystalline forms, one being a mirror image of the other. *See also* optical isomerism.

enantiotropic [Chemistry] Substances that exist in two different physical forms, one being stable below a certain temperature (the transition temperarure), the other above it, *e.g.*, sulphur exists as alpha sulphur at all temperatures below 96°C, above this, the stable form is beta-sulphur.

endergonic [Chemistry] Pertaining to a biochemical reaction in which the final products have more free energy than the reactants. Such reactions require energy to occur.

endocrine gland [Physics] A gland which secretes a hormone. The gland has no duct, the hormone diffuses into the blood stream from the gland either directly or by way of the lymphatic.

endoenzyme [Biochemistry] An enzyme that remains within a living cell and does not diffuse through the cell wall into the surrounding medium.

endosmosis [Chemistry] The inward flow of water into a cell containing an aqueous solution, through a semipermeable membrane, due to osmosis.

endothermic [Chemistry] Denoting a chemical reaction that takes heat from its surroundings. *Compare* exothermic.

endotoxin [Biology] A toxic proudct of bacteria which is associated with the structure of the cell, and can only be obtained by destruction of the cell.

end point [Chemistry] The stage in a titration, usually indication by a change of colour of an indicator, at which a particular reaction has reached completion or the desired stage.

energy [Physics] The capacity for doing work. Energy can only exist in the absence of matter in the form of radiant energy. The derived SI unit of energy is the joule.

energy band *See* band.

energy converter [Chemistry] A substance which can convert the radiant energy of sunlight into chemical, electrical or thermal energy, *e.g.*, silicon, selenium and tellurium. *See also* solar cell.

energy flux [Physics] The rate of flow of energy per unit area. *See* flux.

energy level [Physics] An atom as a whole, or an individual nucleus, can exist only in certain definite states characterized by the energy of the state. Thus, for each different atom or nucleus, there exists a series of energy levels corresponding to these permissible states. The lowest stable energy level of an atom or nucleus is referred to as the ground state; atoms or nuclei at higher energy levels than the ground state are said to be in excited state.

energy metabolism [Biochemistry] Chemical reactions responsible for the energy transformations within the cells.

energy value [Biochemistry] A measure of the heat energy available by the complete combustion of a stated weight of the food; often expressed in joules per kilogram or large calories per lb.. It takes no account of the value of the food from any other point of view, or sometimes even of the suitability of the food for use by the human organism.

engine [Engineering] A device for converting one form of energy into another, especially for converting other forms of energy into mechanical (*i.e.*, kinetic) energy.

enfleurage [Chemistry] Extraction of odoriferous substances of flowers by means of fats and tallow.

enols [Chemistry] Compounds containing the group -CH=C(OH)- in their molecules. *See also* keto-enol tautomerism.

enrichment [Nucleonics] The process of increasing the abundance of a specified isotope in a mixture of isotopes. It is usually applied to an increase in the proportion of U-235, or the addition of Pu-239 to natural uranium for use in a nuclear reactor or weapon.

enthalpy [Chemistry] H. A thermodynamic property of a system defined by $H = U + pV$, where H is the enthalpy, U is the internal energy of the system, p its pressure, and V its volume. In a chemical reaction carried out in the atmosphere the pressure remains constant and the enthalpy of reaction, ΔH, is equal to $\Delta U + p\Delta V$. For an exothermic reaction ΔH is taken to be negative.

entomology [Zoology] A branch of life sciences dealing with the study of insects.

entrainer [Chemistry] An additive to liquid mixtures that are difficult to separate by ordinary distillation.

entrainment [Engineering] The transport of particles (*e.g.*, fine droplets) in a moving stream of a fluid (*e.g.*, the vapour of a boiling liquid).

entropy [Chemistry] S. A measure of the unavailability of a system's energy to do work; an increase in entropy is accompanied by a decrease in energy availability. When a system undergoes a reversible change the entropy *(S)* changes by an amount equal to the energy *(Q)* absorbed by the system divided by the thermodynamic temperature *(T)* at which the energy is absorbed, *i.e.* $\Delta S = \Delta Q/T$. However, all real processes are to a certain extent irreversible changes and in any closed system an irreversible change is always accompanied by an increase in entropy.

In a wider sense entropy can be

interpreted as a measure of a system's disorder; the higher the entropy the greater the disorder. As any real change to a closed system tends towards higher entropy, and therefore higher disorder, it follows that the entropy of the universe (if it can be considered a closed system) is increasing and its available energy is decreasing.

enyme [Chemistry] A hydrocarbon with a double (-ene) and a triple (-yne) bond between carbon atoms of its molecule.

enzyme. [Biochemistry] Any one of a large number of complex organic proteins produced by all cells, that act as a catalyst for chemical reations of biological processes. An enzyme is easily destroyed by heat, and many chemical substances. It needs certain conditions, particulary a suitable pH value before it will act. Enzymes are highly specific in their catalytic behaviour; a given enzyme is effective only for one particular reaction.

enzymolysis. [Biochemistry] The decomposition of a substance catalyzed by an enzyme.

epact [Astronomy] 1. The difference in days between the length of a solar year and a lunar year. 2. The moon's age in days at the start of the calendar year.

ephemeris [Astronomy] A table that gives the predicted positions and the movements of a celestial body such as a planet or comet.

Also an annual publication containing astronomical data.

ephemeris time [Astronomy] Time measured on the basis of the orbital movements of the planets and the moon.

epicentre [Geology] The point on the surface of the earth that lies directly above the focus of an earthquake.

epicycle [Mathematics] A circle whose centre rolls round the circumference of a large circle within slipping. [Astronomy] *See* deferent.

epicyclic gears [Engineering] A system of gears in which one or more wheels move around the outside, or the inside, of another wheel whose axis is fixed.

epidermis [Biology] The outermost layer of cells of an organism.

epidiascope [Optics] An optical projector for projecting an enlarged image of either an opaque object or a transparency upon a screen. Used for illustrating lectures.

epididymis [Physiology] Complexly coiled tube adjacent to the testis were sperm are stored.

epigenesis [Biology] The theory that development proceeds from a structureless cell by the successive formation and addition of new parts which do not pre-exist in the fertilized egg.

epigenetic [Geology] Produced or formed at or near the surface of the earth.

epimerism [Chemistry] A type of optical isomerism in which a molecule has two chiral centres; two optical isomers *(epimers)* differ in the arrangement about one of these centres.

epiphyte [Biology] A plant that grows upon another plant, for position and support only.

epistasis [Biology] The suppression of the effect of one gene by another.

epitaxy [Chemistry] The growth of one crystalline substance on another so that both have the same crystal structure. Epitaxial layers are used in the manufacture of semiconductor devices.

epithelium [Biology] The layer of tissue covering the internal and external surfaces of the body; including the line of vessels and other small cavities; consists of cells joined by small amounts of cementing substances.

epoxy [Chemistry] A compound in which an oxygen atom is bound to two cabron atoms, forming a three-membered ring.

epoxy resins [Materials] Synthetic resins produced by copolymerizing epoxide compounds with phenols. They contain -O- linkages and epoxide groups and are usually viscous liquids. They can be hardened by addition of agents, such as polyamines, that form cross-linkages. Alternatively, catalysts may be used to induce further polymerization of the resin. Epoxy resins are used in electrical equipment and in the chemical industry (because of resistance to chemical attack). They are also used as adhesives.

epsom salt [Chemistry] $MgSO_4 \cdot 7H_2O$. Magnesium sulphate.

equation, chemical *See* chemical equation.

equation of state [Chemistry] An equation that relates the pressure p, volume V, and thermodynamic temperature T of an amount of substance n. The simplest is the ideal gas law :

$$pV = nRT,$$

where R = universal gas constant. Applying only to ideal gases, this equation takes no account of the volume occupied by the gas molecules (according to this law if the pressure is infinitely great the volume becomes zero), nor does it take into account any forces between molecules. A more accurate equation of state would therefore be

$$(p + k)(V - nb) = nRT,$$

where k = a factor that reflects the decreased pressure on the walls of the container as a result of the attractive forces between particles, and nb = the volume occupied by the particles themselves when the pressure is infinitely high. In the *van der Waals equation of state,*

$$k = n^2a/V^2,$$

where a = a constant. This equation more accurately reflects the behaviour of real gases; several others have done better but are more complicated.

equation of time [Astronomy] The difference between mean solar time, as given by a clock, and apparent solar time, *i.e.*, sundial time. The time of rotation of the earth upon its axis is not exactly equal to the time from noon to noon, the difference being caused by the motion of the earth relative to the sun to complete a revolution in one year, and also by the inclination of the ecliptic to the equator.

equatorial axis [Geology] The diameter of the earth described between two points on the equator.

equatorial plane [Astromomy] The plane passing through the equator of the earth, or of other celestial body, perpendicular to its axis of rotation and equidistant from its poles.

equator, terrestrial [Geology] The great circle of the earth, laying in a plane perpendicular to the axis of the earth, that is equidistant from the two poles. *See also* magnetic equator and celestial equator.

equilateral figure [Mathmatics] A figure having all its sides equal in length.

equilibrium [Physics] A state in which a system has its energy distributed in the statistically most probable manner; a state of a system in which forces, influences, reactions, etc., balance each other out so that there is no net change. A body is said to be in *thermal equilibrium* if no net heat exchange is taking place within it or between it and its surroundings. A system is in chemical equilibrium when a reaction and its reverse are proceeding at equal rates (*see also* equilibrium constant). These are examples of *dynamic equilibrium* in which activity in one sense or direction is in aggregate balanced by comparable reverse activity.

equilibrium constant [Chemistry] For a reversible reaction of the type

$$xA + yB \rightleftharpoons zC + wD$$

chemical equilibrium occurs when the rate of the forward reaction equals the rate of back reaction, so that the concentrations of products and reactants reach steady-state values. It can be shown that at equilibrium the ratio of concentrtations.

$$[C]^z[D]^w/[A]^x[B]^y$$

is a constant for a given reaction and fixed temperature, called the equilibrium constant K_c (where the c indicates concentrations have been used)

The equilibrium constant shows the *position* of equilibrium. A low value of K_c indicates that $[C]$ and $[D]$ are small compared to $[A]$ and $[B]$; *i.e.* that the back reaction predominates. It also indicates how the equilibrium shifts if concentration changes. For example, if $[A]$ is increased (by adding A) the equilibrium shifts towards the right so that

[C] and [D] increase, and K_c, remains constant.

For gas reactions, partial pressures are used rather than concentrations. The symbol K_p is then used. Thus, in the example above

$$K_p = P_c^{\ z} \, PD^w / P_A^{\ x} \, P_B^{\ y}$$

It can be shown that, for a given reaction $K_p = K_c (RT)^{\Delta r}$, where Δr is the difference in stoichiometric coefficients for the reaction (*i.e.*, $z + w - x - y$). Note that the units of K_p and K_c depend on the numbers of molecules appearing in the stoichiometric equation. The value of the equilibrium constant depends on the temperature. If the forward reaction is exothermic, the equilibrium constant decreases as the temperature rises; if endothermic it increases (*see also* van't Hoff's isochore).

The expression for the equilibrium constant can also be obtained by thermodynamics; it can be shown that the standard eqilibrium constant K is given by exp. $(-\Delta G / RT)$, where ΔG is the standard Gibbs free energy change for the complete reaction. Strictly, the expressions above for equilibrium constants are true only for ideal gases (pressure) or infinite dilution (concentration). For accurate work activities are used.

equimolecular mixture [Chemistry] A mixture containing substances in equal molecular proportions; *i.e.*, in the ratio of their molecular weights.

equinox [Astronomy] The moment (or, astronomically, the point) at which the sun apparently crosses the celestial equator; the point of intersection of the ecliptic and the celestial equator.

equipartition of energy [Physics] The theory that the energy of gas molecules in a large sample under thermal equilibrium is equally divided among their available degrees of freedom, the average energy for each degree of freedom being $kT/2$, where k is the Boltzmann constant and T is the thermodynamic temperature. The proposition is not generally true if quantum considerations are important, but is frequently a good approximation.

equivalent, electrochemical *See* electrochemical equivalent.

equivalent weight *See* chemical equivalents.

equivocation [Physics] A term used in information theory to indicate the rate of loss of information (per second or per symbol) at the receiving end of a channel of information due to noise.

erbium [Chemistry] Er. A soft silvery metallic element belonging to the lanthanoides; At. No. 68; r.d. 9.066 (293K); m.p. 1770K; b.p. 3141K. It occurs in apatite, gadolinite, and xenotine from certain sources. There are six natural isotopes, which are stable, and twelve artificial isotopes are known. It has been used in alloys for nuclear technology as it is a neutron absorber.

erg [Physics] A unit of work or energy in the c.g.s., system of units; the work done by a force of 1 dyne acting a distance of 1 cm. 1 erg=10^{-7} joule.

erythrocyte [Biology] The cells of the blood that contain haemoglobin and whose function is to transport oxygen through the body. Erythrocytes have no means of propulsion, and in mammals the cells have no nuclei. Human blood contains approximately five million erythrocytes per cubic millimetre. Also known as red blood cells.

escape velocity [Physics] The velocity that a projectile or space probe would need to attain in order to escape from a particular gravitational field. The escape velocity from the surface of a planet (or moon) depends on the planet's (or moon's) mass and diameter. The escape velocity from the earth's surface is about 11200 metres/s and from the moon's surface about 2370 metres/s.

essential amino acid [Biochemistry] An amino acid that an organism is unable to synthesize in sufficient quantities. It must therefore be present in the diet. In man the essential amino acids are arginine, histidine, lysine, threonine, methionine, isoleucine, leucine, valine, phenylalanine, and tryptophan. These are required for protein synthesis and deficiency

leads to retarded growth and other symptoms.

essential element [Biochemistry] Any of a number of elements required by living organisms to ensure normal growth, development, and maintenance. Apart from the elements carbon, hydrogen, oxygen, and nitrogen, plants, animals, and microorganisms all require a range of elements in inorganic forms in varying amounts, depending on the type of organism. The *major elements*, present in tissues in relatively large amounts (greater than 0.005%), are calcium, phosphorus, potassium, sodium, chlorine, sulphur, and magnesium. The *trace elements* occur at much lower concentrations and thus requirements are much less. The most important are iron, manganese, zinc, copper, iodine, cobalt selenium, molybdenum, chromium, and silicon.

essential fatty acids [Biochemistry] Fatty acids that must normally be present in the diet of certain animals, including man. Essential fatty acids, which include linoleic and lionlenic acids, all possess double bonds at the same two positions along their hydrocarbon chain and so can act as precursors of prostaglandins. Deficiency of essential fatty acids can cause dermatosis, weight loss, irregular oestrus, etc.

essential oil [Chemistry] A natural oil with a distinctive scent secreted by the glands of certain aromatic plants. Terpenes are the main constituents. Essential oils are extracted from plants by steam distillation, extraction with cold neutral fats or solvents.

esterases [Biochemistry] Enzymes that control hydrolysis of esters.

ester gums [Materials] Products made by esterification of organic acid in rosin with polyhydric alcohols, especially glycerol. Used in varnishes. Also known as rosin esters.

esterification [Chemistry] The formation of an ester by the chemical reaction of an organic acid with an alcohol.

esters [Chemistry] Organic compounds derived by replacing hydrogen of an acid by an organic radical or group. Many esters are pleasant-smelling liquids and are used for flavouring essences. Many vegetable and animal fats and oils also belong to this class.

estrus [Physiology] The period in female mammals during which ovulation occurs and the animal is receptive to mating; marked by intense sexual urge.

etalon [Optics] An interferometer used for studying fine spectrum lines. It depends upon the interference effects produced by multiple reflection between fixed, parallel, half-silvered glass or quartz plates.

ethanolamines [Chemistry] Organic amines derived from ethanol. They are manufactured by the action of ammonia on ethylene oxide and are used for the absorption of acid gases, and as intermediates in the production of surfactants.

ethanoyl group [Chemistry] The organic group CH_3CO-.

ethenoid plastics [Materials] A class of thermoplastic resins made from substances containing a double bond, *e.g.*, acrylic, styrene and vinyl resins.

ether [Physics] The hypothetical medium that was supposed to fill all space: postulated as a medium to support the propagation of electromagnetic radiations. Once the subject of controversy, now regard as an unnecessary assumption. Also spelled a ether.

ethers [Chemistry] A group of organic compounds with the general formula $R-O-R'$ formed by the condensation of two alcohol molecules. The compound commonly called 'ether' is diethyl ether, $C_2H_5.O.C_2H_5$.

ethoxy [Chemistry] The univalent radical C_2H_5O-.

ethylene-propylene rubber [Materials] EPR. A fully saturated, stereo-regular, synthetic rubber prepared by the solution polymerization of approximately equal proportions of ethylene and propylene. It cannot be cured by sulphur vulcanization but satisfactory vulcanization can be achieved using peroxide curing systems.

ethyl fluid [Chemistry] A solution of tetraethyl, $Pb(C_2H_5)_4$, and dibromoethane, $C_2H_4Br_2$, used as an anti-knock compound in motor fuel.

ethyl group [Chemistry] The univalent alkyl radical-C_2H_5.

ethynylation [Chemistry] The process of making an acetylenic derivative by condensing acetylene (ethyne) with a compound such as an aldehyde; for example, butynediol is produced by the condensation of acetylene with formaldehyde.

etiolation [Botany] The whitening or yellowing of green parts of the plant grown in darkness; occurs due to lack of chlorophyll.

euchlorine [Chemistry] A gaseous mixture of chlorine, Cl_2, and explosive chlorine dioxide, ClO_2.

eudiometer [Engineering] A glass tube for measuring volume changes during chemical reactions between gases.

eugenics [Biology] The study of the genetic control of human populations, with a view to improving their constitution, by selectively encouraging breeding among those people considered by eugenicists to be the most desirable.

europium [Chemistry] Eu. A soft silvery metallic element belonging to the lanthanides; At.No. 63; r.d. 5.245 (293K); m.p. 1095K; b.p. 1802K. It occurs in small quantities in bastanite and monazite. Two stable isotopes occur naturally: europium-151 and europium-153, both of which are neutron absorbers. Experimental europium alloys have been tried for nuclear-reactor parts but until recently the metal has not been available in sufficient quantities.

eutectic crystallization [Chemistry] Simultaneous crystallization of the constituents of an eutectic alloy during cooling of the melt.

eutectic mixture [Chemistry] A solid solution of two or more substances, having the lowest freezing point of all the possible mixtures of the components. This is taken advantage of in alloys of low melting point, which are generally eutectic mixtures.

eutectic point [Chemistry] Two or more substances cabable of forming solid solutions with each other have the property of lowering each other's freezing point attainable, corresponding to the eutectic mixture, is termed the eutectic point.

eutectic temperature [Chemistry] The temperature at the lowest melting point of an eutectic.

evacuate [Engineering] To remove gases and vapours from an enclosure. Also known as exhaust.

evaporation [Physics] The conversion of a liquid into vapour, without necessarily reaching the boiling point.

evaporometer *See* atmometer.

evening star [Astronomy] A misnomer for a planet which can be seen by naked eye just after sunset.

exa- [Scientific Techniques] E. A

prefix used in the metric system to denote 10^{18} times. For example, 10^{18} metres = 1 exametre (Em).

excess electron [Electronics] Electron introduced into a semiconductor by a donor impurity and available for conduction.

exchange current [Electricity] The magnitude of the current flowing through a galvanic cell when it is working in a reversible manner.

exchange force [Physics] The type of force that holds nucleons together in the nucleus of an atom, visualized as the exchange of mesons between the nucleons.

excision [Medicine] Removal of a part by cutting.

excitaton [Physics] A process in which a nucleus, electron, atom, ion, or molecule acquires energy that raises it to a quantum state *(excited state)* higher than that of its ground state. The difference between the energy in the ground state and that in the excited state is called the *excitation energy*.

exciton [Electronics] A non-conduction, non-localized, excited electronic state in a semiconductor. It may be regarded as a bound electron-hole pair, or alternatively as an atomic excitation passed from atom to atom.

exclusion principle *See* Pauli's exclusion principle.

execute [Computers] Generally, to run a compiled or assembled program on the computer.

exergonic [Chemistry] A reaction characterized by the release of energy.

exitance [Physics] M. The radiant or luminous flux emitted per unit area of a surface. The *radiant exitance (M_e)* is measured in watts per square metre (Wm^{-2}), while the *luminous exitance (M_v)* is measured in lumens per square metre (lm m^{-2}). Exitance was formerly called *emittance*.

exocrine gland [Physiology] A structure whose secretion is passed directly or by ducts, such as tear and salivary glands.

exoenzyme [Biochemistry] An enzyme that functions outside the cell that produces it, *e.g.*, pepsin.

exogamy [Biology] Union of gametes from organisms which are not closely related. Also known as outbreeding.

exosmosis [Physics] Passage of a fluid outward through a cell membrane.

exosphere [Geology] The outermost layer of the earth's atmosphere, in which the density is such that an air molecule moving directly outwards has a 50% chance of escaping rather than colliding with another molecule. The exosphere lies beyond the ionosphere and starts about 400 kilometres above the earth's surface.

exothermic [Chemistry] Denoting a chemical reaction that releases heat into its surroundings. *Compare* endothermic.

exotoxin [Biology] A toxin that is produced by bacteria and is passed into the environment of the cell during growth.

expander [Materials] A mixture of lamp black, barium sulphate and some other organic materials that increases the capacity of storage batteries.

expansion of the universe [Astronomy] The widely accepted theory that the universe is expanding, *i.e.*, that clusters of galaxies are receding from each other. It is based upon the evidence of the red shift (*see also* Doppler effect) and the theory of relativity. *See* Hubble's constant.

expectorant [Medicine] A drug which promotes expectoration.

expectoration [Physiology] The elimination of secretion from the respiratory tract by coughing.

explicit function [Mathematics] A variable quantity, x, is said to be an explicit function of y, when x is directly expressed in terms of y.

explosion [Chemistry] A violent and rapid increase of pressure in a confined space. it may be caused by an external source of energy (*e.g.*, heat or by an internal exothermic chemical reaction in which relatively large volumes of gases are produced. Explosions may also occur as the result of the release of internal energy during an uncontrolled nuclear reaction (either fission or fusion or both).

explosives [Materials] Substances that undergo a rapid chemical change, with production of gas, on being heated or struck. The volume of gas produced being very great relative to the bulk of the solid explosive; great pressures are set up when the action takes place in a confined space.

exponent [Mathematics] The number indicating the power of a quantity. Thus the **exponent of** x in x^4 is 4.

exponential [Mathematics] A function that varies as the power of another quantity. If $y = a^x$, y varies exponentially with x. The function e^x is called the *exponential function (see e)*. It is equal to the sum of the *exponential series, i.e.*

$$e^x = 1 + x + \frac{x^2}{2!} + \frac{x^3}{3!} + \dots + \frac{x^n}{n!}$$

exponential experiment [Nucleonics] A nuclear experiment involving a subcritical assembly of fissionable and moderator material.

exponential growth [Biology] The period of bacterial growth during which cells divide at a constant rate.

exposure meter [Optics] Generally, a photocell, used to measure light intensity for the purpose of determining proper camera adjustments.

extender [Materials] An inert substance added to a product (paint, rubber, washing powder, etc.) to dilute it (for economy) or to modify its physical properties.

extensive property [Chemistry] The property of a system, such as internal energy or volume, that changes with the quantity of material in the system. The quantitative value of extensive property equals the sum of the values of the property for the individual constituents.

extinction coefficient [Chemistry] A measure of the amount of light absorbed by a substance in solution. If light of intensity I_1, is passed through a distance d of a solution containing a molecular concentration c of the dissolved substance, so that its intensity is reduced to I_2, then the extinction coefficient is given by :

$$[\log_{10}(I_1/I_2)]/cd$$

extraction [Chemistry] The process of separating a desired constituent from a mixture, by means of selective solubility in an appropriate solvent. [Metallurgy] Also used to describe any process by which a pure metal is obtained from ore.

extraordinary ray *See* ordinary ray.

extrapolation [Scientific Techniques] The process of filling in values or terms of a series on either side of the known values, thus extending the range of values.

extremely high frequencies [Electronics] EHF. Radio frequencies in the range 30,000 to 300,000 megahertz.

extremely low frequencies

[Electronics] ELF. Radio frequencies below 300 hertz.

extrinsic properties [Electronics] The properties of a semiconductor as modified by impurities or imperfections within the crystal.

extrinsic semiconductor [Electronics] A semiconductor in which the carrier density results mainly from the presence of impurities or other imperfections as, opposed to an intrinsic semiconductor in which the electrical properties are characteristic of the ideal crystal.

extrinsic sol [Chemistry] A colloid whose stability is attributed to electric charge on the surface of colloidal particles.

extrusion [Engineering] A process in which a hot or cold semisoft solid substance, such as plastic, is forced through a orifice or a die to produce a continuously formed piece in the desired shape.

eye-piece [Optics] In optical instruments, the lens or system of lenses nearest to the observer's eye; generally used to view the image formed by the objective. Also known as ocular.

eyespot [Botany] A small photosensitive pigment body in certain unicellular algae.

F *See* farad; fluorine.

Faber flaw [Physics] A deformation in the superconducting material which acts as nucleation centre for the growth of a superconducting region.

fabric [Engineering] A textile structure composed of mechanically interlocked fibre or filaments. The word generally refers to wool, cotton or synthetic fibres.

fabrication [Engineering] The assembly of parts into a structure.

face-centred lattice [Crystallography] A lattice whose unit cells are cubes, with lattice points at the centre of each face of the cube, as well as at the vertices.

facet [Materials] The plane surface of a crystal, a cut stone, or other fractured surface.

factice [Materials] A soft material made by reacting sulphur or sulphur chloride with vegetable oils. Also known as vulcanized oil.

factor [Mathematics] A number of quantity is exactly divisible by its factors; thus the factors of 18 (*i.e.* the integral or whole-number factors) are 1, 2, 3, 6, 9, and 18.

factorial [Mathematics] The product of a number and all the consecutive positive whole numbers below it down to 1. Thus, factorial 6, is written as 6! or 6!

$= 6 \times 5 \times 4 \times 3 \times 2 \times 1 = 720.$

faculae [Astronomy] Large bright areas of the photosphere of the sun, whose temperatures are higher than the average of the sun's surface.

facultative cells [Biology] Cells that can live in presence or absence of oxygen.

fading [Electronics] Variations in the field strength of a radio signal, usually, gradual that are caused by changes in the transmission medium.

Fahrenheit scale [Physics] A temperature scale in which (by modern definition) the temperature of boiling water is taken as 212 degrees and the temperature of melting ice as 32 degrees. It was invented by G.D. Fahrenheit, who set the zero at the lowest temperature he know how to obtain in the laboratory (by mixing ice and common salt) and took his own body temperature as 96°F. The scale is no longer in scientific use. To convert to the Celsius scale the formula is $C = 5(F - 32)/9$.

Fajans' and Soddy law [Nucleonics] The emission of an alpha particle during radioactive change produces an element two places to the left in the periodic table; and the emission of a beta particle, however, produces an element one place to the right in the periodic table.

Fajans' rules [Chemistry] Rules indicating the extent to which an ionic bond has covalent character caused by polarization of the ions. Covalent character is more likely if :

(a) the charge of the ions is high;

(b) the positive ion is small or the negative ion is large;

(c) the positive ion has an outer electron configuration that is not a noble-gas configuration.

fall-out [Nucleonics] Radio-active particles deposited from the atmosphere either from a nuclear explosion or from a nuclear accident. *Local fall-out,* within 250 km of an explosion, falls within a few hours of the explosion. *Trophospheric fall-out* consists of fine particles deposited all round the earth in the approximate latitude of the explosion within about one week. *Stratospheric fall-out* may fall anywhere on earth over a period of years. The most dangerous radioactive isotopes in fall-out are the fission fragments iodine-131 and strontium-90. [Chemistry] Hazardous chemicals discharged into and subsequently released from the atmosphere, especially by factory chimneys.

farad [Electricity] F. The SI unit of capacitance, being the capacitance of a capacitor that, if charged with one coulomb, has a potential difference of one volt between its plates. $1 F = 1 CV^{-1}$. The farad itself is too large for most applictions; the practical unit is the microfarad ($10^{-6}F$).

Faraday. *See* Faraday's constant.

Faraday's constant [Electricity] F. The electric charge carried by one mole of electrons or singly ionized ions, *i.e.*, the product of the Avogadro constant and the charge on an electron (disregard-

ing sign). It has the value 9.648 670 X 10^4 coulombs per mole. This number of coulombs is sometimes treated as a unit of electric charge called the *faraday*.

Faraday dark space [Electronics] The nonluminous region that separates the negative glow from the positive column in a cold-cathode discharge tube.

Faraday effect [Optics] The rotation of the plane of vibration of polarized light on traversing an isotropic transparent medium placed in a magnetic field possessing a component in the direction of the light ray. Although originally restricted to light, the Faraday effect is now known to apply to other electromagnetic radiations. Also known as Faraday rotation; Kundt effect.

Faraday's law of electromagnetic induction [Physics] The electromotive force induced in a circuit by a changing magnetic field is equal to the negative of the rate of change of magnetic flux linking the circuit. Also known as law of electromagnetic induction.

Faraday's laws of electrolysis [Chemistry] Two laws describing electrolysis :

(I) The amount of chemical change during electrolysis is proportional to the charge passed.

(II) The charge required to deposit or liberate a mass m is given by $Q = Fmz/M$, where F is the Faraday constant, z the charge of the ion, and M the relative ionic mass.

These are the modern forms of the laws. Originally, they were stated by Faraday in a different form :

(I) The amount of chemical change produced is proportional to the quantity of electricity passed.

(II) The amount of chemical change produced in different substances by a fixed quantity of electricity is proportional to the electrochemical equivalent of the substance.

fast breeder reactor [Nucleonics] A type of fast nuclear reactor that uses highly enriched fuel in the core, fertile material in the blanket, and a liquid metal coolant, having breeding ratio equal to one or more than one.

fast burst reactor [Nucleonics] A nuclear reactor that produces microsecond pulses of fast neutrons which are used in biomedical research.

fast-effect [Nucleonics] Increase in neutrons due to fissions caused by fast neutrons in a thermal reactor.

fast neutrons [Nucleonics] Neutrons obtained as a result of nuclear fission, and have very high energy; used in breeder reactors.

fast reactor [Nucleonics] A nuclear reactor in which fission is produced by fast neutrons, with little or no moderator to slow down the neutrons.

fat [Chemistry] A glycerol ester of higher fatty acids such as stearic or palmitic acid, *e.g.*, lard and tallow. Fats differs from the oils in the respect that former are solid while the latter are liquid at room temperature.

fat dye [Materials] A variety of oil-soluble dyes used for colouring wax products.

fatigue of metals [Metallurgy] The deterioration of metals owing to repeated stresses above a certian critical value; it is accompanied by changes in the crystalline structure of the metal.

fatty acids [Chemistry] Monobasic organic acids having the general formula R.COOH, where R is hydrogen or a group of carbon and hydrogen atoms. The saturated fatty acids have the general formula $CnH_{2n+1}COOH$. Many fatty acids occur in living things, usually in the form of glycerides in fats and oils.

f-block elements [Chemistry] The block of elements in the periodic table consisting of the lanthanoid series (from cerium to lutetium) and the actinoid series (from thorium to lawrencium). They are characterized by having two s-electrons in their outer shell (n) and f-electrons in their inner $(n-1)$ shell.

feberifuge *See* antipyretic.

feed back [Electronics] The coupling of a portion of the output of a circuit or device to its input.

feeder [Electricity] A transmission line used between a transmitter and an antenna. [Engineering] A device for delivering materials to a processing unit.

Fehling's solution [Chemistry] A reagent used to test for sugars, aldehydes etc. It consists of two solutions, one of copper sulphate, and the other of alkaline tartrate, which are mixed just before use. The mixed solution forms red precipitate of cuprous oxide with reducing aldehydes.

feldspar [Minerals] General name for a group of sodium, potassium, calcium, barium, and aluminium silicates; used in pottery, enamel, glass, soaps, abrasives and roofing materials. Also known as feldspad.

feldspar *See* feldspar.

femitrons [Electronics] A glass of field emission microwave devices.

femto- [Scientific Techniques] F. A prefix used in the metric system to denote 10^{-15}. For example, 10^{-15} second = 1 femtosecond (fs).

Fermat's principle [Optics] An electromagnetic wave takes the path that involves least travel time while propagating between any two points. Also known as least-time principle.

ferment [Biochemistry] Any substance that can initiate fermentation. Also known as enzyme.

fermentation [Biochemistry] A form of anaerobic respiration occurring in certain microorganisms, *e.g.*, yeasts. It comprises a series of biochemical reactions by which sugar is converted to ethanol and carbon dioxide. Fermentation is the basis of the baking, wine, and beer industries.

fermi [Scientific Techniques] A unit of length formerly used in nuclear physics. It is equal to 10^{-15} metre. In SI units this is equal to 1 femtometre (fm).

Fermic energy [Physics] The average energy of electrons in a metal equal to 3/5 of Fermi level.

Fermi-Dirac statistics [Physics] The branch of statistical mechanics used with systems of identical particles having the property that their wave function changes sign if any two particles are interchanged. *See* fermions.

Fermi hole [Physics] A region surrounding an electron in a solid in which the energy band theory predicts that the probability of finding other electrons is less.

Fermi level [Physics] The energy level in a solid at which the probability of finding an electron is 1/2. The Fermi level in conductors lies in the conduction band (*see* energy bands), in insulators it lies in the valence band, and in semiconductors it falls in the gap between the conduction band and the valence band. At absolute zero all the electrons would occupy energy levels up to the Fermi level and no higher levels would be occupied.

ferminos [Physics] Particles that conform to Fermi-Dirac statistics. The numbers of fermions are conserved throughout all nuclear interactions, but they are divided into two groups, baryons and leptons, which are distinguished from each other in that members of one group cannot transform into members of the other group. All fermions have spin 1/2.

fermium [Chemistry] Fm. A synthetic transuranic element. At. No. 100. The most stable isotope, ^{257}Fm. has a half-life of only 80 days.

ferrate [Chemistry] 1. A salt of the hypothetical ferric acid, H_2FeO_4. 2. A multiple iron oxide with another oxide, e.g., Na_2FeO_4.

ferric [Chemistry] Containing trivalent iron. Ferric salts are usually yellow or brown in colour.

ferrimagnetism [Physics] The type of magnetism in materials in which the magnetic moments of adjacent atoms are anti-parallel, but of unequal strength, or in which the number of magnetic moments orientated in one direction outnumber those in the reverse direction. Ferrimagnetic materials therefore have a resultant magnetization similar to that of ferromagnetism. Typical ferrimagnetic materials are the ferrites.

ferrite [Chemistry] A salt of the hypothetical 'ferrous acid' that exists in strong alkaline solution only, e.g., $NaFeO_2$. [Metallurgy] The name applied to several types of iron ore. Pure alpha-iron, or solid solutions of alpha-iron is the solvent.

ferrites [Materials] A group of ceramic materials that exhibit the property of ferrimagnetism. They consist of iron oxide to which small quantities of transition metal oxides (e.g., cobalt and nickel oxides) have been added the .ferrites have the formula $MO.Fe_2O_3$ where M is a divalent transition metal ion. By suitable combination of metallic oxides, ferrites can be made which exhibit ferromagnetism, but as they are electrical insulators and therefore do not differ from the effects of eddy currents, they can be used as cores in coils and transformers in electronic equipment at frequencies that would be impossible with ordinary ferromagnetic materials.

ferro- [Chemistry] Prefix denoting iron, especially in name of alloys; e.g., ferromanganese.

ferroaluminium [Metallurgy] An alloy of aluminium (up to 80%) and iron.

ferroboron [Metallurgy] An alloy of iron and boron that is added to steel to make it hard.

ferrochrome [Metallurgy] An alloy of chromium with 30%-40% iron, obtained by the reduction of chromite with carbon in an electric furnace.

ferroelectric [Physics] Dielectric materials that have electrical properties analogous to certain magnetic properties such as hysteresis' *e.g.*, barium titanate and potassium sodium tartrate (Rochelle salt). Ferroelectric materials usually also have piezoelectric properties (*see* piezoelectric effect).

ferroelectric converter [Electricity] A device that transforms thermal energy into electric energy by utilizing the changes in dielectric property of a ferroelectric substance when heated beyond its Curie temperature.

ferroelectricity [Physics] Spontaneous electric polarization in a crystal.

ferromagnetic substances *See* ferromagnetism.

ferromagnetism [Physics] The phenoemnon associated with *ferromagnetic* substances within a certain temperature range where there are net atomic magnetic moments, which line up in such a way that magnetization persists after the removal of the applied field. Below a certain temperature, called the *Curie point* (or Curie temperature) an increasing magnetic field applied to a ferromagnetic substance will cause increasing magnetization to a high value, called the *saturation magnetization*. This is because a ferromagnetic substance consists of small (1-0.1 mm across) magnetized regions called *domains*. The total magnetic moment of a sample of the substance is the vector sum of the magnetic moments of the component domains. Within each domain the individual atomic magnetic moments are spontaneously aligned by *exchange forces*, related to whether or not the atomic electron spins are parallel or antiparallel. However, in an unmagnetized piece of ferromagnetic material the magnetic moments of the domains themselves are not aligned; when an external field is applied those domains that are aligned with the field increase in size at the expense of the others. In a very strong field all the domains are lined up in the direction of the field and provide the high observed magnetization. Iron nickel, cobalt, and their alloys are ferromagnetic. Above the Curie point, ferromagnetic materials become paramagnetic.

ferromanganese [Metallurgy] An alloy of manganese (70%-80%) and iron.

ferroprussiate paper [Materials] A paper used in blue print process to reproduce drawings.

ferrosilicon [Metallurgy] An alloy of silicon (15%) and iron, used in special steels.

ferrous [Chemistry] Containing bivalent (divalent) iron; more loosely, pertaining to iron. Ferrous salts are generally pale green in colour.

fertile material [Nucleonics] Isotopes that can be transformed into fissile material by the absorption of neutrons, *e.g.*, thorium-232; uranium-238.

fertilization [Biology] The union of two sexually dissimilar gametes to form a zygote.

fertilizers [Materials] Materials put into the soil to provide compounds of elements essential to plant life; more particularly nitrogen phosphorus, and potassium.

fetus [Biology] The unborn offspring after it has largely completed its embryonic development; from the third month of pregnancy to birth in man. Also spelled foetus.

fibreglass [Materials] Trademark for a variety of products made of glass fibres or glass flakes including insulating glass wool and reinforced plastics.

fibre, optical [Materials] An extremely fine drawn glass fibre of high purity that will transmit laser light impulses with high fidelity.

fibre optics [Optics] The technique of transmiting light through long, thin and flexible fibres of glass, plastic or other transparent materials.

fibrin [Biochemistry] An insoluble blood protein precipitated in the blood of vertebrates in the form of a meshwork of fibres during the process of clotting. It is formed when thrombin acts upon fibrinogen.

fibrinogen [Biochemistry] A soluble protein found in the blood of vertebrates that causes clotting of the blood by the action of the enzyme thrombin as a result of which fibrin is formed.

fidelity [Electronics] A measure of the frequency response of a sound-producing system. 'High fidelity' systems are usually taken to be those that are capable of reproducing frequencies up to 12000 hertz without any distortion.

field coil [Electricity] A coil of wire used for magnetizing an electromagnet, *e.g.*, as in a dynamo.

field effect diode [Electronics] A semiconductor diode in which charge carriers are of only one polarity.

field emission [Electronics] The emission of electrons from an unheated surface as a result of a strong electric field existing at that surface.

field-emission microscope [Electronics] A type of microscope for observing the surface structure of a solid. A high negative voltage ($>10kV$) is applied to a metal tip placed at the centre of a spherical fluorescent screen in a vacuum. Field emission from the tip produces electrons, which create an enlarged image on the screen. As resolution is limited by the vibrations of the metal atoms, the tip is usually cooled with liquid helium.

field lens [Optics] The lens in the eye-piece system of optical instruments farthest from the eye.

field magnet [Electricity] A magnet that provides a magnetic field in the dynamo, electric motor, or other electrical machine.

filament [Electricity] A thin thread. In incandescent electric light bulbs and thermionic valves, the filament is a wire of tungsten or other metal of high melting point, which is heated by the passage of an **electric current** [Electronics] A cathode made of resistance wire, through which an electric current is sent to produce high temperature required for emission of electrons.

file [Computers] A collection of information that has a describable structure, allowing all, or part, of it to be retrieved from the store (or backing storage) on demand.

filler [Materials] A solid substance added to synthetic resins, paints, and rubbers, either to modify their properties or to reduce their cost.

film [Chemistry] A thin layer of a substance formed on the surface of a liquid or at the interface between two immiscible liquids, usually only a few molecules thick. [Materials] A flexible strip (usually cellulose acetate or a polyester) coated with a light sensitive emulsion.

film badge [Nucleonics] A badge containing a masked photographic film worn by workers in contact with ionizing radiations to indicate the extent of their exposure to these radiations.

filter [Chemistry] A device for separating solids or suspended particles from liquids. It consists of a porous material (*e.g.*, filter-paper) through the pores of which only liquids and dissolved substances can penetrate. [Physics] A material or device inserted in the path of an electromagnetic radiation to alter its frequency distribution.

filter capacitor [Electricity] A capacitor used in a power supply filter system to provide a low reactance path for a.c. without affecting d.c.

filtrate [Chemistry] A clear liquid after filtration; a substance that has been filtered, and contains no suspended matter.

filtration [Chemistry] The process of separating solids from liquids by passing them through a filter.

finder [Optics] A small low-powered telescope having a wide angle lens, fixed parallel to the axis of a large telescope so that the object to be observed may be located and set in the field of vision of the large telescope.

fineness of gold [Metallurgy] The quantity of gold in an alloy expressed as parts per thousand. Thus gold with a fineness of 900 is in alloy containing 90% gold. *See also* carat.

fine structure [Physics] Closely spaced spectral lines arising from transitions between energy levels that are split by the vibrational or rotational motion of a molecule or by electron spin. They are visible only at high resolution. *Hyperfine structure,* visible only at very high resolution, results from the influence of the atomic nucleus on the allowed energy levels of the atom.

fire [Chemistry] A chemical reaction accompanied by the evolution of heat, light, and flame. It is generally applied to the chemical combination with oxygen of carbon and other elements constituting the substance being burnt.

fireclay [Minerals] Clay consisting principally of aluminium oxide, Al_2O_3, and silica, SiO_2, which will only soften at high temperatures and is therefore used as a refractory material.

fire-damp [Geology] An explosive mixture of methane (CH_4) and air, formed in coal mines; due to decomposition of coal or other carbonaceous matter.

fire extinguishers [Engineering] Devices for extinguishing fires in their early stages by the ejection of a fire-inhibiting substance, such as water, carbon dioxide, or chemical foam.

First law of thermodynamics *See* laws of thermodynamics.

Fischer-Tropsch process [Chemistry] A process for the manufacture of hydrocarbon oils from coal, lignite, or natural gas. The process essentially consists of the hydrogenation of carbon monoxide in the presence of catalysts to form hydrocarbons and steam.

fissile material [Nucleonics] Isotopes that are capable of undergoing nuclear fission. Sometimes the term is restricted to isotopes that are capable of undergoing fission upon impact with a slow neutron (*e.g.*, ^{233}U, ^{235}U, ^{239}Pu).

fission [Biology] Process of asexual reproduction in which an organism divides into almost equal parts. [Nucleonics] *See* nuclear fission.

fission bomb *See* atomic bomb.

fission fuel *See* nuclear fuel.

fission neutron [Nucleonics] A neturon produced as a result of nuclear fission.

fission-track dating [Nucleonics] A method of estimating the age of glass and other mineral objects by observing the tracks made in them by the fission fragments of the uranium nuclei that they contain. By irradiating the objects with neutrons to induce fission and comparing the density and number of the tracks before and after irradiation it is possible to estimate the time that has elapsed since the object solidified.

fixation *See* nitrogen fixation.

fixed air [Chemistry] Former name for carbon dioxide, CO_2.

fixed alkali [Chemistry] Former name for potassium or sodium carbonate, to distinguish them from volatile alkali, ammonium carbonate.

fixed bias [Electronics] A constant value of bias voltage which is independent of signal strength.

fixed oil [Materials] A non-volatile oil occurring in plants. These oils are generally edible, *e.g.*, coconut oil.

fixed stars [Astronomy] Heavenly bodies termed fixed because they do not appear to alter their relative positions on the celestial sphere.

flagellates [Biology] Micro-organisms furnished with one or more slender, whiplike processes termed flagella.

flame [Chemistry] A hot luminous mixture of gases undergoing combustion. The chemical reactions in a flame are mainly free-radical chain reactions and the light comes from fluorescence of excited molecules or ions or from incandescence of small solid particles (*e.g.*, carbon).

flame spectrum [Spectroslopy] An emission spectrum obtained by evaporating substances in a nonluminous flame.

flame test [Chemistry] A simple test for metals, in which a small amount of the sample (usually moistened with hydrochloric acid) is placed on the end of a platinum wire and held in a Bunsen flame. Certain metals can be detected by the colour produced: barium (green), calcium (brick red), lithium (crimson), potassium (lilac), sodium (yellow), strontium (crimson red).

flammable [Materials] A substance capable of supporting combustion.

flash bomb [Engineering] A bomb that illuminates the ground for night aerial photography.

flash burn [Medicine] Tissue injury resulting from exposure to high-intensity radiant heat.

flash dry [Chemistry] The rapid evaporation of moisture from a porous medium by a sudden reduction in pressure or by placing the material in an updraft of warm air.

flash magnetization [Physics] Magnetization of a ferromagnetic substance by a current impulse of short duration.

flash over [Electricity] An electric discharge around or over the surface of an insulator.

flash pasteurization [Biochemistry] A pasteurization technique in which a heat-labile liquid such as milk, is briefly subjected to temperature around 383K.

flash photolysis [Chemistry] A technique for studying free-radical reactions in gases. The apparatus used typically consists of a long glass or quartz tube holding the gas, with a lamp outside the tube suitable for producing an intense flash of light. This dissociates molecules in the sample creating free radicals, which can be detected spectroscopically.

flash point [Chemistry] The lowest temperature at which a substance give off sufficient inflammable vapour to produce a momentary flash when a small flame is applied.

flavoproteins [Biochemistry] Yellow conjugated proteins in which the prosthetic group is either flavin mononucleotide or flavin adenine dinucleotide. Flavoproteins are enzyme of the dehydrogenase type.

Fleming's rule [Electricity] If the fore-finger, second finger, and thumb of the right hand are extended at right angles to each other, the forefinger indicates the direction of the flux, the second finger the direction of the E.M.F., and the thumb the direction of motion in an electric generator. If the left hand is used the digits indicate the conditions obtaining in an electric motor.

flint [Minerals] Natural variety of impure silica, SiO_2. [Materials] 'Flints' of lighters are composed of pyrophoric alloys of metals such as cerium and iron.

flint glass [Materials] A variety of heavy and brilliant glass containing lead silicate; used for optical purposes.

flocculation [Chemistry] The process of small particles aggregating into larger clumps.

flocculent [Chemistry] Aggregated in woolly masses; used to describe precipitates.

flotation process [Metallurgy] A technique for the separation of a mixture of sulphide ore from each other and gangue, e.g., of

zinc blende, ZnS and galena, PbS, making use of the surface tension of water. Zinc blende is not easily wetted by water floats, supportted by the surface film of water, while galena sinks. In modern practice, special material are added to the water to cause one of the constituents to float in the froth produced by aerating and agitating the water. See froth flotation.

flowers of sulphur [Chemistry] A fine powder, consisting of very small crystals of sulphur obtained by the condensation of sulphur vapour during distillation of crude sulphur. Also known as sublimed sulphur.

flue [Engineering] A passage for conveying combustion products from fireplace to or through chimney.

flue dust [Metallurgy] Fine particles of alloys or metals emitted along with the gases of metallurgical furnaces.

flue gas [Materials] The gaseous products of combustion from a boiler furnace consisting predominantly of carbon dioxide, carbon monoxide, oxygen, nitrogen and steam. Analysis of the flue gases is used to check the efficiency of furnace.

fluid [Physics] A substance that takes the shape of the vessel containing it; a liquid or gas.

fluidity [Mechanics] The reciprocal of viscosity expresses the ability of a substance of flow. The c.g.s.

units is the reciprocal of the poise, known as the rhe.

fluidization [Chemistry] A technique used in industrial processes, in which a mass of solid particles is brought into a state of suspension by an upward steam of gas blown through it in a reactor. The material in the resultant 'fluidized bed,' which resembles a boiling liquid, is more accessible to chemical reactions, etc., than the same solid material in the static state.

fluid resistance [Mechanics] The force exerted by a liquid or gas opposing the motion of a body through it.

fluorescence [Physics] A form of luminescence in which certain substances (*e.g.*, quinine sulphate solutions, paraffin oil, fluorescein solution) are capable of absorbing light of one wavelength (*i.e.*, colour, when in the visible region of the spectrum) and in its place, emitting light of another wavelength or colour. Unlike phosphorescence, the phenomenon ceases immediately when the source of light is removed.

fluorescent lamp [Electronics] A light source consisting of a glass tube inside of which is coated with a fluorescent material which gets activated by the radiations produced by the ionized mercury vapour.

fluorescent pigment [Chemistry] A pigment capable of absorbing both visible and non-visible electromagnetic radiation and releasing them quickly as energy of desired wavelength.

fluoridation [Chemistry] The addition of minute quantities of fluorides above 1 part per million to drinking water supplies to give protection against caries (decay) in the teeth of growing children.

fluorination [Chemistry] The introduction of fluorine into a compound by substitution or by an addition reaction.

fluorine [Chemistry] F.Element. At.No. 9. At.Wt. 18.9984; placed in group VII A of periodic table. A pale yellowish-green gas; is a member of halogen family, it is the most electronegative element; highly toxic, and corrosive.

fluorocarbons [Chemistry] Compounds obtained by replacing the hydrogen atoms of hydrocarbons by fluorine atoms. Their high stability to temperature makes them suitable for a variety of uses, including aerosol propellants, oils, polymers, etc. They are often known as *freons*. There has been some concern that their use in aerosols may cause depletion of the ozone layer.

fluoroscope [Engineering] A fluorescent screen (*see* fluorescence) for the direct visual observation of X-*ray* images; used diagnostically in medicine.

fluorspar [Minerals] CaF_2. Natural calcium fluoride, consisting of colourless crystals, often coloured due to impurities.

flux [Chemistry] A substance added to assist fusion. [Physics] The rate of flow of mass or energy per unit area normal to the direction of the flow. [Nucleonics] The product of the number of particles per unit volume and their average velocity. *See also* magnetic flux; electric flux; luminous flux.

flux density [Physics] The magnetic flux or luminous flux per unit of cross sectional area. The S.I. unit of magnetic flux density is tesla.

fluxing [Metallurgy] Making of a liquid phase in a ceramic body under heat treatment by melting low-fusion components.

flux mapping [Nucleonics] The process of measuring the radiation flux at various points within a nuclear reactor or some other radiation source.

fluxmeter [Engineering] An instrument for the measurement of magnetic flux.

flyash [Engineering] Fine, non-combustible particles carried in a gas stream from a furnace. It is a mixture of alumina, silica, unburnt carbon, and various metallic oxides.

f-number of a lens [Optics] The ratio of focal length to diameter.

foam [Chemistry] A dispersion of bubbles in a liquid. Foams can be stabilized by surfactants. Solid foams (*e.g,* expanded polystyrene or foam rubber) are made by foaming the liquid and allowing it to set.

foam glass [Materials] A light, opaque, cellular glass formed by adding powdered carbon to crushed glass and firing the mixture.

foam metal [Metallurgy] Cast metal with tiny gas bubbles evenly distributed throughout the body of the metal.

focal distance *See* focal length.

focal length [Optics] The distance from the optical centre or pole to the principal focus of a lens or spherical mirror. The focal length of a spherical mirror is half of its radius of curvature. Also known as focal distance.

focal plane [Optics] A plane perpendicular to the axis of an optical system and passing through the focus of the system.

focus [Optics] The point at which converging rays, usually of light, meet (real focus); or a point from which diverging rays appear to be directed (virtual focus). The 'principal focus, of a lens or spherical mirror is the point on the principal axis through which rays of light parallel to principal axis pass or appear to pass.

foetus *See* fetus.

fog [Physics] The effect caused by the condensation of water droplets upon particles of dust, soot, etc.

food additive [Chemistry] A substance added to food-stuffs during processing to improve colour, flavour, texture or storing qualities, *e.g.*, food colours, food preservatives, antioxidants, emulsifiers etc.

food preservation [Chemistry] The prevention of chemical decomposition and the development of harmful bacteria in foods. It is generally effected by the sterlization of the food (*i.e.*, by the destruction of bacteria in it) by heating in sealed vessels, *i.e.*, canning; or by making the conditions unfavourable for the development of bacteria, by pickling, drying, freezing, smoking, etc.

forbidden band [Electronics] A range of unallowed energy bands for an electron in a solid.

force [Physics] An external agency capable of altering the state of rest or motion in a body; measured in newtons (SI units), dynes (c.g.s. units), or poundals (f.p.s. units). The force F, required to produce an acceleration, a, in a mass, m, is given by $F=ma$. If m is in kilograms, a in m s^{-2}, F will be in newtons.

forensic science [Sci] The application of science for discussion, debate, argumentative or legal purposes.

forge [Metallurgy] To convert a hot metal into desired shapes by applying compressive forces.

formalin [Chemistry] A 40% solution of formaldehyde, used as a disinfectant.

format [Computers] The specific arrangement of data on a printed page, punched card to meet established presentation requirements.

form oil [Materials] An oil utilized on the contact surface of wooden or metal concrete forms to prevent concrete from sticking.

formula [Chemistry] A way of representing a chemical compound using symbols for the atoms present. Subscripts are used for the numbers of atoms. The *molecular formula* simply gives the types and numbers of atoms present. For example, the molecular formula of ethanoic acid is $C_2H_4O_2$. The *empirical formula* gives the atoms in their simplest ratio; for ethanoic acid it is CH_2O. The *structural formula* gives an indication of the way the atoms are arranged. Commonly, this is done by dividing the formula into groups; ethanoic acid can be written $CH_3.CO.OH$ (or more usually simply CH_3COOH). Structural formula can also show the arrangement of atoms or groups in space.

formula weight [Chemistry] The relative molecular mass of a compound as calculated from its molecular formula.

formyl [Chemistry] The univalent group $O=CH-$, derived from formic acid.

fortification [Chemistry] Addition of nutrients to food prouducts that are normaly not present in it, for example addition of vitamin D to milk.

fortin barometer [Engineering] A mercury barometer that is used in conjunction with various correction tables, enables accurate measurements of atmospheric pressure.

FORTRAN [Computers] A set of procedure oriented computer languages used generally for scientific or algebraic applications.

forward bias [Electronics] A bias voltage applied to a p-n junction in the direction that causes a large current flow.

fossil [Geology] The remains of an organism preserved in rocks in the earth's curst. Usually only the hard parts (bones, shells, etc.) are so preserved, but occasionally remains of organisms having no hard parts have been recognized.

fossil fuel [Geology] Coal, oil, and natural gas, the fuels used by man as a source of energy. They are formed from the remains of living organisms and all have a high carbon or hydrogen content. Their value of fuels relies on the exothermic oxidation of carbon to form carbon dioxide (C + O_2 ---> CO_2) and the oxidation of hydrogen to form water
$$H_2 + \tfrac{1}{2}O_2 \rightarrow H_2O$$

Fourier series [Mathematics] An expansion of a periodic function as a series of trigonometric functions. Thus,

$$f(x) = a_o + (a_1\cos x + b_1\sin x) + (a_2\cos 2x + b_2\sin 2x) + ...,$$

where a_o, a_1, b_1, b_2, etc; are constants, called Fourier coefficients.

fourth dimension [Physics] Time in the theory of relativity in which space and time are conceived as particular aspects of a four-dimensional world.

fraction [Chemistry] Any portion of a mixture characterized by closely similar properties. [Mathematics] An expression which is the product of a real number or complex number with multiplicative inverse of a real or complex number.

fractional crystallization [Chemistry] The separation of mixture of dissolved substances by making use of their different solubilities. The solution containing the mixture is evaporated until the least soluble component crystallizes out.

fractional distillation [Chemistry] The separation of a mixture of several liquids that have different boiling points, by collecting separate 'fractions' boiling at different temperatures.

fractionating column [Engineering] A long vertical column, containing rings, plates, or bubble caps, that is attached to a still. As a result of internal reflux a gradual separation takes place between high and low boiling 'fractions, of a liquid mixture.

fractionation [Chemistry] Separation of a mixture in successive stages, each stage removing from the mixture some proportion of one of the substances, as by differential solubility in water-solvent mixtures.

fragrance [Chemistry] An odorant used to impart a pleasant smell to shaving lotion, tooth pastes, face creams and other men's accessories.

frame of reference [Physics] A set of reference axis for defining the position of a point or body in space. A frame of reference in four-dimensional continum consists of an observer, a coordinate system, and a clock to correlate positions with times.

francium [Chemistry] Fr. Element. A radioactive alkali metal, At. No. 87. At. Wt. 233. Fr. 223 is the longest lived isotope, having a half-life of 21 minutes.

frangible [Mechanics] Breakable, brittle, or fragile.

Frasch process [Engineering] A process for extracting sulphur from deep deposits. A series of concentric pipes is sunk down to the level of sulphur deposits superheated steam is forced down to melt the sulphur, which is then forced to the surface by compressed air blown through the central pipes.

Fraunhofer diffraction [Optics] Diffraction of a parallel beam of light observed at an effectively infinite distance from the diffracting object, usually with an aid of lenses which collimates the light before diffraction and focus it at the point of observation.

Fraunhofer lines [Optics] Dark lines in the continuous spectrum of the sun, caused by the absorption of certain wavelengths of the white light from the hotter regions of the sun by chemical elements present in the cooler chromosphere surrounding the sun.

free carbon [Metallurgy] Elemental carbon present in a metal in uncombined state.

free charge [Electricity] Electric charge which is not bound to a fixed and definite site in a solid.

free electron [Physics] An electron not constrained and restricted to a particular site, but is free to move under the influence of electric or magnetic fields.

free energy [Chemistry] A measure of a system's ability to do work. The *Gibbs free energy* (or *Gibbs function*), G, is defined by $G = H - TS$, where G = the energy liberated or absorbed in a reversible process at constant pressure and constant temperature (T), H = the enthalpy and S entropy of the system. Changes in Gibbs free energy. ΔG, are useful in indicating the conditions under which a chemical reaction will occur. If ΔG is positive the reaction will only occur if energy is supplied to force it away from the equilibrium position (*i.e.* when $\Delta G = 0$). If ΔG is negative the reaction will proceed spontaneously to equilibrium.

The *Helmholtz free energy* (or *Helmholtz function*), F is defined by $F = U - TS$, where U = internal energy. For a reversible isothermal process, ΔF represents the useful work available.

free radical [Chemistry] A group of atoms which possesses at least one unpaired electron and may exist independently for short periods (short-lived free radicals) during the course of a chemical reaction, or for longer periods (free radical of long life) under special conditions.

free space wave [Physics] An electromagnetic wave propagating in a vacuum, free from boundary effects.

freeze drying [Chemistry] A process of drying heat-sensitive substances, such as food or blood plasma, by freezing and then removing the frozen water by volatilization at low pressure and temperature.

freezing [Physics] Change of state from liquid to solid; it takes place at a constant temperature (freezing point) for any given substance under a given pressure.

freezing mixtures [Chemistry] Certain salts, that, when dissolved in water or mixed with crushed ice, produce a considerable lowering of temperature. The action depends upon absorption of heat of solution by the dissolving salt; in the case of mixtures in contact with ice, the melting point of ice is lowered in the presence of a dissolved substance; latent heat of fusion of ice is absorbed, and the salt dissolves in the melting ice.

freezing point [Chemistry] The temperature at which a liquid and a solid are in equilibrium, *i.e.*, the temperature at which liquid solidifies. For a pure substance, the freezing point is the same as its melting point.

French chalk [Materials] Finely ground talc which has a characteristic soft and soapy feel.

French polish [Materials] Shellac dissolved in methylated spirit.

Frenkel defect *See* defect.

freons [Materials] Trade name for a group of polyhalogenated hydrocarbons containing fluorine and chlorine, *e.g.*, trichlorofluoromethane; used as coolants and refrigerants.

frequency [Physics] v.f. The number of cycles, oscillations, or vibrations of a wave motion or oscillations in unit time, usually one second. In a wave motion the frequency is equal to the velocity of propagation divided by the wavelength. The derived SI unit of frequency is hertz (H_z).

frequency band [Physics] A range of frequencies of electromagnetic radiations falling within prescribed limits.

frequency modulation [Electronics] FM. The type of radio transmission system in which the frequency of a carrier wave is modulated rather than its amplitude (as in amplitude modulation). It provides a method of transmission free from 'static' interference.

frequency monitor [Electronics] An instrument for indicating the amount of deviation of the carrier frequency of a transmitter from its assigned valve.

frequency of a vibrating string [Physics] The fundamental frequeney, *F,* of a stretched string of length a under tension *S,* is given by

$$F = \sqrt{S}/\pi d/2ra$$

where *r* is the radius of the string and *d* its density.

fresnel [Physics] A unit of frequency equal 10^{12} hertz.

Fresnel biprism [Optics] A flat triangular prism which has two very acute angle and one very obtuse angle; used to observe interference of light from a slit passing through the two halves of the prism.

Fresnel diffraction [Optics] A form of diffraction in which the light source or the receiving screen, or both, are at a finite distance from the diffracting system. *Compare* Fraunhofer diffraction.

Fresnel frings [Optics] One of the series of dark and light bands which appears near the edge of a shadow in Fresenel diffraction.

Fresnel lens [Optics] An optical lens whose surface consists of a number of smaller lenses so arranged that they give a short focal length; used in headlight, searchlights, etc.

Fresnel mirrors [Optics] Two plane mirrors which are inclined to each other on an order of a degree and used to observe the interference of light from a slit and is reflected from both mirrors.

friable [Materials] Easily crumbled , rubbed or pulverized into powder.

friction [Mechanics] The force that offers resistance to relative motion between surfaces in contact. *See* tribology.

Friedel-Craft's reaction [Chemistry] Originally the synthesis of aromatic hydrocarbons by reacting alkyl halides with benzene derivatives in the presence of anhydrous aluminium chloride as a catalyst. It is now extended to include the addition of alkenes to, and the condensation of alcohols with, aromatic hydrocarbons in the presence of such catalysts as anhydrous ferric chloride, gallium chloride, boron trifluoride, and hydrogen fluoride.

frit [Materials] A ground glass used in making glazes and enamels. Finely powdered glass may be called a frit.

frost [Physics] A covering of ice produced by the sublimation of water vapour on objects colder than 0°C.

frosted glass [Materials] Glass that has been etched with sand, or appears to have been so treated.

frother [Chemistry] Substances that

make longer lasting air bubbles by reducing surface tension; used in froth flotation process.

froth flotation [Metallurgy] The separation of a mixture of finely divided minerals by agitaing them in a froth of water and a frothing agent, so that some of the components of the mineral float and others sink. The process can be made selective by adjusting the nature of the froth with the addition of suitable surface active agents.

fructose [Biochemistry] $C_6H_{12}O_6$. A sweet soluble crystalline hexose, m.p. 275-277K; occurs in sweet ripe fruits, in the nectar of flowers, and in honey. Also known as fruit sugar; laevulose.

frustum [Mathematics] Any part of a solid figure cut off by a plane parallel to the base, or lying between two parallel planes.

F-star [Astronomy] A star of yellowish colour with a surface temperature of about 7000K.

fuel [Chemistry] A substance used for producing heat energy, either by means of the release of its chemical energy by combustion (*see* fossil fuels) or its nuclear energy by nuclear fission.

fuel cell [Electricity] A cell in which the chemical energy of a fuel is converted directly into electrical energy. The simplest fuel cell is one in which hydrogen is oxidized to form water over porous sintered nickel electrodes. A supply of gaseous hydrogen is fed to a compartment containing the porous cathode and a supply of oxygen is fed to a compartment containing the porous anode; the electrodes are separated by a third compartment containing a hot alkaline electrolyte, such as potassium hydroxide. The electrodes are porous to enable the gases to react with the electrolyte, with the nickel in the electrodes acting as a catalyst. At the cathode the hydrogen reacts with the hydroxide ions in the electrolyte to form water, with the release of two electrons per hydrogen molecule:

$$H_2 + 2OH^- \longrightarrow 2H_2O + 2e^-$$

At the anode, the oxygen reacts with the water, taking up electrons, to form hydroxide ions :

$$\tfrac{1}{2}O_2 + H_2O + 2e^- \longrightarrow 2OH^-$$

The electrons flow from the cathode to the anode through an external circuit as an electric current. The device is a more efficient converter of electric energy than a heat engine, but it is bulky and requires a continuous supply of gaseous fuels. Their use to power electric vehicles is being actively explored.

fuel element [Nucleonics] A fabricated rod, or other shape which consists of or contains the fissionable fuel for use in a nuclear reactor.

fuel oil [Materials] Any liquid product burned to generate heat exclusive of oils with a flash point below 37.7°C; includes furnace oils, stove oils.

fuel pellet [Nucleonics] A form of nuclear fuel element consisting of flat sheet of fuel which is generally a sandwitch of uranium fuel protected by metallic cladding.

fugacity [Chemistry] F. A thermodynamic function used in place of partial pressure in reactions involving real gases and mixtures. For a component of a mixture, it is defined by $d\mu = RTd(\ln f)$, where μ is the chemical potential. It has the same units as pressure and the fugacity of a gas is equal to the pressure if the gas is ideal.

fumes [Chemistry] Smoke or air consisting of solid particles generated by condensation from the gaseous state.

fumigation [Chemistry] The destruction of bacteria, insects, and other pests by exposure to toxic chemicals in gaseous state to poisonous gas or smoke.

fundamental constants [Scientific Techniques] Those parameters that do not change throughout the universe. The charge on an electron, the speed of light in free space, the Planck constant, the gravitational constant, the electric constant, and the magnetic constant are all thought to be examples. Also known as universal constants.

functional model [Computers] A software subroutine used by a simulator that emulates the behavior of a physical circuit of circuit component in response to an input stimulus.

fundamental frequency [Physics] The lowest frequency at which a system vibrates freely.

fundamental tone [Physics] The component tone of lowest pitch in a complex tone.

fundamental units [Scientific Techniques] A set of independently defined units of measurement that forms the basis of a system of units. Such a set requires three mechanical units (usually of length, mass, and time) and one electrical unit; it has also been found convenient to treat certain other quantities as fundamental, even though they are not strictly independent. In the metric system the centimetre-gram-second (c.g.s.) system was replaced by the metre-kilogram-second (m.k.s) system; the latter has now been adapted to provide the basis for SI units. In British Imperial units the foot-pound-second (f.p.s.) system was formerly used.

fundus [Biology] The bottom or the base of an organ; the part of a hollow organ farthest from its opening.

fungi [Biology] Simple plants that contain no chlorophyll. They may consist of one cell or many cellular tube-like threads. They feed on dead or living organisms, and cause disease of plants and of some animals, also cause decay of food, fabrics, and timber. Certain fungi are used in breuring and baking and for the production of antibiotics.

fungicide [Materials] A substance capable of destroying harmful fungi, such as moulds and mildews.

furnace [Engineering] An enclosed chamber in which heat is liberated and transferred directly or indirectly to the substances for the purpose of effecting a physical or chemical change.

furnace oil [Materials] Distillate fuel oil primarily used for heating purposes.

fuse [Electricity] A safety device containing a piece of wire of low melting point which melts if too high electric current is passed through it. This breaks the circuit in which fuse is placed. [Chemistry] To melt.

fused salt [Chemistry] A salt in the molten state.

fused oil [Materials] A volatile mixture of *iso*-amyl, butyl, propyl, and heptyl alcohols; obtained as a by product in alcoholic fermentation of starch, grains, or fruits.

fusible alloys [Metallurgy] Alloys of low melting point; generally eutectic mixtures of metals of low melting point such as bismuth, lead, tin, and cadmium. Fusible alloys having a melting point a little above the boiling point of water; are used in the construction of automatic sprinklers, heat from a fire melting the metal and releasing a spray of water.

fusion [Chemistry] A change of state of a substance from solid to liquid. Also known as melting. [Nucleonics] *See* nuclear fusion.

fusion bomb [Nucleonics] An explosive bomb that derives energy as a result of nuclear fusion.

fusion fuel [Nucleonics] A substance that can generate energy in a fusion reaction, such as deuterium, tritium, helium-3.

fusion mixture [Chemistry] A mixture of anhydrous sodium and potassium carbonates, Na_2CO_3 and K_2CO_3.

fyrel [Metallurgy] A flexible multifilament fibre made of nickel-chromium alloy that withstands heat upto 2500°K; used in space applications.

fyrex [Chemistry] Naturally occurring ammonium sulphate; soluble in water; used in flameproofing textiles, wood, and fibres.

g [Mechanics] Symbol for value of the acceleration of free fall; equal to 9.80665 metres per second.

gadolinium [Chemistry] Gd. A soft silvery metallic element belonging to the lanthanoides; At.No. 64; r.d. 7.901 (293K), m.p. 1585K; b.p. 3546K. There are seven stable natural isotopes and eleven artificial isotopes are known. Two of the natural isotopes, gadolinium-155 and gadolinium-157, are the best neutron absorbers of all the elements.

gaging [Nucleonics] The measurement of thickness, density, or quantity of material that absorbs a certain amount of radiation.

gain [Electronics] The increase in signal power produced by an amplifier; usually expressed as ratio of output to input voltage, power, or current, expressed in decimels.

galactic centre [Astronomy] The gravitation centre of the Milky Way galaxy; the sun and other stars of the galaxy revolve about this centre.

galactic noise [Astronomy] Radio frequency noise that originates outside the solar system and is strongest in the direction of the Milky Way.

galactic radiation [Astronomy] Radiation emanating from the Milky Way.

galaxy [Astronomy] A large aggregate of stars, gas, and dust, containing thousands of millions of stars, forming a stellar system. The solar system as a part of a galaxy called Milky Way. There are a large number of galaxies unevenly placed in space.

galena [Minerals] Natural lead sulphide, PbS, the principal ore of lead; Used as semiconductor in crystal rectifiers. Also known as lead glance.

Galilean telescope [Optics] A type of refracting telescope consisting of a convex lens as objective and a diverging (concave) lens as an eye-piece. It forms errect images.

gallium [Chemistry] Ga. A soft silvery metallic element belonging to group IIIA of the periodic table; At.No. 31; r.d. 5.90; m.p. 302.9K; b.p. 2676K. The two stable isotopes are gallium-69 and gallium-71; there are eight radioactive isotopes, all with short half-lives. The metal has only a few minor uses (*e.g.,* as an activator in luminous paints) but gallium arsenide is extensively used as a semiconductor in many applications. Gallium corrodes most other metals because it repidly diffuses into their lattices.

galvanic [Electricity] Pertaining to electricity flowing as a result of chemical action.

galvanic battery [Electricity] A galvanic cell, or a combination of two or more such cells for producing electric energy.

galvanic cell [Electricity] An electrolytic cell in which electric energy is produced as a result of chemical reactions.

galvanic corrosion [Metallurgy] Electro-chemical corrosion associated with the current in a galvanic cell caused by dissimilar metals in an electrolyte because of difference in potential of two metals.

galvanism [Biology] The use of

galvanic current for biological or medical purposes.

galvanize [Metallurgy] Deposition of zinc to the surface of metals by the process of hot dipping.

galvanometer [Engineering] An instrument for detecting, comparing, measuring small electric currents, but not usually calibrated in amperes; it requires calibration when an actual current measurement is needed. Working of galvanometers usually depends upon the magnetic effect produced by an electric current. *See* ammeter; ballistic galvanometer.

gamete [Biology] A reproduction cell, usually haploid and sexually differentiated. The female gamete (or ovum) unites with the male gamete (or spermatozoan) during fertilization to produce a zygote, which develops into a new individual. Also known as germ cell.

gametocyte [Biology] A cell that undergoes meiosis to form gametes.

gamma emission [Nucleonics] A transition between two energy levels of a nucleus in which a gamma rays is emitted. Also known as gamma decay.

gamma heating [Nucleonics] Heating of a substance due to absorption of gamma rays.

gamma-iron [Metallurgy] An allotropic form (*see* allotropy) of iron, which is nonmagnetic and

exists between 1173K and 1673K. *See* austenite.

gamma rays [Nucleonics] Gamma radiation. γ-rays. Electromagnetic radiation of the same nature, but shorter wavelength than X-rays (10^{-10}-10^{-13} metre). Ray emitted by the nuclei of radioactive atoms during decay. Gamma rays are emitted in quantized units called photons.

gamogony [Biology] 1. Spore formation by multiple fission in sporozoans. 2. Sexual reproduction.

ganglion [Physiology] A group of nerve cell bodies, usually located outside the brain and spinal cord.

gangrene [Medicine] A form of tissue death usually occurring due to insufficient blood supply.

gangue [Geology] The useless stony minerals that occur along with a metallic ore.

garnet [Minerals] A group of minerals of varying composition, mainly double silicates of calcium with other metals. Several varieties are red in colour, and are used as gems.

gas [Physics] A substance whose physical state (the gaseous state) is such that it always occupies the whole of the space in which it is contained. In a perfect gas, the atoms and molecules would move freely but in a real gas they are subjected to small intermolecular forces (Van der Waals forces).

gas alarm [Engineering] A signal

system which warns mine workers of dangerous concentration level of methane.

gas black [Chemistry] Fine carbon particles formed by partial combustion of natural gas. Also known as carbon black; channel black.

gas carbon [Chemistry] A hard deposit consisting of fairly pure carbon, found on the walls of the retorts used for the destructive distillation of coal in the manufacture of coal-gas. It is a good conductor of electricity, and it is used for making carbon electrodes. Also known as retort carbon.

gas chromatography [Chemistry] Gas-liquid chromatography. A very sensitive method of analysing the components of a complex mixture of volatile substances. The apparatus consist of a long narrow tube, packed with an inert support material of uniform particle size (*e.g.* diatomaceous earth) that has been coated with a non-volatile liquid, called the stationary phase, the whole tube and its content being maintained in a thermostatically controlled oven. The sample to be analysed is carried through the tube by a noble gas (*e.g.*, argon) so that the progress through the tube of various components of the mixture is selectively interfered with the stationary phase, some components passing through the tube more rapidly than others. A detector measures the electrical or thermal conductivity, or some

other characteristic property, of the gas leaving the column, differences being recorded on a strip chart, which indicates peaks corresponding to the various components. The instrument is calibrated by analysing samples of known compositions.

gas cleaning [Engineering] Removing pollutants or contaminants from gases.

gas constant [Physics] R. In the gas equation, $PV = RT$, the gas constant, R, equals 8.31434 joules per kelvin per mole or 1.9858 calories per degree celcius per mole.

gas-cooled reactor [Nucleonics] A nuclear reactor in which the coolant is a gas like air, carbon dioxide, or helium.

gas discharge [Electronics] Conduction of electricity in a gas due to movement of ions produced by collision between gas molecules and electrons.

gas laser [Optics] A laser whose active material is gas enclosed in quartz or glass tube.

gas law [Physics] Any law relating the pressure, volume, and temperature of a gas. *See* Boyle's law; Charle's law; Gay-Lussac's law; ideal gas law.

gas mantle [Engineering] A structure composed of the oxides of thorium (90%) and cerium (1%) made by impregnating a combustible fabric with a solution of the nitrates of the metals, and decomposing the nitrates by heat.

gas maser [Physics] A maser in which microwave radiation interacts with gas molecules; used in highly stable oscillator applications, as in atomic clocks.

gas mask [Engineering] A device for protecting the face and breathing organs against poisonous 'gases'. The air is drawn through a layer of activated carbon, which adsorbs vapours, and also through a filter pad, which retains solid particles of smoke. Such an arrangement is effective against war 'gases' and smokes, but not against gases of low molecular weight such as carbon monoxide or coal-gas. Sometimes also known as respirator.

gasohol [Materials] A mixture of unleaded gasoline and 10-15% ethyl alcohol used as fuel in internal combustion engines.

gas oil [Materials] A liquid petroleum distillate with viscosity and boiling range between kerosine and lubricating oil, usually includes diesel fuel, heating oils and light-fuel oils.

gasoline [Materials] A mixture of volatile hydrocarbons used as a fuel in internal combustion engines having an octane number of atleast 60. The major components are branched chain paraffins, cycloparaffins, and aromatics. Also known as petrol.

gas thermometer [Engineering] An apparatus for measuring temperature by the alteration in pressure produced by temperature changes in a gas kept at constant volume or by the alteration in volume of a gas kept at constant pressure.

gastrectomy [Medicine] Surgical removal of all or part of the stomach.

gastric acid [Biochemistry] Hydrochloric acid secreted by partictal cells in the fundus of stomach.

gastric juice [Physiology] The digestive fluid secreted by gastric glands; contains gastric acid and enzymes.

gastritis [Medicine] Inflammation of the stomach.

gastroenterology [Medicine] The branch of medicine concerned with the study of stomach and intestines.

gas turbine [Engineering] An engine that converts the chemical energy of a liquid fuel into mechanical energy by internal combustion, the gaseous products of which are expanded through a turbine; used as the power plant in aeroplanes (both turbo-propeller and turbo-jet driven), locomotives, and experimentally in motorcars; also used as an auxiliary power plant in electrical generating stations.

gate [Electronics] 1. A circuit with only one output, but which has more than one input and which can be activated by various combinations of input signals. 2. A signal that activates a circuit for a predetermined time or until another signal is received. 3. A device for selecting portions of a wave, either on a time or amplitude basis. 4. The electrode in a field-effect transistor that controls the flow of current through the channel.

gauss [Physics] The C.G.S. system unit of magnetic flux density. If a magnetic field of 1 oersted intensity exists in a medium of unit magnetic permeability, *e.g.*, air, then the induction will be 1 gauss; equal to 1 maxwell per square centimetre or 10^{-4} weber per square metre (*i.e.*, 1 tesla).

Gaussian elimination [Mathematics] A method for solving systems of linear equations based on manipulation of the matrix representing those equations.

Gauss's law [Electricity] The total electric flux of a closed surface in an electric field is 4π times the electric charge within that surface.

gate array [Electronics] A chip containing many unconnected gates of the same type in a regular pattern.

Gay-Lussac's law [Chemistry] When gases combine, they do so in simple ratio by volume to each other, and to the gaseous products, measured under the same conditions of temperature and pressure.

Geiger counter [Nucleonics] An instrument for the detection of ionizing radiations (chiefly alpha, beta, and gamma rays), capable of registering individual particles or photons. It consists normally of a fine wire anode surrounded by a coaxial cylindrical metal cathode, in a glass envelope containing gas at low pressure. A potential difference, of about 1000 volts, is maintained between the two electrodes. The ions produced in the counter by an incoming ionizing particle are accelerated by the applied potential difference towards their appropriate electrodes, causing a momentary drop in the potential between the latter. This voltage pulse is then passed on to various electronic circuits by means of which it can, if desired, be made to work as counter. Also known as Geiger-Muller counter.

Geissler tube [Electronics] A tube for showing the luminous effects of a discharge of electricity through various rarefied gases, consists of a sealed glass tube containing platinum electrodes.

gel [Chemistry] A colloidal solution that has set to a jelly, the viscosity being so great that the solution has the elasticity of a solid. The formation of a gel is attributed to a mesh-like structure of the disperse phase or colloid, with the dispersion medium circulating through the meshwork.

gelatin [Biochemistry] A complex protein formed by the hydrolysis of collagen in animal cartilages and bones, by boiling with water; soluble in water, the solution has the property of setting to a jelly; used in photography, as an adhesive, textile size. Also spelled as gelatine.

gelation [Physics] The process of freezing. *See also* regelation of ice. [Chemistry] The formation of a gel from a sol.

gel filtration [Chemistry] A chromatographic procedure for the separation of a mixture of molecules on the basis of size, through the capacity of porous polymers to exclude solutes above a certain size.

gelignite [Materials] An explosive consisting of a mixture of nitroglycerin, cellulose nitrate, potassium nitrate and wood pulp.

gem [Chemistry] Designating molecules in which two functional groups are attached to the same atom in a molecule. For example, chloral hydrate, $CCl_3CH(OH)_2$, is a gem diol in which both hydroxyl groups are on the same carbon atom.

gene [Biology] A chromosomal segment that codes for a single polypeptide chain or RNA molecule. It is a basic unit of inheritance, comprising part of a chromosome which controls an individual inherited characteristic of an organism and which is capable of mutation as a unit. The gene is regarded as being a particular molecular configuration of the nucleic acids at a particular point on the length of a chromosome. There is considerable evidence to support the belief that genes function by controlling the manufacture of specific proteins in cells.

gene library [Biology] A random collection of DNA fragments that includes all the genetic information of a given species; sometimes called a shotgun collection.

generation time [Nucleonics] The average lapse of time between the creation of a neutron by nuclear fission and a subsequent fission produced by that neutron.

generator [Electricity] A machine that converts mechanical energy into electrical energy. Also known as electric generator; dynamo.

gene splicing [Biology] The enzymatic attachment of one gene or part of a gene to another; also removal of nitrons and splicing of exons during mRNA synthesis.

genetic code [Biology] The code by which inherited characteristics are handed from generation to generation. The code is expressed by the molecular configuration of the chromosome of cells. Chromosomes consist of deoxyribonucleic acid (DNA) and protein, the code-bearing material being the DNA. Four different nitrogenous bases (adenine, cytosine, guanine, and thymine) occur in the nucleotides of DNA, and it appears that the sequence of three of these bases constitutes a unit of the genetic code, in that each sequence of three bases codes for one of the twenty different amino acids that go to make up the enzymes that control the characteristics of a cell. Chromosomes, which almost always exist in the nuclei of cells, tranfer their coded information to the cytoplasm of these cells (where the enzyme proteins are assembled in units called ribosomes) by way of a 'messenger' nucleic acid (ribonucleic acid).

genetic drift [Biology] The tendency, within small interbreeding populations, for heterozygous gene pairs to become homozygous for one allele or the other by chance rather than by selection.

genetic equilibrium [Biology] The situation in which the distribution of alleles in a population is constant in successive generations (unless altered by selection or mutation).

genetic information [Biology] The hereditary information contained in a sequence of nucleotide bases in chromosomal DNA or RNA.

genetic map [Biology] A diagram showing the relative sequence and position of specific genes along a chromosome molecule.

genetics [Biology] The branch of life sciences that deals with the study of heredity, variation, development, and evolution.

genome [Biology] All the genes of an organism or individual.

genotype [Biology] The fundamental hereditary constitution, assortment of genes, of any given organism.

genus [Biology] A rank in taxonomic classification in which closely related species are grouped together.

geocentric [Astronomy] Having the earth as a centre; measured from the centre of the earth.

geochemistry [Geology] The scientific study of the chemical composition of the earth. It includes the study of the abundance of the earth's elements and their isotopes and the distribution of the elements in environments of the earth (lithosphere, atmosphere, biosphere, and hydrosphere).

geodesic [Mathematics] Pertaining to the geometry of curved surfaces. A 'geodesic line', also called a 'geodesic', is the shortest distance between two points on a curved surface.

geology [Geology] The scientific study of the earth's crust.

geomagnetism [Physics] The study of the magnetic field associated with the earth.

geometry [Mathematics] The mathematical study of the properties and relations of lines, surfaces, and solids in space.

geophysics [Geology] The study of the earth and its atmosphere by physical methods.

geothermal energy [Geology] Heat within the earth's interior that is a potential source of energy. Volcanoes, geysers, hot springs, and fumaroles are all sources of geothermal energy.

geotropism [Botany] A growth response toward or away from the earth; the influence of gravity on growth of plants.

germ [Biology] Any microorganism, especially pathogens.

germanium [Chemistry] Ge. Element. At. No. 32. At.Wt. 72.59; placed in group IV A of periodic table. A brittle white metal; m.p. 1210K; b.p. 3103K. Used in semiconductors, alloys and glass.

German silver [Metallurgy] An alloy of copper, zinc, and nickel in varying proportions, approximating to 50% copper, 25% nickel, and 25% zinc.

germicide [Materials] A substance capable of destroying bacteria.

germination [Botany] The beginning or the process of development of a spore or seed.

gerontrology [Physiology] The scientific study of aging processes in biological systems, especially in human beings.

gestation period [Biology] The period in mammals (human) from fertilization to birth.

getter [Engineering] A substance used for removing the last traces of air or other gases in attaining a high vacuum, *e.g.*, magnesium metal is used in thermionic valves; after exhausting and sealing the valve a small amount of magnesium left in the valve is vapourized by heat and combines chemically with any remaining oxygen and nitrogen.

GeV [Electricity] Abbreviation for giga electron-volt, *i.e.*, 10^9 electron-volts.

ghost [Physics] False lines appearing in a line spectrum due to imperfections in the ruling of the diffraction grating used.

ghost image [Electronics] An undesired duplicate image at the right of the desired image on a television screen due to multi-path effect, thus a reflected signal travelling over a longer path arrives slightly later than the desired signal.

giant planets [Astronomy] The planets Jupiter, Saturn, Uranus, and Neptune.

giant star [Astronomy] A class of highly luminous stars having low density, and a diameter 10 to 100 times that of the sun.

gibberellins [Biochemistry] A group of plant hormones, that promote the growth of plant stems and fruit, and have other beneficial effects.

gibbous [Astronomy] The shape of the moon or a planet when it is more than half-phase, but less than full phase.

Gibbs' function *See* free energy.

giga- [Scientific Techniques] G. Prefix denoting a thousand million (10^9).

gigabit [Physics] One billion bits. Also known as billibit.

gigaelectron volt [Physics] A unit of energy equal to 10^9 electron volts or $(1.60210 \pm 0.00007) \times 10^{-10}$ joule.

gilbert [Physics] The c.g.s. unit of magnetomotive force in electromagnetic units, equal to $10/4\pi$ ampere-turns.

gilding [Metallurgy] Covering with a thin layer of metallic gold, often by electrolysis.

gillion [Scientific Techniques] 10^9, one thousand million.

glacial acetic acid [Chemistry] Pure acetic acid; solid crystalline acetic acid below its freezing point (289.7K).

gland [Physiology] A structure which produces a substance essential and vital to the existence of the organism.

glass [Materials] A hard brittle amorphous mixture made by fusing silicates, sometimes borates and phosphates with certain basic oxides and then rapidly cooled to prevent crystallization. Ordinary soda glass is made by melting together sand (silica), sodium carbonate, and lime. Glass for special purposes may contain lead, potassium, barium or other metals in place of the sodium, and boric oxide in place of the silica. *See* crown glass; flint glass; hard glass.

glass enamel [Materials] A finely ground flux of mainly lead borosilicate blended with coloured ceramic pigments. Different varieties give characteristics of acid resistance or alkali resistance.

glass fibre [Materials] Fine glass fibres, usually less than a quarter of a micrometre in diameter, that are woven into a cloth and impregnated with various resins. Owing to their high tensile strength and corrosion resistance these materials are used as acoustic, electrical, or thermal insulating material and as a reinforcing material in laminated plastics.

glassine [Materials] A thin transparent, and very flexible paper obtained by excessive beating of the paper-pulp.

glass, metallic [Materials] Metal alloys having amorphous structure similar to that of silica glass; obtained by rapid cooling of molten alloys; are harder and more resistant to corrosion.

glass, optical *See* optical glass.

glass transition [Engineering] The change, characteristic of many rubbers and other polymers, from a plastic or rubbery to a glassy or brittle state.

glass wool [Materials] A material consisting of very fine glass threads, resembling cotton wool; used for filtering and absorbing corrosive liquids.

Glauber's salt [Chemistry] $Na_2SO_4.-10H_2O$. Crystalline sodium sulphate.

glaze [Engineering] A mixture similar to porcelain enamel used for covering pottery.

globular clusters [Astronomy] Self-contained, approximately spherical clusters of about one hundred thousand stars; some hundred of these clusters are known to be distributed about the centre of the Milky Way, and although they appear to be outside the galaxy, they are believed to be gravitationally associated with it.

globule [Astronomy] A black volume of cosmic dust viewed against the darker background of bright nebulae.

globular protein [Biochemistry] Any protein that is soluble in aqueous medium and in which polypeptide chain is lightly folded in three dimensions to yield a globular shape.

globulins [Biochemistry] Groups of proteins soluble in dilute solutions of mineral salts, such as sodium chloride. They generally contain glycine and coagulate when heated. They occur in many animal and vegetable tissues and fluids; *e.g.*, lactoglobulin in milk, serum globulin in blood, vegetable globulins in seeds. Globulins are the main proteins of antibodies.

glove box [Engineering] A metal box that provides protection to workers who have to manipulate radioactive materials or that enables the manipulation of substances requiring a dust-free, sterile, or inert atomsphere. Manipulation is carried out by means of gloves fitted to parts in the walls of the box.

glow discharge [Electricity] A silent discharge of electricity through a gas at low pressure in an electron tube, characterized by several regions of luminous glow and a voltage drop in the vicinity of the **cathode** discharge.

glucogenic amino acids [Biochemistry] Amino acids whose carbon chain can be metabolically converted into glucose or glycogen.

gluconeogenesis [Biochemistry] The biosynthesis of new carbohydrates from noncarbohydrate precursors.

glucose *See* dextrose.

glucose oxidase [Biochemistry] An enzyme that catalyzes the oxidation of glucose to gluconolactone.

glucosidase [Biochemistry] An enzyme that hydrolyses glucoside.

glucosides [Biochemistry] A class of compounds containing the cyclic form of glucose, in which the hydrogen of the hemiacetal hydroxyl has been replaced by an alkyl or aryl group.

glue [Materials] A general name for adhesives, particularly those made by extracting hides, bones, cartilariges, etc., of animals with water.

gluten [Biochemistry] A mixture of proteins (gliaden and glutelin) found in wheat flour.

glycerides [Chemistry] Esters of glycerol with organic acids. Animal and vegetable fats are mainly composed of triglycerides of fatty acids, such as stearic, palmitic, and oleic, a molecule of such a triglyceride being derived by the combination of one molecule of glycerol with three fatty acid molecules.

glycerin (e) *See* glycerol.

glycogen [Biochemistry] A non reducing, white, amorphous polysaccharide carbohydrate formed from glucose in the liver and other organs of animals, serving as a sugar reserve.

glycogenesis [Biochemistry] The metabolic formation of glycogen from glucose.

glycogenolysis [Biochemistry] The metabolic breakdown of glycogen.

glycolipids [Biochemistry] Complex lipids that consist of compounds of fatty acids with carbohydrates and contain nitrogen but no phosphoric acid; found in brain tissues.

glycolysis [Biochemistry] The metabolic conversion of glucose into lactic acid or pyruvic acid by a series of enzyme-catalyzed reaction that occur in living organisms.

glycoproteins [Biochemistry] Complex proteins that contain at least one carbohydrate group. Also known as glucoproteins.

glycosides [Biochemistry] Ether-type compounds, derived from sugars and hydroxy compounds. If the latter component in a glycoside is a non-sugar, it is called an a glycone. Glycosides in which the sugar is glucose are called glucosides. Glycosides occur widely in plants.

glyptal resins [Materials] A class of synthetic resins obtained by the reaction of polyhydric alcohols with polybasic organic acids or their anhydrides; *e.g.*, glycerol and phthalic anhydride; used chiefly for surface coatings.

gnotobiotics [Biology] The study of germ-free life, especially in experimental conditions in which animals are inoculated with specific strains of microorganisms.

goblet cell [Biology] A cell that secretes mucus onto a surface or into a cavity.

gold [Chemistry] Au. Element. At. No. 79. At.Wt. 196.967; placed in group IB of periodic table. A bright yellow soft metal that is extremely malleable and ductile; m.p. 1337.5K; b.p. 3080K, not corroded by air, is unattacked by most acids, but dissolves in aqua regia. It occurs mainly as the free metal; most compounds are unstable and are easily reduced to gold. Used in jewellery, dentistry, gilding, anodes, alloys, photography and solders.

gold-leaf electroscope *See* electroscope.

gold number [Chemistry] A measure of the amount of protective colloid which must be added to a standard red gold sol mixed with sodium chloride solution to prevent the solution from causing the sol to coagulate, as indicated by a change in colour from red to blue.

Goldschmidt process [Metallurgy] The preparation of metals from their oxides by aluminothermic reduction.

Golgi body [Biology] A complex membranous organelle of eubaryotic cells; functions in secretion to the exterior and formation of the plasma membrane.

Golgi cell [Biology] 1. A nerve cell with long axons. 2. A nerve cell with short axons that branch repeatedly and terminate near the cell body.

gonad [Biology] A gamete-producing gland; an ovary or testis.

goniometer [Engineering] An instrument for the measurement of angles between crystal faces.

Gouy balance [Engineering] Device for the measurement of diamagnetic and paramagnetic susceptibilities of substances.

governor [Engineering] A device for regulating the speed of an engine or machine, on the principle of negative feedback, so that its speed is kept constant under all conditions of leadig.

graduation [Engineering] The marking that indicates the scale of an instrument, *e.g.*, the stem of a thermometer is graduated in degrees.

graft [Biology] The transplantation of an organ or tissue in plants and animals. Grafts may be from one place to another on the same individual (autograft or between different individuals. A graft between individuals of the same

species is called homograft **or** allograft; between individuals **that** are genetically identical **(as** between identical twins) it is an isograft, or syngraft; and between different species it is a heterograft or xenograft.

Graham's Law [Chemistry] The rate of diffusion of a gas is inversely proportional to the square root of its density. Also known as law of gaseous diffusion.

gram-atomic weight [Chemistry] The atomic weight of an element expressed in grams; *i.e.*, the atomic weight on a scale on which the atomic weight of carbon-12 is taken as exact 12.

gram-equivalent [Chemistry] The equivalent weight of an element or compound expressed in grams on a scale in which carbon-12 has an equivalent weight of 12 grams in those compounds in which its formal valency is 4.

gram-ion [Chemistry] The sum of the atomic weights of the atoms in an ion expressed in grams.

gram-molecular volume [Chemistry] The volume occupied by a gram-molecular weight of a chemical in the gaseous state at 273K and 760 millimetres of pressure.

gram-molecular weight [Chemistry] The molecular weight of a compound expressed in gram; *i.e.*, the molecular weight on a scale on which the atomic weight of carbon-12 isotope is taken as exact 12. Also known as gram molecule.

gram-negative [Biology] of bacteria, decolourizing and staining with counterstain when treated with Gram's stain.

Gram's method [Biology] A method of staining and classifying bacteria in which gentian violet is used to stain a bacterial smear. If the bacteria retain the violet dye, after washing with a solution of iodine and potassium iodide in water (Gram's solution) and counterstaining with safranine, they are said to be 'gram positive'. If they do not retain the dye they are said to be 'gram negative.'

gram weight [Mechanics] A unit of force, the pull of the earth on the gram mass; it varies slightly in different localities, depending on the value of g, the acceleration of free fall at the given place. Force expressed in grams, weight=force in dynes divided by the appropriate value of g at the place under consideration. A force of 1 gram weight= approximately 981 dynes. Also known as gram force.

granite [Geology] Any of a class of heterogeneous igneous rocks, containing quartz, feldspar, and other minerals.

graph [Mathematics] A diagram, generally plotted between axes at right angles to each other, showing the relation of one variable quantity to another. [Scientific Techniques] A suffix applied to instrument that automatically record or write down observations; *e.g.*, spectrograph.

graphite [Minerals] A natural allotropic form of carbon, used for pencil leads, in electrical apparatus, and as a lubricant for heavy machinery. Also used as a moderator in nuclear reactors. Also known as blacklead; plumbago.

graphite-moderated reactor [Nucleonics] A nuclear reactor in which graphite is the main moderating material.

graphite resistor [Electricity] A resistor made of carbon for resistance heating and is suitable for any temperature.

graphitic carbon [Metallurgy] Carbon in iron or steel present in the form of graphite.

graticule [Optics] A scale, or network of fine wires, in the eyepiece of a telescope or microscope.

grating *See* diffraction grating.

gravimetric analysis [Chemistry] A branch of quantitative chemical analysis. The amount of a substance present is determined by converting it, by a suitable chemical reaction, into some other substance of known chemical composition, which can be readily isolated, purified, and weighed.

gravitation [Physics] The mutual attraction between all bodies in the universe.

gravitational collapse [Astronomy] The implosion of a star or other astronomical body from an initial size to a size hundreds or thousands times smaller.

gravitational constant [Physics] G. The fundamental constant that appears in Newton's law of gravitation. It has the value $6.664 \times 10^{-11} \, \text{Nm}^2 \, \text{kg}^{-2}$.

gravitational displacement [Physics] The gravitational field strength times the gravitational constant. Also known as gravitational flux density.

gravitational field [Physics] The region in space in which a test body would experience a gravitational force; quantitatively, the gravitational force per unit mass on the body at a particular point.

gravitational force [Physics] The force on a body due to its gravitational attraction on other bodies.

gravitational interaction [Physics] The interaction between all massive particles. It is the weakest of all known interactions, being some 10^{40}th times weaker than the electromagnetic interaction. *Compare* strong interaction and weak interaction.

gravitational law [Mechanics] Every particle in the universe attracts every other particles with a force directly proportional to the product of the masses of the inversely proportional to the square of the distance between the two particles them. Thus, the force of attraction between two massses m_1 and m_2, separated by a distance of d, is given by

$$F = G m_1 m_2 / d^2,$$

where G is the gravitational constant.

gravitational mass [Physics] The mass of a body as it determines the force it experiences in a gravitational field; equal to inertial mass according to the equivalence principle.

gravitational red shift [Physics] A displacement of spectral lines toward the red when the gravitational potential at the observer of the light is greater than at its source.

graviton [Physics] A hypothetical particle or quantum of gravitational energy (*see* gravitation), which not yet been observed. If it exists, it is expected to have zero rest mass and charge, and spin 2.

gravity [Physics] The gravitational attraction at the surface of a planet or other celestial body. As gravity is proportional to the mass of the planet or satellite and inversely proportional to the square of the distance from its centre, the gravity on a planet or satellite in terms of the earth's gravity is given by

$$(d_p/d_e)^2/M_p,$$

where M_p is the mass of the planet in earth masses, and d_p and d_e are the diameters of the planet and earth respectively.

gray [Nucleonics] Gy. The derived SI unit of absorbed dose of ionizing radiation. The energy in joules absorbed by one kilogram of irradiated material.

grease [Materials] A semi-solid lubricant composed of emulsified petroleum oils and soluble hydrocarbon soaps.

great circle [Geology] A circle obtained by cutting a sphere by a plane passing through the centre, *e.g.*, regarding the earth as a sphere, the equator is a great circle, as are all the meridians of longitude. On the earth's surface, an apparent straight line joining any two points is an arc of great circle, *i.e.*, a geodesic.

great year [Astronomy] The period of one complete cycle of the equinoxes around the ecliptic, about 25,800 years. Also known as platonic year.

green [Optics] The hue evoked in an average observer by monochromatic radiation having a wavelength in the approximate range from 492 to 577 nanometres; however the same sensation can be produced in a variety of other ways.

greenhouse [Botany] Glass enclosed and climate controlled structure in which out-of-season plants are cultivated and protected.

greenhouse effect [Physics] 1. The effect produced inside a greenhouse : ultraviolet radiation is admitted to the greenhouse through its glass roof and is absorbed by the contents. The infrared radiation emitted by the contents cannot escape through the glass and the temperature of the interior rises. 2. A similar effect that applies to the whole earth. Shortwave solar radiation passes through the atmosphere but atmospheric carbon dioxide absorbs the longwave radiation emitted by the earth. Thus solar energy is trapped by the earth's atmosphere.

green laser [Optics] A laser which uses mercury and argon to generate a green line at 5225°A, corresponding to the wavelength that is easily transmitted through seawater.

green vitriol [Chemistry] $FeSO_4$. $7H_2O$. Ferrous sulphate crystals. Also known as copperas.

Greenwich apparent time [Astronomy] Local apparent time at the Greenwich meridian.

Greenwich mean time GMT [Astronomy] Mean solar time at the meridian of Greenwich.

grid [Electricity] A metal plate with holes used as storage cell or battery as a conductor and support for the active material. [Electronics] An electrode placed between the cathode and anode of an electron tube, which has one or more openings through which electrons or ion can pass controls the flow of electrons from cathode to anode

grid bias [Electronics] A direct current voltage applied between the cathode and the control grid of a thermionic valve, which determines its operating conditions.

grid control [Electronics] The control of anode current of an

electron tube by control of the control grid potential with respect to the cathode of the tube.

grid current [Electronics] Flow of electrons towards a positive grid, in an electron tube.

grid metal [Metallurgy] An alloy of lead and antimony (5-12%) and sometimes containing small amounts of tin; used for grids in storage batteries.

grid voltage [Electronics] The voltage between the grid and the cathode of an electron tube.

Griess reagent [Chemistry] A reagent used to test for nitrous acid; consists of a solution of sulphanilic acid, α-naphthyla-mine and acetic acid in water.

Grignard reagents [Chemistry] A class of organometallic compounds of magnesium, with the general formula RMgX, where R is an organic group and X a halogen atom (*e.g.* CH_3MgCl, $C_2H_5 MgBr$, etc.). They actually have the structure $R_2Mg.MgCl_2$, and can be made by reacting a haloalkane with magnesium in ether; they are rarely isolated but are extensively used in organic synthesis, when they are made in one reaction mixture. Grignard reagents have a number of reactions that make them useful in organic synthesis. With methanal they give a primary alcohol

$$CH_3MgCl + HCHO \rightarrow CH_3CH_2OH$$

Other aldehydes give a secondary alcohol

$$CH_3CHO + CH_3MgCl \dashrightarrow (CH_3)_2 CHOH$$

With alcohols, hydrocarbons are formed

$$CH_3MgCl + C_2H_5OH \rightarrow C_2H_5 CH_3$$

Water also gives a hydrocarbon

$$CH_3 MgCl + H_2O \rightarrow CH_4$$

grog [Materials] Fired refractory substance that is used in the manufacture of products which must withstand high temperature.

Grossbery units [Biology] A unit for the standardization of lipase.

ground potential [Electricity] Zero potential with respect to ground or earth.

ground state [Physics] The most stable energy state of a nucleus, atom, or molecule. The normal state of an atom when its circum-nuclear electrons move in orbits such that the energy of the atom is a minimum.

ground waves [Physics] Electromagnetic radiation of radio frequencies that travel more or less directly from transmitting aerial to receiving aerial, that is without reflection from the ionosphere. Also known as surface waves.

group [Chemistry] 1. The set of elements that have similar chemical properties and constitute a vertical column in the periodic table. 2. A number of covalently bonded atoms that form part of a compound and have characteristic properties. methanol, for example, consists of the methyl group (CH_3^-) and the hydroxyl group ($-HO$).

grout [Materials] A fluid mixture of cement and water, or mixture of cement, sand, and water.

growth [Physiology] An irreversible increase in size and or weight. It involves cell division and cell expansion through synthesis of new materials, and is closely related to subsequent development processes.

growth hormone [Biochemistry] Any hormone that regulate growth in plants and animals.

G-star [Astronomy] A star of spectral type G with surface temperatures 5500-4200 K for gaints, and 6000-5000K for dwarfs.

guano [Materials] Phosphate and nitrogen-rich matter, partially decomposd excrement of seabirds; used as fertilizer.

gum arabic [Materials] A water soluble, yellowish gum obtained from certain varieties of acacia. Used in food and pharmaceutical products and as an adhesive. Also known as gum acacia.

gun-cotton *See* nitrocellulose.

gun-metal [Metallurgy] A class of copper alloys containing about 90% copper, 8%-10% tin, and 2% to 4% zinc.

Gunn effect [Electronics] Development of rapidly fluctuating current in a small block of semiconductor when a constant voltage above a critical value is applied to contacts on opposite faces.

gunpowder [Materials] An explosive consisting of a mixture of potassium nitrate, sulphur, and charcoal.

gutta-percha [Materials] A material very similar to rubber, obtained from the latex of certain Malayan trees; chemically, it consists of the trans-form of polyisoprene. It is thermoplastic at about 343K and resembles unvulcanized rubber.

guttation [Botany] Loss of water as liquid from plant surface. The water droplets appear on leaves as a result of root pressure.

G-value [Nucleonics] The number of molecules produced decomposed for each 100 electron volts absorbed by a substance from ionizing radiation.

gynecology [Medicine] The branch of medical science dealing with diseases of women, especially those affecting the sex organs.

gypsum [Minerals] $CaSO_4.2H_2O$, Natural hydrated calcium sulphate, that loses three quarters of its water of crystallization when heated to 393K; and becomes Plaster of Paris.

gyration [Mechanics] Motion around a fixed axis or centre.

gyro-compass [Physics] A compass that does not make use of magnetism, and is therefore not effected by magnetic storms, etc.; it consists of a universally mounted spinning wheel that has a rigidity of direction of axis and plane of rotation relative to space; the rotation being electrically maintained. Also known as gyroscopic compass.

gyromagnetic effect [Physics] The rotation induced in a body by a change in its magnetization, or the magnetization resulting from a rotation.

gyromagnetic ratio [Physics] γ. The ratio of the angular momentum of an atomic system to its magnetic moment. The inverse of the gyromagnetic ratio is called the *magnetomechanical ratio*.

gyrostablizer [Engineering] Gyroscope used to stablize ships and air planes.

gyrotron [Electronics] A device which detects the motion of a system by measuring the phase distortion that occurs when a vibrating tuning-fork is moved.

gyttja [Geology] A fresh water anaerobic mud containing an abundance of organic matter; capable of supporting aerobic life.

Haber process [Chemistry] An industrial process for producing ammonia by reaction of nitrogen with hydrogen :

$$N_2 + 3H_2 \longleftrightarrow 2NH_3$$

The reaction is reversible and exothermic, so that a high yield of ammonia is favoured by low temperature (*see* Le Chatelier's principle). However, the rate of reaction would be too slow for equilibrium to be reached at normal temperatures, so an optimum temperature of about 720K is used, with a catalyst of iron containing potassium and aluminium oxide promoters. The higher the pressure the greater the yield, although there are technical difficulties in using very high pressures. A pressure of about 250 atmospheres is commonly employed.

habitat [Ecol] The part of physical environment in which animals and plants live and grow.

hadron [Physics] Any elementary particles which has strong interaction. Hadrons include the baryons and mesons as well as the recently discovered psi particle.

haem [Biology] An iron-containing molecule that binds with proteins as a cofactor or prosthetic group to form the *haemoproteins*. These are haemoglobin, myoglobin, and the cytochromes. Essentially, haem comprises a porphyrin with its four nitrogen atoms holding the iron(II) atom as a chelate. This iron can reversibly bind oxygen (as in haemoglobin and myoglobin) or (as in the cytochromes) conduct electrons by conversion between the iron(II) and iron (III) series. Also spelled heme.

haematin [Biology] An iron-containing constituent of haemoglobin.

haematinic [Biology] Any substance which is required for the production of red blood cell and its constituents.

haematite [Minerals] Fe_2O_3. Natural ferric oxide. A valuable ore of iron.

haematoloy [Biology] The science dealing with composition, formation, diseases, and functions of blood. Also spelled hematology.

haemocynin [Biology] A blue copper-containing blood pigment, and is the second most abundant blood pigment after haemoglobin and function similarly in acting as oxygen-carrier in the blood. Also spelled hemocynin.

haemocyte [Biology] A blood cell especially in invertebrates. Also spelled hemocyte.

haemoglobin [Biology] One of a group of globular proteins occurring widely in animals as oxygen carriers in blood. Vertebrate haemoglobin comprises two pairs of polypeptide chains (forming the *globin* protein) with each chain folded to provide a bindig site for a haem group. Each of the four haem groups binds one oxygen molecule to form *oxyhaemoglobin*. Dissociation occurs in oxygen-depleted tissues; oxygen is released and haemoglobin is reformed. The haem groups also bind other inorganic molecules, including carbon monoxide. This binds more strongly than oxygen and competes with it (hence its toxicity). In vertebrates, haemoglobin is contained in the red blood cells (erythrocytes). Also spelled hemoglobin.

haemolysis [Biology] The release of haemoglobin from red bloodcells due to rupture of cell membrance. Also spelled hemolysis.

hafnium [Chemistry] Hf. Element. At. No. 72. At.Wt. 178.49; placed in group IV B of periodic table. A rare metal, melting point 2500K; boiling point above 4875K, used in the manufacture of tungsten filaments and as a neutron absorber in nuclear reactors.

hair salt [Minerals] $Al_2(SO_4)_3$. $18H_2O$. Natural aluminium sulphate.

half cell [Electricity] An electrode in contact with a solution of ions, forming part of a cell. Various types of half cell exist, the simplest consisting of a metal electrode immersed in a solution of metal ions. The commonest gas half cell is the hydrogen half cell. Half cells can also be formed by a metal in contact with an insoluble salt or oxide and a solution. The calomel half cell is an example of this. Half cells are commonly referred to as electrodes.

half-life [Nucleonics] The time taken for the activity of a radioactive isotope to decay to half of its original value, that is for half of the atoms present to disintegrate. Half-lives vary from fractions of a second to millions of years. Symbol $t_{1/2}$.

half-period zones [Optics] The division of a wave front into elements of area or zones such that secondary wavelets (see Huygens' construction) reaching a given point ahead of the wave from adjacent zones differ in phase by half a period, or π. This construction is used in theoretical investigations of Fresnel diffraction in simple cases.

half-wave [Electricity] Pertaining to half of one cycle of a wave.

half-wave amplifier [Electronics] A magnetic amplifier whose total induced voltage has a frequency equal to the power supply frequency.

half-wave antenna [Electronics] An antenna whose electrical length is half the wavelength being transmitted or received.

half-wave plate [Optics] A plate of double refracting material cut parallel to the optic axis and of such a thickness that a phase difference of π or $180°$ is introduced between the ordinary ray and the extraordinary ray for light of a particular wavelength. The half-wave plate is chiefly used to alter plane of vibration of plane-polarized light.

halide [Chemistry] A binary compound of one of the halogen elements (fluorine, chlorine,

bromine, or iodine); a salt of the hydride of one of these elements.

Hall effect [Electricity] If an electric current flows in a wire placed in a strong transverse magnetic field, a potential difference is developed across the wire, at right angles to both the magnetic field and the wire.

Halley's comet [Astronomy] A bright comet that takes about 76 years to orbit the sun; last seen in 1986 it is next due to be visible in 2062.

Hall mobility [Electronics] The product of the conductivity and the Hall constant for a conductor or semiconductor, or, it is the velocity of the carriers under the influence of an electric field of 1 volt per metre. Also known as drift mobility.

halo [Optics] A luminous ring sometimes observed surrounding the sun or the moon; caused by the refraction of light by ice crystals in the atmosphere.

haloforms [Chemistry] The four compounds with formula CHX_3, where X is a halogen atom. They are *chloroform* ($CHCl_3$), and by analogy, *fluoroform* (CHF_3), *bromoform* ($CHBr_3$), and *iodoform* (CHI_3).

halogenation [Chemistry] The introduction of halogen atoms into a compound by addition or substitution.

halogens [Chemistry] A group of elements in the periodic table (group VIIB): fluorine (F), chlorine (Cl), bromine (Br), iodine (I), and astatine (At). All have a characteristic electron configuration with outer ns^2np^5 electrons.

halophyte [Botany] A plant that grows in soils with a high concentration of salts, as found in salt marshes, *e.g.*, species of spartina.

Hamiltonian [Physics] H. A function used to express the energy of a system in terms of its momentum and positional coordinates. In simple cases this is the sum of its kinetic and potential energies.

hand rule [Physics] The rule that, when grasping a conductor in the right hand with the thumb pointing in the direction of the current, the fingers will then open in the direction of lines of flux.

haploid [Biology] Having a set of single (unpaired) chromosomes; *e.g.*, gametes.

hard acid [Chemistry] A Lewis acid of low polarizability, small size, and high positive oxidation state; it does not have easily excitable outer electrons, *e.g.*, H^+, Li^+, Mg^{2+}, Al^{3+}.

hard base [Chemistry] A Lewis base having high polarizability and low electronegativity, it is easily oxidised, or possess lowlying empty orbitals, *e.g.*, OH^-, H_2O, F, CN^-, OCH_3^-.

hardened steel [Metallurgy] Steel hardened by quenching from high temperatures.

hard glass [Materials] A variety of glass containing potassium and high silica content; used for making glassware. Also known as Bohemian glass.

hard iron [Metallurgy] Iron or steel which is not easily magnetized by induction.

hard radiation [Physics] Radiation whose particles or photons have a high energy, and thus can penetrate through a large number of materials including metals.

hardware [Computers] The physical, tangible, and permanent components of a computer or data-processing system.

hardwater [Chemistry] Water that does not form an immediate lather with soap, owing to the presence of calcium, magnesium, and iron compounds dissolved in the water. The addition of soap produces an insoluble scum consisting of salts of these metals with the fatty acids of the soap, until no more is left in solution. Hardness is divided into two types : 1. Temporary hardness, due to hydrogen carbontes (bicarbonates) of the metals. These enter the water by the passage of the water, containing dissolved carbon dioxide, over solid carbonates (chalk or limestone deposits, etc.). Such hardness is removed by boiling, the soluble bicarbonates being decomposed into the insoluble carbonates, carbon dioxide and water. 2. Permanent hardness, due to sulphates of the metals. This is destroyed by the addition of washing-soda, sodium carbonate, which precipitates the insoluble carbonates. All hardness may be destroyed by passing the water through zeolites.

harmonic [Physics] An oscillation having a frequency that is a simple multiple of a *fundamental* sinusoidal oscillation. The fundamental frequency of a sinusoidal oscillation is usually called the *first harmonic*. The *second harmonic* has a frequency twice that of the fundamental and so on.

harmonic antenna [Electronics] An antenna whose electrical length is an integral multiple of a half-wavelength at the operating frequency of the receiver or transmitter.

harmonic oscillator [Electronics] An oscillator circuit whose output voltage is a sine-wave function of time. Also known as sinusoidal oscillator; sine-wave oscillator.

harmonic motion *See* simple harmonic motion.

harmonic series [Mathematics] A series in which the reciprocals of the terms are in arithmetical progression, *e.g.*, $1 + \frac{1}{2} + \frac{1}{3} + \frac{1}{4} \dots$

hartree [Physics] Atomic unit of energy equal to e^2/l_o, where e is the charge of an electron and l_o is the atomic unit of length. It is equal to 4.360×10^{-18} joule or 27.21 eV.

hazardous material [Materials] Any material which, if improperly handled, can be damaging to health and well-being of human being. These may be poisons or toxic agents, corrosive chemicals, flammable materials, explosives, and radioactive substances.

heat [Physics] Energy possessed by a substance in the form of kinetic energy of atomic or molecular translation, rotation, or vibration. The heat contained by a body is the product of its mass, its temperature, and its specific heat capacity; it is expressed in joules (SI units), or calories (c.g.s. units).

heat capacity [Physics] C. When the temperature of a system is increased by an amount dT as a consequence of the addition of a small quantity of heat dQ, the quantity dQ/dT is called the heat capacity. In practice, it is the heat in joules required to raise the temperature of a body or system by 1 K.

heat exchanger [Engineering] Any device that transfers heat from one fluid to another without allowing the fluids to come into contact with each other.

heat flux [Physics] The amount of heat transferred across a surface of unit area in unit time.

heat of adsorption [Chemistry] The increase in enthalpy when 1 mole of a substance is adsorbed upon another at constant pressure.

heat of combustion [Chemistry]

The amount of heat evolved by 1 mole of a substance when it is burned in oxygen.

heat of cooling [Chemistry] Increase in enthalpy during cooling of a system at constant pressure, resulting from an internal change such as allotropic transformation.

heat of decomposition [Chemistry] The change in enthaply during the decomposition of 1 mole of a compound into its elements at constant pressure.

heat of dilution [Chemistry] The increase in enthalpy accompanying the addition of a specified amount of solvent to a solution at constant pressure.

heat of dissociation [Chemistry] The increase in enthalpy when molecules break apart or valence linkage rupture, at constant pressure.

heat of formation [Chemistry] The increase in enthalpy resulting from the formation of 1 mole of a substance from its elements at constant pressure.

heat of neutralization [Chemistry] The change in enthalpy when 1 mole of an acid or base is exactly neutralized. For all strong acids or bases, its value is approximately 57500 joules (13700 calories).

heat of reaction [Chemistry] The quantity of heat given out or absorbed in a chemical reaction, usually per mole of reacting substance. *See* Hess's law.

heat of solution [Chemistry] The change in enthalpy when 1 mole of a substance is dissolved in a large volume of water or any other specified solvent at constant pressure.

heat pump [Engineering] A machine for extracting heat from a fluid that is at a slightly higher temperature than its surroundings. A heat pump can be used to raise the temperature of 'low temperature heat', so that it can be usefully employed.

heat radiation *See* infrared radiation.

heat shield [Engineering] The shielding surface that protects a spacecraft from excessive heating on re-entering the earth's atmosphere.

heavy hydrogen *See* deuterium.

heavy water [Nucleonics] D_2O. Deuterium oxide. Water in which the hydrogen is replaced by deuterium; present in natural water to the extent of about 1 part in 5000. The term is also used when referring to water that contains appreciably more D_2O or HDO than natural water. Heavy water is used as a moderator in some nuclear reactors.

heavy water reactor [Nucleonics] A nuclear reactor in which heavy water is used as moderator and sometimes also as coolant.

hectare [Mechanics] Metric unit of area; 10000 square meters, 2.471 05 acres.

hecto [Scientific Techniques] Symbol h. A prefix used in the metric system to denote 100 times. For example, 100 coulombs = 1 hectocoulomb (hC).

Heisenberg's uncertainty principle See uncertainty principle.

heliocentric [Astronomy] Measured from the centre of the sun.

helium [Chemistry] He. Element. At. No. 2. At.Wt. 4.0026. A noble gas that is placed in zero group of the periodic table; occurs in certain natural gases occluded in radioactive ores (e.g., monozite, pitchblende), and in the atmosphere (1 part in 200000); non-inflammable and very light, it is valuable for filling airships and balloons.

helium I. [Chemistry] The phase of liquid helium which is stable at temperatures above the lambda point (2.186K) and has the properties of normal liquid, except low viscosity; also known as quantum fluid.

helium II. [Chemistry] The phase of liquid helium on cooling helium I below its lambda point; stable between absolute zero and about 2.186K; has remarkable property of superfluidity, extremely high thermal conductivity and viscosity approaching almost zero.

hematology See haematology.

heme See haem.

hemicelluloses [Biochemistry] A class of polysaccharides (mainly pentosans) that occur in cell walls of plants associated with cellulose and lignin.

hemihydrate [Chemistry] A compound that has one molecule of water of crystallization for every two molecules of the compound; e.g., $2CaSO_4.H_2O$.

hemocyte See haemocyte.

hemoglobin See haemoglobin.

hemolysis See haemolysis.

henry [Physics] H. The derived SI unit of self-and mutual inductance (see self-induction; mutual induction). An inductance in a closed circuit such that a rate of change of current of 1 ampere per second produces an induced E.M.F. of 1 volt.

Henry's law [Chemistry] The mass of a gas dissolved by a definite volume of liquid at constant temperature is directly proportional to the pressure. From this it, follows that the volume of a gas absorbed by a given volume of liquid at constant temperature is independent of the pressure. The law holds only for sparingly soluble gases at low pressure.

heparin [Biochemistry] A substance that prevents blood clotting by neutralizing prothrombin and stopping the action of thrombin; occurs in a variety of tissues, most abundantly in liver.

hepatectomy [Medicine] Surgical removal of the liver or a part of it.

hepatic [Physiology] Pertaining to the liver.

herbaceous [Botany] Pertaining to, or having the characteristics of a herb; nonwoody.

herbarium [Botany] A collection of plant specimens, pressed and mounted on paper, or placed in liquid preservatives, and systematically arranged.

herbicides [Materials] Substances that kill plants or inhibit their growth. Selective herbicides affect only particular plant types, making it possible to attack weeds growing among cultivated plants.

herbicolous [Biology] Living on herbs.

herbivore [Biology] An animal that eats only vegitation.

hereditary [Biology] Of or pertaining to heredity or inheritance.

hereditary disease [Medicine] A genetically determined illness transmitted from parent to child.

heredity [Biology] The sum of genetic endowment obtained from parents.

hermaphroditism [Biology] A state characterized by the presence of both male and female sex organs in the same organism.

hernia [Medicine] Abnormal protrusion of an organ or other body part through its containing wall.

hertz [Physics] H_z. The derived SI units of frequency defined as the frequency of periodic phenomenon of which the periodic time is one second; equal to 1 cycle per second. 1 kilohertz (kHz) = 10^3 cycles per second.

Hertz effect [Electronics] Increase in the length of a spark induced across a spark gap when the gap is irradiated with ultraviolet light.

hertzian waves [Physics] Electromagnetic radiation covering a range of frequency from above 3×10^{10} hertz to 1.5×10^5 hertz, corresponding to long radio waves of 2000 meters. Also known as wireless waves; radio waves.

Hess's law [Chemistry] If a chemical reaction is carried out in stages, the algebraic sum of the amounts of heat evolved in the separate stages is equal to the total amount of heat evolved when the reaction occurs directly; a consequence of the law of conservation of energy as applied to thermochemistry. Also known as law of constant heat summation.

hetero [Chemistry] Chemical prefix denoting different.

hetero atom [Chemistry] An odd atom in the ring of a heterocyclic compound. For instance, nitrogen is the hetero atom in pyridine.

heteroazeotrope [Chemistry] Liquid mixture that is not completely miscible in all proportions in the liquid phase, yet does not form an azeotrope.

heteroblastic [Biology] Arising from different tissues or germ layers, in referring to similar organs in different species.

heterocyclic compounds [Chemistry] Organic compounds containing a ring structure of atoms in the molecule, the ring including atoms of elements other than carbon, *e.g.*, pyridine, C_5H_5N, having a molecule consisting of 5 carbon atoms and one nitrogen atom in a closed ring, with a hydrogen atom attached to each carbon atom.

heterodyne [Physics] A beat effect (*see* beats) produced by superimposing two waves of different frequency; used extensively in radio receivers in which the received wave is combined with a wave generted within the receiver. The two combining waves produce an intermediate frequency, which is amplified and then demodulated. *See* superheterodyne.

heterogametic sex [Biology] The sex with dissimilar sex chromosomes, one (the *y*-chromosome) being shorter than the other (the *x*-chromosome).

heterogamy [Biology] Sexual reproduction involving the union of two gametes which differ in size and structure, *e.g.*, egg and sperm.

heterogeneous [Chemistry] Not of a uniform compostion; showing different properties in different portions.

heterogeneous chemical reaction [Chemistry] A chemical reaction in which reactants and products are present in different phases.

heterogeneous radiation [Physics] Radiation having a number of different frequencies, different particles, or different particle energies.

heterograft *See* graft.

heterolytic fission [Chemistry] The breaking of a chemical bond so that oppositely charged ions are formed, *e.g.*, $HCl = H^+ + Cl^-$. Compare homolytic fission.

heterotroph [Biology] An organism which cannot synthesize its own food from inorganic materials and therefore must live either at the expense of autotrophs or upon breakdown of decaying organic matter.

heterozygous [Biology] Possessing two different allele for a given character in the corresponding loci of homologous chromosomes.

heuristic [Matematics] Denoting a method of solving a problem for which no alogrithm exists. It involves trial and error, as in iteration.

hexagonal system [Crystallography] A crystal system that has three equal axes is perpendicular to the other three.

high energy compound [Chemistry] A compound that undergoes a large decrease in free energy upon hydrolysis under standard conditions.

higher order language [Computers] HOL A computer language (such as FORTAN or LISP) requiring fewer statement than machine language and usually substintially easier to use and read.

high fidility [Electronics] Audio reproduction that closely approximates the sound of the original performance. Also known as hi-fi.

high frequency [Physics] HF. Radio frequencies between 3000 and 30000 kilohertz.

high frequency welding [Engineering] A method of welding thermoplastic materials in which the heat required to fuse the surfaces toge 'her is generated by the application of radio frequency electromagnetic radiation. Also known as radio frequency welding.

high-pass filter [Electronics] A filter that transmits all frequencies above a given cutoff frequecny and attenuates all others.

high Q [Electronics] A characteristic so that a component has a high ratio of reactance to effective resistance.

high-speed steel [Metallurgy] A very hard steel containing 12%-22% tungsten, with chromium vanadium, molybdenum and small amounts of other elements which remain hard and tough at red heat; used for tools which remain hard even at red heat.

high tension *See* high voltage.

Hill reaction [Biochemistry] The oxygen evolution and photoreduction of an artificial electron acceptor by a chloroplast preparation in the absence of carbon dioxide.

histochemistry [Biology] The study of the distribution of the chemical constituents of tissues by means of their chemical reactions.

histology [Biology] The study of the structure of the tissues and organs of living being.

histolysis [Biology] Disintegration of organic tissue.

histones [Biochemistry] The group of five basic proteins associated with the chromosomes of all eukaryotic cells.

hodoscope [Electronics] An apparatus for tracing the path of a charged particle (usually a cosmic ray particle).

hole [Physics] A vacant electron energy state near the top of an energy band in a solid. Also known as electron hole.

hole conduction [Electronics] Conduction in a semicon ductor when electrons move into holes under the influence of applied voltage and thereby create new holes.

holmium [Chemistry] Ho. Element. At. No. 67. At. Wt.164. 93. m.p. 174K; b.p. 2993K, a lanthanide placed in group III B of periodic table.

holocellulose [Biochemistry] All the carbohydrate components of a cellulose raw material.

holoenzyme [Biochemistry] A catalytically active complex made up on an apoenzyme and a coenzyme. The former is responsible for the specificity of the holoenzyme while the latter determines the nature of reaction.

hologram [Engineering] The intermediate photographic record that contains the information for reproducing a three-dimensional image by holography.

holography [Physics] A technique of reproducting three-dimensional images without cameras or lenses using photographic film and coherent light. A beam of cohrent light from a laser is split in two by a semitransparent mirror, so that one beam can be diffracted by the object to be reproduced onto a photographic film or plate. The other beam falls directly onto the film or plate. The two beams form interference patterns on the plate thus forming the hologram. The fine speckled pattern on the plate contains information characteristic of the wave fronts themselves, rather than of the light intensities as in normal photography. To reproduce the image the hologram is illuminated by coherent light (usually of the same wavelength as the original beam). The hologram acts as a diffraction grating and produces two sets of diffracted waves, which form equal angles with the plate. One set of waves forms a real image on a screen or photographic plate, while the other forms a three-dimensional virtual image.

holophytic [Biology] Pertaining to the type of nutrition in which complex organic molecules are synthesized from inorganic molecules using light energy.

holozoic [Biology] Designating organisms that feed on other organisms or solid organic mater.

homeomorphous [Chemistry] Having the same crystalline form but different chemical composition.

homeopathy [Medicine] A system of medicine that treats disease by administering to the patient small dose of drugs which produce the signs and symptoms of the disease in a healthy person.

homocentric [Optics] Pertaining to rays which have the same focal point, or which are parallel. Also known as stigmatic.

homocyclic compounds [Chemistry] Organic compounds the molecules of which contain a ring structure of atoms of the same kind (usually carbon), *e.g.*, benzene, C_6H_6.

homodesmic [Crystallography] Pertaining to a crystal, having atoms bonded in a single way.

homogeneous chemical reaction [Chemistry] A chemical reaction in which reactants, catalyst, and products are in the same phase.

homogeneous radiation [Physics] Radiation having an extremely narrow band of frequenices, or a beam of particles each having same energy.

homograft *See* graft.

homologous pair [Physics] In Spectrographic analysis an homologous pair consists of the particular spectral line (*See* line spectrum) utilized in the determination of the concentration of an element and an internal standard line, such that the ratio of the intensities of the radiation producing the lines remain unchanged with variations in the conditions of excitation.

homologous series [Chemistry] A series of chemical compounds of uniform chemical type, showing a regular gradation in physical properties, and capable of being represented by a general molecular formula, the molecule of each member of the series differing from the preceding one by a definite constant group of atoms, *e.g.*, the alkanes are represented by the general formula C_nH_{2n+2}.

homologues [Chemistry] Member of the same homologous series; *e.g.*, methane, CH_4, and ethane, C_2H_6.

homology [Chemistry] The state in series of organic compounds that differ from each other by a CH_2 such as methane series C_nH_{2n-2}, in which there is a similarity between compounds in the series and a gradation in the properties.

homolysis [Chemistry] The breaking of a chemical bond so that neutral atoms or radicals are formed for example, $A : B \rightarrow A \cdot + \cdot B$ Also known as homolytic fission

homomorphs [Chemistry] Chemical molecules that are similar in size and shape, but not necessarily having any other characteristics in common.

homopolar [Electricity] Having equal distribution of charge, *i.e.*, electrically symmetical.

homopolar bond [Chemistry] A covalent bond whose total dipole moment is zero, *e.g.*, H-H bond in H_2 molecule.

Hooke's law [Mechanics] Within the elastic limit, a strain is proportional to the strees producing it.*See* elasticity; elastic modulus.

horizontal component [Physics] B_o.The horizontal component of the earth's magnetic field. *See* magnetism, terrestrial.

horizontal polarization [Electronics] Transmission of radio waves so that the electric lines of force are horizontal, while the magnetic lines of force are vertical.

horizontal sweep [Electronics] The sweep of electron beam from left to right across the screen of a cathode-ray tube.

hormone [Biochemistry] A chemical messanger liberated by a certain type of gland (endocrine gland) and transported into the blood to a specific organ, where it acts to control growth, metabolism, sexual reproduction, and other body processes. Hormones may be steriods, polypeptides, or amines. They are recognised by specific molecules or receptors in the cells of target organs. These receptors are usually proteins located at the membrane or in the cytoplasm. The hormones exert their effects on enzymes or on nucleic acids.

hormone receptor [Biology] A specific hormone-binding site on the cell surface or within the cell.

horn silver [Minerals]   AgCl. Natural silver chloride, an important ore of silver. Also known as cerargyrite, chlorargyrite.

horology [Scientific Techniques] The science of measuring time and the constructing instruments for time-measurement.

horsepower [Physics] h.p. British unit of power; work done at the rate of 550 foot-pounds per second. 1 h.p. = 745.7 watts.

hot atom [Nucleonics] An atom that has very high kinetic or internal energy as a result of nuclear process such as beta decay or neutron capture.

hot cathode [Electronics] A cathode in which electron or ionemission is produced by heat. Also known as thermionic cathode.

hot-wire instruments [Electricity] An electrical measuring instrument (ammeter or voltmeter) that depends upon the expansion, or change in resistance, of a wire heated by the passage of an electric current.

hubble [Astronomy] A unit of astronomical distance equal to 10^9 light years or 9.4605×10^{24} metres.

Hubble effect *See* red shift.

hue [Optics] The characteristic of a colour such as red, green, yellow that is determined by its wavelength.

hum [Electronics] An electrical disturbance occurring at the power supply frequency or its harmonics.

humidification [Engineering] The process of increasing the amount of water vapour in a gas.

humidifier [Engineering] A device for supplying moisture to the air and for maintaining desired humidity conditions.

humidistat [Engineering] An instrument that measures and controls relative humidity. Also known as hydrostat.

humidity [Physics] A measure of the water vapour present in the air. It may be expressed in terms of relative humidity, or the absolute humidity.

humus [Materials] The dark brown colloidal matter present in soil as a result of animal, microbial and vegetable decomposition. It is an important source of nutrients for plants.

Hund's rules [Physics] Two rules giving the order in energy of the atomic states formed by equivalent electrons : I Electrons tend to avoid as far as possible being in the same orbital. II Electrons in the different orbitals of the same energy have parallel spins.

Huygen's construction [Optics] Each point of a wave front may be regarded as a new source of sec-ondary wavelets. Knowing the position of the wave front at any given time, the construction en-ables its position to be deter-mined at any subsequent time.

Huygen's principle [Optics] I. The resultant displacement at any point due to the superposition of any system of waves is equal to the sum of the displacement of the individual waves at that point. This principle forms the basis of the theory of light inter-ference. II. The principle that each point on a light wavefront may be regarded as a source of secondary waves, the envelope of these secondary waves determin-ing the position of the wavefront at a later time.

hydrate [Chemistry] A compound containing combined water. It is generally applied to salts con-taining water of crystallization.

hydration [Chemistry] The process in which water molecules be-come attached to ions or mole-cules.

hydraulic cement [Materials] Cement that hardens in contact with water.

hydraulic press [Mechanics] An application of Pascal's law; a device whereby a force applied by a piston over a small area is transmitted through water to another piston having a large area; by this means very great forces may be obtained.

hydraulics [Mechanics] The prac-tical application of hydrodynam-ics to engineering.

hydrazide [Chemistry] An organic compound of general formula $RCO-NH-NH_2$.

hydrazo group [Chemistry] The bivalent radical $-HN=NH-$.

hydride [Chemistry] A binary compound with hydrogen.

hydroboration [Chemistry] The process of making organoboranes by the additing of a compound with B-H bond to an unsaturated hydrocarbon.

hydrocarbons [Chemistry] Organic compounds that contain only carbon and hydrogen. They are classified as either aliphatic or aromatic compounds (or a com-bination of both). Hydrocarbons may be either saturated or un-saturated compounds.

hydrochloride [Chemistry] A salt formed when an organic base (*i.g.* an alkaloid) combines with hydrochloric acid. The salt so formed is usually more soluble than the base.

hydrocooling [Engineering] Re-moving heat from freshly har vested fruits and vegetables by immersion in ice water, then while still wet, subjected to vacuum. The evaporation of water chills the fruits and vege-tables.

hydrocracking [Chemistry] Crack-ing of petroleum or its products in the presence of hydrogen and a suitable catalyst.

hydrodistillation [Chemistry] Removal of essential oils from plant components by using high temperature steam.

hydrodynamics [Mechanics] The mathematical study of the forces, energy, and pressure of liquids in motion.

hydroelectricity [Electricity] Electrical energy obtained from water power, the latter being used to drive a dynamo.

hydroforming [Chemistry] The use of hydrogen at high temperature and pressure, in presence of catalyst to convert olefinic hydrocarbons to branched chain paraffins to yield high-octane gasoline.

hydrogasification [Chemistry] Production of gaseous or liquid fuels by direct addition of hydrogen to coal.

hydrogel [Chemistry] A colloidal gel in which water is the dispersion medium for example, coagulated silicic acid.

hydrogen [Chemistry] H. The lightest element. At. No. 1. At. Wt. 1.008; m.p. 14K, b.p. 20.2 K; placed in group I A of periodic table. A colourless, odourless, tasteless gas, which forms diatomic molecules. It is used in the oxyhydrogen burner, as a reducing agent, in the manufacture of synthetic ammonia and of synthetic oil, and for hydrogenation of oils. Three isotopes of hydrogen are known; the two 'heavy' isotopes, deuterium and tritium, are of great importance in nuclear physics.

hydrogenase [Biochemistry] Enzyme that catalyzes the oxidation of hydrogen.

hydrogenation [Chemistry] Subjecting to the chemical action of, or causing to combine with hydrogen.

hydrogenation of coal [Chemistry] The manufacture of artificial mineral oil from coal by the action of hydrogen. This depends on causing the carbon in coal to combine with hydrogen to form hydrocarbons. See Bergius process; Fischer-Tropsch Process.

hydrogenation of oils [Chemistry] Artificial hardening of liquid animal and vegetable oils by the action of hydrogen. Liquid fats and oils contain a high percentage of liquid triolein, $C_{57}H_{110}O_6$ by the action of hydrogen in the presence of finely divided nickel as catalyst; the result being a hard fat of higher melting point .

hydrogen 'bomb [Nucleonics & Nuclear physics] A device in which heavy hydrogen nuclei, under intense heat and pressure, undergo an uncontrolled, selfsustaining fusion reaction to produce an explosion.

hydrogen bond [Chemistry] A weak electrostatic chemical bond that forms between covalently bonded (see valence, eletronic theory of) hydrogen atoms and a strongly electronegative atom with a lone pair of electrons (e.g. oxygen, nitrogen, fluorine). The.hydrogen bond is of enormous importance in bio-chemical processes, especially the N-H.....N bond, which enables complex proteins and nucleic acids to be built up.

hydrogen half cell [Chemistry] A type of *half cell in which a metal foil is immersed in a solution of hydrogen ions and hydrogen gas is bubbled over the foil. The standard hydrogen electrode, used in measuring standard *electrode potentials, uses a platinum foil with a 1.0 M solution of hydrogen ions, the gas at 1 atmosphere pressure, and a temperature of 298K. It is written Pt(s) $lH_2(g)$, $H^+(aq)$, the effective reaction being $H_2 \rightarrow 2H^+ +2e^-$.

hydrogen ion [Chemistry] A positively charged hydrogen atom; a proton. The general properties of acids in solution are due to the presence of hydrogen ion.

hydrogen ion concentration [Chemistry] The number of grams of hydrogen ions per litre of solution. It is useful as a measure of the acidity of a solution and in this context is usually expressed in terms of pH which is $\log_{10}1/[H^+]$, where $[H^+]$ is the hydrogen ion concentration. As pure water at ordinary temperature dissociates slightly into hydrogen ions and hydroxyl ions.($H_2O \rightleftharpoons H^+ + OH^-$), the concentration of each type of ion being 10^{-7} mole per litre, the pH of pure water will be $\log_{10}1/10^{-7}=7$; this figure is accordingly taken to represent neutrality on the pH scale. If acid is added to water its hydrogen ion concentration will increase and its pH will therefore decrease.

Thus a pH below 7 indicates acidity and similarly a pH in excess of 7 indicates alkalinity.

hydrogen maser [Physics] A maser in which hydrogen gas is used for providing an output signal with a high degree of stability and spectral purity.

hydrogenolysis [Chemistry] A reaction in which hydrogen causes a chemical change that is similar to role of water in hydrolysis.

hydrolases [Biochemistry] A class of enzymes that control hydrolysis, *e.g.*, esterases, proteases.

hydrolith [Materials] C_aH_2. Calcium hydride. A substance that is decomposed by water and used for the production of hydrogen, according to the equation. $CaH_2+2H_2O \longrightarrow Ca(OH)_2+2H_2$. [Geology] A rock that is free of organic material.

hydrology [Geology] The study of water with reference to its occurrence and properties in the hydrosphere and atomsphere.

hydrolysis [Chemistry] The chemical reaction of a substance by water, the water itself being decomposed; the reaction is of the type $AX + H_2O \rightarrow A(OH) + HX$. Salts of weak acids, weak bases, or both, are partially hydrolyzed in solution; esters may be hydrolyzed to form an alcohol and acid.

hydrolytic enzyme [Biochemistry] A catalyst that acts like a hydrolase.

hydrolytic process [Chemistry] A reaction in which water effects a double decomposition with another compound, hydrogen going to one compound and hydroxyl part to another.

hydrometallation [Chemistry] A chemical reaction involving the addition of M-H bond across a double bond, where M stands for boron, aluminium, silicon, tin and many other elements.

hydrometer [Engineering] A direct reading instrument for measuring the density or relative density of liquids.

hydronic radiation [Physics] A form of electromagnetic radiation used for communication under water.

hydronium ion [Chemistry] The oxonium ion, H_3O^+, *i.e.*, a proton, H^+ combined with a molecule of water.

hydrophilic [Chemistry] Having an affinity for water.

hydroponics [Chemistry] Having no affinity for water; waterrepellent.

hydroponics [Botany] Cultivation of plants without the use of soil, using instead solutions of those mineral salts that a plant normally extracts from the soil.

hydrophyte [Botany] A Plant that grows in highly wet environment either completely equatic or rooted in water or mud but with stems and leaves above water.

hydrosol [Chemistry] A colloidal solution, as distinct from a hydrogel, water being the solvent.

hydrosphere [Geology] The watery portion of the earth's crust, comprising the oceans, seas, and all other waters. Composition by weight is given as : oxygen 85.85%, hydrogen 10.68% chlorine 2.01%, sodium 1.1%, magnesium 0.15%, not more than 0.05% of any other element being present. The chief constitutents are water, sodium chloride, and magnesium chloride.

hydrostat *See* humidistat.

hydrostatics [Mechanics] The mathematical study of forces and pressures in liquids at rest.

hydrosulpha [Chemistry] A salt formed when an organic base (*e.g.*, an alkaloid) combines with sulphuric acid. The salt so formed is usually more soluble than the base.

hydrosulphide [Chemistry] A compound containing the univalent HS^- group.

hydroxide [Chemistry] A compound containing the OH^-, hydroxyl group, *e.g* sodium hydroxide, NaOH.

hydroxyl group [Chemistry] The univalent-OH group present in electrovalently bonded form in inorganic alkalies and in covalently bonded form in alcohols and phenols.

hygiene [Medicine] The science that deals with the principles and practices of good health.

hygro [Physics] Prefix denoting moisture, humidity, *e.g.*, hygrometer.

hygrodeik [Physics] A wet and dry bulb hygrometer with a chart attached, which enables the relative humidity to be obtained directly from the readings of the two thermometers.

hygrometer [Physics] Any instrument designed to measure the relative humidity of the atmosphere.

hygroscope [Physics] An instrument for showing variations of relative humidity of the air.

hygroscopic [Chemistry] Having a tendency to absorb moisture.

hyper- [Scientific Techniques] Prefix denoting over, above, beyond.

hyperbola [Mathematics] The plane curve obtained by intersecting a circular cone of two nappes with a plane parallel to the axis of the cone.

hypercapina [Medicine] Excessive amount of carbon dioxide in the blood.

hyperchromatism [Medicine] Excessive pigmation in the skin.

hyperfine structure [Physics] A splitting of spectral lines due to spin of the atomic nucleus or to the occurrence of a mixture of isotopes in the element.

hypergolic [Chemistry] Capable of igniting spontaneously upon contact.

hyperlipemia [Medicine] Excessive amount of fats in the blood.

hypermetropia [Medicine] A defective vision resulting from too short an eyeball so that unaccommodated rays focus behind the retina, hence a person is unable to *see* near objects distinctly. It is corrected by the use of convex spectacle lenses. Also known as long-sightedness.

hyperons [Physics] A group of elementary particles, belonging to the class called baryons, which have greater mass than the neutron but very short lives. All baryons that are not nucleons are known as hyperons, but as all hyperons decay into nucleons they can be regarded as excited nucleons. For each hyperon there is a corresponding anti-particle.

hyperproteinemia [Medicine] Excess of protein present in the blood.

hypersonic [Physics] Pertaining to frequencies above 500 megahertz.

hypersonics [Physics] Production and utilization of sound waves of frequencies above 500 megahertz.

hypertonic [Chemistry] A solution is said to be hypertonic with respect to another if it has a greater osmotic pressure.

hypnotic [Chemistry] A substance producing sleep. Also known as somnificant; soporific.

hypo [Materials] $Na_2S_2O_3.5H_2O$.

Sodium thiosulphate. Formerly incorrectly called 'sodium' hyposulphite'; used in photography.

hypochlorite [Chemistry] A salt of hypochlorous acid. Hypochlorites of sodium, potassium, and calcium are used as disinfectants and for bleaching, by virtue of their oxidizing properties.

hypocycloid [Mathematics] The figure traced by a point on the circumference of a circle that rolls, without slipping, round the inside of a large fixed circle.

hypothesis [Scientific Techniques] A supposition put forward in explanation of observed facts and that cannot be tested by further controlled experiments.

hypotonic [Chemistry] A solution is said to be hypotonic with respect to another if it has a smaller osmotic pressure. It has a lower concentration of solute molecules and higher concentration of solvent molecules.

hypsometer [Engineering] An apparatus for measuring atmospheric pressure to ascertain elevations by the determination of the boiling point of liquid.

hysteresis [Electronics] An oscillator effect wherein a given value of an operating parameter may result in multiple values of output power or frequency. [Physics] The dependence of the state of a system on its previous history, generally in the form of a lagging of a physical effect behind its cause.

hysteresis loop [Physics] The closed curve followed by a mathematical displaying hysteresis (such as ferromagnet) on a graph of a driven variable (such as magnetic flux) versus the driving variable (such as magnetic field)

hysteresis loss [Physics] The energy converted to heat in a material because of magnetic or other hysteresis, accompanying cyclic variation of magnetic field or other driving variable.

Hz *See* hertz.

iceberg [Geology] A large mass of detached land ice floating in sea or standard in shallow water.

ice nucleus [Physics] Any particle which can act as a nucleus (cental point) in the formation of ice crystals in atmosphere.

ice point [Physics] The temperature at which there is equilibrium between ice and water at standard atmospheric pressure (*i.e.* the freezing or melting point under standard conditions) It was used as a fixed point ($0°$) on the Celsius scale, but the kelvin and the International Practical Temperature Scale are based on the triple point of water.

iclathyology [Biology] A branch of vertebrate zoology which deals with the study of fishes.

ideal crystal [Chemistry] A crystal whose lattice is perfectly regular and contains no foreign atoms or ions or other defects or imperfections.

ideal gas [Physics] A hypothetical gas that obeys the *gas laws exactly. An ideal gas would consist of molecules that occupy negligible space and have negligible forces between them. All collisions made between molecules and the walls of the container or between molecules and other molecules would be perfectly elastic, because the molecules would have no means of storing energy except as translational kinetic energy. Also known as perfect gas.

ideal solution [Chemistry] A solution that confirms to Raoult's law over all ranges of temperature and concentration, and exhibits no change of internal energy on mixing and no attractive forces between components. *See* Raoult's law.

igneous [Geology] Pertaining to rocks which have formed from a molten mass.

ignition [Chemistry] The action of setting fire to something. Initiating combustion by raising the temperature of the reactants to the ignition temperature.

ignition temperature [Chemistry] The lowest temperature to which a substance must be heated before combustion can take place. Also known as ignition point; autogenous ignition temperature.

illuminance [Optics] *Ev.* The amount of light falling on unit area of a surface per second, the derived SI unit of illuminance is the lux (lumen per square metre),

illumination Oil [Materials]. Any oil such as kerosine suitable for burning to provide illumination.

ilmenite [Minerals] $FeTiO_3$. Natural ferrous titanate; an ore of titanium.

image [Optics] A representation of a physical object formed by a lens, mirror, or other optical instrument. An image is formed when rays of light, starting from the same point on an object, meet at a point which is the same point on the image as on the object. True sharp images are formed only with rays of a single wavelength passing fairly close and parallel to the principal axis of the lens or mirror. The image is said to be real if the rays of light can be produced on a screen. The image is said to virtual if the rays of light starting from a point, after reflection or refraction appear to diverge from a second point. All virtual images are errect and cannot be taken on a screen.

image converter [Electronics] A device for converting an image formed by nonvisible radiation (such as infrared or ultraviolet radiation) into a visible image.

imaginary numbers [Mathematics] Numbers with negative squares; thus −1 is an imaginary number, denoted by i; $i^2 = -1$.

imbibition [Chemistry] The phenomenon in which a substance absorbs a liquid and swells, but does not necessarily dissolve in the liquid. The process in reversible, the substance contracting on drying.

imides [Chemistry] Organic compounds containing the group -
−CO.NH.CO-(the *imido group*).

imido group *See* imides.

imines [Chemistry] Compounds containing the group-NH- (the *imino group*) joined to two other groups; *i.e.*secondary amines.

imino group *See* imines.

immersion cleaning [Metallurgy] Surface cleaning of a metal by dipping into a cleaning liquid.

immersion coating [Engineering] Applying material to the surface of an object by dipping into liquid.

immersion objective [Optics] An optical microscope objective in which the front surface of the lens is immersed in a liquid on the cover glass of the microscope specimen slide. Cedar-wood oil or sugar solution is frequently used. It has the same refractive index as the glass of the slide, so that the object is effectively immersed in it. The presence of the liquid increases the effective aperture of the objective, thus increasing the resolution.

immiscible [Chemistry] Incapable of being mixed to form a homogeneous substance; it is usually applied to liquids; *e.g.*, oil and water are immiscible.

immittance [Electricity] A term used to denote both impedance and admittance.

immune [Biology] Safe from attack; protected against a disease by an acquired immunity.

immune clearance [Biochemistry] The rapid removal of antigen introduced into the body of an immune individually as result of its complexing with antibody.

immune haemolysin [Biology] A substance formed in the blood in response to an injection of erthocytes from another species.

immune protein [Biochemistry] Any anitbody.

immune response [Biochemistry] A response to the introduction of antigen into the body involving production of specific antibodies or lymphocytes, which combine with antigen. It is basic mechanism of active immunity.

immmune serum [Biochemistry] Blood serum obtained from an immunized individual and carrying antibodies.

immunity [Biology] The ability of plants and animals to withstand harmful infective agents and toxins. It may be due partly to a number of non-specific mechanisms, such as phagocotosis or an imprevious skin. In vertebrates it is largely the result of a specific mechanism whereby certain substances antibodies or lymphocytes present in the body com bine with a particular foreign substance (antigen) introduced specifically acquired immunity includes passive immunity, where the antibody has been derived from another individual and active immunity. where the antibody is produced following stimulation with antigen (*e.g.*, by vaccination or by exposure to infection).

immunization [Biology] The process of making an animal resistant to infection or harmful agents. *See* immunity.

immunoglobin [Biochemistry] An antibody protein generated to a specific antigen.

immunological tolerance [Biology] The failure of antibody response to an antigen, usually one to which the animal has been exposed previously.

impact [Mechanics] A forceful collision between two bodies which is sufficient to cause an appreciable change in the momentum of the system on which it acts.

impedance [Electricity] Z. The quantity that measures the opposition of a circuit to the passage of a current and therefore determines the amplitude of the current. In a b.c.circuit this is the resistance (X) also has to be taken into account, according to the equation : $Z^2 = R^2 + x^2$, where Z is the impedance. The *complex impedance* is given by $Z = R + iX$, where $i = \sqrt{-1}$. The ratio of the imaginary part, the reactance, to the real part is an indication of the difference in phase between the voltage and the current.

impermeable [Scientific Techniques] Not permitting the passage of fluids.

implantation [Medicine] Placing a tissue transplant in depth in a body.

implicit function [Mathematics] A variable quantity, x, is said to be an implicit function of y, when x and y are connected by a relation that is not explicit. *See* explicit function.

implosion [Chemistry] The sudden reduction of pressure by chemical or physical reaction which cause an inrushing of the surrounding medium. [Physics] The inward collapse of an evacuated vessel.

impulse [Physics] A force acting during a very short time; given (for a constant force) by the product of the magnitude of the force and the time during which the force acts; it is equal to the total change of momentum produced by it.

impurity [Chemistry] An undesirable foreign material in a pure substance. [Electronics] A substance that, when diffused into a semiconductor in small amounts, either provides free electrons or accepts electrons from it.

incandescence [Physics] The emission of light by a substance as a result of raising it to a high temperature. An *incandescent lamp* is one in which light is emitted by an electrically heated filament. *See* electric lighting.

inching *See* jogging.

inclination [Physics] 1.The angle between the earth's magnetic field and the horizontal at a given point on the earth's surface. 2.The angle between the orbital plane of a planet, satellite, or comet and the plane of the earth's ecliptic.

inclined plane [Mechanics] A plane surface at an angle to some force or reference line.

inclinometer [Engineering] An instrument for measuring the angle of inclination that an aircraft makes with the horizontal.

incoherent light [Optics] Electromagnetic radiation not all of the same phase, and possibly also consisting of various wave-lengths.

incoherent waves [Physics] Waves having no fixed phase relationship.

incompressibility [Mechanics] Property of a substance which maintains its original volume under increased pressure.

incretion [Physiology] An internal secretion.

incubator [Engineering] A box designed to maintain a constant internal temperature by the use of a thermostat; used for rearing

chickens and prematurely born infants, and in bacteriology for cultivation of bacteria.

india ink [Materials] Black ink containing a suspension of carbon and blue binder. Also known as Chinese ink; sumi ink.

Indicator A substance that, by a sharp colour change, indicates the completion of a chemical reaction. It is frequently used in volumetric analysis. Indicators for titrations of acids and alkalies are usually weak organic acids or bases, yielding ions of a different colour from the unionized molecules.

indium [Chemistry] In. Element. At. No. 49. At. Wt. 114.82; placed in group III A of periodic table. a soft silvery-white metal, m.p. 429.5K; b.p. 2353 K. The uses of the metal are specialpurpose electroplates and some special fusible alloys. Several semiconductor compounds are used, such as InAs, InP, and InSb. With only three electrons in its valency shell, indium is an electron acceptor.

Induced current [Electricity] Electric current in a conductor caused by e.m.f. set up a changing magnetic field surrounding it.

induced dipole [Electricity] An induced dipole produced by application of an electric field.

induced enzyme [Biochemistry] An enzyme that is not made by the cell (*i.e.,* is repressed) unless it is induced by its substrate or a closely related compound.

induced fission [Nucleonics] & Nuclear physics] Fission taking place only as a result of bombardment of nucleus with neutrons, gramma rays or other carriers of energy.

induced fit [Biochemistry] A change in the shape of an enzyme to confirm to the structure of its substance.

induced radioactivity [Nucleonics & Nuclear physics] Radioactivity induced in naturally stable elements by bombarding them with neutrons or other high energy particles. Also known as aritificial radioactivity.

inducer [Biochemistry] A molecule capable of inducing the synthesis of a given enzyme; usually the enzyme's substrate.

inductance [Electricity] The property of an electric circuit or component that causes an e.m.f.to be generated in it as a result of a change in the current flowing through the circuit (*self inductance*), or of a change in the current flowing through a neighbouring circuit with which it is magnetically linked (*mutual inductance*). In the case of self inductance, L, the e.m.f., E, generated is given by $E = -L\, dI/dt$, where I is the instantaneous current and the minus sign indicates that the e.m.f. induced is in opposition to the change of current. In the case of mutual inductance, M, the e.m.f. E_1, induced in one circuit is given by $E_1 = -M.\, dI_2 /dt$, where I_2 is the instantaneous current in the other circuit.

induction, charging [Electricity] A process of electrically charging an insulated conductor, using the force due to another nearby charge to separate the positive and negative charge existing on the conductor.

induction coil [Physics] A device that produces a series of highvoltage pulses by means of electromagnetic induction. It consists of a coil of wire with only a few turns, would on an iron core and surrounded by another coil with many more turns. When the current in the first coil is interrupted suddenly, a large e.m.f. is induced in the second. A pulse current in the first coil induces a large pulsed e.m.f. in the second.

induction, electromagnetic *See* electromagnetic induction.

induction heating [Electricity] A form of heating in which electrically conducting material is heated as a result of the electric current induced in it by an alternating magnetic field.

induction, magnetic *See* magnetic induction.

induction motor [Electricity] A type of electric motor in which an alternating current supply fed to the primary winding set up a flux causing electrical currents to be induced in the secondary winding of the motor. The interaction between these currents and the flux causes the motor to rotate.

induction period [Chemistry] The time of acceleration of a chemical reaction from zero to maximum rate.

inductive effect [Chemistry] The effect of a group of atom of a compound in pulling electrons twards itself or in pushing them away. Inductive effects can be used to explain some aspects of organic reactions. For instance, electron-withdrawing groups, such as $-NO_2$, $-CN$, $-CHO$, $-COOH$, and the halogens substituted on a benzene ring, reduce the electron density on the ring and decrease its susceptibility to further (electrophilic) substitution. Electron-releasing groups, such as -OH, $-NH_2$, $-OCH_3$, and $-CH_3$, have the opposite effect.

inelastic collision [Physics] A collision in which some of the kinetic energy of the colliding bodies is converted into internal energy in one body so that kinetic energy is not conserved. In collisions of macroscopic bodies some kinetic energy is turned into vibrational energy of the atoms, causing a heating effect.

inert [Chemistry] Not easily affected by chemical reaction.

inert gases *See* noble gases.

inertia [Physics] The tendency of a body to preserve its state of rest or uniform motion in a straight line.

inertial flow [Mechanics] The flow in which no external forces are exerted on a fluid.

inertial frame [Physics] A frame of reference in which bodies move in straight lines with constant speeds unless acted upon by external forces, *i.e.* a frame of reference in which bodies are not accelerated. Newton's laws of motion are valid in an inertial system but not in a system that itself accelerated with respect to such a frame.

inertial mass [Mechanics] The mass of a body as determined by its momentum (in accordance with the law of conservation of momentum), as opposed to 'gravitational mass', which is determined by the extent to which it responds to the force of gravity. The acceleration of a falling body increases in proportion to its gravitationai mass and decreases in proportion to its inertial mass. Since all falling bodies have the same constant acceleration it follows that the two types of mass must be equal.

intertial wave [Mechanics] Any wave motion in which no form of energy other than kinetic energy is present.

inert-pair effect [Chemistry] An effect seen especially in groups III and IV of the periodic table, in which the heavier elements in the group tend to form compounds with a valency two lower than the expected group valency. It is used to account for the existence of thallium (I) compounds in group III and lead (II) in group IV. In forming compounds, elements in these groups promote an electron from a filled s-level state to an empty p-level. The energy required for this is more than compensated for by the extra energy gain in forming two more bonds. For the heavier elements, the bond strengths or lattice energies in the compounds are lower than those of the lighter elements. Consequently the energy compensation is less important and the lower valence states become favoured.

infection [Medicine] Invasion of body by a pathogenic organism, with or without disease manifestation.

inferior planet [Astronomy] A planet that circles the sun in an orbit that is smaller than the earth's orbit.

infestation [Medicine] The state or condition of having animal parasites on or in the body.

infinitesimal [Scientific Techniques] Vanishingly small but not zero. Infinitesimal changes are notationally made in the calculus, which is sometimes called the *infinitesimal calculus.*

infinity [Mathematics] _ A quantity that is greater than any assignable quantity.

inflammability *See* flammability.

inflammation [Medicine] Local tissue response to injury charac-terized by swelling, pain or redness.

inflection [Mathematics] A point on a curve at which the tangent changes from rotation in one direction to rotation in the opposite direction. If the curve $y=f(x)$ has a stationary point $dy/dx=0$, there is either a maximum, minimum, or inflection at this point. If $d^2y/dx^2 = 0$, the stationary point is a point of inflection.

informational molecules [Biochemistry] Molecules containing information in the form of specific sequences of different building blocks; include proteins and nucleic acids.

infrared radiation [Physics] Electromagnetic radiation possessing wavelengths between those of visible light and those of radio wave, *i.e.*, from approximately 0.75 micrometre to 1000 micrometres. Infrared radiation has the power of penetrating fog or haze, which would scatter ordinary visible light. Infrared radiation is produced by vibrational or rotational motion of molecules.

infrared stars [Astronomy] Celestial bodies whose principal emission is infrared radiation. They are believed to consist of stars surrounded by dust clouds. In some cases the light from the central star penetrates the dust so that it can be seen with optical telescopes.

infrasonic [physics] Having a frequency below the frequency of audible sound waves, *i.e.*, a frequency of less than about 20 hertz.

infrasound [Physics] Vibrations or pressure waves in a medium with frequencies below about 16 hertz. The ear distinguishes such vibrations as a series of pulses, rather than a continuous sound.

infusible [Chemistry] Difficult to melt; having a very high melting point.

ingestion [Biology] The act of food intake by the animals through mouth.

ingot iron [Metallurgy] A variety of relatively pure iron produced in an open hearth furnace.

inhibitor [Chemistry] A substance capable of retarding or stopping a chemical reaction.

initiation factors [Biochemistry] Specific protein required to initiate synthesis of a polypeptide by ribosomes.

injection moulding [Engineering] A process by which thermoplastic articles are moulded. The thermoplastic material is softened in a heated chamber and then injected under pressure through an orifice into a cool closed mould.

inorganic chemistry [Chemistry] The study of the chemical reactions and properties of elements and their compounds. Inorganic chemistry usually includes the study of elemental carbon, its oxides, metal carbonates, and sulphides, while all other carbon compounds are studied under organic chemistry.

input [Computers] The information that is delivered to a data processing device. [Electronics] The power or signal fed into an electrical or electronic device.

insecticide [Materials] A substance used for killing insect pests.

insemination [Physiology] The introduction of sperm into vagina.

in situ [Scientific Techniques] In the original location.

insolation [Physics] The solar radiation that is received at the earth's surface per unit area. It is related to the solar constant, the duration of daylight, the altitude of the sun, and the latitude of the receiving surface. It is measured in MJm^{-2}.

insoluble [Chemistry] Not capable of forming a solution (especially in water, unless some other solvent is specified). It is a relative term, since most substances have been shown to dissolve in water to some extent.

instantaneous frequency [Physics] The rate of change of phase of an oscillation, expressed in radians per divided by 2π.

instantaneous value [Scientific Techniques] The value of any varying quantity at a specified instant.

insulating oil [Materials] A mixture of chlorinated hydrocarbons and fluorinated hydrocarbons used in switches and transformers as insulator and colling medium. Also known as electrical oil.

insulation [Electricity] The prevention of the passage of electricity, or heat, by conduction.

insulator [Electricity] A non-conductor of electricity or heat.

integral [Mathematics] 1.Consisting of whole numbers or integers. 2.A mathematical function obtained by the process of integration.

integral calculus [Mathematics] The branch of the calculus making use of the processes of integration. It is used for calculating areas and volumes and for other problems concerned with summation of infinitesimally small elements.

integrated circuit [Electronics] A microelectronic circuit incorporated into a chip of semiconductor, usually crystalling silicon (a silicon chip). Integrated circuits consist of whole systems rather than single components, and are used in modern computers and a large number of electronic devices. They are also used in other industries (e.g.cars, radios, etc.) in which small reliable electronic control circuits are required.

integration [Mathematics] A mathematical process used in the calculus; the inverse process of differentiation. It gives a method of finding the area enclosed by curves, and of finding solutions

to other problems involving the summation of infinitesimals.

intensifier [Materials] A substance used to increase the density or contrast of an image on a photographic film or plate. It is usually a compound from which a metal (e.g.silver, lead, uranium, etc.) can be deposited.

interface [Chemistry] A specification or protocol for how data should be formatted and sequenced to pass between two adjacent parts of the net work; the interface specification may also describe maintenance procedures, performance, and other chaaracteristics; the interface may be implemented in a network as hardware and software.

interface [Chemistry] The surface that separates two chemical phases.

interference [Physics] The interaction of two or more wave motions affecting the some part of a medium so that the instantaneous disturbances in the resultant wave are the algebric sum of the instantaneous disturnbances in the interfering waves.

interferometer [Optics] Any instrument that divides a beam of light into a number of beams and re-unites them to produce interference; Used for the accurate determination of wavelengths of light; the testing of prisms and lenses; the examination of the hyperfine structure of spectrum lines; measurement of the diameters of stars and the determination of the number of light waves of a certain wavelength in the standard metre.

intergalactic matter [Astronomy] The matter between the galaxies.

intergalactic space [Astronomy] The space between galaxies, in which intergalactic matter may occur.

intermediatary metabolism [Biochemistry] In the cells, the enzyme-catalyzed reactions that extract chemical energy from nutrient molecules and utilize it to synthesize and assemble cell components.

intermediate [Chemistry] A compound used in an intermediate step in the manufacture of the final product by chemical synthesis.

intermediate frequency [Physics] In superheterodyne radio receivers, the carrier wave frequency of the incoming radio wave is changed to a fixed intermediate frequency by heterodyne action, for ease of amplification before detection.

intermediate neutrons [Nucleonics] Neutron with kinetic energies between those of epithermal and fast neutrons, *i.e.*, between 100 to 100, 1000 electron-volts.

intermediate reactor [Nucleonics & Nuclear Physics] A nuclear reactor in which chain reaction is sustained mainly by intermediate neutrons.

intermetallic compound [Metallurgy] A compound in which two or more metals are held together by metallic bonds. They occur in some alloys.

intermolecular force [Chemistry] The attractive force between two molecules. If energy is a function of the distance between the centres of the molecules, the intermolecular force is the negative gradient of potential energy.

internal combustion engine [Engineering] An engine in which energy supplied by a buring fuel is directly transformed into mechanical energy of the controlled combustion of the fuel in an enclosed cylinder behind a piston.

internal conversion [Chemistry] A process in which an excited atomic nucleus decays to the *ground state and the energy released is transferred by electromagnetic coupling to one of the bound electrons of that atom rather than being released as a photon.

internal energy [Physics] U. The total of the kinetic energies of the atoms and molecules of which a system consists and the potential energies associated with their mutual interactions. It does not include the kinetic and potential energies of the system as a whole nor their nuclear energies or other intra-atomic energies. The value of the absolute internal energy of a system in any particular state connot be measured; the significant quantity is the change in internal energy, ΔU. For a closed system (*i.e.*, one that is not being replenished from outside its boundaries) the change in internal energy is equal to the heat absorbed by the system (Q) from its surroundings, less the work done (W) by the system on its surroundings, *ø.e.* $\Delta U = Q\text{-}W$. See also energy; heat; thermodynamics.

internal resistance [Electricity] The resistance with a source of electricity. In the case of a cell, when the current is supplied, the potential difference between the terminals is lower than the e.m.f. The internal resistance r is :

$$r = (E\text{-}V)\ I$$

where E is the e.m.f., V the postential difference between terminals, and I the current.

internal stress [Mechanics] The stress within a solid material, *i.g.*, alloys, metals, glass, etc., as a result of treatment, cold working, or non-uniform molecular structure of that material.

international Candle [Optics] The former unit of huminous intensity. A point source emitting light uniformly in all directions at one-tenth of the rate of the Harcourt pentane lamp burning under specified conditions.

interplanetary probe [Engineering] An instrumented spacecraft that flies close to or lands upon a planet.

interplanetary space [Astronomy] The space between the sun and the planents within the solar system.

interpolation [Scientific Techniques] The process of filling in intermediate values or terms of a series between known values or terms.

interstellar [Astronomy] Between the stars.

interstellar matter [Astrons] Clouds of hydrogen atoms or molecules, mixed with a small proportion of dust, that exist between stars. The density of these clouds is very low, ranging between some 10^7 and 10^9 atoms per m^3 (compared to about 10^{25} molecules per m^3 for a perfect gas at S.T.P.).

interstellar space [Astronomy] The space between stars within a galaxy, in which interstellar matter may occur.

interstitial [Chemistry] A crystal defect in which an additional atom or ion is situated between the normal sites in a crystal lattice.

interstitial compound See *compound, interstitial.*

intoxication [Medicine] 1. Poisoning. 2. The state produced by excessive intake of ethyl alcohol.

intrinsic conductivity [Electronics] The conductivity of a semiconductor or metal in which impurities and structural defects are absent or are very small.

intrinsic semiconductor [Electronics] A semiconductor in which the concentration of charge carriers is characteristic of the material itself rather than of the content of impurities and structural defects of the crystal.

invasion [Medicine] The process by which microorganisms enter the body.

inverse functions [Mathematics] If $y = f(x)$ and a function can be found so that $x = g(y)$, then $g(y)$ is said to be the inverse function of $f(x)$. If y is a trigonometrical function of the angle x, say $y = \sin x$, then x is the inverse *trigonometrical function* of y, written $x = $ arc sin y or $\sin^{-1} y$.

inverse square law [Physics] A relationship in which the effect of a source (*e.g.*, light, electricity, gravity, magnet) is inversely proportional to the square of the distance from the source.

inversion [Chemistry] A chemical reaction involving a change from one optically active configuration to the opposite configuration. The Walden inversion is an example. *See* nucleophilic substitution.

inversion temperature *See* Joule-Thomson effect.

invertase [Biochemistry] An enzyme contained in yeast that converts cane-sugar into glucose and fructose. *See* inversion of cane-sugar. Also known as sucrase.

inverter [Electricity] A device for converting direct current into alternating current.

invert sugar [Chemistry] A mixture of glucose and fructose in equal proportions, obtained by the inversion of cane-sugar.

in vitro [Biology] Pertaining to biological reactions taking place in an artificial apparatus.

in vivo [Biology] Pertaining to biological reactions taking place in a living cell or organism.

involute [Biology] Being coiled, curled, or rolled in at the edge. [Mathematic] The curve formed when a piece of string is unwound from, or wound on to, another curve.

iodine [Chemistry] I Element At. No. 53. At.Wt. 126.90; m.p. 386.5K; b.p. 456.5K; placed in group VII A of the periodic table. A blackish-grey, crystalline solid. It is essential to the functioning of the thyroid gland; lack of iodine in the diet is a cause of gotire; used in medicine, chemical analysis, and photography. The radioisotope ^{131}I, (half-life 8.6 days) is used in the treatment and diagnosis of disorders of the thyroid gland.

iodine number [Chemistry] A measure of the degree of unsaturation (content of double bonds) of a product, such as an oil or fat; it is expressed in grams of iodine absorbed by 100 gm of the given substance. Also known as iodine value; Jiubl number.

ion [Chemistry] An electrically charged atom or group of atoms. Positively charged ions (cations) have fewer electrons than is necessary for the atom or group to be electrically netural; while negative ions (anions) have more.

ion chamber *See* ionization chamber.

ion current [Physics] The electric current set up due to motion of ions.

ion engine [Engineering] An engine that provides thrust by expelling high velocity ions; used for propelling space vehicles.

ion exchange chromatography [Chemistry] A chromatographic method based on the ability of polymers to absorb ionized solutes reversibly. It can be carried out in columns as well as on sheets.

ion exchange resins [Materials] Certain substances have the power of acting on solutions containing ions, such as solutions of salts, and replacing some of the ions by others; *e.g.*, in a typical cation exchange (base exchange) action, when hard water is passed through a suitable ion exchange resin or a zeolite, the calcium ions in the water are replaced by sodium ions. In anion exchange resins acid radicals or anions are exchanged similarly. Ion exchange resins have many impor-

tant industrial and analytical uses in addition to water softening.

ion exclusion [Chemistry] The process in which a synthetic resin of the ion exchange type absorbs nonionized solutes such as sugar, while it does not absorb ionized solutes that are also present in a solution in contact with the resin.

ionic bond [Chemistry] Bonds formed by transfer of electrons; for example, the calcium atom has an electron configuration of $[Ar]4S^2$, *i.e.* it has two electrons in its outer shell. The chlorine atom is $[Ne]3s^23p^5$, with seven outer electrons. If the calcium atom transfers two electrons, one to each chlorine atom, it becomes a Ca^{2+} ion with the stable configuration of an inert gas (Ar). At the same time each chlorine, having gained one electron, becomes a Cl^- ion, also with an inert-gas configuration (Ar). The bonding in calcium chloride is the electrostatic attraction between the ions. Also known as electrovalent bond.

ionic crystal [Chemistry] A crystal in which the lattice sites are occupied by charged ions held together primarily by their electrostatic interaction.

ionic product [Chemistry] The product of the concentrations of ions present in a given solution taking the stoichiometry into account. For a sodium chloride solution the ionic product is $[Na^+][Cl^-]$; for a calcium chloride solution it is $[Ca^{2+}]$ $[Cl^-]^2$. In pure water, there is an equilibrium with a small amount of self-ionization :

$$H_2O \rightarrow H^+ + OH^-$$

The equilibrium constant of this dissociation is given by

$$K_w = [H^+][OH^-]$$

since the concentration $[H_2O]$ can be taken as constant. K_w is referred to as the ionic product of water. It has the value 10^{-14} mol^2 dm^{-6} at 298K. In pure water (*i.e.*, no added acid or added alkali) $[H^+] = [OH^-] = 10^{-7}$ mol dm^{-3}. *See also* solubility product.

ionic strength [Chemistry] A measure of the intensity of the electrical field due to the ions in a solution of an electrolyte. It is defined as half of the sum of the terms obtained by multiplying the molarity of each ion by the square of its valence.

ionization [Chemistry] The formation of ions by an atom or a compound.

ionization current [Electronics] The electric current produced by the movement of ions or electrons in an electric field as a result of ionizing radiation.

ionization potential [Chemistry] I. The minimum energy required to remove an electron from a specified atom or molecule to such a distance that there is no electrostatic interaction between ion and electron. Originally defined as the minimum potential through which an electron would have to fall to ionize an atom, the ionization potential was measured in volts. It is now, however, defined as the energy to effect an ionization and is conveniently measured in electron volts (although this is not an SI unit).

ionizing radiation [Physics] Radiation that is capable of causing ionization, either directly or indirectly. Electrons and alpha particles are considerably more effective in this respect than neutrons or gamma-rays.

ion-microprobe analysis [Chemistry] A technique for analysing the surface composition of solids. The sample is bombarded with a narrow beam of high-energy ions. Ions ejected from the surface by sputtering are detected by mass spectrometry. The technique allows quantitative analysis of both chemical and isotopic composition for concentrations as low as a few parts per million.

ion mobility [Electricity] The velocity of an ion in an unit electric field.

ionomer resins [Materials] Synthetic resins cross-linked (*see* crosslinkage) through ionized carboxyl groups in their macromolecules. Although they have the usual properties of crosslinked polymers, they can be processed like thermoplastic resins.

ionosphere [Physics] The region of the earth's upper atmosphere in which free electrons arising from ionization occur, mainly as a result of ultraviolet radiation and X-ray from the sun. The ionosphere is useful in that it enables intercontinental radio transmission round the curved surface of the earth to be achieved, as a result of its property of reflecting electromagnetic waves of radio frequencies (*see* sky wave); but it is an obstacle to radio astronomy because it reflects a large proportion of the radiation that arrives from extra-terrestrial sources.

ionospheric wave *See* sky wave.

ion pair [Nucleonics & Nuclear physics] A positive ion and an equal-charge negative ion, usually an electron, that is produced by the action of radiation upon a neutral atom or molecule.

ion pump [Electronics] A high vacuum pump in which gas is removed from a system by ionizing the atoms or molecules and adsorbing the resulting ions on a surface, usually of a metal.

iridium [Chemistry] Ir. Element. At. No. 77. At.Wt. 192.2; placed in group VIII of the periodic table. A rare metal resembling, and occurring together with, platinum; m.p. 2683K, b.p. 4403K; it is extremely hard and resistant to chemical action. Alloys of iridium are used for numerous purposes where extreme hardness and where a high melting point are required.

iris [Optics] 1. The coloured part of the eye of vertebrates. 2. A diaphragm forming an adjustable opening over a lens in an optical instrument.

iron [Chemistry] Fe (Ferrum). Element. At. No. 26. At.Wt. 55.847; placed in group VIII of the periodic table. A white magnetic metal, m.p. 1808K b.p. 3023K. Physical properties are greatly modified by the presence of small amounts of other metals and of carbon; one of the most widely used metals, and plays a vital role in biological proceses.

irradiance [Physics] *E.* The radiant flux per unit area reaching a surface; in SI units it is measured in watts per square metre (Wm⁻^{2}). Irradiance refers to electromagnetic radiation of all kinds, whereas illuminance refers only to visible radiation.

irradiation [Physics] Exposure to any form of radiation, often exposure to ionizing radiation is implied.

irrational number [Mathematics] A number that cannot be expressed as the ratio of two integers. An irrational number may be a surd, such as $\sqrt{2}$ or $\sqrt{3}$, which can be expressed to any desired degree of accuracy but cannot be assigned an exact value. Alternatively, it may be a transcendental, such as π or e.

irreversible reaction [Chemistry] A chemical reaction that proceeds to completion; the resulting products do not react to form the original substances.

irritability [Physiology] The property of living organisms that enables them to respond to external stimuli.

isentropic process [Physics] Any process that takes place without a change of entropy. The quantity of heat transferred, δQ, in a reversible process is proportional to the change in entropy, δS, *i.e.* $\delta Q = T\delta S$, where T is the thermodynamic temperature. Therefore, a reversible adiabatic process is isentropic, *i.e.*, when $\delta Q = 0$, δS also equals 0.

isinglass [Materials] A product containing about 90% gelatin, made from the dried swimming bladders of fish; used for clarifying alcoholic beverages.

isoantigen [Biochemistry] A type of antigen that induces antibody production in members of a species that have different genetic constitutions.

isobar [Physics] A line connecting points having equal atmospheric pressure. [Nucleonics Two or more atoms having same number of nucleons in their nuclei, but differing in atomic numbers and chemical properties, *e.g.*, C-14 and N-14 are isobars.

isobits [Computers] Binary digits having the same value.

isochore [Physics] A line that graphically represents the rela-

tionship between the pressure and the temperature of a liquid or gas, the volume of the system being kept constant.

isochromatic [Optics] Pertaining to a variation of certain quantities related to light, in which the colour or wavelength of the light remains same.

isochronon [Engineering] A clock designed to maintain very accurate time.

isocline [Geology] A line connecting points of equal angle of magnetic dip.

isocyanate [Chemistry] A salt or ester of isocyanic acid; a compound containing the -N=C=O group.

isodiapheres [Nucleonics & Nuclear physics] Nuclides in which the difference between the number of neutrons and protons is the same, *e.g.*, a nuclide and its decay product after it has emitted an alpha-particle are isodiapheres.

isodimorphism [Chemistry] The phenomenon of a dimorphous substance being isomorphous (*see* isomorphism) with another dimorphous substance in both its forms.

isodynamic line [Geology] A line passing through point of equal horizontal intensity of the earth's magnetic field (*see* magnetism, terrestrial).

isoelectric [Electricity] Pertaining to a constant electric potential.

isoelectric point [Chemistry] The pH value at which a substance or system (*e.g.*, a protein solution) is electrically neutral; at this value electrophoresis does not occur when a direct electric current is applied.

isoelectronic [Physics] Pertaining to molecules or ions having the same number of electrons *e.g.* nitrogen (N_2) and carbon monoxide (CO).

isoenzymes *See* isozymes.

isogamy [Biology] Reproduction resulting from the union of two gametes that are identical in size and structure.

isogonal line [Geology] A line passing through points of equal magnetic declination.

isogonism [Chemistry] A type of isomorphism in which two substances having little or no chemical resemblance but have the same crystalline form.

isogony [Biology] Growth of parts at such a rate as to maintain relative size differences.

isograft *See* graft.

isoimmunization [Medicine] Development of anti-Rh agglutins in the blood of an Rh-negative person who has been given an Rh-positive transfusion, or who is carrying an Rh-positive foetus.

isokom [Chemistry] A line joining points of equal viscosity on a phase diagram.

isolation [Chemistry] Indentification and separation of a pure substance which is present in very low amount in a complex mixture.

isomegethic solution [Chemistry] Solutions formed of solute molecules of almost same size.

isomerism [Chemistry] The existence of chemical compounds (*isomers*) that have the same molecular formulae but different molecular structures or different arrangements of atoms in space. In *structural isomerism* the molecules have different molecular structures: *i.e.*, they may be different types of compound or they may simply differ in the position of the functional group in the molecule. Structural isomers generally have different physical and chemical properties. In *stereoisomerism*, the isomers have the same formula and functional groups, but differ in the arrangement of groups in space. Optical isomerism is one form of this (*see* optical activity). Another type is *cis-trans isomerism* (formerly *geometrical isomerism*), in which the isomers have different positions of groups with respect to a double bond or central atom. [Nucleonics & Nuclear physics] In nuclear physics, nuclei having the same atomic number and the same mass number, but which exist in different energy states, are said to be isomeric, *e.g.*, a nucleus in its ground state and a nucleus in a metastable excited state are isomers.

isometric [Crystallography] Referring to a system of crystallization in which the axes are at right angles to each other. [Physics] A line on a graph (isometric line) showing change of temperature with pressure, when the volume is kept constant.

isomorphism [Chemistry] The existence of two or more substances *(isomorphs)* that have the same crystal structure, so that they are able to form solid solutions.

iso-spin *See* isotopic spin.

isosterism [Chemistry] The phenomenon in substances having molecules with the same number of atoms and the same total number of electrons; this leads to similarity in physical properties, *e.g.*, carbon dioxide, CO_2, and nitrous oxide, N_2O.

isotherm line [Physics] A line connecting points at an equal temperature in graphs of thermodynamic relation.

isothermal process [Physics] Any process that takes place at constant temperature, *e.g.*, the isothermal expansion of a gas. Compare adiabatic.

isotones [Nucleonics & Nuclear physics] Atoms whose nuclei contain the same number of neutrons but have a different atomic number.

isotonic solutions [Chemistry] Solutions having the same osmotic pressure, being of the same molar concentration for example, human blood and physiological salt solution.

isotope exchange reaction [Chemistry] A chemical reaction in which interchange of atoms of a given element between two or more chemical forms of the element occurs, the atoms in one form being isotopically lebelled so as to distinguish them from atoms in the other form.

isotopes [Chemistry] Atoms of the same element that have the same atomic number, *i.e.*, the same position in the periodic table, but different masses. The difference in mass arises due to the presence of one or more extra neutron, in the nucleus.

isotopic number [Chemistry] Neutron excess. The difference between the number of neutrons in an isotope and the number of protons.

isotopic spin [Nucleonics & Nuclear physics] A quantum number, I, used to work out the properties of groups of elementary particles when the members of the groups are indentical in all respects except that of electric change; *e.g.*, the nucleon has isotopic spin, $I = +^1/_2$, and its two states, the proton and the neutron are then described as different orientations of that spin in a fictitious 'isotopic space'. The word 'spin' is not intended to imply any conventional image of rotation in this context, it is used in analogy to angular momentum to which the concept of isotopic spin bears a close formal resemblance. Isotopic spin is conserved in all strong nuclear interactions. Also known as iso-spin; isobaric spin.

isotopic tracer [Chemistry] An isotope of an element, either radioactive or stable, a small amount of which is incorporated into a sample material (the carrier) in order to follow the course of that element through a process. Also known as isotopic label; label; tag.

isotropic [Biology] Having a tendency of equal growth in all directions. [Physics] having identical property in all directions.

isotropic radiator [Physics] An energy source that radiates energy uniformly in all directions.

isotropy [Physics] The quality of a property which does not depend on the direction along which it is measured or of a medium or entity whose properties do not depend on the direction along which they are measured.

isozymes [Biochemistry] Different molecular forms of proteins with the same enzymatic activity. They differ from each other in their substrate affinity, in their maximum activity, or in regulatory properties. Also known as isoenzymes.

item [Computers] A set of adjacent digits, bits, or characters which is treated as a unit and conveys a single unit of information.

iteration [Computers] The process of repeating a sequence of instruction with minor modifcations between successive repetitions.

ivory black [Materials] Animal black formed by calcining ivory; used as a pigment.

izal [Materials] An emulsion of tar oils, widely used as a disinfectant and deodorant.

J

jack [Electricity] A connecting device into which a plug can be inserted to make circuit connections. [Engineering] A portable device for lifting heavy loads through small heights, operated by a lever, a screw, or a hydraulic press.

jacket [Nucleonics & Nuclear physics] A thin container for one or more fuel slugs, used to prevent the fuel from escaping into coolant of a reactor.

Jahn-Teller effect [Chemistry] The distortion of symmetrical structures resulting from partially filled electronic energy levels. Any non-linear molecular system possessing degenerate electronic state will be unstable and will undergo distortion to form a system of lower symmetry and lower energy and thus will remove degeneracy.

jamming [Electronics] Radiation or re-radiation of electromagnetic waves so as to impair the usefulness of a specific segment of the radio spectrum that is being used by the enemy for communication or radar.

japanning [Metallurgy] The finishing of metal objects with a glossy, black baking paint or varnish that primarily consists of a hard asphalt base.

jar [Electricity] A unit of capacitance equal to about 1.11265×10^{-9} farad; this unit is now obsolete.

jasper [Minerals] A coloured dense, opaque to translucent form of natural silica, SiO_2 containing iron oxide impurities. Also known as jaspis.

javelle water [Materials] A solution containing potassium hypochlorite, $KOCl$; made by the action of chlorine on a cold solution of potassium hydroxide, KOH; used for bleaching.

jelly [Chemistry] A modified form of word 'gel'; a colloidal suspension that has set.

jet [Materials] A very hard, lustrous form of natural carbon, allied to coal. [Mechanics] A strong well defined stream of compressible fluid, either gas or liquid, issuing from an orifice or nozzle.

jet coal [Geology] A hard, lustrous, black variety of lignite occurring in isolated masses in bituminous shale. Also known as black amber.

jet engine [Engineering] A gas turbine that produces a stream of hot gas enabling an aircraft to be propelled through the air by reaction propulsion. Air taken in at the front of the engine is compressed by a radial compressor. The compressed air then enters the combustion chambers providing the oxidant for the combustion of the liquid fuel. The energy released expands the gas and accelerates it rearwards, some of the energy of the gas being used to drive a turbine, which in trun operates the compressor. After leaving the turbine the gas passes to the rear jet nozzle producing forward thrust by reaction on the structure of the jet tube.

jet fuel [Materials] Hydrocarbons with a flash point of 325K; used as fuel in jet air craft.

jet propulsion *See* reaction propulsion.

jeweller's rouge [Materials] Red powdered haematite, iron (III) oxide, Fe_2O_3. It is a mild abrasive used in metal cleaners and polishes.

jigging [Engineering] A gravity method which separates mineral from gangue particles by utilizing an effective difference in settling rate through a periodically dilated bed.

job [Computers] A unit work to be done by computer.

job step [Computers] A unit of work from the viewpoint of the user.

jogging [Electricity] Quickly repeated closing and opening of an electrical circuit to produce small movements of the driven machine. Also known as iching.

Josephson effects [Electronics] Electrical effects observed when two superconducting materials (at low temperature) are separated by a thin layer of insulating material (typically a layer of oxide less than 10^{-8} m thick). If normal metallic conductors are separated by such a barrier it is possible for a small current to flow between the conductors by the tunnel effect. If the materials are superconductors (*see* superconductivity), several unusual phenomena occur :

(a) A supercurrent can flow through the barrier; *i.e.*, it has zero resistance.

(b) If this current exceeds a critical value, this conductivity is lost; the barrier then only passes the 'normal' low tunnelling current and a voltage develops across the junction.

(c) If a magnetic field is applied below the critical current value, the current density changes regularly with distance across the junction. The net current through the barrier depends on the magnetic fie'd applied. As the

field is increased the net current increases from zero to a maximum, decreases to zero, increases again to a (lower) maximum, decreases, and so on. If the field exceeds a critical value the superconductivity in the barrier vanishes and a potential difference develops across the junction.

(d) If a potential difference is applied across the junction, a high-frequency alternating current flows through the junction. The frequency of this current depends on the size of the potential difference.

joule [Mechanics] J. The derived SI unit of work and energy. The work done when the point of application of a force of one newton is displaced through a distance of 1 metre in the direction of the force. The joule is also the work done per second by a current of 1 ampere flowing through a resistance of 1 ohm. 1 joule=10^7 ergs=1 watt second =1 newton metre = 0.238846 calorie.

Joule effect [Physics] The heating effect produced by the flow of current through a resistance.

Joule's equivalent *See* mechanical equivalent of heat.

Joule heat [Electricity] The heat evolved when current flows through a medium having electrical resistance, as given by Joule's laws.

Joule's laws [Physics] The internal energy of a gas at constant temperature is independent of its volume. Joule's law is obeyed strictly only by a perfect gas, real gases show deviations from it. [Electricity] The heat produced by an electric current I, passing through a conductor of resistance R, for a time t, is equal to $I^2 Rt$. If I is measured in amperes, R in ohms, and t in seconds, the heat produced is measured in joules.

Joule-Thomson effect [Physics] When a gas expands through a porous plug, a change of temperature occurs, proportional to the pressure difference across the plug. The temperature change is due partly to a departure of the gas from Joule's law, the gas performing internal work in overcoming the mutual attractions of its molecules and thus cooling itself; and partly to deviation of the gas from Boyle's law. The latter effect can give rise to either cooling or heating, depending upon the initial temperature and pressure difference used. For a given mean pressure, the temperature at which the two effects balance, resulting in no alteration of temperature, is called the 'inversion temperature'. Gases expanding through a porous plug below their inversion temperature are cooled, otherwise they are heated. Also known as Joule-Kelvin effect.

jumper [Electricity] A short length of conductor used to make a connection between two points

or terminals in a circuit or to provide a path around a break in a circuit.

junction [Electronics] A region of transition between two different semiconducting regions in a semiconducor device, such as a junction, or between a metal and a semiconductor.

junction diode [Electronics] A rectifier based upon a semiconductor. Also known as junction rectifier.

junction laser [Optics] A laser in which a junction in a semiconductor serves as the source of coherent laser beam.

junction rectifier *See* junction diode.

junction transistor [Electronics] A transistor in which emitter and collector barriers are formed between semiconductor regions of opposite conductivity type.

jupiter [Astronomy] Largest planet in the solar system, and fifth in the order of distance from the sun, having twelve small satellites, with its orbit between those of mars and saturn; diameter 142800 kilometres. Mean distance from the sun 778.34 million kilometres. Sidereal revolution period 11.86 years; mass approximately 318.4 times that of the earth; surface temperature probably about -150 C.

juvenile hormone [Biochemistry] A hormone secreted by endocrine . glands associated with the brain in insects that prevents metamorphosis into the adult form and maintains the presence of larval characteristics.

juvenile water [Geology] Water in earth's crust that has come from magma.

juxtaposition [Scientific Techniques] The act of placing side by side.

K-A age [Geology] The radioactive age of a rock determined from the ratio of potassium-40 to argon-40 present in the rock.

K-A decay [Nucleonics & Nuclear physics] Radioactive decay of potassium-40 to argon-40 as a result of an orbital electron capture by potassium-40. The ratio of potassium-40 to argon-40 is used to determine the age of rocks.

kainite [Minerals] $MgSO_4$. $KCl.3H_2O$. A natural hydrated double salt of potassium and magnesium that contains about 30% of potassium chloride.

kaolin [Minerals] Any of a group of clay minerals, including kaolinite, nacrite, dickite, and anauxite.

karyo- [Biology] A prefix denoting the nucleus of a cell or its contents; *e.g.*, 'karyotype', the sum of morphological characteristics of the chromosomes of a cell.

karyolysis [Biology] Dissolution of a cell nucleus.

karyokinesis [Biology] The phenomena involved in division of the nucleus in mitosis.

katabolism *See* catabolism.

katharometer [Engineering] A device for measuring thermal conductivity, especially as a detector in gas chromatography. Also known as thermal conductivity cell.

K-band [Physics] A band of radio frequencies extending from 10,900 to 36,000 megahertz, corresponding to wavelengths of 2.75 to 0.834 cm.

K-capture [Nucleonics & Nuclear physics] A type of beta interaction in which a nucleus captures an electron from K-shell and emits a neutrino.

keep [Engineering] Short bars of soft iron or steel placed across the pole of a permanent magnet to prevent permanent magnets from losing their magnetism.

kelp [Materials] Sea-weed or its ash, used as a source of iodine.

kelvin [Physics] K. The SI unit of thermodynamic temperature defined as the fraction 1/273.16 of thermodynamic temperature of the triple point of water, *i.e.*, the triple point of water contains exactly 273.16 kelvins. The units of kelvin and celsius (centigrade) temperature interval are identical. A temperature expressed in degrees celsius is equal to the temperature in kelvins ⁻< 273.15 C. This is true both for thermodynamic temperatures and on the International Practical Temperature Scale. The name 'degree kelvin' (Symbol^{O}K) was discontinued by international agreement in 1967.

Kelvin effect *See* Thomson effect.

Keplerian motion [Astronomy] Orbital movement of a body about another that is not disturbed by the presence of a third celestial body.

Kepler's laws [Astronomy] Three laws that describe the motions of planets in their orbits. I. The planets describe elliptical orbits with the sun at one of the focus of the ellipse. II. The line between a planet and the sun sweeps out equal areas in equal times. III. The square of the period of a planet's orbit is proportional to the cube of the semi-major axis of the ellipse.

keratin [Biochemistry] Any of a group of fibrous proteins occurring in hair, feathers, hooves, and horns. Keratins have coiled polypeptide chains that combine to form supercoils of several polypeptides linked by disulphide bonds between adjacent cysteine amino acids.

kerenel [Physics] A positively charged nucleus lacking the outer most valence electrons.

kerosine [Materials] A refined petroleum fraction used as a fuel for domestic purposes, jet engines, and as a base for insecticides, its components are mostly paraffinic and naphthenic hydrocarbons in the C_{10} to C_{14} range. Also known as kerosene; paraffin oil; lamp oil.

Kerr cell [Optics] A glass cell filled with a dielectric liquid, such as nitrobenzene, which contains two electrodes placed between two polarizing mediums. Light can only pass through the cell if the two planes of polarization are parallel.

Kerr effect [Optics] When plane-polarized light is reflected from a highly polished pole of an electromagnet the light becomes elliptically polarized. Similarly, if a beam of light is passed through certain transparent liquids or solids to which a potential difference is applied, the plane of polarization of the light is rotated through an angle that depends upon the magnitude of the applied potential difference. This effect is made use of in the Kerr cell.

ketals [Chemistry] Organic compounds, similar to acetals, formed by addition of an alcohol to a ketone. If one molecule of ketone (RR"CO) reacts with one molecule of alcohol R"OH, then a *hemiketal* is formed. The rings of ketose sugars are hemiketals. Further reaction produces a full ketal (RR"C(OR")$_2$).

keto-enol tautomerism [Chemistry] A form of tautomerism in which a compound containing a -CH_2-

CO- group (the *keto form* of the molecule) is in equilibrium with one containing the -CH=C(OH)- group (the enol). It occurs by migration of a hydrogen atom between a carbon atom and the oxygen on an adjacent carbon. See isomerism.

ketone body [Chemistry] Any of the three organic compounds named acetoacetic acid (3-oxobutanoic acid, CH_3COCH_2COOH), β-hydroxybutyric acid (3-hydroxybutanoic acid, $CH_3CH(OH)CH_2COOH$), and acetone or propanone (CH_3COCH_3) produced by the liver as a result of the metabolism of body fat deposits. Ketone bodies are normally used as energy sources by peripheral tissues.

ketones [Chemistry] A series of organic compounds having the general formula $RR''C=O$, where R and R'' are univalent hydrocarbon radicals, *e.g.*, acetone, dimethyl ketone, $(CH_3)_2C=O$.

ketose [Biochemistry] A monosaccharide that contains a carbonyl group at other than a terminal position.

ketosis [Biochemistry] A condition in which the ketone body concentration of the blood, tissues, and urine is abnormally high.

keV [Electricity] The symbol for 1000 electron-volts.

key [Computers] A data item that identifies the data record. [Electricity] A manual switch used for transmitting code signals.

kieselguhar [Geology] A mass of hydrated silica (SiO_2) formed from skeletons of minute plants known as diatoms. It is a very porous and absorbent material, used for filtering and absorbing various liquids; in the manufacture of dynamite and in other industries. Also known as diatomaceous earth; infusorial earth.

killed steel [Metallurgy] Completely deoxidised steel obtained by addition of aluminium or silicon, in which the reaction between carbon and oxygen during solidification is suppressed.

killed vaccine [Medicine] A suspension of killed microorganisms used as antigens to produce immunity.

kiloton [Physics] A unit used to specify the yield of a fission or fusion bomb, equal to explosive power of 1000 metric tons of trinitrotoluene (TNT).

kilovar [Electricity] A unit of 1000 volt amperes reactive.

kilowatt-hour [Electricity] kWh. The commercial unit of electrical energy. It is equivalent to a power consumption of 1000 watts for 1 hour.

kinase [Biochemistry] An enzyme catalysing phosphorylation of acceptor molecule by ATP. An enzyme that activates the inactive form of other enzymes.

kinematics [Mechanics] The branch of mechanics concerned with the phenomena of motion without reference to mass or force. Kinematics deals with motion from the standpoint of measurement and precise description, while dynamics is concerned with the causes or laws of motion.

Kinematic viscosity [Physics] Symbol v. The ratio of the viscosity of a liquid to its density. The SI unit is $m^2 s^{-1}$.

kinetic energy [Mechanics] The energy a body possess by virtue of its motion. The kinetic energy of a mass m, moving with velocity v, is $+mv^2/2$. The energy will be in joules if m is in kilograms and v is in metres per second. (In c.g.s units it will be in ergs.) The kinetic energy of rotation of a body. whose moment of inertia about an axis is I, and whose angular velocity about this axis, is w, is $I\omega^2/2$. Again the energy will be in joules if I is in kg m^2 and w is in radians per second.

kinetics [Chemistry] The study of the rates at which chemical reactions proceed. [Mechanics] The dynamics of material bodies.

kinetic theory of gases [Physics] A theory which attempts to explain the behaviour of physical systems especially gases on the assumption that gases consist of molecules in ceaseless motion in space, the kinetic energy of the molecules depending upon the temperature of the gas; the molecules are considered to be perfectly elastic particles that collide with each other and with the walls of the containing vessel (*see* elastic collision). The pressure exerted by a gas on the walls of the vessel is due to the collisions of the molecules with it.

kinin [Biochemistry] One of the group of polypeptides produced in blood and tissues and acting on blood vessels, smooth muscles, and certain nerve endings.

kink [Engineering] A lightened loop in a wire roop resulting in permanent deformation and damage to the wire.

kink instability [Physics] In a thermonuclear reaction experiment, a type of hydromagnetic instability in the magnetically confined plasma resulting from a local deformation of the plasma. The kink tends to grow because the magnetic lines of forces of the self-induced confining field are crowded on the concave side of the kink.

Kirchhoff's equations [Physics] Equations which state that the partial derivative of the change of enthalpy (or of internal energy) during a reaction, with respect to temperature, at constant pressure (or volume) equals the change in heat capacity at constant pressure (or volume).

Kirchhoff's laws [Electricity] A set of rules for calculating unknown currents, resistances, and voltages in an electric circuit, I. In any network of wires the algebraic sum of the electric currents that meet at a point is zero. II. The algebraic sum of the electromotive forces in any closed circuit of mesh is equal to the algebraic sum of the products of the resistances of each portion of the circuit and the currents flowing through them.

kish [Metallurgy] A variety of free graphite that floats on the surface of molten cast iron as it cools.

kjeldahl's method [Chemistry] A quantitative method of determining the nitrogen content of an organic compound. The compound is decomposed with concentrated sulphuric acid to convert the nitrogen into ammonium sulphate. The sulphate is estimated by adding excess alkali, distilling the ammonia into a standard acid solution, and measuring the excess acid by titration.

klystron [Electronics] An associated electron beam tube used to generate or amplify electromagnetic radiation in the microwave region, by velocity modulation; used as amplifier or oscillator.

knock [Engineering] Ignition of a portion of the gasoline in the cylinder head of an automobile engine due to spontaneous oxidation rather than to spark. It causes serious power loss, especially in high compression engines. See also octane number.

knot [Physics] A unit of speed equal to 1 nautical mile per hour.

(Approximately 1.15 statute miles per hour,) or 1.852 km per hour, or 0.5144 metres per second.

knowedge base [Computers] Artificial intelligence data bases that are not merely files of uniform content, but are collections of facts inferences, and procedures corresponding to the types of information needed for problem solution.

Kohlrausch's law [Chemistry] if a salt is dissolved in water, the conductivity of the (dilute) solution is the sum of two values - one depending on the positive ions and the other on the negative ions.

Kondo effect [Electricity] The large anomalous increase in the resistance of certain dilute alloys of magnetic materials in non-magnetic hosts as the temperature is lowered.

Kopp's law [Chemistry] The law that for solids the molal heat capacity of a compound at room temperature and pressure approximately equals the sum of the heat capacities of the elements in the compound.

Krebs cycle [Biochemistry] A complex cycle of enzymatic reactions involving oxidation of two carbon acetyl unit to carbon dioxide and water to provide energy for storage in the form of high energy phosphate bonds. Also known as citric acid cycle; tricarboxylic acid cycle.

Kroll process [Metallurgy] A process for obtaining titanium or zirconium from their ores by producing the tetrachloride of the metal reducing it under reduced pressure or by reacting it with magnesium at red heat.

krypton [Chemistry] Kr. Element At. No. 36, At. Wt. 8380 m.p. 116.5K, b.p. 920.8K. A colourless noble gas element; placed in zero group of the periodic table; used to fill luminescent electric tubes.

Kundt effect *See* Faraday effect.

Kundt rule [Physics] The rule that the optical absorption bonds of a solution are displaced towards the red when its refractive index increases because of the changes in composition of other causes.

Kupfer nickel [Geology] A naturally occurring form of nickel arsenide, NiAs; an important ore of nickel.

Kurrol's salt [Chemistry] $NaPO_3(IV)$. A crystalline high temperature form of sodium meta phosphate obtained by seeding the melt at 823K.

L

l- [Chemistry] Prefix indicating that a compound is levorotatory. A minus sign (-) is now preferred.

L [Chemistry] Prefix indicating the left-handed enantiomer of an optical isomer.

label [Computers] A data item that serves to identify a data record. [Nucleonics & Nuclear physics] A stable form of radioactive nuclide used to investigate some process, such as chemical reaction. *See also* isotopic tracer.

labelled compound [Chemistry] A compound in which a stable atom has been replaced by a radioactive isotope of that atom in order to investigate the chemical process(s).

labile [Chemistry] Describing a chemical compound in which certain atoms or groups can easily be replaced by other atoms or groups. The term is applied to coordination complexes in which ligands can easily be replaced by other ligands in an equilibrium reaction.

lachrymator [Chemistry] A gas that is strongly irritant to the eyes. Also known as lacrimator; tear gas.

lacquer [Materials] A protective or decorative coating agent that primarily dries by evaporation of solvent, rather than by oxidation or polymerization. It contains a substantial amount of cellulose derivative such as cellulose nitrate, cellulose acetate, or cellulose butyrate.

lactams [Chemistry] Organic compounds containing a ring of atoms in which the group -NH.CO.-forms part of the ring. Lactams can be formed by reaction of an $-NH_2$ group in one part of a molecule with a -COOH group in the other to give a cyclic amide. They can exist in an alternative tautomeric form, the *lactim* form in which the hydrogen atom on the nitrogen has migrated to the oxygen of the carbonyl to give -N=C(OH)-. The pyrimidine base uracil is an example of a lactam.

lactase [Biochemistry] An enzyme present in intestinal juices and mucosa which catalyzes the production of glucose and galactose from lactose.

lactation [Physiology] Secretion of milk by mammary glands.

lactometer [Engineering] A hydrometer used to check the specific gravity, *i.e.*, the amount of water present in milk.

laevorotatory [Chemistry] Rotating or deviating the plane of vibration of polarized light to the left (observer looking against the oncoming light). *See* optical activity.

lake [Chemistry] In dyeing, a coloured insoluble substance formed by the chemical combination of a soluble dye with a mordant.

lambda [Mechanics] A unit of volume equal to 10^{-6} litre or 10^{-9} cubic metre.

lambda particle [Nucleonics & Nuclear Physics] An elementary particle; classified as a hyperon,

that has no charge and is 2183 times heavier than an electron.

lambda point [Physics] λ. The temperature of 2.186 K below which helium becomes a superfluid.

lambert [Optics] A unit of luminance. The luminance of a uniform diffuser of light that emits one lumen per square cm.

Lambert's law [Optics] The illuminance of a surface upon which the light falls normally from a point source is inversely proportional to the square of the distance between the surface and the source. If the normal to the surface makes an angle θ with the direction of the rays, then the illuminance is proportional to $\cos\theta$.

lamellar solids [Chemistry] Solid substances in which the crystal structure has distinct layers (*i.e.*, has a layer lattice). The micas are an example of this type of compound. Intercalation compounds are lamellar compounds formed by interposition of atoms, ions, etc., between the layers of an existing element or compound. For example, graphite is a lamellar solid.

laminar flow [Mechanics] The flow of a fluid that closely follows the shape of a streamlined surface without turbulence.

laminate [Materials] A sheet of material made of several different bonded layers.

laminated iron [Metallurgy] Thin sheets of iron (or, more frequently of its alloy) used for cores of transformers instead of solid iron cores, in order to reduce losses due to eddy currents.

lamp-black [Materials] An allotropic form of carbon obtained by burning low-grade heavy oils in insufficient air, in a closed system; used as black pigment. Also known as soot.

lanthanide contraction [Chemistry] A phenomenon encountered in lanthanides; the radii of the atoms of the members of $4f$ series decrease slightly with the increase in atomic number. It is seen that from La^{3+} to Lu^{3+}, the ionic radius shrinks from 1.15 $A°$ to 0.93 $A°$ This is due to imperfect mutual shielding effect of f electrons because $4f$ orbitals are much diffused, hence the inward pull experienced by the $4f$ electrons increases. The contraction effect is of sufficient magnitude to cause the elements which follow in the third transition series to have sizes very similar to those of the second row of transition elements, e.g., Hf^{4+} and Zr^{4+} have almost indentical ionic radii being 0.80 $A°$ and 0.81$A°$ respectively.

lanthanides [Chemistry] A series of elements in the periodic table, generally considered to range in proton number from cerium (58) to lutetium (71) inclusive. The lanthanides all have two outer s-electrons (a $6s^2$ configuration), follow lanthanum, and are classified together because an increasing proton number corresponds to increase in number of $4f$ electrons. In fact, the $4f$ and $5d$ levels are close in energy and the filling is not smooth. The outer electron configurations are as follows :

57 lanthanum (La) $5d^1 6s^2$

58 cerium (Ce) $4f^1 5d^1 6s^2$ (or $4f^2 6s^2$)

59 praseodymium (Pr) $4f^3 6s^2$

60 neodymium (Nd) $4f^4 6s^2$

61 promethium (Pm) $4f^5 6s^2$

62 samarium (Sm) $4f^6 6s^2$

63 europium (Eu) $4f^7 6s^2$

64 gadolinium (Gd) $4f^7 5d^1 6s^2$

65 terbium (Tb) $4f^9 6s^2$

66 dysprosium (Dy) $4f^{10} 6s^2$

67 holmium (Ho) $4f^{11} 6s^2$

68 erbium (Er) $4f^{12} 6s^2$

69 thulium (Tm) $4f^{13} 6s^2$

70 ytterbium (Yb) $4f^{14} 6s^2$

71 lutetium (Lu) $4f^{14} 5d^1 6s^2$

Note that lanthanum itself does not have a $4f$ electron but it is generally clasified with the lanthanides because of its chemical similarities, as are yttrium (Y) and scandium (Sc). Scandium, yttrium and lanthanum are d-block elements; the lanthanides and actinoids make up the f-

block. Also known as lanthanoids; lanthanons; rare-earths.

lanthanum [Chemistry] La. Element At. No. 57, At. Wt. 138.91,
m.p. 1191K, b.p. 3737K; placed
in III B group of the periodic
table. Its compounds are used in
electronic devices and as catalyst.

Laplace equation [Mathematics]
The partial differential equation:

$$\delta^2 u/\delta x^2 + \delta^2 u/\delta y^2 + \delta^2 u/\delta z^2 = 0$$

It may also be written in the form
$\Delta^2 u = 0$, where δ^2 is called the
Laplace operator.

lard [Materials] Purified internal
fat of the hog that mainly
contains stearin, palmitin, and
olein.

Larmor precession [Physics] The
orbital motion of the electrons
about the nucleus of an atom is
usually such as to give the atom
a resultant angular momentum
and a magnetic moment. These
two properties cause the atom to
precess (see precessional motion)
about the direction of any applied magnetic field. This is
known as Larmor precession.

larvicide [Materials] A chemical
agent used to kill the eggs and
larva of insects.

laser [Optics] Light Amplification
by Stimulated Emission of Radiation. A device that produces a
beam of coherent or monochromatic light as a result of photon-
stimulated emission. Such beams
have extremely high energy.

Materials capable of producing
this effect are certain high-purity
crystals and semiconductors. It
works on essentially the same
principle as the maser, except
that the 'active medium consists
of, or is contained in, an optically
transparent cylinder with a reflecting surface at the other. The
stimulated waves make repeated
passages up and down the cylinder, some of them emerging as
light through the partially reflecting end.

Lasers have also been constructed
using a mixture of helium and
neon to produce a continuous
beam. Lasers are used in industry
for cutting diamonds, in flash
photolysis, spectroscopy medicine, and surgery. Also known as
optical maser.

laser amplifier [Electronics] A
laser which is used to increase
the output of another laser. Also
known as light amplifier.

laser beam [Optics] A narrow
beam of coherent, powerful, and
nearly monochromatic electromagnetic radiation emitted by a
laser.

lasing [Optics] Production of visible or infrared light waves
having nearly a single frequency
pumping or exciting electrons
into high energy states in a laser.

latent heat [Physics] The quantity
of heat absorbed or released in
an isothermal transformation of
phase.

latent heat of fusion [Physics] The

heat required to convert unit mass of a solid to a liquid at the same temperature.

latent heat of vapourization [Physics] The heat required to convert unit mass of liquid to vapour at the same temperature; measured in joules per kilogram. The corresponding molar latent heats are measured in joules per mole.

lateral inversion [Optics] The change produced in the image formed by a plane mirror in which the side are reversed.

latex [Materials] 1. A milky fluid produced by certain plants; the most important is that obtained from the rubber tree consisting mainly of a colloidal suspension of rubber globules in a watery liquid 2. An analogous emulsion or suspension of a synthetic rubber or similar polymer.

latitude [Astronomy] The angular distance of a point from the equator measured upon the curved surface of the earth.

lattice [Crystallography] The regular network of fixed points in three dimensional space about which molecules, atoms, or ions vibrate in a crystal. [Nucleonics & Nuclear physics] In a nuclear reactor, a structure consisting of discrete bodies of fissile and non-fissile material (especially moderator), arranged in a regular geometrical pattern.

lattice constant [Crystallography] A parameter defining the unit cell of a crystal lattice, that is, lenth of one of the edges or an angle betwen the edges.

law [Scientific Techniques] A regularity which applies to all members of a broad class of phenomena.

law of conservation of charge *See* conservation of charge.

law of conservation of energy *See* conservation of energy.

law of conservation of mass *See* conservation of mass.

law of conservation of momentum *See* conservation of momentum.

law of constant angles [Crystallography] The law that the angles between the faces of a crystal remain constant as the crystal grows.

law of constant composition   *See* Laws of chemical combination.

law of constant heat summation ι *See* Hess's law.

law of cooling *See* Newton's law of cooling.

law of corresponding states [Chemistry] For two substances, any two ratios of pressures, or volume to their respective critical properties are equal, the third ratio must equal the other two.

law of electric charges [Electricity] The law that like charge repel, and unlike charges attract each other.

law of electromagnetic induction *See* Faraday's laws of electromagnetic induction.

law of electrostatic attraction *See* Coulomb's law.

law of gravitation *See* Newton's law of gravitation.

law of isomorphism *See* Mitcherlich's law.

law of mass action [Chemistry] The law of mass action states that the rate at which a chemical reaction takes place at a given temperature is proportional to the product of the active masses of the reactants. The active mass of a reactant is taken to be its molar concentration.

For example, for a reaction

$$A + B \longrightarrow C$$

the rate is given by

$$R = k[A][B]$$

where k is the rate constant.

law of partial pressure *See* Dalton's law of partial pressure.

laws of motion *See* Newton's laws of motion.

laws of reflection *See* reflection of light, laws of.

laws of refraction *See* refraction, laws of.

laws of thermodynamics [Physics] Three laws of thermodynamics are : I. Heat is a form of energy, and the total amount of energy of all kinds in an isolated system is conservation of energy. II. 1. There is a preferred direction for any process. 2. It is impossible for a cyclic process to take heat from a cold reservoir and convert it into work without at the same time transferring heat from a hot to cold reservoir. 3. It is impossible to construct a machine which is able to convey heat by a cyclic process from one reservoir at lower temperature to another at higher temperature unless work is done on the machine by some outside agency. III. The entropy of a pure perfectly crystalline substance is zero at the absolute zero of temperature.

L band [Electronics] A band of radiofrequencies extending from 390 to 1550 megahertz, corresponding to wavelengths of 76.9 to 19.37 centimetres.

L capture [Nucleonics & Nuclear physics] A type of generalized beta interaction in which a nucleus captures an electron from L shell.

leaching [Chemistry] Washing out a soluble constituent from a mixture.

lead [Chemistry] Pb Element. At. No. 82. At.Wt. 207.19, placed in group IV A of the periodic table. A soft, bluishwhite metal, m.p. 600.5K, b.p. 2013K, is used in the lead accumulator; in alloys and in plumbing; compounds are used in paint manufacture and in petrol additives. Also known as plumbum.

lead-chamber process [Chemistry] A process for the manufacture of sulphuric acid by the action of nitrogen dioxide, NO_2, on sulphur dioxide, SO_2, to give nitric oxide, NO, and sulphur trioxide, SO_3. The former reacts with oxygen from the air to give NO_2 again; the SO_3 combines with water to give sulphuric acid, the process being carried out in large lead chambers. The process is now obsolete and has been replaced by the contact process.

lead equivalent [Nucleonics & Nuclear Physics] A measure of the absorbing power of a radiation screen, expressed as the thickness of a lead screen in millimetres that would afford the same protection as the material being considered.

lead pigment [Chemistry] Compounds of lead used in paints to give colour; examples are white leads; basic lead carbonate; lead sulphate; lead chromate; and lead oxide.

leakage current [Electricity] The flow of direct current through a poor dielectric in a capacitor. [Electronics] The alternating current that pass through a rectifier without being rectified.

Le Chatelier principle [Chemistry] If a system in equilibrium is subjected to a external force the system tends to react in such a way as to oppose the effect of the applied force.

lecithins [Biochemistry] Naturally occurring complex lipids essentially consisting of glycerides in which one of the acyl groups is replaced by a phosphorycholine group; they are chemically similar to fats, but additionally contain nitrogen and phosphorus.

Leclanche cell [Electricity] A primary voltaic cell consisting of a carbon rod (the anode) and a zinc rod (the cathode) dipping into an electrolyte of a 10-20% solution of ammonium chloride. Polarization is prevented by using a mixture of manganese dioxide mixed with crushed carbon, held in contact with the anode by means of a porous bag or pot; this reacts with the hydrogen produced. This wet form of the cell, has an e.m.f. of about 1.5 volts. Leclanche cells are widely used for many purposes which require an intermittent current. The common dry cell is a special form of Leclanche cell.

LED See ligth-emitting diode.

left-hand rule See Fleming's rules.

Lenard rays [Electronics] Cathode rays produced in air by a tube that has a thin glass or metallic foil window at the opposite end of the cathode, through which the electron beam can pass into atmosphere.

lens [Optics] Any device that causes a beam of rays to converge or diverge on passing through it. The optical lens is a portion of a transparent refracting medium (*see* refraction of light), usually glass, bounded by two surfaces, generally curved. Such lenses are classified according to the nature of the surfaces into bi-concave, bi-convex, plano convex, etc. The centres of the spheres of which the lens surfaces are considered to form a part are termed the centres of curvature; the line joining these is the optical axis, the optical centre is a point on the axis within the lens; all rays passing through this point emerge without deviation. A parallel beam of light incident on a lens is made to converge (convex lens) or diverge (concave lens). The point of divergence is called a principal focus. Regarding all distances as being measured from the optical centre, and taking all distances as positive when measured in a direction opposite to that of the incident light, the distances of the object and image from the lens are given by the formula $1/v - 1/u = 1/f$, where u and v are the distances from the lens of object and image respectively and, f is the focal length, *i.e.*, the distance of the focus from the lens.

lenticular [Optics] Pertaining to a lens, especially a bi-convex lens, or resembling such a lens in shape.

Lenz's law [Physics] When an electric circuit and a magnetic field move relatively to each other, the electric current induced in the circuit will have a magnetic field opposing the motion. *See* induction, electromagnetic.

lepton [Nucleonics & Nuclear Physics] A collective name for electrons, muons, and neutrinos. They form a class of elementary particles that react by the electromagnetic interaction and the weak interaction but are insensitive to the strong interaction. All leptons have spin, but antileptons have spin in an opposite direction to that of their corresponding particles.

lethal gene [Physics] *See* lethal mutation.

lethal mutation [Biochemistry] Mutation of gene to a yield a totally defective gene product unable to sustain the life of organism. Also known as lethal gene.

leucocidin [Biochemistry] A toxic substance released by certain bacteria which destroys leucocytes.

leucocyte [Biochemistry] A type of blood cell that has a nucleus but no pigment. White cells are larger and less numerous than red cells (about 6000-9000 per cubic millimetre of blood). They are important in defending the body against disease because they devour bacteria and produce antibodies. They are all capable of amoeboid movement.

There are several types of leucocytes. They can be divided into two groups, granulocytes and agranulocytes, according to presence or absence of granules in the cytoplasm. Leucocytes have a short life span and are continuously produced in the myeloid tissue of the red marrow. Also known as white blood cell; white corpuscle.

leucoplast [Biochemistry] A colourless plastid, *i.e.*, one not containing chlorophyll or any other pigment.

leucosin [Biochemistry] One of the structurally stable scleroproteins found in wheat and other cereals.

leucosis [Medicine] An excess of white blood cells.

lever [Physics] A rigid bar that may be turned freely about a fixed point of support, the fulcrum used to multiply force or motion.

leverage [Mechanics] The multiplication of force or motion achieved by a lever.

levigate [Engineering] To separate fine powder from coarse material by suspending the fine material in a liquid.

levo form [Chemistry] An optical isomer that induces levorotation, *i.e.*, rotates the plane of polarization of plane polarized light to the left.

levulose [Biochemistry] Levorotatory D-fructose.

Lewis acid [Chemistry] A substance that can accept an electron pair from a base; $AlCl_3$, BF_3, SO_3 are some of the examples of Lewis acids.

Lewis base [Chemistry] A substance that can donate an electron pair; *e.g.*, NH_3, OH^-, H_2O.

libration [Physics] Any oscillatory rotational motion such as of a molecule in a solid, or that of the moon, which does not have enough energy to make complete rotation. Due to libration, about 58% of the moon's surface can be seen from the earth.

life cycle [Biology] The functional and morphological stages through which an organism passes between two successive primary stages.

ligand [Chemistry] A single atom or a group of atoms attached to a central metal atom in a coordination compound. In a ligand one, two, or more atoms may be attached to the central metal atom, and it is referred to correspondingly as a uni-, bi-, or multidentate ligand, *e.g.*, NH_3, CN^-, F^-, EDTA.

ligase [Biochemistry] Any of a class of enzymes that catalyse the formation of covalent bonds using the energy released by the cleavage of ATP. Ligases are important in the synthesis and repair of many biological molecules, including DNA.

light [Optics] Electromagnetic

radiation able to be detected by the human eye, ranging approximately from 4000 A°(extreme violet) to 7700A° (extreme red). Light is produced by surfaces at temperatures above 900 K, however the majority of radiation emitted is infrared.

light-emitting diode [Electronics] (LED). A device used to display figures (digital display), etc., in calculators and other equipment giving a visual display. It consists essentially of a semiconductor diode, made from such materials as gallium arsenide, in which light is emitted at a *p-n* junction when electrons and holes recombine. The light emitted is proportional to the bais colour depends on the type of material used.

light metal [Engineering] A metal or alloy of specific gravity less than 3 that is strong enough for construction purposes, *e.g.*, aluminium, magnesium and their alloys.

lightmeter [Engineering] A small device for measuring illumination.

lightning [Electricity] An electric discharge in the form of spark or flash between two charged clouds, or between a cloud and the earth.

lightning conductor [Electricity] A conductor of electricity connected to earth and ending in one or more sharp points attached to the higher part of a building. It provides a direct path of low resistance to earth.

light pen [Computers] An input/output computer device used with a visual display unit. When pointed at a cathode ray tube it can sense whether or not the spot is illuminated.

light quantum *See* photon.

light-sensitive [Electronics] Having photoconductive, photoemissive or photovoltaic characteristics.

light water reactor [Nucleonics & Nuclear Physics] A nuclear reactor that uses ordinary water as moderator.

light-year [Astronomy] An astronomical measure of distance; the distance travelled by light in one year; equal to 9.4650×10^{15} m.

lignin [Biochemistry] A complex organic material that occurs in the woody tissues of plants, often combined with cellulose.

lignite [Geology] A brownish-black, natural deposit resembling coal, which contains a higher percentage of hydrocarbons than ordinary coal; it is probably of more recent origin. Also known as brown coal.

lime [Chemistry] Specifically calcium oxide (CaO); more generally any of the various chemical and physcial forms of quicklime, hydrated lime, and hydraulic lime.

lime, air-slaked [Chemistry] Lime which has absorbed carbon dioxide from atmosphere, It consists of a powder composed of calcium carbonate and calcium hydroxide.

lime, chlorinated [Chemistry] CaCl-(OCl). A white powder obtained by treating slaked lime with chlorine; used as deodorizer, disinfectant, and bleaching agent. Also known as bleaching powder.

lime hydrated [Chemistry] $Ca(OH)_2$. Calcium hydroxide; obtained by treating quick lime with water.

lime hydraulic [Chemistry] A variety of calcined limestone which, when pulverized, absorbs water without swelling or heating and gives a cement that hardens under water. The limestone burned for this purpose usually contains 10-17% silica, alumina and iron and 40-45% lime and magnesia.

lime-nitrogen [Materials] Trade name for calcium cynamide. Also known as nitrolime.

limestone [Minerals] $CaCO_3$. Natural calcium carbonate.

lime-water [Chemistry] A solution of calcium hydroxide, $Ca(OH)_2$, in water.

limit [Mathematics] Limiting value. A function of a variable quantity x, written $f(x)$, approaches a limiting value k as x approaches a value a, if the difference $k-f(a+\delta)$ may be made smaller than any assignable value by making δ sufficiently small. [Physics] The lines appearing in the line spectrum of any element can be ground into definite series. The shortest wavelength of any such series is called the limit of the series. At this series limit, the lines crowd closer and closer together from the long wavelength side.

lindane [Chemistry] $C_6H_6Cl_6$. Trade name for gamma isomer of $1,2,3,4,5,6$-hexa-chlorocyclohexane. Also known as gamma-benzenehexachloride; used as pesticide.

Linde process [Engineering] A process for producing liquid air, based on the Joule-Thomson effect. Air is compressed and expanded through a nozzle, which causes it to cool. The cool air is passed through a counter-current heat exchanger to reduce the temperature of the incoming air. Eventually the temperature is reduced sufficiently to liquefy the air.

line [Computers] The basic unit of information exchange, consisting of one or more information elements, between a cache and main memory. Also known as block.

linear absorption coefficient [Physics] a. A measure of a medium's ability to absorb radiation, but not to scatter or diffuse it. It is given by $\phi/\phi_0 = e-al$, where ϕ_0

is the initial radiant flux or luminous flux and ϕl is the flux after it has travelled a distance l through the medium.

linear accelerator [Nucleonics & Nuclear Physics] A device that accelerates electrons, protons, or heavy ions in a straight line by the action of alternating voltages. Also known as linac.

linear actuator [Engineering] A device that converts some kind of power such as electric or hydraulic, into linear motion.

linear amplifier [Electronics] An amplifier in which the changes in output current are directly proportional to changes in applied input voltage.

linear expansion [Physics] Expansion of a body in one direction.

linear polarization See plane polarization.

line defect. See defect.

line of collimation [Optics] The imaginary line through the optical centre of the objective glass and the cross-hair intersection in the diaphragm, in a telescope.

line of force [Physics] An imaginary line in a field of force (such as electric, magnetic, or gravitational) whose tangent at any point gives the direction of the field at that point. Also known as line of flux; flux line.

line profile [Astronomy] A curve that indicates the internal vari-

ation in intensity of a spectral line of a celestial body.

line spectrum [Physics] A spectrum (emission or absorption) consisting of definite single lines, each corresponding to a particular wavelength; it is characteristic of an element in the atomic state.

lipase [Biochemistry] An enzyme secreted by the pancreas and the glands of the small intestine of vertebrates that catalyses the breakdown of fats into fatty acids and glycerol.

lipids [Biochemistry] A class of organic compounds that are esters of fatty acids and are insoluble in water but soluble in many organic solvents. They are usually divided into three groups : (a) 'Simple lipids', which include fats and oils as well as waxes; (b) 'Compound lipids', which include phospho-lipids and glycol-ipids; (c) 'Derived lipids', of which the most important are the steroids. Also known as lipoids.

lipochrome [Biochemistry] Any fat soluble pigment that occurs in natural fats, such as carotenoid. Also known as chromolipid.

lipoclastic [Biochemistry] Applied to enzymes having the power of hydrolyzing (see hydrolysis) fats into fatty acids and glycerin; e.g., lipase. Also known as lipolytic, fatsplitting.

lipolysis [Biochemistry] The splitting of component fatty acids from a lipid, *i.e.*, part of the process of catabolism liquid molecules. Lipolysis is affected in the body, largely in the gut, by the lipase enzyme.

lipoprotein [Biochemistry] A protein that includes a lipid in its structure.

lipotropic agent [Biochemistry] A substance that helps to regulate the metabolism of fats and cholesterol in animals, because of its affinity for fats and oil; *e.g.*, inositol.

Lipowitz alloy [Metallurgy] A fusible alloy, m.p. 338-343K, consisting of 50% bismuth, 27% lead, 13% tin, and 10% cadmium.

liquation [Physics] The separation of a solid mixture by heating till one of the constituents melts and can be drained away.

liquefaction [Physics] A change in phase of a substance to the liquid state, especially of a substance which is a gas at normal pressure and temperature; *e.g.*, liquefaction of air.

liquefaction of gases [Physics] The conversion of a gaseous substance into a liquid. This is usually achieved by one of four methods or by a combination of two of them :

(1) by vapour compression, provided that the substance is below its critical temperature;

(2) by refrigeration at constant pressure, typically by cooling it with a colder fluid in a counter-current heat exchanger;

(3) by making it perform work adiabatically against the atmosphere in a reversible cycle;

(4) by the Joule-Thomson effect (*see also* Linde process).

liquid [Physics] A state of matter intermediate between a solid and a gas, in which the molecules are relatively free to move with respect to each other but are restricted by cohesive forces to the extent that the liquid maintains a fixed volume. Liquids assume the shape of the vessel containing them, and are only slightly compressible.

liquid crystal [Materials] A substance that flows like a liquid but has some order in its arrangement of molecules. *Nematic crystals* have long molecules all aligned in the same direction, but otherwise randomly arranged. *Cholesteric* and *smectic* liquid crystals also have aligned molecules, which are arranged in distinct layers. In cholesteric crystals, the axes of the molecules are parallel to the plane of the layers; in smectic crystals they are perpendicular.

liquid-drop model [Nucleonics & Nuclear Physics] A model of the atomic nucleus in which the nucleons are regarded as being analogous to the molecules in a

liquid, the interactions between which maintain the droplet shape by surface tension. The model has been useful in the theory of nuclear fission.

liquid poison [Nucleonics & Nuclear Physics] A neutron absorbing liquid that can be injected into the cooling system of a nuclear reactor; used to shut down a nuclear reactor.

lithium [Chemistry] Li. Element. At. No. 3. At.Wt. 6.939, placed in IA group of the periodic table. A light, silvery-white alkali metal, m.p. 453K, b.p. 1613K. It is the lightest solid known. Chemically it resembles sodium, but is less reactive.

lithophone [Materials] A mixture of zinc sulphide, ZnS, and barium sulphate, $BaSO_4$. Used in paints as a non-poisonous substitute for white lead.

lithosphere See earth's crust.

litmus [Chemistry] A soluble purple substance of vegetable origin; it is turned red by acids and blue by alkalies; used as an acid-base indicator.

litre [Scientific Techniques] Unit of volume in the metric system. Formerly defined as the volume of 1 kilogram of pure water at 4°C and 760 mm pressure (which is equivalent to 1000.028 c.c.). However, is not used for high precision measurements. For approximate purposes 1 litre=1000 c.c., though this practice is now deprecated.

lixivation [Chemistry] The extraction of soluble material from a mixture by washing with water.

local cell [Electricity] A galvanic cell resulting from differences in potential between adjacent areas on the surface of a metal immersed in an electrolyte.

local oscillator [Electronics] The oscillator in a heterodyne or superheterodyne radio receiver that produces the radio frequency oscillation with which the received wave is combined.

locus [Biology] The particular fixed point of a gene in a chromosome. [Mathematics] The locus of a point is the line that can be drawn through adjacent positions of the point, thus tracing out the path of the point in space.

lodestone [Minerals] Fe_3O_4. A magnetic variety of natural iron oxide. Also known as magnetite.

logarithmic scale [Scientific Techniques] A scale of measurement in which an increase of one unit represents a tenfold increase in the quantity measured (for common logarithms).

longitude [Geology] The angle that the terrestrial meridian through the geographical poles and a point on the earth's surface makes with a standard meridian (usually through Greenwich) is the longitude of the point.

[Astronomy] In astronomy, the 'celestial longitude' is the angular distance of a celestial body from the vernal equinox along the ecliptic, measured through 360° towards the east.

longitudinal [Scientific Techniques] Pertaining to lengthwise dimension.

longitudinal wave [Physics] Wave in which the vibration or displacement takes place in the direction of propagation of the waves; e.g., sound waves.

long sight [Medicine] A defect of vision resulting from too short an eye-ball so that unaccommodated rays focus behind the retina. Also known as farsightedness; hypermetropia; hyperopia.

Lorentz transformation [Mathematics] A set of equations for correlating space and time coordinates in two frames of reference, especially at relativistic velocities.

Loschmidt's number [Chemistry] The number of particles per unit volume of an ideal gas at STP. It has the value $2.687\ 19 \times 10^{25}$ m^{-3}.

loudness of sound [Physics] The maguitude of the physiological sensation produced by sound. As the ear responds differently to different frequencies, the loudness of a sound will depend to a certain extent on its frequency. However, loudness can be roughly correlated with the cube root of the intensity of sound, and different levels can be conveniently compared by the units decibel and phon.

low frequency [Physics] LF. A frequency in the range 30-300 kilohertz in the radio spectrum.

LSI [Electronics] Large-scale integration; from 1000 to 10000 transistors in a circuit.

lubricant [Materials] A substance used to reduce friction between parts or objects in relative motion.

luciferase [Biochemistry] An enzyme occurring in fire flies that catalyzes the oxidation of luciferin.

luciferin [Biochemistry] A complex albumin present in some animals, which under the influence of enzyme luciferase, emits heatless light when combined with oxygen. When luciferin comes in contact with ATP, a chemical reaction occurs which causes luminescence.

lumen [Optics] lm. The SI unit of luminous flux equal to the flux emitted by a uniform point source of 1 candela in a solid angle of 1 steradian.

luminance [Optics] L. The luminous intensity of any surface in a given direction per unit projected area of the surface, viewed from that direction. It is given by the equation $L=dI/(dA\cos\theta)$, where I is the luminous intensity and θ is the angle between the line of sight and the normal to the surface area *A* being considered. It is measured in candela per square metre.

luminescence [Optics] The emission of light from a body from any cause other than high temperature. It is caused by the emission of photons when an excited atom returns to the ground state. Fluorescence and phosphorescence are particular cases of luminescence.

luminescent dye [Materials] A dye that is made luminous by excitation with an outside energy source; used in luminous paint.

luminescent screen [Electronics] The screen in a cathode-ray tube, which becomes luminous when bombarded by an electron beam and maintains its luminosity for appreciable time period.

luminosity [Optics] The property of emitting light. [Astronomy] The amount of light emitted by a star, irrespective of its distance from the earth, usually expressed as a magnitude.

luminous flux [Optics]ϕ_v. A measure of the rate of flow of light, *i.e.*, the radiant flux in the wavelength range 380-760 nanometres, corrected for the dependence on wavelength of the sensitivity of the human eye. It is measured by reference to emission from a standard source, usually in lumens.

luminous intensity [Optics] I_v. A measure of the light-emitting ability of a light source, either generally or in a particular direction. It is measured in candelas.

luminous pigment [Materials] A pigment that absorbs light energy and radiates visible light when exposed to ultraviolet light.

Luna program [Astronomy] A series of soviet space probes launched for flight missions to the moon.

lunar crust [Astronomy] The outer layer of the moon.

lunar day [Astronomy] The time interval between two successive crossings of the meridian by the moon.

lunar month [Astronomy] The period of revolution of the moon about the earth.

lunar orbit [Astronomy] Orbit of a space craft around the moon.

lunation [Astronomy] The time period between two successive new moons, equal to 29 days 12 hours and 44 minutes.

lute [Materials] A substance such as clay or cement, for packing a joint or coating a porous surface to produce imperviousness to gas or liquid.

lutetium [Chemistry] Lu. Element. At. No. 71. At.Wt. 174.97., a lanthanide, placed in group IIIB of periodic table m.p. 1925K, b.p. 3675K.

lux [Optics] lx. The SI unit of illuminance equal to the illumination produced by a luminous flux of 1 lumen distributed uniformly over an area of 1 square metre.

lyddite [Materials] An explosive consisting of picric acid (trinitrophenol, $C_6H_2OH(NO_2)_3$, mixed with 10% nitrobenzene and 3% vaseline.

lye [Chemistry] A solution of potassium or sodium hydroxide.

Lyman series [Physics] A series of lines that occurs in the ultraviolet region of the spectrum of hydrogen.

lyophilic [Chemistry] Having an affinity for a solvent ('solvent-loving'; if the solvent is water the term *hydrophilic* is used).

lyophobic [Chemistry] Lacking any affinity for a solvent ('solvent-hating'; if the solvent is water the term *hydrophobic* is used).

lysin [Biochemistry] A substance, particularly antibodies, capable of lysing a cell.

lysine [Biochemistry] $NH_2(CH_2)_4$ CH-(NH_2)-COOH. An essential crystalline soluble amino acid, obtained from many proteins by hydrolysis.

lysis [Medicine] The dissolution or destruction of cells (especially blood cells or bacteria) by a class of antibodies called lysins.

lysogeny [Botany] The formation of an inter cellular space in plants by dissolution of cells.

lysol [Materials] A mixture of the cresols dissolved in soft soap; used as a disinfectant.

lysozyme [Biochemistry] An antibiotic enzyme found in egg white; acts as a mild antiseptic.

lytic infection [Biology] Penetration of a host cell by lytic phage.

lytic phage [Biology] Any phage that cause host cells to lyse.

laytic reaction [Biology] A reaction that leads to lysis of a cell.

m [Chemistry] Abbreviation for meta, milli.

M [Chemistry] Abbreviation for molarity; mega.

macassar oil [Materials] Any of the several fatty oils or oily preparations used in hairdressing preparations.

Mace [Materials] Trade name for a tear gas containing chloraceto phenone, can cause permanent damage to the eyes.

mace oil [Materials] An essential oil containing pinene and dipentene; used as a flavouring agent.

maceral [Geology] The microscopic organic substances found in coal.

macerate [Chemistry] To soften or break up a fibrous substance by soaking in water at room temperature, often accompanied by mechanical action.

machine [Engineering] A device, used to help in performing work. It is a combination of rigid or resistant bodies having definite motions and capable of performing useful work. The user applies a force to the machine, the machine applies the force to a load of some kind. In general the purpose of a machine is to overcome a large load with a small force.

machmeter [Engineering] An instrument for measuring the speed of an aircraft relative to the speed of sound. *See* mach number.

mach number [Mechanics] The ratio of the speed of a fluid or body to the local speed of sound. The speed of a fluid or body is therefore said to be supersonic if its mach number is greater than unity. *See also* hypersonic.

macro- [Scientific Techniques] Prefix denoting large, in contrast to microsmall.

macrometer [Optics] Instrument having a focusing telescope and two mirrors with which the ranges of distant objects can be estimated.

macromolecular [Chemistry] Consisting of or pertaining to macromolecules; having a very high molecular mass.

macromolecular crystal [Chemistry] A crystalline solid in which the atoms are all linked together by covalent bonds; *e.g.*, Carbon (in diamond), boron nitride, and silicon carbide.

Magellanic clouds [Astronomy] Two small patches of light that appear, from the southern hemisphere, to be detached from the main bright band of stars constituting the Milky Way. These objects are separate galaxies, being two of the smaller members of the local group to which our galaxy belongs.

macroscopic [Scientific Techniques] Large enough to be seen by naked eye.

magic numbers [Nucleonics & Nuclear Physics] The numbers 2, 8, 20, 28, 50, 82, and 126. Atomic nuclei containing these numbers of neutrons or protons have exceptional stability because of comparatively high binding energy.

magma [Geology] The molten rock material from which igneous rocks are formed.

magnalium [Metallurgy] A light alloy, r.d. 2-2.5; consists of aluminium with 5% to 30% magnesium.

magnesite [Minerals] $MgCO_3$.

Natural magnesium carbonate, which occurs in white masses; used in the manufacture of refractories and fertilizers.

magnesium [Chemistry] Mg. Element. At. No. 12. At.Wt. 24.312, placed in group IIA of the periodic table. A light, silvery-white metal m.p. 924K, b.p. 1380K, that tarnishes easily in air. It burns with an intense white flame to form magnesium oxide, MgO; used in lightweight alloy; in photography; incendiary bombs; production of iron, nickel, zinc, titanium, zirconium from their compounds and antiknock gasoline additives. Its compounds are used in medicine. It is essential cofactor for certain phosphate enzymes. High concentration of Mg^{2+} ions, are needed to maintain ribosome structure. It is contained in chlorophyll molecule and is thus essential for photosynthesis.

magnesothermic reduction [Metallurgy] Reduction of oxides to the corresponding metals at high temperatures with the aid of metallic magnesium. It is analogous to aluminothermic reduction.

magnet [Physics] A ferromagnetic substance that has a magnetic field and magnetic moment associated with it. *See also* magnetic domains.

magnetic amplifier [Electronics] A device for the amplification of small direct currents and of low frequency alternating currents. Also known as transductor.

magnetic bias [Physics] A steady magnetic field applied to the magnetic circuit of a relay or other magnetic device.

magnetic bottle [Physics] A magnetic field used in the containment of a plasma during controlled thermonuclear reaction experiments.

magnetic bubble memory [Computers] A form of computer memory in which a small magnetized region of a substance is used to store information. Bubble memories consist of materials, such as magnetic garnets, that are easily magnetized in one direction but hard to magnetize in the perpendicular direction.

magnetic card [Computers] A card with a magnetic surface on which data can be stored by selective magnetization.

magnetic circuit [Physics] A closed path following the lines of force of a magnetic field.

magnetic constant [Physics] Permeability of free space, μ_o. The fundamental constant that has the value 40×10^{-7} henry per metre. It arises as the constant of proportionality in Ampere's law, its value depending on the choice of units. *See also* magnetic permeability.

magnetic declination *See* declination.

magnetic dip *See* inclination.

magnetic dipole [Physics] An object such as a permanent magnet, current loop, or a particle with angular momentum, which experiences a torque in a magnetic field, and itself gives rise to a magnetic field, as if it consisted of two magnetic poles of opposite sign separated by a small distance.

magnetic domains [Physics] The fact that ferromagnetic substances are not necessarily always magnetized led to the theory that they consist of separate domains, each of which is spontaneously magnetized, but the magnetic moments of which may not be aligned. If an external magnetic field is applied to the substance to magnetic moment of the domains (not the domains themselves) are rotated so that they lie parallel to the field; the substance then acts as permanent magnet.

magnetic drum [Computers] A cylinder coated with magnetic material for storing information in a computer, especially in the backing storage.

magnetic elements [Physics] The three quantities, magnetic declination, angle of dip (*see* magnetic dip), and horizontal component, which define completely the earth's magnetic field (*see* magnetism, terrestrial) at any point.

magnetic equator [Physics] A line of zero magnetic dip lying fairly near the geographical equator, Also known as aclinic line; geomagnetic equator.

magnetic field [Physics] A field of force that is said to exist at any point if a small coil of wire carrying an electric current experiences a couple when placed at the point. A magnetic field may exist at a point as a result of the presence of either a permanent magnet or a circuit carrying an electric current, in the neighbourhood of the point.

magnetic field of electric current [Physics] A wire or coil carrying an electric current is surrounded by a magnetic field. The direction of the field relative to the current may be determined by the following simple rule: If a corkscrew, held in the right hand, is turned along the conductor in the direction of the current, the movement of the thumb indicates the direction of the magnetic field produced. The strength of the magnetic field at the centre of a circular coil of wire of radius r, consisting of n turns, in which a current of I amperes is flowing is; $nI/2\pi r$ amperes per metre in SI units or $2\pi nI/10r$ oersted in c.g.s. units.

magnetic field strength [Physics] H. The force that would be exerted on a unit north-pole at a given point in the magnetic field. The strength of a magnetic field measured in amperes per metre (SI units) or oersteds c.g.s.

units). It is given by $H=B/\mu_o-M$, where B is the magnetic flux density, M is the magnetization, and μ_o is the magnetic constant.

magnetic filter [Chemistry] Filteration device in which the filter screen is magnetized to trap and remove fine iron and other magnetic materials from liquids or suspensions.

magnetic fluid [Materials] A mixture of fine iron particles in oil or other liquid; its viscosity increases sharply in a strong magnetic field.

magnetic flux [Physics] ϕ A. measure of quantity of magnetism, taking account of the strength and the extent of a magnetic field. The flux $d\phi$ through an element of area dA perpendicular to B is given by $d\phi = BdA$. The unit of magnetic flux is the weber.

magnetic flux density [Physics] B. The magnetic flux passing through unit area of a magnetic field in a direction at right angles to the magnetic force. It is defined by the effect on a current-carrying conductor in the field. For a field B; $B=F/\theta V$, where F is the force on a charge θ, perpendicular to the field with velocity V. The derived SI unit of magnetic flux density is tesla (weber per square metre). The c.g.s. unit is gauss.

magnetic focusing [Electronics] Focusing a beam of electrons or other charged particles by utilizing the action of a magnetic field.

magnetic force [Physics] The attractive or repulsive force exerted by a magnetic field on a magnetic pole or an electric charge.

magnetic induction [Physics] 1. The induction of magnetism in a body by an external magnetic field. 2. *See* magnetic flux density.

magnetic ink [Materials] Ink containing magnetic particles to permit reading of printed characters by a magnetic character reader as well as by human eye.

magnetic lens [Physics] A magnetic field with axial symmetry, capable of converging beams of charged particles of uniform velocity and of forming images of objects placed in the path of such beams; the field may be produced by solenoids, electromagnets, or permanent magnets.

magnetic mirros [Physics] The regions of high field strength at the end of an externally generated magnetic field used in the containment of the plasma in controlled thermonuclear reaction experiments. Ions that enter these regions of high strength reverse their direction of motion (are reflected) and return to the central region of the plasma in which they are plasma in which they are trapped.

magnetic moment [Physics] *1*. The ratio between the maximum torque

(T_{max}) exerted on a magnet, current-carrying coil, or moving charge situated in a magnetic field and the strength of that field. It is thus a measure of the strength of a magnet or current-carrying coil. 2. In the Sommerfield approach this quantity (also called *electromagnetic moment* or *magnetic area moment*) is the ratio T_{max}/B. 3. In the Kennelly approach the quantity (also called *magnetic dipole moment*) is T_{max}/H. 4. In the case of a magnet placed in a magnetic field of field strength H, the maximum torque T_{max} occurs when the axis of the magnet is perpendicular to the field. In the case of a coil of N turns and area A carrying a current I, the magnetic moment can be shown to be $m = T/B = NIA$ or $m = T./H = «NIA$. Magnetic moments are measured in A m^2. 5. An orbital electron has an orbital magnetic moment IA, where I is the equivalent current as the electron moves round its orbit. It is given by $I = q\omega/2\pi$, where q is the electronic charge and ω is its angular velocity. The orbital magnetic moment is therefore $IA = q\omega A/2\pi$, where A is the orbital area. If the electron is spinning there is also a spin magnetic moment atomic nuclei also have magnetic moments.

magnetic monopole [Physics] A hypothetical unit of magnetic 'charge' analogous to electric charge. No evidence has been found for the existence of a separate magnetic pole, they are always found in pairs.

magnetic permeability [Physics] μ The ratio of the magnetic flux density in a medium to the external magnetic field strength that induces it. The 'relative permeability', μr, is the ratio of the permeability of a substance to the permeability of free space (*see* magnetic constant). For most of the substances μr has a constant small value. When r is less than 1, the material is said to be diamagnetic; if μr is greater than 1, it is paramagnetic. A few substances, notably iron, have very large values of μ, which tend to fall as the field strength increase so that the magnetic flux density tends to a limiting value called the saturation value. Such substances are said to be ferromagnetic.

magnetic pole [Physics] One of the two regions located at the ends of a magnet that generates and respond to magnetic field in much of the same way as electric charge generate and respond to electric field.

magnetic potential [Physics] The work which must be done against a magnetic field to bring a magnetic pole of unit strength from a reference point to the point under consideration. Also known as magnetic scalar potential.

magnetic-resonance imaging (Electronics) A diagnostic imaging technique that detects the variation in density and relaxation times of hydrogen nuclei throughout a patient's body when it is subjected to a strong magnetic field and displays a two-dimensiol image on a CRT.

magnetic scattering [Physics] Scattering of neutrons as a result of the interaction of magnetic moment of the neutrons with magnetic moments of the atoms or other particles.

magnetic shell [Physics] Two layers of magnetic charge of opposite sign, separated by an infinitesimal distance.

magnetic storm [Astronomy] A sudden disturbance in the earth's magnetic field (*see* magnetism, terrestrial) associated with sunspot activity, which affects compasses and radio transmission.

magnetic susceptibility [Physics] *Xm*. The ratio of the magnetization (*M*) produced in a substance to the magnetic field strength (*H*) to which it is subjected, *i.e.*, $X_m = M/H$. The susceptibility is related to the relative permeability, μr (*see* magnetic permeability) by $X_m = 1-μr$. Ferromagnetic materials have high position values of X_m.

magnetic tape [Electronics] Plastic tape coated with a ferromagnetic powder, used in tape recorders and computers. The tape is passed over the gap in a magnetic circuit, which is modulated in accordance with information to be recorded. The tape retains a

record of the modulation, which can be played back through a suitable circuit.

magnetic transducer [Engineering] A device for converting mechanical energy into electrical energy. It consists of a magnetic field including a variable reluctance path and a coil surrounding all or a part of this path, so that variation in reluctance leads to a variation in magnetic flux through the coil and a corresponding induced e.m.f.

magnetism [Physics] A group of phenomena associated with magnetic fields. Whenever an electric current flows a magnetic field is produced; as the orbital motion and the spin of atomic electrons are equivalent to tiny current loops, individual atoms create magnetic fields around them, when their orbital electrons have a net magnetic moment as a result of their angular momentum. The magnetic moment of an atom is the vector sum of the magnetic moments of the orbital motions and the spins of all the electrons in the atom. *See* diamagnetism; ferrimagnetism; ferromagnetism; paramagnetism.

magnetization [Physics] 1. The magnetic moment per unit volume of a magnetized body. It is equal to $B/μ_o-H$, where *B* is the magnetic flux, $μ_o$ is the magnetic constant, and *H* is the magnetic field strength. 2. The process of magnetizing a magnetic material,

magneto [Electricity] A small dynamo provided with a spark-coil, for ignition of petrol vapour in petrol internal-combustion engines. Also known as magno-electric generator.

magnetoelectrcity [Physics] Magnetic techniques for generating electricity, such as in ordinary generators. [Electronics] The appearance of an electric field in certain substances such as chromic oxide (Cr_2O_3), when they are subjected to a static field.

magnetohydrodynamics [Physics] MHD. 1. The study of the behaviour of moving electrically conducting fluids in magnetic fields. 2. A method of generating electricity by subjecting the free electrons in a high velocity flame or plasma to a strong magnetic field. The free-electron concentration in the flame is increased by the thermal ionization of added substances of low ionization potential. These electrons constitute a current when they flow between electrodes within the flame, under the influence of the external magnetic field.

magnetometer [Engineering] An instrument for comparing strengths of magnetic fields, magnetic moments and sometimes also the direction of a magnetic field, such as the earth's magnetic field.

magnetomotive force [Physics] MMF. A quantity analogous to the electromotive force; defined as the circular integral of the magnetic field strength around a closed path.

magnetochemistry [Chemistry] The branch of physical chemistry concerned with measuring and investigating the magnetic properties of compounds. It is used particularly for studying transition-metal complexes, many of which are paramagnetic because they have unpaired electrons. Measurement of the magnetic susceptibility allows the magnetic moment of the metal atom to be calculated, and this gives information about the bonding in the complex.

magneton [Physics] A unit for measuring magnetic moments of nuclear, atomic, or molecular magnets. 1. The *Bohr magneton* μB has the value of the classical magnetic moment of an electron, given by

$$\mu_B = eh/4\pi m_c = 9.274 \times 10^{-24} \text{ A m}^2,$$

where e and m_e are the charge and mass of the electron and h is the Planck constant. 2. The *nuclear magneton*, μ_N is obtained by replacing the mass of the electron by the mass of the proton and is therefore given by

$$\mu_N = \mu_B \, m_e/m_p = 5.05 \times 10^{-27} \text{ A m}^2.$$

magnetosphere [Physics] The region where the earth's (or any celestial body) magnetic field is constrained by the streaming solar wind. It has a shape similar to that of a comet with a blunt nose towards the sun and a long tail always away from the sun. This magnetic cavity effectively shields the earth and its atmosphere from the direct influx of the solar wind.

magnetron [Electronics] A thermionic valve capable of producing high power oscillations in the microwave region 1-40 gigahertz. It is used as a pulsed microwave radiation source for radar.

magnification [Optics] The ratio of the linear dimensions of the final image formed by an optical to the linear dimensions of the object.

magnifying glass [Optics] A convex lens. *See* microscope simple.

magnifying power [Optics] The ratio of the tangent of angle subtended at the eye by the final image to the tangent of angle subtended by the object placed at the least distance of distinct vision (*i.e.*, the shortest distance from the eye at which the object can be seen distinctly).

magno [Metallurgy] An alloy containing about 95.5% nickel and 4.5% manganese; used in the manufacture of incandescent lamps and radio tubes.

magnox [Metallurgy] A magnesium alloy used for sheathing uranium fuel elements in certain types of nuclear reactor.

major planet [Astronomy] Any of the four planets that are larger than earth; Jupiter, Saturn, Neptune, and Uranus.

malachite [Minerals] $CuCO_3$. $Cu(OH)_2$. Natural basic copper carbonate. A bright green mineral.

malignant [Medicine] Pertaining to the growth and proliferation of certain neoplasms which is likely to have a fatal termination if not checked by treatment.

malleability [Metallurgy] The ability to be hammered out into thin sheets, a characteristic property of metals.

malnutrition [Physiology] The state of being poorly nourished, may be caused by inadequate intake of one or more essential nutrients or due to faulty digestion, assimilation or metabolism.

malonyl [Chemistry] The bivalent radical-$OCCH_2CO$-, derived from malonic acid.

malt [Biochemistry] Grain (usually barley) that has been allowed to germinate and then heated and dried.

maltase [Biochemistry] An enzyme occurring in yeast and other organisms that hydrolyzes (*see* hydrolysis) maltose into glucose.

manganese [Chemistry] Mn. Transition element. At. No. 25. At. Wt. 54.938; m.p. 1517K; b.p. 2; placed in group VII B of the periodic table. A greyish-white, hard brittle metal, mainaly used in making alloys and steel.

manganese bronze [Metallurgy] A copper-zinc alloy containing up to 4% manganese, 55-60% copper, and 38-42% zinc.

manganese steel [Metallurgy] A very hard variety of steel containing up to 13% manganese.

manganin [Metallurgy] An alloy containing 80-85% copper, 12-15% manganese, 2-4% nickel. Its electrical resistance is affected only slightly by change in temperature it is therefore used in making resistance coils.

manometer [Engineer] Any instrument used for measuring gaseous pressure.

mantissa [Mathmatic] The decimal, always positive, portion of a common logarithm.

manure [Materials] Animal excreta used to enrich the soil. It helps to form humus and provides nutrients for plants.

mare [Astronomy] One of the large, dark, flat areas on the surface of moon.

margarine [Materials] A butter substitute prepared from purified vegetable and animal fats and oils. Milk is added to a suitable blend of fats; bacterial action in the milk produces a butter-like flavour; vitamins A and D along with suitable salts, emulsifiers and colour are added.

marine biology [Biology] A branch of biology that deals with the study of living organisms which inhibit the sea.

marine microbiology [Biology] The study of microorganisms living in the sea.

Markownikoff rule [Chemistry] In the addition reaction, the additive molecule R-H adds as H and R, with R going to the carbon atom with the lesser number of hydrogen atoms bonded to it.

mars [Astronomy] The planet fourth in distance from the sun with two small satellites, having its orbit between those of earth and jupiter. It is visible to naked eye as a bright red star, except for short periods when it is near its conjunction with the sun. Its diameter is about 6790 km. The mean distance from the sun is about 227.94 million kilometres. Sidereal period ('year')=686.98 days, mass 0.107 that of the earth. Its atmosphere is composed mainly of carbon dioxide, has a pressure of only about 0.01 atmosphere. The polar ice caps are solid carbon dioxide. The day temperature at the equator is about--40°C, dropping to about -70°C at night.

Marsh's test [Chemistry] A sensitive test for arsenic that depends upon the formation of arsine. AsH_3 when arsenic or its compounds are present in a solution evolving hydrogen. When arsine is passed through a narrow, heated tube it is decomposed and leaves a deposit of metallic arsenic.

marsh gas [Chemistry] Combustible gas; mainly consits of methane; formed as a result of decay of vegetation in stagnant water.

martensite [Mechanics] A solid solution of carbon in alpha-iron (*see* iron) formed when steel is cooled too rapidly for pearlite to form from austenite. It is responsible for the hardness of quenched steel.

mascagnite [Chemistry] A mineral form of ammonium sulphate $(NH_4)_2SO_4$

mascon [Geology] A large, high-density mass concentration below the surface of moon; of unexplained origin.

maser [Physics] A device for coherent amplification or generation of electromagnetic waves in which an assembly of atoms or molecules is excited to an unstable energy state, is stimulated by an electromagnetic wave to radiate excess energy at the same frequency and phase as the stimulating wave. When a maser is made to operate at optical frequencies, it is called optical maser, or laser. Derived from Microwave Amplification by Stimulated Emission of Radiation.

mash [Biochemistry] Mixture of malted barley and water used for making wort in brewing operations.

mass [Mechanics] The quantity of matter contained in a body. It is a quantitative measure of the body's resistance to be accelerated. Mass of a body is constant regardless of its location in the universe, whereas weight of a body changes with the change in its distance from the centre of the earth (or other celestial body). The SI unit of mass is kilogram.

mass action law *See* law of mass action.

mass decrement [Chemistry] The difference between the isotopic mass of an isotope and its mass number.

mass defect [Nucl.Sc.] The difference between the mass of a nucleus and the sum of the masses of its constituent nucleons. The energy equivalent to the mass defect, on the basis of the mass-energy equation, must be supplied to a nucleus to split it into its component nucleons.

mass-energy equation [Nucl.Sc.] Mass and energy are mutually convertible under certain conditions. The equation connecting the two qualities in any such transformation is $E=mc^2$, where c is the velocity of light and E is the energy, released when a mass m is completely converted into energy.

mass number [Nucl.Sc.] The integer nearest to the atomic mass of an isotope, *i.e.*, the number of nucleons in the nucleus of an atom.

mass spectrograph [Chemistry] An instrument for the determination of the exact masses of individual atoms, *i.e.*, isotopic masses by photographing the mass spectrum produced.

mass spectrometer [Chemistry] An instrument for obtaining the mass spectrum of a beam of ions by means of suitably disposed magnetic and electric fields. The deflection of any individual ion in these fields depends on the ratio of its mass to its electric charge, m/e. Such a spectrum will appear as a number of lines on a photographic plate, each corresponding to a definite value of m/e.

mass spectrum [Physics] A spectrum obtained with a mass spectrometer or spectrograph in which a beam of ions is arranged in order of increasing charge to mass ratio.

material science [Engineering] The study of nature, behaviour, and use of materials applied to science and technology.

mating [Biology] The meeting of individuals for sexual reproduction.

matrix [Computers] A lattice work of input and output leads with logic elements connected at some of their intersections. [Engineer] A mould for shaping a cast. [Mathematics] An arrangement of mathematical elements into rows and columns according to algebraic rules, in order to solve a set of linear equations. [Metallurgy] The crystalline phase in an alloy, in which the other phases are contained. [Materials] A binding agent used to make and agglomerate mass.

matte [Metallurgy] A mixture of sulphides of iron and copper obtained as an intermediate stage in the smelting of copper.

maxwell [Physics] c.g.s. unit of magnetic flux. The flux through one square centimetre normal to a magnetic field of strength one gauss. 1 maxwell $=10^{-8}$ weber.

means-ends analysis [Computers] A problem-solving approach (used by GPS) in which problem-solving operators are chosse in an interative fashion to reduce the difference between the curent problems-solving state and the goal state.

mean free path [Chemistry] The average distance travelled between collisions by the molecules in a gas, the electrons in a metallic crystal, the neutrons in a moderator, etc. According to the kinetic theory of gases, the mean free path between elastic collisions of gas molecules of diameter d (assuming them to be rigid spheres) is equal to $1/(2npd2)1/2$, where n is equal to number of molecules per unit volume in a gas. As n is proportional to the pressure of the gas, the mean free path is proportional to the pressure of the gas.

mean life (τ)[Chemistry] The average time during which a system, such as an atom, nucleus, or elementary particle, exists in a specified form; for a radionuclide or an excited state of an atom or nucleus; it is the reciprocal of decay constant. Its SI unit is second (s).

mean normal stress [Physics] In a system stressed multiaxially, the algebraic mean of the three principles stresses. Its SI unit is pascal (Pa), while other unit is newton per metre squared (N/m^2).

mean solar day [Astronomy] The duration of one rotation of the earth on its axis with respect to the sun, the length of mean solar day is 24 hours of mean solar time, or 24 hours 3 minutes and 56.555 seconds of mean sidereal time.

mean stress [Physics] The algebraic mean of the maximum and minimum values of a periodically varying stress. its SI unit is pascal (Pa).

mean sun [Astronomy] A ficticious sun conceived to move eastward along the celestial equator at a rate that provides a uniform measure of time equal to average apparent time.

mechanical admittance (Ym) [Physics] The inverse of the mechanical impedance, i.e., the ratio of the velocity to the driving force; a useful quantity when considered as analogous to electrical impedance. Its SI unit is second per kilogram $(s.kg-1)$.

mechanical compliance [Physics] The ratio of the displacement produced to the force producing it in an elastic displacement. Its SI unit is metre per newton (m/N).

mechanical efficiency [Physics] In an engine, the ratio of brake horsepower to indicated horsepower.

mechanical impedance (Z_m) [Physics] The mechanical impedance of a vibrating system is the complex ratio of the force acting in the direction of motion at a point or surface to the velocity at that point or surface ;

$$Z_m = R_m + i\ X_m$$

where, R_m = mechanical resistance, and X_m = mechanical reactance. Its SI unit is newton second per metre.

mechanical mass (m) [Physics] The part of the mass of a particle which is supposed to be an intrinsic property of the particle and not due to the interaction of the particle with itself through the medium of some field. The two types of masses are not experimentally distinguishable, however. Its SI unit is kilogram (kg).

mechanism [Engineering] The portion of a machine which possesses two or more components so arranged that the motion of one compels the motion of the other. [Chemistry] Explanation of the way in which a reaction takes place.

median [Mathematics] 1. A line joining a vertex of a triangle to the mid-point of the opposite side. 2. The middle number in a sequence of numbers.

median lethal dose [Nucl.Sc.] (M)LD$_{50}$. The dose of ionizing radiation that would kill 50% of a large batch of organisms within a specified period.

medicinal paraffin [Medicine] A mineral oil or paraffin of no nutritive value as it is not effected by digestive enzymes and passes through the intestine unchanged. Due to its lubrication properties, it is used as a mild laxative.

medicine [Medicine] 1. Any agent administered for the treatment of a disease. 2. The science or art of healing and treating as distinguished from surgery and obstetrics.

medium [Chemistry] The carrier in which a reaction takes place. [Physics] The entity in which objects exist and phenomena takes place.

mega- [Science Technology] M. Prefix denoting one million or 10^6 times, in metric units.

megahertz [Physics] MHz. 1 million hertz. A measure of frequency equal to 10^6 cycles per second.

megaton bomb [Physics] A nuclear weapon with an explosive power equivalent to one million tons of T.N.T. (approximately 4×10^{15} joules).

meiosis [Biology] A kind of nuclear division, usually two successive cell divisions, which results in daughter cells with the haploid number of chromosomes, one half the number of chromosomes in the original cell.

Meissner effect [Physics] The falling off of the magnetic flux within a superconducting metal when it is cooled to a temperature below the critical temperature in a magnetic field.

melting point [Physics] The constant temperature at which the solid and liquid phase of a substance are in equilibrium at a given pressure. Melting points are normally quoted for standard atmospheric pressure.

memory [Computers] The stored data which can be retrieved.

memory cell [Computers] A single storage element of a memory, along with associated circuits for storing and reading out one bit of information.

Mendeleev's law *See* periodic law.

mendelevium [Chemistry] Md. Transuranic element, At. No. 101. The most stable isotope, ^{256}Md, has a half-life of 60 days.

meniscus [Physics] The curved surface of a liquid in a vessel. If the contact angle between the liquid and the wall of the vessel is less than 90°, the meniscus is concave; if greater, the meniscus is convex.

menstruation [Physiology] The periodic discharge of blood from the uterus for 4 to 5 days once a month in the female from puberty, lasts upto the menopause; governed by hormones from the anterior pituitary gland and the ovaries.

mensuration [Mathematics] The measurement of length, areas, and volumes.

mercuric [Chemistry] A compound of bivalent mercury, i.e., Hg(II).

mercurous [Chemistry] A compound of univalent mercury, i.e., Hg(I), e.g., mercurous chloride, Hg_2Cl_2.

mercury [Astronomy] A planet with its orbit nearest to sun. Mean distance from the sun 57.91 million kilometres. Sidereal period ('year')=87.969 days. Mass 0.054 that of the earth, diameter 4840 kilometres. It has no atmosphere and a day temperature of about 400°C. [Chemistry] Hg. Element. At. No. 80. At. W. 200.59, placed in group II B of the periodic table. A liquid, silvery-white metal, m.p. 234.2K; 629.7k; Used in thermometers, barometers, manome-

ters, and nuclear reactors; its alloys (called amalgams) are used in dentistry.

mercury-cathode cell [Chemistry] Electrolytic cell used to manufacture chlorine and caustic soda from sodium chloride solution.

mercury cell [Electricity] A primary voltaic cell consisting of a zinc anode and a cathode of mercury(II) oxide (HgO) mixed with graphite. The electrolyte is potassium hydroxide (KOH) saturated with zinc oxide, the overall reaction being :

$$Zn + HgO \longrightarrow ZnO+Hg$$

The e.m.f. is 1.35 volts and the cell will deliver about 0.3 ampere-hour per cm^3.

mercury vapour lamp [Electricity] A lamp emitting a strong bluish light by the passage of an electric current through mercury vapour in a bulb. The light is rich in utraviolet radiations; used in artificial sun-ray treatment and in street lighting. *See also* fluorescent.

meridian, celestial [Astronomy] The great circle of the celestial sphere passing through the zenith and the celestial poles, meeting the horizon at points called the north and the south points.

meridian, magnetic *See* magnetic meridian.

meridian, terrestrial [Geology] Meridian of longitude. An imaginary great circle drawn round the earth that passes through both poles.

mesic atom [Physics] An atom in which one of the electrons is replaced by a negative muon or meson orbiting close to or within the nucleus. Also known as mesonic atom.

meso- [Scientific Technology] A prefix indicating that a substance is optically inactive due to intramolecular compensation.

mesomerism *See* resonance.

meson [Physics] Any of a group of unstable elementary particles with strong nuclear interactions and baryon number equal to zero. They are believed to consist of a quark and its antiquark. Positive, negative, and neutral mesons exist; when charged the magnitude of the charge is equal to that of the electron. Mesons are found in cosmic rays and are emitted by nuclei under bombardment by high energy particles. Muons were originally called.mesons, but they are now classified as leptons rather than mesons.

mesonic atom *See* mesic atom.

mesophases [Physics] Phases intermediate between crystalline and liquid phases (*see* liquid crystals; cybotaxis). Three different types are recognized : smectic, nematic, and cholesteric crystals, in accordance with the different arrangements of the molecules in them.

mesophites [Biology] Microorganisms that grow best at temperatures between 298K and 313K, usually will not grow below 276K.

mesophytes [Botany] Common land plants that live in a climate with an average amount of moisture.

mesosphere [Geology] The region of the earth's atmosphere between the stratosphere and the thermosphere, extending from some 40 kilometres above the earth's surface; characterized by temperature that generally falls with altitude.

meta [Chemistry] 1. Denoting positions separated by one atom in a hexagonal ring of atoms, particularly the benzene ring. Abbreviated to m-as a prefix in naming a compound : *e.g.*, m-dinitrobenzene (alternatively, 1, 3-dinitrobenzene). *Compare* ortho; para. 2. A prefix indicating an inorganic acid (or a corresponding salt) of a lower degree of hydration; *e.g.*, metaphosphoric acid, HPO_3, as compared with orthophosphoric acid, H_3PO_4.

metabolic turnover [Biochemistry] The constant steady-state metabolic replacement of cell components.

metabolism [Physiology] The physical and chemical processes taking place in living organisms by which foodstuffs are broken into complex elements (assimilation, anabolism), which are transformed into simple ones (disassinulation, catabolism) and energy is made available for use.

metabolite [Biochemistry] Any substance that takes part in the process of metabolism.

metagenesis [Biology] The phenomenon in which one generation of certain plants and animals reproduces a sexually, followed by a sexually reproducing generation.

metal [Chemistry] Any of a class of chemical elements that are typically lustrous solids that are good conductors of heat and electricity. Not all metals have all these properties (*e.g.*, mercury is a liquid). In chemistry, metals fall into two distinct types. Those of the *s*-and *p*-blocks (*e.g.*, sodium and aluminium) are generally soft silvery reactive elements. They tend to form positive ions and so are described as electropositive. This is contrasted with typical nonmetallic behaviour of forming negative ions. The transition elements (*e.g.*, iron and copper) are harder substances and generally less reactive. They form coordination complexes. All metals have oxides that are basic.

metal fatigue [Mat] A cumulative effect causing a metal to fail after repeated applications of stress, none of which exceeds the ultimate tensile strength. The *fatigue strength* (or *fatigue limit*) is the stress that will cause failure after a specified number (usually 10^7) of cycles.

metallic bond [Chemistry] A chemical bond of the type holding together the atoms in a solid metal or alloy. In such solids, the atoms are considered to be ionized, with the positive ions occupying lattice positions. The valence electrons are able to move freely (or almost freely) through the lattice, forming an 'electron gas'. The bonding force is electrostatic attraction between the positive metal ions and the electrons. The existence of free electrons accounts for the good electrical and thermal conductivities of metals. *See also* energy bands.

metallic crystal [Chemistry] A crystalline solid in which the atoms are held together by metallic bonds. Metallic crystals are found in some interstitial compounds as well as in metals and alloys.

metallic soap [Chemistry] An insoluble salt formed by a heavy metal and a fatty acid such as stearic, oleic, palmitic, lauric or erucic acid (especially salts of lead and aluminium); used for waterproofing textiles and as a drier for paints.

metallizing [Metallurgy] Coating a plastic or similar material with a thin deposit of metal by means of spraying or vacuum deposition.

metalloenzyme [Biochemistry] An enzyme having a metal ion as its prosthetic group.

metallography [Metallurgy] The study of the crystalline structure of metals and alloys.

metalloid [Chemistry] Any of a class of chemical elements intermediate in properties between metals and nonmetals. The classification is not clear cut, but typical metalloids are boron, silicon, germanium, arsenic, and tellurium. They are electrical semiconductors and their oxides are amphoteric.

metallurgy [Metallurgy] The branch of applied science concerned with the production of metals from their ores, the purification of metals, the manufacture of alloys, and the use and performance of metals in engineering practice. *Process metallurgy* is concerned with the extraction and production of metals, while *physical metallurgy* concerns the mechanical behaviour of metals.

metamerism [Chemistry] A type of isomerism exhibited by organic compounds of the same chemical class or type; it is caused by the attachment of different radicals to the same central atom or group, *e.g.*, diethyl either $(C_2H_5)_2O$ and methylpropyl ether, $CH_3OC_3H_7$.

metamict state [Geology] The amorphous state of a substance that has lost its crystalline structure as a result of the radioactivity of uranium or thorium.

metamorphism [Geology] The transformation of the structure or constitution of rocks due to such natural factors as heat and pressure.

metastable phase [Chemistry] The state of supercooled water (*see* super-cooling) or of supersaturated solutions (*see* supersaturation) in which the phase that is normally stable under the given conditions does not form unless a small amount of the normally stable phase is already present. Thus, supercooled water will remain as liquid water below 0°C until a small crystal of ice is introduced.

metastable state [Physics] An excited state (*see* excitation) of an atom or nucleus that has an appreciable life-time.

metathesis [Chemistry] *See* double decomposition.

metazoa [Biology] A subkingdom of multicellular animals whose bodies are composed of specialized cells grouped together to form tissues and that possess a coordinating nervous system.

meteor [Astronomy] A solid body from outer space which enters the earth's atmosphere and shines brightly because of the heat produced by friction with air; small meteors are completely burned to gases. Also known as shooting star.

meteoric ionization [Astronomy] Ionization resulting from collisional interactions of a meteoroid and its vapourization products with the air.

meteorite [Astronomy] Any meteoroid that has fallen to the earth's surface and is not completely burned to gas.

meteorid [Astronomy] Any solid object moving in interplanetary space that is smaller than a planet or asteorid.

meteorology [Science Technology] The science of the weather; the study of such conditions as atmospheric pressure, temperature, wind strength, humidity, from which conclusions as to the forthcoming weather are drawn.

meteor shower [Astronomy] A large number of meteors that enter the earth's atmosphere when the earth's orbit crosses the orbit of a comet, *i.e.*, an orbit that contains either the material of which comets are made or into which they disintegrate.

methylated spirits [Materials] A mixture consisting mainly of ethanol with added methanal (‾9.5%), pyridine (‾0.5%), and blue dye. The additives are included to make the ethanol undrinkable.

metonic cycle [Astronomy] A time period of 235 lunar months, or 19 years 11 days; after which the phases of moon occur on the same days of the month.

metre [Scientific Technology] m. The SI unit of length being the length of the path travelled by light in vacuum during a time interval of $1/(2.99\ 792\ 458 \times 10^8)$ second.

metrology [Physics] The scientific study of measurements.

mho [Electricity] A unit of conductance, equal to conductance between two points of a conductor at a potential difference of 1 volt when a current of 1 ampere is passing through them. The conductance of a conductor in mhos is the reciprocal of its resistance in ohms. Also known as siemen.

mica [Minerals] Any of several silicates of varying chemical composition but with similar physical properties the most important of which are muscovite, $H_2KAI_3(SiO_4)_3$, and phlogopite, $H_2KMg_3Al(SiO_4)_2$. Naturally occurring mica can be split along its cleavages into small thick pieces or thin sheets. Being an excellent insulator and being resistant to high temperatures, mica is used as a dielectric in capacitors, as a support for electrodes in thermionic valves, and for protecting heating elements in irons, etc.

micelle [Chemistry] An electrically charged colloidal particle, usually organic in nature. The term is also applied to casein complex in milk.

micro- [Scientific Technology] 1. Prefix denoting one-millionth or 10^{-6} in metric units. Symbol μ. 2. Prefix meaning 'very small'; on a small scale.

micro-aerophiles [Biology] Microorganisms that can grow in extremely low concentrations of oxygen; can spoil foodstuff intended to be stored in the absence of air, if traces (of oxygen) are left.

microanalysis [Chemistry] Identification, charcterization and analysis of sample ranging from 0.1 to 10 milligrams. This often involves the use of a microscope and still more often of chromatography.

microbe [Biology] A microscopic organism, especially a becterium of a pathogenic nature.

microbicide [Materials] A substance or agent that kills microbes.

microbiology [Biology] The branch of biology concerned with the structure and function of microorganisms, including protozoens, algae, fungi, bacteria, viruses,and rickettsiae.

microcode [Computer] The lowest level of instructions that directly control the inyeraction of a processor's computing elements- that is, machine instructions wired into the hardware that is being controlled.

microelement [Electronics] Capacitor, diode, inductor, resistor, transformer, or other electronic component mounted on a ceramic water which is atleast 0.025 cm thick and about 0.75 cm^2 in area.

micrometer [Engineering] An instrument attached to a microscope or telescope for the accurate measurement of small distances or angles [Mechanics]. A unit of length equal to one millionth of a metre. Also known as micron. Abbreviated μm. Also spelled micrometre.

micron *see* micrometre.

micronutrient [Biochemistry] Trace elements and compounds required by living systems only in very small quantities.

microorganism [Biology] A unicellular organism that can only be seen with the aid of a microscope of high resolution, including bacteria, protozoans, yeasts, viruses and algae.

microphone [Engineering] A device that converts sound energy into electric energy. It contains a transducer which is actuated by sound waves and delivers essentially equivalent electric waves.

microscope, compound [Optics] A device for producing large images of close small objects with a combination of lenses. It consists of two converging lenses called the objective and the eye-piece respectively. The objective, which is nearest the viewed object, forms a real inverted magnified image of the object just inside the focal distance of the eye-piece. This image is viewed through the eye-piece, which then acts as a simple microscope producing an inverted further magnified virtual image. The useful magnification obtainable with an optical microscope is limited by the wavelength of visible light as two points on a microscopic specimen cannot be distinguished from each other if they are not as far apart as half the wavelength of the light used to illuminate them. Thus for magnifications in excess of about 1500, an ultraviolet microscope or an electron microscope must be used.

microscope, electric *See* electric microscope.

microscope, simple [Optics] A convex lens used to produce a virtual image larger than the viewed object. Also known as magnifying glass.

microtome [Engineering] An instrument for cutting thin sections of material, for microscopical examination.

microwave [Physics] An electromagnetic wave having its wavelength between 0.3 and 30 centimetres, corresponding to frequencies of 1-100 gigahertz.

milk of lime [Chemistry] A suspension of lime in water.

microwave frequency [Physics] A frequency of the order of 10^9-10^{11} hertz.

microwave maser [Physics] A maser which emits microwave radiation.

mictic [Biology] Produced by sexual reproduction.

milky way [Astronomy] The faint band of light seen in the sky that results from the combined light of the many stars near the plane of our galaxy.

milli [Scientific Technology] Prefix denoting one thousandth, in metric units. Symbol m.

millibar [Mechanics] A unit of at-

mospheric pressure, used in meterology, 1000 dynes per square centimetre or 100 newtons per square metre.

millimetre wave [Physics] An electromagnetic wave having its wavelength between 1 mm and 1 cm, corresponding to frequencies between 30 and 300 gigahertz.

millimicron [Mechanics] A unit of length equal to one thousandth of a micron, or one billionth of a metre. Abbrev. $m*$. Also known as nanometre.

mineral [Geology] A naturally occurring substance with a characteristic chemcial composition.

mineral acid [Chemistry] A common inorganic acid, such as hydrochloric acid, sulphuric acid, or nitric acid.

mineral oil *See* paraffin oil.

minority carriers [Electronics] In a semiconductor, the type of carrier that constitutes less than half the total number of carriers.

minor planets [Astronomy] Planets smaller than the earth, specifically Mercury, Venus, Mars and Pluto.

mirage [Optics] The reflection of light by a layer of very warm air, which has been heated by the earth. Thus an object and its reflected image are seen, that gives the appearance of a watery surface.

mirror [Optics] A surface that reflects specularly large fraction of the light falling upon it, thus forming images.

mirror image [Optics] The image of an object as viewed in a mirror; reversed in such a way that the image bears to the object the same relation as a right hand to a left.

mirror nuclei [Nucl.Sc.] A pair of atomic nuclei, each of which would be transformed into the other by changing all its neutrons into protons, and vice-versa.

misch metal [Metallurgy] An alloy of cerium (50%), lanthanum (25%), neodymium (18%), praseodymium (5%), and other rare earths. It is used alloyed with iron (up to 30%) in lighter flints, and in small quantities to improve the malleability of iron.

miscible [Chemistry] Capable of being mixed to form a homogeneous substance; usually applied to liquids, e.g., water and alcohol are completely miscible.

mispickel [Minerals] FeAsS. A natural sulphide of iron and arsenic. Also known as arsenopyrite.

missile [Physics] Any object that is designed to be thrown, dropped projected, or propelled, for the purpose of making it strike a target.

mist [Physics] Droplets of water, formed by the condensation of water-vapour on dust particles.

mite [Biology] Small tiny creatures parasites on man and animals, that cause skin troubles, such as scabies.

mitosis [Biology] Nuclear division involving exact duplication and separation of the chromosome threads so that each of the two daughter nuclei carries a chromosome compliment identical to that of the parent nucleus.

Mitscherlich's law [Chemistry] Substances that have the same crystal structure have similar chemical formulae. The law can be used to determine the formula of an unknown compound if it is isomorphous with a compound of known formula. Also known as law of isomorphism

mixer [Electronics] A device having two or more inputs, and a common output, used to combine or separate audio or video signals linearly in desired proportions to produce an output signal.

mixture [Chemistry] A heterogeneous association of substances obtained by mixing two or more different substances which cannot be represented by a chemical formula. The substances can be elements, compounds or materials. Mixtures can be solid, liquid or gaseous. The constituents of the mixture can be separated by suitable physical means. The properties of the mixture are an aggregate of the properties of the constituents, whereas a compound has individual properties, often quite unlike those of the component elements.

m.k.s. system [Scientific Technology] A system of units derived from the metre, kilogram, and second. Now superseded for scientific purposes by the SI units, which are based on the m.k.s. system.

mobility [Physics] Freedom of particles to move either under the influence of fields or forces, or in random motion.

mock silver [Metallurgy] An alloy of aluminium containing 5% copper and 10% tin, or 5% silver and 5% copper.

moderator [Nucl.Sc.] The material used in nuclear reactors to slow down neutrons produced by nuclear fission. These substances consist of atoms of light elements (*e.g.*, deuterium in heavy water, graphite, beryllium to which the neutrons are able to impart some of their energy on collision, without being captured.

modulated carrier [Electronics] Radio frequency carrier wave whose amplitude, phase, or frequency has been varied according to the signal to be conveyed.

modulation [Electronics] The process of varying some characteristic of one wave such as amplitude, frequency, and phase (usually a radio frequecny carrier wave) in accordance with some characteristic of another wave, or vary the velocity of the electrons in an electron beam in some characteristic manner.

modulation depth [Electronics] The difference in brightness between black and white in a CRT display.

modulus [Mechanics] A constant factor or multiplier for the conversion of units from one system to another.

modulus of elasticity [Mechanics] The ratio of the increment of some specified form of stress to the increment to some specified form of strain, such as Young's modulus, the bulk modulus, or the shear modulus. Also known as coefficient of elasticity.

Mohs scale [Minerals] An empirical scale in which each mineral listed is softer than (*i.e.*, is scratched by) all those below it. 1. Talc, 2. gypsum, 3. calcite, 4. fluorite, 5. apatite, 6. orthoclase, 7. quartz, 8. topaz, 9. corundum, 10 diamond.

moisture [Physics] The water vapour content of the atmosphere or the total water present in a given volume of air.

molality [Chemistry] A method of expressing the strength of a solution (*See also* concentration); the number of moles of solute per kilogram of solvent.

molar [Chemistry] Denoting that an extensive physical property is being expressed per amount of substance, usually per mole.

molar heat capacity [Chemistry] Cm. The heat capacity of substance divided by the amount of substance. The amount of heat required to raise the temperature of 1 mole of a substance by 1 kelvin. Expressed in joules per mole per kelvein (SI units).

molarity [Chemistry] Concentration expressed in moles of solute per litre of solvent.

molar solution [Chemistry] An obsolete expression for a solution with a concentration of mole per litre.

molar volume [Chemistry] Vm. The volume occupied by 1 mole of a substance. All gases have approximately equal molar volumes under the same conditions of temperature and pressure. At 760 mm Hg and 0°C the molar volume of a perfect gas is 22.415 litres per mole. Also known as gram molecular volume.

mold [Engineering] A cavity which imparts its shape to a fluid or malleable substance. [Biology] Any of the various wooly fungus growths. They grow best in a moist environment at about 298 K even in absence of light. Also spelled mould.

mole [Chemistry] Symbol mol. The SI unit of amount of substance. It is equal to the amount of substance that contains as many elementary units as there are atoms in 0.012 kg of carbon-12. The elementary units may be atoms, molecules, ions, radicals, electrons, etc., and must be specified. 1 mole of a compound has a mass equal to its relative molecular mass expressed in grams.

molecular association [Chemistry] The formation of dimers or polymers from a single species as a result of some intermolecular forces, such as hydrogen bonding.

molecular attraction [Chemistry] A force which attracts the molecules toward each other.

molecular binding [Chemistry] The force which holds the molecules at some site on the surface of a crystal.

molecular biology [Biology] The study of the structure of the molecules that approaches the subject of life at molecular level.

molecular compounds [Chemistry] Chemical compounds formed by the chemical combination by two or more complete molecules, e.g., the hydrates of salts.

molecular concentration [Chemistry] The concentration of a solution expressed in terms of moles in a given volume.

molecular distillation [Chemistry] The evaporatoin of molecules from a surface with least damage to their chemical composition at pressures of about 10^{-2} mm Hg, and their subsequent condensation under such conditions that their mean free path is of the same order as the distance between the heated and cooled surfaces. Used for distilling heat-sensitive organic compounds such as vitamins and glycerides.

molecularity [Chemistry] The number of molecules involved in forming the activated complex in a step of a chemical reaction. Reactions are said to be unimolecular, bimolecular, or trimolecular according to whether 1, 2, or 3 molecules are involved.

molecular orbital [Chemistry] A wave function describing an electron in a molecule.

molecular sieves [Materials] Highly porous aluminosilicate adsorbents, containing pores (lattice vacancies) of uniform size, that are selective in their action with respect to molecules of a particular size and character.

molecular volume [Chemistry] The volume occupied by one mole of a substance; equal to its molecular weight divided by its density.

molecular weight [Chemistry] The ratio of the average mass per molecule of a specified isotopic composition of a substance to 1/12 of the mass of an atom of C-12. The sum of the atomic weights of all the atoms that comprise a molecule. Also known as relative molecular mass.

molecule [Chemistry] One of the fundamental units forming a chemical compound; the smallest part of a chemical compound that can take part in a chemical reaction. In most covalent compounds, molecules consist of groups of atoms held together by covalent or coordinate bonds. Covalent substances that form macromolecular crystals have no discrete molecules (in a sense, the whole crystal is a molecule). Similarly, ionic compounds do not have single molecules, being collections of oppositely charged ions.

mole fraction [Chemistry] x. A measure of the amount of a component in a mixture. The mole fraction of component A is given by $X_A = n_A/N$, where n_A is the amount of substance of A (for a given entity) and N is the total amount of substance of the mixture (for the same entity).

mollusca [Biology] Any of a varied group of soft-bodied animals, most of which have a hard protective shell. Snails, slugs, oysters, sqid, and octopuses are all mollusca.

molybdenum [Chemistry] Mo. Element At. No. 42. At.Wt. 95.94; placed in group VIB of the periodic table. A hard white metal m.p. 2883K, b.p. 5833K; used for special steels and alloys; compounds are used as pigments for printing inks, paints; and ceramics catalysts; and solid lubricants.

moment, magnetic [Physics] *See* magnetic moment.

moment of force [Chemistry] *See* torque.

moment of intertia [Mechanics] The moment of inertia I of a body about any axis is the sum of the products of the mass, dm, of each element of the body and the square of r, its distance from the axis. $I = \Sigma r^2 \, dm$.

momentum [Mechanics] The product of the mass and the velocity of a body. For speeds approaching that of light, the variation of mass with velocity of the body must be used in the expression for the momentum.

momentum conservation [Physics] *See* conservation of momentum.

monazite [Minerals] A yellow or brown rare-earth mineral containing phosphates of cerium, thorium, and other rare earths, with some occluded helium.

Mond's process [Metallurgy] A process for the extraction of nickel by the action of carbon monoxide, CO, on the impure metal. This gives nickel carbonyl, $Ni(CO)_4$, a gas that decomposes when heated to 423-473K into pure nickel and carbon monoxide, the latter being used again.

monel metal [Metallurgy] A corrosion-resistant alloy of copper (25%-35%), nickel (60%-70%) and small amounts of iron, manganese, silicon, and carbon.

monitor [Computers] To supervise a program during its execution, usually by means of a diagnostic routine.

monobasic acid [Chemistry] An acid having one atom of acidic hydrogen in a molecule; an acid giving rise to only one series of salts, *e.g.*, nitric acid, HNO_3.

monochromatic light [Optics] Light consisting of vibrations of the same or nearly the same frequency; light of one colour.

monoclinic [Chemistry] Relating to crystals that have three unequal axes with one oblique intersection.

monohydrate [Chemistry] A crystalline compound having one molecule of water per molecule of compound.

monomer [Chemistry] A molecule (or compound) that joins with others in forming a dimer, trimer, or polymer.

monomorphic [Biology] Having or exhibiting only a single form.

monopropellant [Materials] A propellant which combines fuel and oxidiser in one compound or mixture, *e.g.*, gunpowder, and nitromethane.

monosaccharides [Biochemistry] A group of carbohydrates consisting chiefly of sugars having a molecular formula, $C_6H_{12}O_6$ (hexoses) or $C_5H_{10}O_5$ (pentoses); unlike the polysaccharides, they cannot be hydrolyzed to give further simpler sugars.

monotropic [Chemistry] Existing in only one stable physical form, any other form obtainable being unstable under all conditions.

monovalent [Chemistry] Having a valence of one. Also called univalent.

monosoon [Climate] A wind system in which prevailing summer and winter winds are reversed in direction. In India, the summer monsoon brings heavy rainfall.

month [Astronomy] I. The 'solar month' is one twelfth of a solar year. The 'calendar month' is any of the twelve divisions of the year according to the Gregorian calendar II. The 'lunar month' is the time taken for the moon to complete one orbit of the earth. III. The 'synodic month' is the period between two successive phase of the moon, equal to 29.5306 days. IV. The 'sidereal month' is the moon's period with respect to successive conjunctions with a star, equal to 27.3217 days. V. The 'anomalistic month' is the moon's period between two successive perigees, equal to 27.5546 days. VI. The 'draconic month' is the moon's period with respect to two successive similar nodes, equal to 27.2122 days.

moon [Astronomy] 1. The only satellite of the earth. It is cold and reflects light from the sun. Mean distance from the earth 384400 kilometres; mass 0.0123 that of the earth, *i.e.*, about 7.35 X 10^{22} Kg; diameter 3476 kilometres. It is devoid of water or an atmosphere. Man first set foot on the moon on July 20th, 1969. 2. A natural satellite of any planet.

mordant [Chemistry] A substance used in certain dyeing processes. Mordants are often inorganic oxides, or salts, which are absorbed on the fabric. The dyestuff then forms a coloured complex with the mordant, the colour depending on the mordant used as well as the dyestuff.

morphology [Biology] The study of the form and structure of organisms.

mortar [Materials] A building material consisting mainly of cement, lime and sand that hardens on exposure through chemical action between the ingredients and atmospheric carbon dioxide.

mosaic [Electronics] A device for the electrical storage of the optical image. It usually consists of a sheet of mica one side of which is covered with mutally insulated particles of a photoemissive material, each of which is capacitively coupled through the mica to a conducting coating on the reverse side. This conducting coating, called the signal plate, is the output electrode from which the electrical signal representing the optical image is obtained. [Nucl.Sc.] A photomicrograph of a track in an emulsion, prepared from a number of photographs of consecutive fields of view and reconstructed as though the track lay in one plane.

mosaic gold [Chemistry] SnS_2. Crystalline stannic sulphide consisting of shining, golden-yellow scales.

Moseley's law [Chemistry] The frequencies of the lines in the X-ray spectra of the elements are related to the atomic numbers of the elements. If the square roots of the frequencies of corresponding lines of a set of elements are plotted against the atomic numbers a straight line is obtained.

Mossbauer effect [Chemistry] The emission and absorption of gamma rays by certain nuclei, bound in a crystal, without loss of energy through nuclear recoil, with the result that radiation emitted by one such nucleus can be absorbed by another. The mossbauer effect has been used to test the predictions of the theory of relativity and to investigate the properties of the solid state and the nature of magnetism.

mother-liquor [Chemistry] A solution from which substances are recovered by evaporation or crystallization.

motor [Electricity] A device for converting other form of energy into mechanical energy. The most common forms are the internal combustion engine and the electric motor.

mould [Engineering] A vessel into which a hot liquid is poured to solidify in the shape of the mould. Powders can be used instead of liquids and then pressure and heat are applied to convert powder into a solid of the shape of mould. [Biology] Simple fungus consisting of a mass of fine fluffy threads that cover the dead organic matter. Also spelled mold.

M-shell [Physics] The third layer of electrons about the nucleus of an atom, having electrons characterized by the principal quantum number 3.

MSI [Computers] medium-scale intergration: from 100 transistors to 1000 transistors in a circuit.

M-star [Astronomy] Stars having surface temperatures of about 3000 K to 3400 K.

mucilage [Chemistry] A plant product obtained from roots, seeds, or other parts of plants by extraction with cold or hot water.

mucoproteins [Biochemistry] Proteins that contain a carbohydrate group.

multienzyme system [Biochemistry] A sequence of related enzymes participating in a given metabolic pathway.

multiple star [Astronomy] A system of stars consisting of three or more components held together by gravitation.

multiplet [Physics] A line in a spectrum formed by two or more closely spaced lines and resulting from small differences of energy level in the atoms or molecules. [Nucl.Sc.] A group of related elementary particles that differ only in electric charge.

multiplication constant [Nucl.Sc.] The 'effective' multiplication constant of a nuclear reactor is the ratio of the average number of neutrons produced by nuclear fission per unit time, to the total number of neutrons absorbed or leaking out in the same time. *See* subcritical and supercritical.

multiplicity [Physics] The number of energy levels into which an atom or nucleus splits as a result of coupling between orbital angular momentum and spin angular momentum. [Nucl.Sc.] The number of elementary particles in a multiplet.

muon [Physics] An elementary particle with a mass 207 times that of an electron; it exists in negatively and positively charged forms designated μ^+ and μ^- respectively. It was originally so called as it was classified as a meson. However as these particles have spin 1/2, they are now classified as leptons. Also known as * meson.

mustard gas [Materials] A highly poisonous gas, $(ClCH_2\ CH_2)_2S$; dichlorodiethyl sulphide. It is made from ethene and disulphur dichloride (S_2Cl_2), and used as a war gas.

mutagenic agent [Materials] Any of a number of chemicals able to induce mutation in DNA and in living cells, *e.g.*, nitrous oxide, dimethyl sulphate, and ethylmethane sulphonate. Also known as mutagen.

mutarotation [Optics] A change in the optical rotation of a substance.

mutation [Biology] A change in one or more of the bases in DNA, which results in the formation of an abnormal protein. Mutations are inherited only if they occur in cells that give rise to the gamets; sematic mutation may give rise to cancers and chimaeras. Mutations result in

new allelic forms of a gene and hence new variations upon which natural selection can act. The natural rate of mutation is low, but the mutations frequency can be increased by such factors as ionizing radiations and mutagenic chemicals.

mutual conductance [Biology] *See* transconductance.

mutual induction [Physics] The induction of an E.M.F. in a circuit due to a changing current in a separate circuit with which it is magnetically linked. The induced E.M.F. is proportional to the rate of change of the current in the second circuit, the constant of proportionality being called the coefficient of mutual induction, or the mutual induction. The derived SI unit of mutual inductance is the henry.

mycetozoa [Biology] A class of organisms that exhibit both animal and plant characters during their life cycle.

mycology [Botany] The branch of botany that deals with the study of fungi.

mycosis [Medicine] An infection or a disease caused by a fungus.

mycotoxin [Materials] Any poisonous substance produced by molds or fungi.

mydriasis [Medicine] Prolonged dilation of the pupil of the eye.

myology [Medicine] The study of muscles in normal as wall as diseased state.

myokinase [Biochemistry] An enzyme present in muscle and other tissues that catalyses the reaction, $2ADP \rightarrow ATP + AMP$.

myopia [Medicine] A condition of eye in which the image of a distant object is formed before the retina of the eye when accomodation does not occur. It can be corrected by using concave spectacle lenses. Also known as nearsightedness.

myxedema [Medicine] A condition which results due to deficiency of thyroxin secretion in an adult; characterized by a low metabolic rate and decreased heat production.

myxoxanthin [Biochemistry] Caroteniod pigment in algae with vitamin A activity.

n-[Chemistry] Chemical prefix for normal (straight-chain) hydrocarbon compounds.

N *See* neutron number; newton; normality.

nacre [Materials] A form of calcium carbonate secreted by the epithelial cells in the mentle of oyster. Also known as mother of pearl.

nacreous pigment [Materials] A pigment containing guanine crystals obtained from fish scales or skin; which produces a pearly

lustre. It can be applied as surface coatings, as in stimulated pearls, or incorporated in plastics.

nadir [Astronomy] The point on the celestial sphere vertically below the observer, or $180°$ from the zenith.

name reaction [Chemistry] A chemical reaction, usually organic that is identified by the name of its discoverer, *e.g.*, Friedel-Crafts; Cannizarro; Diels-Alder; Hofmann etc.

nano [Scientifi Technology] *n.* Prefix meaning one thousand millionth, *i.e.*, 10^{-9} units; 1 nanogram $=10^{-9}$ g.

nanometre *See* millimicron.

napalm [Materials] A soap of aluminium which is a mixture of oleic and naphthenic fatty acids; used to gelatinize oil or gasoline for use in napalm bombs or flame throwers.

naphtha [Materials] A mixture of hydrocarbons in various proportions, obtained from paraffin oil, coal-tar, etc.; used as a gasoline ingredient, solvent for paints and rubber, and cleaning solvent.

narcotic [Medicine] Producing sleep, stupor, insensibility, coma, or convulsions.

nascent [Chemistry] Pertaining to an atom or simple compound at the moment of its coming into being, when it may have greater activity than its normal state; *e.g.*, when zinc reacts with dilute sulphuric acid, the hydrogen evolved is nascent at the moment of formation.

native [Chemistry] Pertaining to an element found in the earth in free state and not combined in a mineral, *e.g.*, gold and sulphur are found uncombined in deposits; they occur as native elements.

natural [Chemistry] Occurring in nature; not artificially prepared.

natural aging [Metallurgy] Spontaneous aging at room temperature of a supersaturated metallic solid solution.

natural frequency [Physics] The frequency of free oscillation of any system.

natural gas [Materials] A mixture of gaseous hydrocarbons of low molecular weights predominantly methane, often containing other gases, issuing from the earth in some localities, more particularly near deposits of mineral oil. Its composition is about 85% methane, 10% ethane, and the rest being propane, butane, and nitrogen; used as a fuel (alone or mixed with coal gas) and as a source of intermediates for organic synthesis.

natural glass [Geology] An amorphous, vitreous inorganic material that has solidified from magma too quickly to crystallize.

natural-language processing [Comp.] Processing of natural language (English, for example) by a computer, or for other purpose such as language translation.

natural radioactivity [Nucl.Sc.] Radioactivity exhibited by naturally occurring radionuclides. *See also* radioactivity.

natural science [Science Technology] Collectively, the branches of science dealing with objectively measurable phenomena pertaining to transformations and relationships of energy and matter; includes biology, chemistry, and physics.

near stars [Astronomy] Stars in the celestial neighbourhood of the sun, sometimes taken as those 22 stars within 13 light-years of the sun.

nebula [Astronomy] A cloudy, luminous patch in the heaven that consists of a galaxy of stars, or of materials from which such galaxies are being formed.

necrosis [Medicine] Death of a part of the body due to absence of blood supply, injury, disease, or other pathologic state.

Neel temperature [Physics] The temperature above which an antiferromagnetic substance becomes paramagnetic. The susceptibility increases with temperature reaching a maximum at the Neel temperature, after which it abruptly declines.

negative catalysis [Chemistry] A catalytic reaction which is slowed down by the presence of a catalyst.

negative crystal [Optics] A uniaxial crystal in which the extra ordinary wave travels faster than the ordinary wave, such as calcite.

negative feedback [Electronics] Feedback in which a portion of the output is fedback 180° out of phase with the input signal, resulting in a decrease of amplification. Also known as inverse feedback; reverse feedback.

negative glow [Electronics] The luminous flow in a glow discharge cold-cathode tube occurring between the cathode dark space and the Faraday dark space.

negative pole [Physics] The south seeking pole of a magnet.

nematic crystals [Physics] Liquid crystals in which the molecules are not arranged in layer but all their axes are parallel. *See also* cholesteric crystals; smetic crystals.

nematicide [Materials] A chemical used to destroy plant parasitic nematodes.

nematodes [Biology] Worm like creatures that have two sexes an intestinal canal, many of them are parasitic to man, *e.g.*, guinea worms, filarial worms.

nematology [Biology] The systematic study of nematodes.

neodymium [Chemistry] Nd. Element. At. No. 60. At.Wt. 144.24. m.p. 1289K, b.p. 3341K; a member of the rare-earth group of elements; placed in group IIIB of the periodic table.

neon [Chemistry] Ne. Element. At. No. 10. At.Wt. 20.183; m.p. 24.4K, b.p. 27.1K; placed in zero group of periodic table. A colourless, odourless, invisible gas; present in trace amounts in the atmosphere. A discharge of electricity through neon at low pressure produces an intense orange-red glow : used for neon signs.

neoplasm [Biology] New growth of abnormal tissue in plants or animals; a tumour, which may be either benign or malignant.

neoprene [Chemistry] Trans-polychloroprene. $(CH_2CHCClCH_2)_n$. A synthetic rubber having a high tensile strength and better heat and ozone resistance than natural rubber.

neper [Scientific Technology] A unit for expressing the ratio of two voltages, currents, or analogous quantities equal to the natural logarithm of the ratio of the quantities. 1 neper 8.686 decibels.

nephelometer [Optics] An instrument for measuring turbidity of liquids, or scattering of light by particles in suspensions, and visual range through the medium.

nepheloscope [Engineering] An instrument for producing clouds in the laboratory by condensation or expansion of moist air.

nephology [Physics] The study of clouds.

nephoscope [Engineering] A grid-like instrument for determining the speed of celestial objects (including clouds) by observation of time of transit.

neptune [Astronomy] The outermost of the four giant planets, it is 30 astronomical units away from the sun, and the sideral revolution period is 164.8 years. The diameter of neptune is about 44800 kilometres. Its orbit lies between those of uranus and pluto. The surface temperature is about 70K and the dense atmosphere consist mainly of methane and hydrogen. It has two satellites.

neptunium [Chemistry] Np. A radioactive metallic transuranic element belonging to the actinoids; At. No. 93. The most stable isotope, neptunium-237, has a half-life of 2.2×10^6 years and is produced in small quantities as a by-product by nuclear reactors. Other isotopes have mass numbers 229-236 and 238-241. The only other relatively long-lived isotope is neptunium-236 (half-life 5×10^3 years).

Nernst effect [Physics] If a temperature gradient is maintained across an electrical conductor (or semiconductor) that is placed in a transverse magnetic field, a potential difference will be produced across the conductor.

Nernst heat theorem [Physics] The entropy change for chemical reactions involving crystalline solids, is zero at the absolute zero of temperature.

nerve cell *See* neuron.

nerve gas [Materials] Any of the war gases that attack the nervous system, especially the nerves controlling respiration. They are colourless, odourless, tasteless, liquids of low volatility, and are absorbed rapidly through eyes, lungs, or skin; they are lethally toxic to higher animals and man, *e.g.*, isopropylphosphonofluoride.

nervous system [Physilogy] A co-ordinating and integrating system which functions in the adaptation of an organism to its environment; in vertebrates the system consists of the brain, brainstem, spinal cord, cranial, and peripheral nerves, and ganglia.

Nessler's reagent [Chemistry] A solution of potassium mercuric iodide, $K_2[HgI_4]$ used as a test reagent for ammonia, with which it forms a brown colouration or precipitate.

net list [Comp] A database that details the logic elements and the interconnects among them.

neurology [Medicine] The study of the anatomy, physiology, and disorders of nervous system.

neuron [Biology] A nerve cell including the cell body, axon, and denrites. Also known as a nerve cell.

neurotoxin [Toxin] A poison that attacks the nervous system and acts as depressant.

neutral [Chemistry] Neither acid nor alkali , containing equal numbers of hydroxyl and hydrogen ions and having a pH of 7. [Electricity] A body having no net electric charge.

neutral eqilibrium [Mechanics] Equilibrium such that if the system is disturbed a little, there is no tendency for it to move, further, nor to return.

neutral flame [Chemistry] A flame produced by burning fuel in air which is neither oxidising nor reducing.

neutralization [Chemistry] The addition of acid to alkali, or vice-versa, till neither is in excess and the solution is neutral, *i.e.* has pH equal to 7.

neutral temperature [Electricity] The temperature of the hot junction of a thermocouple at which the electromotive force round the circuit is a maximum and the rate of change of E.M.F with temperature is a minimum.

neutral wave [Physics] Any wave whose amplitude does not change with time.

neutretto [Physics] A meson with zero electric charge.

neutrino [Nucl.Sc.] A stable elementary particle with zero electric charge and zero rest mass, but with spin 1/2. It was originally postulated to preserve the laws of conservation of mass and energy and conservation of momentum. The existence of the particle has since been established experimentally, and it is known to exist in two forms : one associated with the beta decay

process and the other with the muon. Both forms have antiparticles.

neutron [Nucl.Sc.] An elementary particle that is a constituent of all atomic nuclei except that of normal hydrogen. The neutron has no electric charge and a mass only very slightly greater than that of the proton (1.67492 X 10^{-27} kilogram). Free neutrons have high penetrating power, hence they have high damaging effect on living tissue. Neutrons directly emitted from atomic nuclei are termed as fast neutrons.

neutron excess *See* isotopic number.

neutron flux [Nucl.Sc.] A measure of the number of neutrons passing through unit area in unit time.

neutron number [Nucl.Sc.] N. The number of neutrons in an atomic nucleus; it is equal to the mass number minus the atomic number.

neutron star [Astronomy] A star that is supposed to occur in the final stage of stellar evolution; it consists of a superdense mass mainly of neutrons, and has a strong gravitational attraction from which only neutrinos and high energy photons could escape so that the star is invisible.

neutron temperature [Physics] A concept used to express the energies of neutrons that are in thermal equilibrium with their surroundings, assuming that they behave like a monatomic gas. The neutron temperature T, on the Kelvin scale, is given by $T = 2E/3k$, where E is the neutron energy and k is the Boltzmann constant.

newton [Mechanics] N. The SI unit of force, being the force required to give a mass of one kilogram an acceleration of 1 ms^{-2}.

Newtonian fluid [Mechanics] A fluid that obeys Newton's law of viscosity, i.e, the viscosity is independent of the rate of shear or the velocity gradient. The tangential force, F, between two parallel layers of fluid is given by

$$F=\eta S.dv/dx$$

where S is the area of the fluid layers, dx is the distance between them, and dv is their velocity. η is a constant called the coefficient of viscosity. A large number of liquids obey Newton's law.

Newtonian mechanics [Mechanics] A system of mechanics based on Newton's laws of motion in which mass and energy are considered as separate, conservative, mechanical properties. It provides an accurate means of determining the motions of bodies possessing ordinary velocities.

Newtonian telescope [Optics] A reflecting telescope in which the light reflected from a concave mirror is reflected again by a plane mirror making an angle of 45° with the telescope axis, so that it passes through a hole in the side of the telescope containing eye piece.

Newton's alloy [Metallurgy] A fusible alloy that melts at about 368K composed of 50% bismuth, 31% lead, and 19% tin. Also known as Newton's metal.

Newton's law of cooling [Physics] The rate at which a body loses heat to its surroundings is proportional to the temperature difference between the body and its surroundings. It is an empirical law true only for small differences of temperature.

Newton's law of gravitation [Mechanics] The law that every two bodies in the universe attract each other with a force that acts along the line joining them, and has a magnitude proportional to the product of their masses and inversely proportional to the square of the distance between them. Also known as law of gravitation.

Newton's laws of motion [Mechanics] Three fundamental laws on which classical dynamics is based. I. Every body continues in its state of rest or uniform motion in a straight line except in so far as it is compelled by external forces to change that state. Also known as law of inertia. II. Rate of change of momentum is proportional to the applied force, and takes place in the direction in which the force acts. III. To every action there is an equal and opposite reaction.

Newton's metal *See* Newton's alloy.

Newton's rings [Optics] Coloured rings that may be observed round the point of contact of a convex lens and a plane reflecting surface. They are caused by the interference effects that occur between light waves reflected at the upper and lower surfaces of the air film separating the lens and the flat surface.

nickel [Chemistry] Ni. Transition element. At. No. 28. At.Wt. 58.71; m.p. 1723K, b.p. 3113K; placed in group VIII of the periodic table. A silvery-white magnetic metal, resembling iron, that resists corrosion; used for nickel-plating, in coinage, for alloys such as nickel steel, nickel silver, platinoid, constantan, nichrome, and as a catalyst. It is present in foods and in animal and human tissues, though it is not essential for plants, but improves the growth of many plants.

nickel-iron accumulator [Electricity] A secondary cell having a positive plate of nickel oxide and a negative plate of iron both immersed in an electrolyte of potassium hydroxide. The reaction on discharge is

$$2NiOOH.H_2O + Fe \rightarrow 2Ni(OH)_2$$
$$+ Fe(OH)_2.$$

the reverse occurring during charging. Each cell gives an e.m.f. of about 1.2 volts and produces about 8×10^4 J per kilogram during each discharge. Also known as Edison cell.

nickel plating [Electricity] Depositing a thin layer of metallic nickel by an electrolytic process.

nickel silver [Metallurgy] A group of alloys of copper, nickel, and zinc in varying proportions, containing up to 30% nickel. A typical composition is 60% copper, 20% nickel, 20% zinc. Also known as German silver.

nickel steel [Metallurgy] Steel containing up to 9% nickel as a major alloying element.

nicol prism [Optics] An optical device, constructed from a crystal of calcite, used for obtaining plane polarized light.

night blindness *See* nyctalopia.

niobium [Chemistry] Nb. Element. At. No. 41. At.Wt. 92.906, placed in group VB of the periodic table. A rare grey metal, m.p. 2741K, b.p. 5015K. Small quantities in stainless steel preserve the steel's corrosion resistance at high temperatures. Also known as columbium.

nit [Optics] A unit of luminance equal to one candela per square metre.

nitration [Chemistry] Introduction of the nitro group, $-NO_2$, into organic compounds by replacing a hydrogen on carbon atom by the use of nitric acid or a mixture of nitric acid and sulphuric acid.

nitre [Chemistry] KNO_3; potassium nitrate. Saltpetre.

nitride [Chemistry] Binary compound of nitrogen with a metal, such as Mg_3N_2.

nitriding [Chemistry] A process of case hardening in which a ferrous alloy is heated in an atmosphere of ammonia or in conatact with nitrogenous material to produce surface hardening by absorption of nitrogen without quenching.

nitrification [Chemistry] The treatment of substance with nitric acid. [Biology] The process of conversion, by the action of bacteria, of nitrogen compounds from animal and plant waste and decay, into nitrates in the soil.

nitrile [Chemistry] R-C=N. Cyanide derived by the removal of a water molecule from an acid amide.

nitrile rubber [Materials] Any of a group of synthetic rubbers formed by polymerizing butadiene and acrylonitrile; the structure of polymer is $-CH_2CH=CHCH_2 CH_2CH(CN)-$. It is highly resistant to oils, fuels, and aromatic solvents.

nitrite [Chemistry] A compound either organic or inorganic, containing $-NO_2$ radical.

nitro- [Chemistry] Chemical prefix showing the presence of the univalent radical $-NO_2$.

nitrochalk [Materials] A mixture, of calcium carbonate, $CaCO_3$, and ammonium nitrate, NH_4NO_3; used as a fertilizer.

nitrogen [Chemistry] N. Element. At. No. 7. At.Wt. 14.0067; m.p. 63.2K, b.p. 77.3K; placed in group VA of the periodic table. An odourless, invisible, chemically inactive gas, forming approximately 78% of the atmosphere. The element is vital to living organisms, forming an essential part of proteins and nucleic acids.

nitrogenase system [Biochemistry] A system of enzymes capable of reducing atmospheric nitrogen into ammonia in the presence of ATP

nitrogen balance [Medicine] The difference between nitrogen intake (as protein) and total nitrogen excretion for an individual. Growing children and convalescents are in positive nitrogen balance; patients with wasting diseases are in negative balance. Also known as nitrogen equilibrium.

nitrogen cycle [Chemistry] The circulation of nitrogen and its compounds between organisms and the environment. Inorganic nitrogen compounds in the soil are taken in by plants, and are combined by the plants with other elements to form nucleic acids and proteins, the latter being the form in which nitrogen can be utilized by the higher animals. The result of animal waste and decay is to bring the nitrogen that the animals had absorbed back into the soil in the form of simpler nitrogen compounds. Bacterial action of various kinds converts these nitrogenous compounds into **compounds**

suitable for use by plants again. In addition to this main circulation, a certain amount of atmospheric nitrogen is 'fixed' (*i.e.*, combined) by the action of bacteria associated with the roots of leguminous plants and by the action of electric discharge in the atmosphere (which converts nitrogen and oxygen into nitric oxide, which further reacts with oxygen to form nitrogen dioxide, that dissolves in water to form nitric acid). Apart from uptake by plants, nitrate is also lost from the soil by denitrification and by leaching. The increasing application of fertilizers in agriculture has now become another important factor in the nitrogen cycle.

nitrogen equilibrium *See* nitrogen balance.

nitrogen fixation [Chemistry] The formation of nitrogenous compounds from atmospheric nitrogen. In nature this may be achieved by electric discharge in the atmosphere or by the activities of certain microorganisms. The symbiotic bacteria, Rhizobium species are associated with leguminous plants, forming characteristic nodules on their roots. The bacteria supply the legume with nitrate, while the legume supplies the bacteria with carbohydrate. Free-living bacteria and some sulphur-bacteria, some blue-green algae, and some yeast fungi have shown the property to fix nitrogen. In industry, nitrogen is fixed by converting it into ammonia (Haber process), nitric acid (Birkeland -Eyde process), calcium cynamide, or oxides of nitrogen.

nitrogenous base [Chemistry] A basic compound containing nitrogen. The team is used especially of organic ring compounds, such as adenine, guanine, cytosine, and thymine, which are constituents of nucleic acids. *See* amine salts.

nitrogen solution [Chemistry] A mixture of ammonium nitrate and ammonia solution; used to neutralize super-phosphate in fertilizer manufacture.

nitroparaffin [Chemistry] Any organic compound in which one or more hydrogens of an alkane is replaced by a nitro, *i.e.*, $-NO_2$ group, such as nitromethane.

nitroso- [Chemistry] The univalent radical MNO in organic compounds. ·

nitrosyl ion [Chemistry] The univalent ion NO^+.

nn junction [Electronics] In a semiconductor, a region of transition between two regions having different properties in *n*-type semiconducting material.

nobélium [Chemistry] No. Transuranic element. At. No. 102. The most stable isotope, ^{250}No, has a half-life of 3 minutes. It is produced in a cyclotron by bombarding curium with nuclei of carbon-13 accelerated to high energies.

noble gases [Chemistry] Members of the group zero of the periodic table, *i.e.*, helium, neon, argon, krypton, xenon, and radon. These gases (except helium) have 8 electrons in their outer-most shell, and are monoatomic in nature. With limited exceptions, they are chemically inert. Also known as inert gases.

noble metal [Metallurgy] A metal such as silver, gold, platinum, palladium, iridium, rhodium, ruthenium and osmium or an alloy that does not corrode or tarnish in air or water, and is not easily attacked by acids. From the chemical point of view, unreactive metals are low in the electromotive series.

nodal points [Optics] Two points on the axis of a lens system, such that if the incident ray passes through one, travelling in a given direction, the emergent ray passes through the other in a parallel direction.

nodes [Physics] Points of zero amplitude in a system of standing waves. *See also* antinodes. [Astronomy] Two points at which the orbit of a celestial body intersects the ecliptic. [Botany] The points on leaf insertion on stem. At the apex of the stem the nodes are very close together but become separated in older regions of the steam by intercalary growth [Electronics] The junction points in a net work. [Mathematic] Points on a curve or surface that can have more than one tangent.

nodical month [Astronomy] The average period of revolution of the moon about the earth with respect to moon's ascending node, a period of 27 days 5 hours 5 minutes and 35.8 seconds.

noise [Acous] Unwanted sound. [Electricity] An effect observed in amplifying circuits due to the amplification, together with the input signal, of spurious voltages arising from such causes as the vibration of certain components, the random motion of the electrons constituting the current in the conductors, etc.

noise pollution [Acous] Excessive noise in the human environment.

no-load current [Electricity] The current which flows in a network when the output is open circuited.

nomogram [Mathematics] An alignment chart arranged so that the value of a variable can be found, without calculation, from the values of one or two other variables which are known. Also known as nomograph.

non-electrolytes [Chemistry] Substances that do not yield ions in solution and therefore form solutions of low electrical conductivity.

non-essential amino acid [Biochemistry] Amino acids of proteins that can be made by humans and other vertebrates from simpler precursors, and thus not required in the diet.

non-ferrous metal [Metallurgy] Any metal other than iron or its alloys.

nonhaeme-iron proteins [Biochemistry] Proteins containing iron but no porphyrin groups.-

nonhypergolic [Chemistry] Not capable of igniting spontaneously upon contact.

nonideal gas [Chemistry] A gas whose molecules have significant interaction, more than that needed to bring about the equilibrium.

nonlinear [Physics] Pertaining to a phenomenon or relation which is other than directly or inversely proportional to a given variable.

nonmetals [Chemistry] Any of a number of elements whose properties, bonding characteristics, and electronic configuration differ markedly from those of metals. In general, nonmetals have low conductivity, high electronegativity, and high ionization energy.

nonstoichiometric compound [Chemistry] A chemical compound in which the elements do not combine in simple ratios. For example, rutile (titanium(IV) oxide) is often deficient in oxygen, having such a formula as $TiO_{1.8}$. Also known as Berthollide compound.

nor [Chemistry] Chemical prefix to the name of a compound indicating one methyl group less, *e.g.*, noradrenalin contains a methyl group less than adrenalin.

norite [Chemistry] Activated carbon used to decolourize solutions.

normality [Chemistry] An obsolescent method of expressing concentrations of solutions; the number of gram-equivalents of reagent per litre of solution. Thus, a solution containing 5 gram-equivalents per litre is 5N solution.

normalizing [Metallurgy] A heat treatment applied to steel in order to relieve internal stresses. It involves heating above a critical temperature and cooling in air.

normal salt [Chemistry] A salt in which all the acid hydrogen atoms have been replaced by a metal, or the hydroxide radicals of a base are replaced by an acid radical; for example, $CaCO_3$; $Al(NO_3)_3$; $Ca_3(PO_4)_2$.

note [Physics] A sound that has a single fundamental frequency.

nova [Astronomy] A star that ejects a small part of its material in the form of a gas cloud. During the process the star becomes 5000 to 10,000 times more luminous than it was before the outburst. 'Dwarf' novae increase their luminosity by a factor of only 10 to 100. Novae appear to be one of a pair of binary stars.

N.T.P. [Scientific Technology] Normal temperature and pressure of 1.01325×10^5 pascals and a temperature of 273K; standard conditions under which volumes of gases are compared. Also known as S.T.P.

n-type conductivity [Electronics] The conductivity in a semiconductor caused by a flow of electrons, where as *p*-type conductivity is caused by a flow of holes.

nuclear barrier [Physics] The region of high potential energy through which a charged particle must pass on entering or leaving an atomic nucleus. Also known as potential barrier.

nuclear battery [Nucl.Sc.] A primary battery in which the energy of radioactive material is converted into electric energy by solar cells or other energy converters.

nuclear breeder [Nucl.Sc.] A nuclear reactor in which more fissionable material is formed in each generation than is used up in fission process.

nuclear charge [Physics] The positive electric charge on the nucleus of an atom. When expressed in units equal to the charge on the nucleus; this is numerically equal to the atomic number of the element, *i.e.*, to the number of protons in the nucleus, and also equal to the number of electrons surrounding the nucleus in the neutral atom. *See* atom, structure of.

nuclear device [Nucl.Sc.] A nuclear explosive used for peaceful purposes, tests, or experiments.

nuclear energy [Nucl.Sc.] Energy released during a nuclear reaction either fission or fusion as the result of the conversion of mass into energy. Nuclear energy is released in nuclear reactors and nuclear weapons. Also known as atomic energy.

nuclear fission [Nucl.Sc.] A nuclear reaction in which a heavy atomic nucleus (*e.g.* uranium) splits into two approximately equal parts, at the same time emitting neutrons and releasing very large amounts of nuclear energy. Fission can be spontaneous or it may be caused by the impact of a neutron (*see* chain reaction), an energetic charged particle, or a photon (photofission). *See also* nuclear reactor and nuclear weapon.

nuclear force [Nucl.Sc.] The attractive force that acts between nucleons when they are extremely close together (closer then 10^{-15} m). The nuclear force replaces the repulsive electromagnetic interaction between protons at such proximities and holds the nucleons together in the atomic nucleus (*see* exchange forces). The precise nature of the nuclear force is not exactly known.

nuclear fuel [Nucl.Sc.] A fissionable, or fertile isotope with a reasonably long half-life that undergoes nuclear fission or nuclear fusion in a nuclear reactor, a nuclear weapon, or a star.

nuclear fuel element [Nucl.Sc.] A rod, tube, plate, or other mechanical shape or form into which a nuclear fuel is fabricated for use in nuclear reactor. The word "element" is used in engineering rather than in chemical sense.

nuclear fusion [Nucl.Sc.] A nuclear reaction between light atomic nuclei as a result of which a heavier nucleus is formed and a large quantity of nuclear energy is released, *e.g.*, the fusion of two deuterium nuclei to form a tritium nucleus and a proton is accompanied by an energy release of 4 MeV (D+D=T+p) +4MeV). For fusion to be possible the reacting nuclei must possess sufficient kinetic energy to overcome the electrostatic field that surrounds them. The temperature associted with fusion reactions are therefore extremely high. Fusion reactions occur on earth during the explosion of a hydrogen bomb (*see* nuclear weapons) and during controlled thermonuclear reactions. Fusion reactions are believed to be the source of the energy of stars (including the sun).

nuclear isomers [Nucl.Sc.] Atoms of an element of the same mass but possessing different rates of radioactive decay.

nuclear magnetic resonance [Physics] A type of radio-frequency spectroscopy, based on the magnetic field generated by spinning of electrically charged atomic nuclei. This nuclear magnetic field is caused to interact with a very strong magnetic field. Data obtained in this way provide valuable information concerning nuclear properties. As the orbital electrons 'shield' the nucleus to a certain extent from the applied magnetic field, at a given frequency nuclei in different electronic (*i.e.* chemical) environments will responds at slightly different values of the applied field. Also known as NMR spectroscopy.

nuclear medical imaging [Computers] The diagnostic technique to produce digital images: the patient is injected with a radioctive tracer elememt that collects in the tissue to be imaged;its distribution is recorded with a gamma camera.

nuclear pile [Physics] *See* nuclear reactor.

nuclear power [Nucl.Sc.] Electric or electromotive power produced from a unit in which the primary energy source is a nuclear reactor.

nuclear radiation [Nucl.Sc.] A term used to denote alpha particles, electrons, neutrons, photons, and other particles which emanate from the atomic nucleus as a result of radioactive decay and nuclear reactions.

nuclear reaction [Nucl.Sc.] Any reaction involving a change in the nucleus of an atom, as distinct from a chemical reaction, which only involves the orbital electrons. Such reactions occur in radioactive elements, and in stars as thermonuclear reactions. Nuclear reactions are represented by enclosing within a bracket the symbols for the incoming and outgoing particles or quanta (separated by a comma), the initial and final nuclides being shown outside the bracket. Thus the reaction :

$$^{14}N_7 + {}^4He_2 \rightarrow {}^{17}O_8 + {}^1H_1$$

is represented as : $^{14}N(\alpha,p)^{17}O$.

nuclear reactor [Nucl.Sc.] A device containing fissionable material in sufficient quantity and so arranged as to be capable of maintaining a controlled, self-sustaining nuclear fission chain reaction. The nuclear fuel used in a reactor consists of fissile material, which undergoes fission, as a consequence of which two nuclides of approximately equal mass are produced together with two or three neutrons and a considerable quantity of energy. These neutrons cause further fissions so that chain reaction develops; in order that the reaction should not get out of control, its progress is regulated by neutron absorbers (see control rods), only sufficient free neutrons being allowed to exist in the reactor to maintain the reaction at a constant level. The essential components of a modern nuclear reactor are : (a) The core, composed of metals or ceramic-clad rods containing enough fissionable material, (b) A source of neutrons to initiate reaction, such as mixture of polonium and beryllium. (c) A moderator to reduce the energy of fast neutrons for more effi-

cient fission; materials such as heavy water, graphite, beryllium are used. (d) A coolant to remove generated heat; water is generally used, which is converted to steam in heat exchangers and used to drive turbines. (e) A control system to absorb neutrons rapidly when their concentration becomes too high; rods of boron or cadmium having high capture cross-reactions, are used for this purpose. (f) Adequate shielding, remote control equipment, and appropriate instrumentation are also required for safety and efficient operation of the nuclear reactor. Also known as atomic pile; atomic reactor; fission reactor; nuclear pile.

nuclear transmutation [Nucl.Sc.] The changing of atoms of one element into those of another by suitable nuclear reactions.

nuclear weapon [Nucl.Sc.] Weapon in which the explosive power is derived from nuclear fission or a combination of nuclear fission and nuclear fusion.

nuclease [Biochemistry] An enzyme that catalyses the splitting of nucleic acids to nucleotides, nucleosides, or other components of nucleosides.

nucleation [Chemistry] The process by which crystals are formed from liquids, supersaturated solutions, or saturated vapours. Crystals originate on a minute trace of a foreign substance acting as a nucleus.

nucleic acids 320

nucleic acids [Biochemistry] Organic acids whose molecules consist of chains of alternating sugar and phosphate units, with nitrogenous bases attached to sugar units. The fundamental units of nucleic acids are nucleotides. The nucleic acids are polynucleotides in which the nucleotides are linked by phosphate bridges. There are two types of nucleic acids; deoxyribonucleic acid (DNA), and ribonucleic acid (RNA). DNA is found in the cell nucleus and is the essential constituent of chromosomes. DNA is responsible for the reproductive power and hereditary characteristics. RNA is found in the nucleus and the cytoplasm and plays the vital role in protein synthesis.

nuclein [Biochemistry] Any of a poorly defined group of nucleic acid protein complexes occurring in cell nuclei.

nucleoid [Biochemistry] The region of a bacterium or blue-green algae, containing DNA and not enclosed by membranes. It may be assoicated with mesosome during cell divisions.

nucleolus [Biochemistry] A small spherical body containing nucleoprotein, one or more of which occur in the nucleus of biological cells. Also known as plasmosome.

nucleon [Physics] A constituent of the atomic nucleus, *i.e.*, a proton or a neutron. The nucleons can transform into each other through β-decay.

nucleonics [Engineering] The technology based on phenomena of atomic nucleus such as fission, fusion, and radioactivity; includes nuclear reactors, various applications of radio isotopes, and radiation detection devices.

nucleon number [Physics] *See* mass number.

nucleophile [Chemistry] An ion or molecule that can donate electrons. Nucleophiles are often oxidizing agents and Lewis bases. They are either negative ions (*e.g.*F⁻) or molecules that have electron pairs (*e.g.* NH₃). In organic reactions they tend to attack positively charged parts of a molecule.

nucleoprotein [Biochemistry] Compounds of nucleic acids and proteins found in cell nuclei principally in the form of chromosomes. Viruses consist almost entirely of nucleoproteins.

nucleoside [Biochemistry] A molecule formed from a nitrogenous base (purine or pyrimidine) and a pentose sugar, *e.g.*, adenosine, which consists of adenine and D-ribofuranose.

nucleotide [Biochemistry] The compound formed by condensation of a nitrogenous base (purine, pyrimidine, or pyridine) with a sugar (ribose or deoxyribose) and phosphoric acid. They are found free in cells as adenosine triphosphate and as part of various coenzymes; they also occur in the form of polynucleotide chains as nucleic acids.

nucleus [Engineering] A vital central point, especially a particle of matter that acts as a centre for the condensation of water vapour in mist or as a centre for the formation of crystals. [Chemistry] A characteristic ring of atoms in a molecule that retains its identity in chemical changes; *e.g.*, the benzene nucleus of six carbon atoms in the benzene ring. [Nucl.Sc.] The central core of an atom, containing protons and neutrons, except for hydrogen which has a nucleus of one proton only. The nucleus of the neutral atom has positive charge which is exactly equal to the total charge of electrons surrounding it. [Biology] A small body of dense materials, covered with a membrance, in the cytoplasm of a cell. It controls all the activities of the cell, and without it, the cell dies. The nucleus contains nuclear sap (a liquid) and chromosomes.

nuclide [Nucl.Sc.] The nucleus of an atom of a specific isotope, characterized by its atomic number, mass number, and its energy state. An 'isotope' refers to a type of atom while 'nuclide' refers to its nucleus.

nutation [Astronomy] An irregular periodic oscillation of the earth's poles. It causes an irregularity of the precessional circle traced by the celestial poles and results from the varying distances and relative directions of the sun and the moon.

nutrient [Physiology] Any sub-stance that is essential to the life and growth of animals and plants, either as such or as transformed by chemical or enzymatic reactions.

nutrification [Physiology] Addition of nutrients to a food either to replace those lost in processing, or to provide nutrients that are not normally present in the food, or to bring the food into conformity with a specific standard for that food.

nutrition [Physiology] All the process by which the living organism receives and utilizes the materials necessary for survival, growth, and repair of worn-out tissues.

nyctalopia [Medicine] Inability to *see* in dimlight due to deficiency of vitamin A. Also known as night blindness.

Nylander reagent [Chemistry] A solution of potassium sodium tartrate, potassium or sodium hydroxide, and bismuth subnitrate in water; used to test for sugar in urine.

nylon [Materials] Any of various synthetic polyamide fibres having a protein-like structure formed by the condensation between an amino group of one molecule and a carboxylic acid group of another.

nylon 6 [Materials] Nylon made by polycondensation of caprolactum.

nylon 66 [Materials] Nylon made by the condensation of hexamethylene diamine with sebacic acid.

nymph [Biology] Any immature larval stage of various hemimetabolic insects.

nymphomania [Physiology] Excessive sexual desire on the part of a woman. Also known as hysteromania.

nystagmus [Medicine] Involuntary moment of the eyeballs.

oasis [Geology] An isolated fertile area, usually limited in extent and surrounded by desert, and marked by vegetation and a water supply.

obduction [Medicine] The act of performing a post-mortem examination to discover the cause of death. Also known as autopsy.

objective [Optics] A lens or system of lenses nearest the object in a telescope, compound microscope or any other optical system.

objected-oriented programming [Computers] A programming approach foused on objects that communicate by message passing. An object is considered to be a package of information and descriptions of oroducers that can manipulate that information.

obliteration [Medicine] Complete removal of an organ or other body part by disease or surgical excision.

obstetrics [Medicine] The branch of medicine dealing with the study of pregnancy, labor, and puerperium.

obstipation [Medicine] Constipation that is difficult to relieve.

obstruction [Medicine] Occlusion or stenosis of hollow viscera, ducts and vessels.

obturaction [Medicine] The closing of an opening or passage.

obtuse angle [Mathematics] An angle greater than 90° and less than 180°.

occlusion [Crystallography] 1. The trapping of small pockets of liquid in a crystal during crystallization. [Physics] The absorption of a gas by a solid such that atoms or molecules of the gas occupy interstitial positions in the solid lattice. Palladium, for example, can occlude hydrogen.

occultation [Astronomy] The disapperance of the light or radio emission from celestial body when another is interposed between it and the observer.

ochre [Minerals] A yellow, brown, or red form of natural hydrated form of ferric oxide, Fe_2O_3, containing various impurities; used as pigments.

octane number [Engineering] A number that provides a measure of the ability of a fuel to resist knocking when it is burnt in a spark-ignition engine. It is the percentage by volume of iso-octane (C_8H_{18} : 2,2,4-trimethylpentane) in a blend with normal heptane (C_7H_{16}) that matches the knocking behaviour of the fuel being tested in a single cylinder four-stroke engine of standard design.

octant [Mathematics] The portion of a circle cut off by an arc and two radii at $45°$; one-eighth of the area of a circle.

octave [Physics] The interval between two musical notes, the fundamental components (*see* quality of sound) of which have frequencies in the ratio two to one.

octet [Physics] A stable group of eight electrons that constitutes the outer electron shell of an atom of a noble gas (except helium whose only electron shell contains two electrons).

odd-even nucleus [Nucl.Sc.] A nucleus that has an odd number of protons and an even number of neutrons.

odd-odd nucleus [Nucl.Sc.] A nucleus that has an odd number of both protons and neutrons.

odometer [Engineering] An instrument used to measure distance traversed, as of a vehicle.

odontology [Medicine] A branch of science that deals with the study of formation, development, and abnormalities of teeth.

odorant [Materials] A substance having a distinctive odor which is deliberately added to essentially odorless materials to provide warning of their presence.

oedema [Medicine] Excess fluid in the body indicated by pitting of the subcutaneous tissues when pressure is applied with the finger; may be caused by cardiac, renal, or hepatic failure and by starvation.

oersted [Physics] The unit of magnetic field strength or magnetic intensity in c.g.s, electromagnetic units defined as the magnetic intensity as a point where a force of 1 dyne acts upon a unit magnetic pole at that point, *i.e.*, the intensity 1 cm from a unit magnetic pole.

oestrogens [Biochemistry] Female sex hormones of which the most important are the sterols **oestradiol** ($C_{18}H_{24}O_2$), **oestrone** ($C_{18}H_{22}O_2$), and **oestriol** ($C_{18}H_{24}O_3$); urinary excretion of these hormones increases throughout normal pregnacy.

ohm [Electricity] *. The derived SI unit of resistance defined as the resistance between two points of a conductor when a constant difference of potential of 1 volt, applied between these two points, produces in the conductor a current of 1 ampere.

Ohm's law [Electricity] The law that the ratio of the potential difference between the ends of a conductor and the current flowing in the conductor is constant. This ratio is termed the resistance of the conductor. For a potential difference of V volts and a current of I amperes, the resistance, R, in ohms is equal to V/I.

oil [Materials] Any of various viscous liquids that are generally immiscible with water. Natural plant and animal oils are either volatile mixtures of terpenes and simple esters (*e.g.* essential oils) or are glycerides of fatty acids. Mineral oils are mixtures of hydrocarbons (*e.g.* petroleum).

oil black [Materials] A variety of carbon-black made from oil, usually an aromatic type petroleum oil.

oil cake [Materials] A mass of oilseeds (*e.g.* linseed, cottonseed) from which the oil has been expelled in a press (expellers) or extracted by a solvent (extractions) : used as cattle food.

oil gas [Materials] A heating gas made from petroleum oil vapours and steam by high temperature cracking.

oil paint [Materials] A paint made with a vegetable oil as the filmogen.

olefins [Chemistry] A family of unsaturated, chemically active hydrocarbons with general formula CnH_{2n}, that contains one carbon carbon double bond. Also known as olefines; alkenes.

oleum [Chemistry] Sulphuric acid containing sulphur trioxide in excess over the formula H_2SO_4; *e.g.*, '20% oleum' contains 20%

SO_3, and 80% H_2SO_4. It is extremely corrosive; used in industrial nitration. Also known as fuming sulphuric acid.

olfaction [Physiology] The act of smelling.

olfactory [Physics] 1. Pertaining to the sense of smell. 2. The function of smelling.

oligomer [Chemistry] A polymer having comparatively few monomer units in the molecule.

oligomeric protein [Biochemistry] A protein having two or more polypeptide chains.

oligosaccharide [Biochemistry] Several monosaccharide groups joined together by glucoside bonds.

omega-minus, Ω [Physics] A negatively charged elementary particle, classified as a hyperon and having a mass 3276 times that of an electron.

omegatron [Engineering] An instrument in which ions are caused to move in spiral paths by the application of an electric field at right angles to a constant magnetic field. As the angular frequency of rotation of the ions depends upon their charge to mass ratio, it is possible by this means to separate ions of different isotopes. The instrument may be used for the absolute determination of atomic masses and for isotopic and chemical analysis.

oncology [Medicine] The study of the causes, development, characteristics, and treatment of tumors.

ondoscope [Electronics] A glow discharge tube used to detect high-frequency radiation, as in the vicinity of a radar transmitter.

-one [Chemistry] Chemical suffix indicating a ketone =C=O.

-onium [Chemistry] Chemical suffix indicating a compelx cation, such as hydronium ion, (H_3O^+).

ontogeny [Biology] The origin and development of an organism from conception to adulthood. The complete developmental history of the individual organism.

oocyte [Biology] An egg before the completion of maturation.

oology [Biology] A branch of zoology, concerned with the study of eggs, especially bird eggs.

ooplasm [Biology] Cytoplasm of an egg.

opacity [Optics] The extent to which a medium is opaque. Numerically the reciprocal of the transmittance.

opaque [Optics] Not permitting a wave motion (*e.g.* light, sound, *x*-rays) to pass. Usually applied to light; not transparent or translucent. *See* opacity.

open clusters [Astronomy] One of the grouping of stars that have a common motion through space and are concentrated along the central plane of Milky way. The open clusters are much less densely populated with stars than the globular clusters, containing only some hundreds of stars interspersed with gas and dust clouds.

operand [Computers] Any one of the quantities entering into or arising from an operation.

operon [Biology] A group of genes whose function is to control the synthesis of the individual enzymes that act together as one enzyme system. One of the genes in an operon, known as the 'operator gene', starts and stops the activity of the complete operon.

ophthalmology [Medicine] The study of the anatomy, physiology, and diseases of the eye.

opposition [Astronomy] A planet having its orbit outside that of the earth is in opposition when the earth is in a line between the sun and the planet.

optical activity [Chemistry] The property possessed by some substances and their solutions of rotating the plane of vibration of polarized light. The amount of this rotation is proportional to distance the light travels in the medium, and to the concentration of the solution.

optical anisotropy [Optics] The behaviour of a medium, or of a single molecule, whose effect on electromagnetic radiation depends on the direction of the propagation of radiation.

optical axis [Optics] The line passing through the optical centre and the centre of curvature of a spherical mirror or lens. Also known as principal axis.

optical brightner [Chemistry] A colourless fluorescent organic compound which absorbs ultraviolet light and emits it as visible blue light.

optical centre [Optics] A point, situated for all practical purposes at the geometrical centre of a thin lens, through which an incident ray passes without being deviated.

optical crystal [Chemistry] Any natural or synthetic crystal such as sodium chloride, calcium fluoride, silver chloride, used for infrared and ultraviolet optics, piezoelectric effects, and short-wave radiation detection.

optical density [Optics] The degree of opacity of a translucent medium expressed by $\log I_o/I$, where I_o if the intensity of the incident ray, and I is the intensity of transmitted ray.

optical frequency [Physics] A frequency comparable to that of electromagnetic waves in the optical region, above about 3×10^{11} hertz.

optical glass [Materials] A type of glass which is free from imperfections, such as unmelted particles, bubbles, which would affect its transmission of light.

optical isomerism [Chemistry] The occurrence of a compound in two different forms, one a mirror image of the other. The two

forms have similar properties in all respects except for their optical activity, which is different. It occurs in compounds that have on asymmetric carbon atom. Optical isomerism is a form of stereoisomerism. Also known as enantiomorphism.

optically flat [Optics] A surface is said to be optically flat if the irregularities do not exceed the wavelength of light.

optical maser [Physics] *See* laser.

optical rotary dispersion [Chemistry] (ORD) The effect in which the amount of rotation of plane-polarized light by an optically active compound depends on the wavelength. A graph of rotation against wavelength has a characteristic shape showing peaks or troughs.

optical temperature [Astronomy] The temperature of a celestial body as calculated from its light radiation.

optical twinning [Crystallography] Simultaneous growth of two crystals which are same except that the structure of the one is that mirror image of the other. Also known as chiral twinning.

optic axis [Optics] The direction in a doubly refracting crystal in which light is propagated without double refraction.

optics [Physics] The study of phenomena associated with the generation, transmission, and detection of electromagnetic radiation in the spectral range extending from long-wave edge of the X-ray region to the short-wave edge of the radio region, or in the wavelength from about 1 nanometre to about 1 millimetre.

optics, geometrical [Physics] The branch of optics built up on the laws of reflection, and refraction and assuming the rectilinear propagation of light; it involves no consideration of the physical nature of light. It is mainly concerned with the formation of images by mirrors and lenses.

o-ray [Optics] The ray which vibrates at right angles to the optic axis in uniaxial crystals.

orbit [Astronomy] The path of one heavenly body around another as a result of their mutual gravitational attraction. Particularly the path of the planets around the sun, or the moon (or artificial satellites) around the earth. [Physics] The path of an electron around the nucleus of an atom. *See* orbital electron.

orbital [Physics] 1. The space dependent part of the Schrodinger wave function of an electron in an atom or molecule in an approximation such that each electron has a definite wave function, independent of the other electrons. 2. A space in which there can be one or two electrons but not more. The space is where there is probability of finding one or both electrons and it varies over the space. The positions and motion of an electron in an orbital cannot be described, only

the probability can be described. The orbitals are classified as *-s, -p, -d, -f* according to their spectral properties.

orbital electron [Physics] An electron contained within an atom; it may be thought of as orbiting around the nucleus, in a manner analogous to the orbit of a planet around the sun. An orbital electron has a high probability of being in the vicinity (at distances of the order of 10^{-10} metres or less) of a particular nucleus, but has only a very little probability of being within the nucleus itself.

orbital velocity [Astronomy] The velocity of a satellite or spacecraft that enables it to orbit round the earth or other celestial body. A synchronous orbit round the earth requires an orbital velocity of about 3200 metres per second.

order [Chemistry] A classification of chemical reactions in which the order is described as first, second, third, or higher, according to number of molecules which appear to enter into reaction.

ordinary ray [Optics] When a ray of light is incident upon a crystal that exhibits double refraction so that the direction of the ray makes an angle with the optic axis of the crystal, the ray splits into two rays. One of these obeys the ordinary laws of refraction and is called the ordinary ray. The other is the 'extraordinary ray'.

ore

328

ore [Geology] A naturally occurring rock or mineral from which a desired product (usually a metal) can be extracted economically.

organic acid [Chemistry] An organic compound that is able to give up a proton to a base; *i.e.*, one that contains one or more carboxyl groups or in some cases hydroxyl groups (*e.g.*, phenol).

organic base [Chemistry] A molecule or ion possessing a lone pair of electrons that can be used for coordination (*see* valence, electronic theory of) with a proton. The common organic compound that fulfil this condition owe their basic character to an oxygen or nitrogen atom.

organic chemistry [Chemistry] The study of the composition, reactions, and properties of organic compounds; the chemistry of carbon compounds excluding the metal carbonates and the oxides and sulphides of carbon.

organism [Biology] Any animal, plant or inividual constituted to carry out all life functions.

organisol [Chemistry] A dispersion of very finely divided resin particles in organic liquid mixture which cannot dissolve the resin at normal temperatures.

organometallic compounds [Chemistry] Organic compounds in which the molecules contain a carbon atom linked directly to a metal atom; *e.g.*, methylsodium,

$NaCH_3$, ferrocene.

organosilicon compounds [Chemistry] Chemical compounds in which silicon atoms play the part of carbon atoms in organic compounds; *e.g.*, silanes (general formula Si_nH_{2n+2}) are the organosilicon analogues of alkanes.

organosol [Chemistry] Colloidal dispersion of any insoluble substance in an organic liquid.

ormulu [Metallurgy] An alloy of copper, zinc, and tin in various proportions; generally containing 50% copper.

ortho- [Scientific Technology] Prefix denoting right, straight, correct. [Chemistry] 1. Denoting adjacency in position in a hexagonal ring of atoms, particularly the benzene ring. Abbreviated to *o*- as a prefix in naming a compound, *e.g.*, *o*-dichlorobenzene (alternatively, 1, 2-dichlorobenzene). *Compare* meta; para. 2. Prefix inciating an inorganic acid (or a corresponding salt) of a higher degree of hydration; *e.g.*, orthophosphoric acid, H_3PO_4, as compared with metaphosphoric acid, HPO_3.

orthochromatic film [Materials] A photographic film sensitive to green in addition to blue and violet light, thus giving a more accurate representation of colours in monochrome than ordinary film.

orthohydrogen [Chemistry] Hydrogen molecules in which the spins of the two constituent atoms are parallel.

osazones [Chemistry] Derivatives formed by reaction of aldehydes and ketones with phenylhydrazine.

oscillator [Mechnanics] A device for producing sonic or ultrasonic pressure waves in a medium. [Electronics] An electronic circuit that converts energy from a direct-current source to periodic varying electric output in the form of alternating current; usually consists of thermionic valves or transistors coupled with a suitable resonant circuit.

oscilloscope [Electronics] An electronic instrument that displays electric signals as patterns on the screen of a cathode ray tube. With no input signal the oscilloscope displays a horizontal straight line. This is formed by a spot of light sweeping continually across the screen. When an input signal is connected, this deflects the spot vertically, hence the spot traces out a pattern of the input signal.

osmium [Chemistry] Os. Trasition. element At. No. 76. At.Wt. 190.2, placed in group VIII of the periodic table. A hard, white, crystalline metal. The heaviest substance known; m.p. 3318K; b.p. 5300K. It occurs together with platinum; used in alloys with platinum and iridium.

osmometer [Engineering] An instrument for measuring osmotic pressures.

osmosis [Chemistry] The passage of water (or other solvent) through a membrance that will permit the passage of the solvent but not of dissolved substances. There is a tendency for solutions separated by such a membrane to become equal in molecular concentration; thus water will flow from a dilute to a concentrated solution, the solutions tending to become more nearly equal in concentration.

osmotic pressure [Chemistry] The applied pressure required to prevent the flow of a solvent across a membrane which offers no obstruction to passage of the solvent, but does not allow passage of solute, and which separates a solution from the pure solvent. The osmotic pressure of a dilute solution is analogous to gaseous pressure; a substance in solution, if not dissociated exerts the same osmotic pressure as the gaseous pressure it would exert if it were a gas at the same temperature and occupying the same volume.

osteology [Medicine] The study of anatomy and structure of bones.

Ostwald's dilution law [Chemistry] A law relating the dissociation constant, K, and the degree of dissociation (or ionization), x, of weak electrolyte of concentration c moles per litre. This law states that for a binary electrolyte

$$K = cx^2/(1-x)$$

out breeding [Chemistry] *See* exogamy.

output [Computers] The data produced by a data-processing operation. 2. [Electronics] The current, power, voltage, driving force, or information which a circuit or device delivers.

oxidase [Chemistry] One of the enzymes which catalyze the oxidation of many substances by free oxygen, *e.g.,* tyrosinase, lactase, uricolase.

oxidation [Chemistry] In its original use, oxidation meant simply combination with oxygen. Its use has, however, been considerably widened to cover a great many processes similar to oxidation, such as chlorination, and other processes of combination with strongly nonmetallic elements, which add electrons readily. In fact the term oxidation in its broadest sense means simply a chemical reaction whereby electrons are removed from one or more of the atoms of a substance. It is, of course, most frequently accompanied by a simultaneous process, in the same reaction, whereby another substance or substances gain the electrons and thus undergo reduction; therefore, calling the process oxidation, under these circumstances, simply means that it is this part (*i.e.*, loss of electrons) of the particular process that is of greatest interest. Removal of hydrogen atoms (each having one electron) from a hydrogen-containing organic compound (dehydrogenation) is a form of oxidation; it is effected by a catalytic reaction with air or oxygen, as in the oxidation of alcohols to aldehydes. Free radical chains play an important part in the oxidation of such organic compounds as rubber, drying oils.

oxidation number [Chemistry] For a given element, the number of electrons it can transfer to another element with which it combines. Conversely, it is the number of electrons necessary to add to or subtract from a combined atom to restore it to its original uncombined (elemental) state. The oxidation number may vary for the same element; and for different elements it may be positive or negative, depending on the number of valence electrons in the outer shell. Thus, there is a close relation between oxidation number (electrons added or subtracted) and the valence, or combining power, of an element. While using the concept of oxidation number, following rules must be observed :

(*i*) The oxidation number of an element is taken to be zero.

(*ii*) The oxidation number of a monoatomic ion is reckoned to be equal to its electric charge.

(*iii*) The oxidation number for a covalent compound of known structure is, for each atom, the charge left on that atom when each shared pair of electrons is assigned to the more electronegative of the two atoms which share it, *e.g.,* for H_2O, H is $+1$ and oxygen -2.

(*iv*) The sum of oxidation numbers of the atoms making up a molecule of a stable compound is zero. Thus for $K_2Cr_2O_7$ (potassium dichromate) each potassium gives $+1$, each chromium $+6$, and each oxygen -2, giving $2 + 12 - 14 = 0$.

oximetry [Physics] Measurement of the degree of oxygen saturation of the circulating blood.

oxo-Prefix indicating the presence of oxygen in a chemical compound.

oxo acid [Chemistry] An acid in which the acidic hydrogen atom(s) are bound to oxygen atoms. Sulphuric acid is an example: the two acidic hydrogens are on the MOH groups bound to the sulphur.

oxonium ion [Chemistry] An ion of the type R_3O^+, in which R indicates hydrogen or an organic group. The hydroxonium ion, H_3O^+, is formed when acids dissociate in water.

oxo process [Chemistry] An industrial process for making aldehydes by reaction between alkanes, carbon monoxide, and hydrogen (cobalt catalyst using high pressure and temperature).

oxygen [Chemistry] O. Element. At. No. 8. At.Wt.15.9994, m.p. 58.7K; b.p. 90.1K; placed in group VI A of the periodic table. An odourless, invisible gas; the most abundant of all the elements in the earth's crust including the seas and the atmosphere; it forms approximately 20% of the atmosphere by volume.

oxygenase [Biochemistry] An enzyme catalyzing a reaction in which oxygen is introduced into an acceptor molecule.

oxygenate [Chemistry] To treat, infuse or combine with oxygen.

oxygen debt [Physics] The accumu-

lation of lactic acid in muscles during violent exercise.

ozone [Chemistry] An allotropic form of oxygen (O_3) in which each molecule contains three oxygen atoms. It is formed by passing silent electric discharge through dry oxygen. Ozone can be characterized by its pungent odour and powerful oxidising power. It is used as an oxidant, bleach, and water purifier, and to treat industrial wastes.

ozone cloud [Physics] A limited region in which total ozone content of the ozonosphere is greater than the normal.

ozonides [Chemistry] 1. A group of compounds formed by reaction of ozone with alkali metal hydroxides and formally containing the ion O_3^-. 2. Unstable compounds formed by the addition of ozone to the C=C double bond in alkenes. *See* ozonolysis.

ozoniser [Engineering] Apparatus that converts oxygen into ozone, by subjecting the oxygen to an electric-brush discharge.

ozonization [Chemistry] The process of treating, impregnating, or combining with ozone.

ozonolysis [Chemistry] A reaction of alkenes with ozone to form an ozonide. It was once used to investigate the structure of alkenes by hydrolysing the ozonide to give aldehydes or ketones. For instance

$$R_2C:CHR^1 \rightarrow R_2CO + R^1CHO$$

These could be identified, and the

structure of the original alkene determined.

ozonosphere [Physics] The layer in the upper atmosphere in which there is an appreciable ozone concentration and in which ozone plays an important role in the radiative balance of the atmosphere by absorbing a large proportion of the sun's ultraviolet radiation. Ozonosphere lies roughly between 10-50km from the surface of the earth with maximum ozone concentration at about 20-25 km. Also known as ozone layer.

pacemaker [Medicine] An electric device used to regulate the pace of the heart beat.

pachymeter [Engineering] An instrument that measures the thickness of materials, such as a sheet of paper

packet switching [Computers] A method of digital communication in which messages are divided into packets of a bit size determined by the needs of the transmission network, and are transferred to their destination in a store-and-forward manner over multiple "Virtual" circuits, which are dedicated to the connection only for the duration of the packets transmission.

packing density [Computers] 1. The number of devices (such as logic circuits) or integrated circuits per unit area of a silicon chip 2. The quantity of information stored in a specified space of a storage system associated with a computer, *e.g.* bits per cm of magnetic tape.

packing fraction [Nucl.Sc.] The quantity $(M-A)/A$, where M is the atomic mass and A is the atomic number of the atom. Packing fractions are usually expressed as parts per 10,000.

packing index [Crystallography] The volume of ion divided by the volume of the unit cell in the crystal.

packing radius [Crystallography] One-half of the smallest approach distance of atoms or ions.

page crossers [Computers] References to memory that span the boundary between two pages.

page fault [Computers] A trap that occurs when data requested by a processing is not found in main memory.

page table [Computers] A data structure that translates virtual addresses into real addresses.

paint [Materials] A thoroughly dispersed mixture having a viscosity ranging from a thin liquid to a semisolid paste and consisting of (i) a drying oil, synthetic resin, or any other film forming material, called the binder, (ii) a solvent or thinner; and (iii) an organic or inorganic pigment. The binder and the solvent are collectively called the vehicle. Paints are used to provide an adherent coating that imparts colour to and often protects the surface.

paint vehicle [Materials] The liquid constituent of paint that consist of volatile solvent or thinner and a film-forming material.

pair production [Physics] . The conversion of a photon into an electron and a positron when the photon traverses a strong electric field, such as that surrounding a nucleus or an electron.

paleontology [Biology] The study of the life of the past as recorded by the fossil remains.

palladium [Chemistry] Pd. Transition element. At. No. 64. At. wt 106.4, m.p. 1824K; b.p. 3413K; metallic element placed in group VIII of the periodic table; used in alloys and as a catalyst.

pallas [Astronomy] An asteriod located between Mars and Jupiter, with a diameter about 480 kilometres.

panary fermentation [Biochemistry] Yeast fermentation of dough in bread-making.

panchromatic film [Materials] A photographic film sensitive to light of all colours including red, thus giving a more accurate representation of colours in monochrome than orthochromatic film.

pancreas [Physoilogy] A gland in the abdomen that produces pancreatic, insulin and glucagon.

panoramic display [Electronics] A display that simultaneously shows the relative amplitudes of all signals recieved at different frequencies.

pantography [Engineering] System for transmitting and automatically recording radar data from an indicator to a remote point.

papain [Biochemistry] An enzyme, found in the fruit and leaves of the pau-pau tree that is capable of digesting proteins. Used for softening meat for human consumption.

paper chromatography [Chemistry] A type of chromatography in which the mobile phase is liquid and the stationary phase is a strip of porous paper. A drop of the mixture is placed at one edge of the paper and eluted with the solvent. The components are separated by the rates at which they move across the paper with the solvent. Indentification can be done by indicators or by their fluorescence in ultraviolet radiation.

paper electrophoresis [Chemistry] A variation of paper chromatography in which an electric current is applied to the ends of the electrolyte impregnated absorbent paper, thus moving chargeable molecules of the unknown sample toward the respective electrodes.

papite [Materials] A poisonous gas

consisting of a mixture of acrolein and stannic chloride.

para-[Chemistry] 1. Prefix designating a benzene compound in which two substituents are in the 1, 4 positions, *i.e.* directly opposite each other, on the benzene ring. The abbreviation *p-* is used; for example, *p*-xylene is 1, 4-dimethylbenzene. *Compare* ortho-; meta-. 2. Prefix denoting the form of diatomic molecules in which the nuclei have opposite spins.

parabiosis [Biology] Experimental joining of two individuals to study the effect of one individual upon the other.

parabola [Mathematics] A curve traced out by a point that moves so that its distance from a fixed point, the focus, is equal to its distance from a fixed straight line, the directrix. The equation of a parabola with its vertex at the origin and its axis along the x-axis is: $y^2 = 4ax$, where a is the distance from the origin to the focus.

parabolic reflector [Optics] A concave reflector, the section of which is parabola; used for producing a parallel beam of electromagnetic radiation when a source is placed at its focus, or for collecting and focusing an incoming parallel beam of radiation. If the radiation is light the reflector is usually called a parabolic mirror, but with microwave or radio frequency radiation (*see* radio telescope) it may be called a 'disc aerial'.

parachlor [Physics] The product of the molecular mass and the fourth root of surface tension, divided by the difference between the density of the liquid and density of the vapour in equilibrium with it. It is generally constant over wide ranges of temperature.

paraffin [Chemistry] Any of saturated aliphatic hydrocarbons of the alkane series C_nH_{2n+2}.

paraffin oil *See* kerosine.

paraffin wax [Materials] A white translucent solid hydrocarbon mixture melting to a colourless liquid in the range 323-333K used for candles, waxed paper, waterproofing, and polishes.

parahydrogen [Chemistry] Hydrogen molecules in which the spins of two constituent atoms are antiparallel.

parallactic ellipse [Astronomy] An annual apparent elliptical course of the celestial body on the celestial sphere about its mean position, caused due to elliptical orbital motion of the earth.

parallax [Optics] The difference in direction, or shift in the apparent position, of a body when viewed from different positions.

parallel circuit [Electricity] An electric circuit in which the elements, branches (having elements in series), or components are connected between two points, with one of the two ends of each component connected to each point. For resistors in parallel, the resulting resistance R is given by :

$$1/R = 1/R_1 + 1/R_2 + 1/R_3 + \ldots$$

For capacitors in parallel, the capacitance of the combination is given by :

$$C = C_1 + C_2 + C_3 + \ldots$$

For cells in parallel, the e.m.f. is equal to the largest value of e.m.f. of all the cells.

parallel processing [Computers] Simultaneous processing, as opposed to the sequential processing in a conventional(Von Neumann) type of computer architecture.

paralysis [Medicine] Complete or partial loss of motor or sensory function.

paramagnetic material [Physics] A material within which an applied magnetic field is increased by the alignment of electron orbits. As the temperature increases the paramagnetism decreases.

paramagnetism [Physics] A property exhibited by substances which, when placed in a magnetic field, are magnetized parallel to the field to an extent proportional to the field (except at very) low temperatures or in extremely large magnetic fields. Substance possessing a magnetic permeability slightly greater than unity, *i.e.* possessing a small positive magnetic susceptibility, are said to be **paramagnetic**. The atoms of a **paramagnetic** substance possess a permanent magnetic moment due to unbalanced electron spins or unbalanced orbital motions of the electrons around the nucleus.

parameter [Mathematics] In two-dimensional analytical geometry it is often convenient to express the variables *(x, y)* each in terms of a third variable t, such that x and y are functions of t; $x=f(t)$, $y=f(t)$. The equations are termed parametric equations, and t is a parameter. [Science Technology] A variable that can be kept constant while the effect of other variables is investigated.

parametric amplifier [Electronics] An amplifier of microwaves that depends on the periodic variation, by an alternating voltage, of the reactance of a thermionic valve or semiconductor device.

paramorphism [Geology] The property of a mineral whose internal structure has changed without change in composition or external form.

parasite [Biology] An organism that lives in or on another organism of different species from which it obtains its food and shelter.

parasitic capture [Nucl.Sc.] The absorption of a neutron by a nuclide that does not result in a nuclear fission or the production of a useful artificial element.

parastate [Physics] A state of a diatomic molecule in which the spins of the nuclei are anti-parallel.

parasymbiosis [Botany] The condition when two organisms grow together without harming one another.

parhelion [Astronomy] Either of the two coloured luminous spots that appear at points 22° on both sides of the sun and at the same elevation as the sun. Also known as sun dog; mock sun.

parison [Metallurgy] An unformed mass of molten glass from which finished products are manufactured.

parkerizing [Metallurgy] Trade name of the process in which iron or steel is coated with a phosphate layer to inhibit the rust formation.

Parke's process [Metallurgy] A process for recovering silver from lead. 1-2% molten zinc is added to lead-silver mixture, and the mixture is heated above the melting point of zinc. A scum containing mostly silver forms on the surface. The process is repeated several times to obtain complete recovery of silver.

parking orbit [Astronomy] A temporary earth orbit during which the space vehicle is checked out and its path carefully measured to determine the amount and time of increase in velocity required to send it into final orbit or into space in the desired manner.

parsec [Astronomy] An astronomical unit of distance, corresponding to a parallax of one second of arc and equal to 3.258 light-years, or 3.0857×10^{13} kilometres.

parthenogenesis [Biology] The development of an unfertilized egg into an adult organism; common among honeybees and wasps.

partial differential [Mathematics] The infinitesimal change in a function consisting of two or more variables when one of the variable changes and the others remain constant. If $z = f(x,y)$, $*z/*x$ is the partial differential to z with respect of x, while y remains unchanged.

partial pressures [Biology] See Dalton's law of partial pressures.

particle [Physics] A very small piece of solid material, that is considered to have mass but not size and volume, that is, it is like a point. Atoms and molecules are considered as particles in this sense. A particle may be considered as a small subdivison of matter, ranging in diameter of a few Angstroms.

particle accelerator [Biology] See accelerator.

partition coefficient [Chemistry] The ratio of the concentrations of a single solute in two immiscible solvents, at equilibrium. The partition coefficent is the equilibrium constant for the process, usually written so that the concentration of the solute in the more soluble phase is the numerator.

parton [Physics] A basic particle,

such as a quark, from which other elementary particles are formed.

parts per million [Scientific Technology] *ppm*. A way of describing small concentrations and means exactly what the term says. mgm per kg is also *ppm*.

parturient [Medicine] In labor; *i.e.*, giving birth.

parturition [Medicine] The process of giving birth.

pascal [Mechanic] Pa. The derived SI unit of pressure equal to 1 newton per square metre.

Pascal's law of fluid pressures [Mechanics] The law which states that confined fluid transmits pressure uniformaly in all the directions when pressure is applied externally.

Paschen series [Physics] A series of lines that occurs in the infrared region of the spectrum of hydrogen whose wave numbers are given by the $R_H[(1/9)-(1/n^2)]$, where R_H is Rydberg constant for hydrogen, and *n* is any integer greater than 3.

Paschen's law [Physics] The breakdown or 'sparking potential' for a pair of parallel electrodes situated in a gas, *i.e.*, the potential that must be applied between them for sparking to occur, is a function only of the pressure of the gas and the separation of the electrodes.

pass [Computers] A complete cycle of reading, processing, and writing in a computer.

passive [Electronics] Denoting an electronic component, such as capacitor, that does not amplify a signal.

passive element [Electricity] The component of an electric circuit that is not a source of energy, such as a resistor, inductor, or a capacitor.

passivity [Chemistry] A property exhibited by iron, chromium and many other metals, involving the loss of their normal chemical activity in an electro-chemical system or in a corrosive environment after treatment with strong oxidising agent like nitric acid, and when oxygen is evolved upon them during electrolysis, forming an oxide layer.

Pasteur effect [Biochemistry] Inhibition of fermentation by passing abundant oxygen to replace anerobic conditions.

pasteurization [Biochemistry] The application of heat for a specified time to a liquid food or beverage for the purpose of killing or inactivating disease causing organisms.

pathogen [Medicine] A disease-producing agent, usually refers to living organisms.

pathology [Medicine] The study of the causes, nature, and effects of diseases and other abnormalities.

Pauli exclusion principle [Physics] Each electron moving round the nucleus of a neutral atom can be characterized by values of four quantum numbers. The **principle**

states that no two electrons in a neutral atom can have the same set of all the four quantum numbers.

p-band [Electronics] A band of radio frequencies extending from 225 to 390 megahertz, corresponding to wavelengths of 133.3 to 76.9 centimetres.

p-block elements [Chemistry] The block of elements in the periodic table consisting of the main groups III (B to Tl), IV (C to Pb), V (N to Bi), VI (O to Po), VII (F to At) and 0 (He to Rn). The outer electronic configurations of these elements all have the form ns^2np^x where $x = 1$ to 6. Members at the top and on the right of the p-block are nonmetals (C, N, P, O, F, S, Cl, Br, I, At). Those on the left and at the bottom are metals (Al, Ga, In, Tl, Sn, Pb, Sb, Bi, Po). Between the two, from the top left to bottom right, lie an ill-defined group of metalloid elements (B, Si, Ge, As, Te).

pearlite [Metallurgy] A microconstituent of iron or steel consisting of alternate layers of ferrite and cementite.

peat [Minerals] An early stage in the formation of coal vegetable matter. It is an accumulation of semicarbonized residue of plants and is used as fuel.

pectins [Biochemistry] A class of complex polysaccharides occurring in plants and fruits. Solutions have the power of setting to a jelly; this is probably responsible for the 'setting' of jams.

pediatrics [Medicine] The branch of medicine that deals with the growth and development of the child with the care, treatment, and prevention of diseases, injuries and defects of children.

pedology [Medicine] The science of the study of the physiological as well as the psychological aspects of childhood.

pelagic [Biology] An organism which inhabits open water, as in midocean.

p-electron [Physics] An atomic electron that has an orbital angular momentum quantum number of unity.

Peltier effect [Electronics] When an electric current flows across the junction between two different metals or semiconductors, a quantity of heat, proportional to the total electric charge the junction, is evolved or absorbed, depending on the direction of the current. This effect is due to the existence of an electromotive force at the junction.

pencil lead [Materials] A mixture of graphite and clay in various proportions, to give different degrees of hardness.

pentode [Electronics] A thermionoic valve containing five electrodes : a cathode, an anode or plate, a control grid, and (between the two latter) two other grids called the screen grid and the suppressor grid.

pentosans [Biochemistry] Polysaccharides that yield pentoses on hydrolysis.

penumbra [Optics] Half-shadow, formed when an object in the path of rays from a large source of light cuts off a portion of the light.

pentolite [Materials] An highly explosive mixture of pentaerythritol tetranitrate and trinitrotoluene in equal quantities.

pepsin [Biochemistry] A digestive enzyme produced in the stomach that converts proteins into peptones; it acts only in an acidic medium.

peptidase [Biochemistry] An enzyme that attacks peptide linkages and splits off amino acids.

peptide [Chemistry] Any of a group of organic compounds comprising two or more amino acids linked by *peptide bonds*. These bonds are formed by the reaction between adjacent carboxyl (-COOH) and amino (-NH_2) groups with the elimination of water (*see* illustration). *Dipeptides* contain two amino acids, *tripeptides* three, and so on. Polypeptides contain more than ten and usually 100-300. Naturally occurring *oligopeptides* (of less than ten amino acids) include the tripeptide glutathione and the pituitary hormones vasopressin and oxytocin, which are octapeptides. Peptides also result from protein breakdown, *e.g.*, during digestion.

peptones [Biochemistry] Organic substances produced by the hydrolysis of proteins by the action of pepsin in the stomach. They are soluble in water, and are absorbed by the body.

per- [Chemistry] Prefix indicating that a chemical compound contains an excess of an element, *e.g.*, a peroxide.

percussion cap [Chemistry] A device used in fire-arms. It consists of small copper cylinder containing mercuric fulminate or other violent explosive that will explode on being struck, thus initiating the explosion of the main charge.

perfect ges *See* ideal ges.

perforation [Scientific Technology] Any hole made by boring, punching, or piercing.

perfume [Materials] A blend of pleasently odorous substances obtained from essential oils of flowers, leaves, fruits, roots, or wood of a wide variety of plants, either by solvent extraction or by steam distillation.

periastron [Astronomy] The coordinates and time when the two stars of a binary star system are nearest to each other in their orbits.

pericentre [Physics] That point on any orbit nearest to the centre of attraction.

periclase [Materials] MgO. A natural or calcined magnesium oxide used as a lining and maintenance material for basic oxygen steel-making furnaces and other refractories.

pericynthion [Astronomy] The time of, or the point of the nearest approach of a satellite in lunar orbit to the moon's surface. Opposite of apocynthion.

perigee [Astronomy] The moon, the sun, or an artificial earth satellite are said to be in perigee when they are at their least distance (in orbit) from the earth. Opposite of apogee.

perihelion [Astronomy] The time of or the point of, the nearest approach of a planet to the sun. Opposite of aphelion.

perimeter [Mathematics] The distance all round a plane figure; *e.g.* the perimeter of a circle is its circumference.

period [Physics] If any quantity is a function of the time, and this function repeats itself exactly after constant time intervals T, the quantity is said to be periodic, and T is called the period of the function.

periodic current [Astronomy] The current produced by the tidal influence of moon and sun or by any other oscillatory forcing function.

periodic damping [Physics] Damping which is less than critical damping.

periodic function [Mathematics] A function $f(x)$ of a real or complex variable is periodic with period T if $f(x+T) = f(x)$ for every value of x.

periodic law [Chemistry] The law that properties of the chemical elements and their compounds are a periodic function of their atomic weights. A modern statement of the law is : The electronic configurations of the atoms of the elements vary periodically with their atomic number consequently all properties of the elements that depend on their atomic structure tend also to change with increasing atomic number in a periodic manner.

periodic motion [Mechanics] Any motion that repeats itself identically at regular intervals.

periodic table [Chemistry] An arrangement of the chemical elements by symbols written in sequence in the order of atomic number (or atomic weight) and arranged in horizontal rows (periods) and vertical columns (groups) to illustrate the occurrence of similarities in the properties of the elements as a periodic function of the sequence.

periodic wave [Physics] A wave whose displacement has a periodic variation with distance or time, or both.

peripherals [Computers] Device connected to the central processing unit or the high-speed store of a computer. Forming part of the hardware, they include backing storage, input and output devices, online equipment, visual display units, etc.

peritectic point [Chemistry] Incongruent melting point identical with discontinuity on liquidus curve.

peritectic process [Chemistry] Process of melting of polyconstituent phase in which both liquid phase and another solid phase are formed; new phases are of different chemical composition.

permafil [Chemistry] A mixture in which the liquid completely undergoes polymerization and hardens without any evaporation.

permanent dipole [Physics] Electric dipole existing independently of presence of external electric field.

permanent dipole moment [Physics] Dipole moment existing independent of the presence of external electric field. Its SI unit is coulomb metre (C.m.).

permanent gases [Physics] An historic classification that included those gases which are very difficult to liquefy (such as nitrogen and oxygen) and which were thought to be non-liquefiable; i.e., that cannot be liquefied by pressure alone at normal temperatures (i..e.., a gas that has a critical temperature below room temperature). The distinction has vanished since all the gases have been liquefied because of the advance in low temperature techniques.

permeability [Physics] 1. It is the capacity of a membrane or other material to allow another substance to penetrate or pass through it; or specifically, the quantity of a specified gas or other substance which passes through under specified conditions. 2. The ratio of magnetic flux density, B, in a medium to the external magnetic field strength, H, i.e.,

$$\mu = B/H$$

Its SI unit is henry per metre (H/m).

It is also called absolute permeability

permease [Biochemistry] Any of a group of enzymes which mediate the phenomenon of active transport.

permendur [Chemistry] A magnetic alloy composed of equal parts of iron and cobalt and has an extremely high permeability when saturated.

permenorm alloy [Chemistry] An alloy containing equal amounts of nickel and iron; used as magnetic core material and in magnetic amplifiers.

permittivity (ε) [Physics] The ratio of the electric displacement, D, in a dielectric medium to the applied electric field strength, E, i.e.,

$$\varepsilon = D/E.$$

It indicates the degree to which the medium can resist the flow of electric charge and is always greater than unity. Its SI unit is farad per metre (F/M). It is also called absolute permittivity.

peroxides [Chemistry] A group of compounds containing the O_2^{2-} ion. They are notionally derived from hydrogen peroxide, H_2O_2, but these ions do not exist in aqueous soution due to to extremely rapid hydrolysis to OH⁻.

peroxide number [Chemistry] Measure of millimoles of peroxide taken up by 1000 grams of fat or oil; used to measure rancidity.

perpetual motion [Physics] The concept of a machine that, once set in motion, will go on forever without receiving energy. It is impossible to make a machine that will go on for ever and be able to do work, *i.e.*, create energy without receiving enery from outside. To do so would contravene the first two laws of thermodynamics.

persistence [Electronics] The rate of decay of luminance of a CRT display after the stimulus is removed.

persistent current [Physics] A magnetically induced current that flows undiminished in a superconducting material or circuit.

Persoz reagent [Materials] A mixture of zinc chloride and zinc oxide in water; used to detect the presence of silk with wool (only silk dissolves in it).

perturbation [Astronomy] Deviation in the motion of the planets from their true elliptical orbits, as a result of their gravitational attractions for each other.

pesticide [Materials] Any substance that destroys pests; the term includes insecticides, herbicides, rodenticides, miticides, etc. Also known as biocide.

peta [Scientific Technology] Perfix denoting one thousand million; 10^{15}. Symbol P, *e.g.*, Pm$=10^{15}$ metres.

petrifaction [Geology] The fossilization process where by an organic structure, such as a tree, changes into a stony or mineral structure. It is generally caused by dissolved hydrated silica, SiO_2, penetrating into the pores and gradually losing its water.

petrochemicals [Chemistry] Chemical substances derived from petroleum or natural gas.

petrogensis [Geology] The branch of petrology dealing with the origin of rocks, particularly igneous rocks. Also known as petrogeney.

petrol [Materials] A complex mixture consisting mainly of hydrocarbons, such as hexane, heptane, and octane; other fuels and special ingredients are often added. Also known as gasoline.

petrolatum [Materials] A purified mixture of hydrocarbons consisting of a semi-solid whitish or yellowish mass. Also known as vaselline; petroleum jelly.

petroleum [Materials] A naturally occurring complex mixture of paraffinic, cycloparaffinic, and aromatic hydrocarbons, containing low percentages of sulphur and trace amounts of nitrogen and oxygen compounds. This after distillation yields combustible fuels, petrochemicals, and lubricants. It is said to have originated from both animal and plant sources from 10 to 20 million years ago.

petroleum ether [Materials] A mixture of the lower hydrocarbons of the alkane series consisting mainly of pentane and hexane; b.p.303-343K

petroleum jelly *See* petrolatum.

petrology [Geology] The branch of geology concerned with the study of the origin, structure, and composition of rocks; principally igneous and metamorphic rocks.

pewter [Metallurgy] An alloy of approximately 4 parts of tin to 1 part of lead, with small amounts of antimony.

Pfund series [Physics] A series of lines in the infrared spectrum of atomic hydrogen whose wave numbers are given by $R_H[1/25-1/n^2]$, where R_H is the Rydberg constant for hydrogen, and n is any integer greater than 5.

pH *See* pH scale.

phagocyte [Biochemistry] A blood cell (particularly a leucocyte) that can engulf a foreign particle or bacterium and debris in the tissues.

phantom [Nucl.Sc.] A volume of material approximating as closely as possible the density and effective atomic number of living tissue, used in biological studies involving radiation.

pharmacognosy [Medicine] The science of crude drugs.

pharmacolite [Minerals] CaH-$(AsO_4).2H_2O$. A white to greyish mineral composed of hydrous acid arsenate of calcium. Also known as arsenic bloom.

pharmacology [Medicine] The study of the action of chemical substances upon animals and man.

pharmacophere [Biochemistry] The portion of a molecule of a substance that is regarded as determining the special physiological action of the substance.

pharmacy [Medicine] 1. The preparation and dispensing of drugs and medicine. 2. The place of dispensing drugs.

phase [Chemistry] A separate part of a heterogeneous body or system that is homogeneous throughout, has definable boundaries and can be separated physically from other phases, *e.g.*, a mixture of ice and water is a two-phase system, while a solution of salt in water is a system of one phase.[Physics] Points in the path of a wave motion are said to be points of equal phase if the displacements at those points at any instant are exactly similar; *i.e.* of the same magnitude and varying in the same manner.

phase angle [Physics] The difference in phase between two sinusoidally varying quantities. The displacement x_1 of one quantity at time t is given by $x_1 = a\sin\omega t$, where ω is the angular frequency and a is the amplitude. The displacement x_2 of a similar wave that reaches the end of its period T, a fraction β of the period before the first is said to lead the first quantity by a time βT. The value of x_2 is then given by $x_2 = a\sin(\omega t + \phi)$. ϕ is called the phase angle and it is equal to $2\pi\beta$.

phase diagram [Chemistry] A graphical representation showing the relations between various phases in a chemical system, and the effects of composition and conditions (temperature, pressure) on them.

phase modulation [Electronics] Modulation of the phase angle of a sinusoidal carrier wave. The phase angle of a sinusoidal carrier wave. The phase of the modulated wave differs from that of the carrier by an amount proportional to the instantaneous value of the modulating wave.

phase rule [Chemistry] For any system at equilibrium, the relationship $P + F = C + 2$ holds, where P is the number of distinct phases, C the number of components, and F the number of degrees of freedom of the system.

phases of the moon [Astronomy] The various shapes of the illuminated surface of the moon as seen from the earth (new moon, first quarter, quarter, full moon third quarter); due to variations in the ralative positions of earth, sun, and moon.

phase velocity [Physics] V_p The speed of propagation of a pure sine wave $Vp = \lambda f$, where λ is the wavelength and f is the frequency. The value of the phase speed depends on the nature of the medium through which it is travelling and may also depend on the mode of propagation. For electromagnetic waves travelling through space the phase speed c is given by $c^2 = 1/\varepsilon_o\mu_o$. where ε_o and μ_o are the electric constant and the magnetic constant respectively.

phenocopy [Biology] The nonhereditary alteration of a phenotype to a form imitating a mutant trait; caused by external conditions during development.

phenotype [Biology] The characteristics possessed by an individual organism as a result of the interaction of its inherited characteristics (*See* genotype) with its environment. 2. A group of organisms having the same phenotype.

pheromone [Biochemistry] Chemical substance secreted by an organism to the external environment that elicit a behavioural response from other organisms of the same species, especially substances that act as sex attractants.

phloem [Botany] A type of vascular tissue in plants; transports organic nutrients both up and down the stem or root.

phon [Physics] A unit of loudness used in measuring the intensity of sounds. The loudness, in phons, of any sounds is equal to the intensity in decibels of a sound of frequency 1000 hertz that seems as loud to the ear as the given sound.

phonon [Physics] The quantum of thermal energy in the lattice vibrations of a crystal. For the vibrational frequency f, the magnitude of the phonon is hf, where h is Planck's constant.

phosphagen [Biochemistry] In excitable tissues, an energy-storing compound containing a high-energy phosphate group, usually in enzymatic equilibrium with the terminal phosphate of ATP.

phosphodiester [Biochemistry] A molecule that contains two alcohols esterified to one molecule of phosphoric acid, which thus serves as a bridge between them.

phospholipids [Biochemistry] Compound lipids that contain phosphoric acid groups and nitrogenous bases. They are found in brain tissue and in egg yolk. Also known as phosphatides.

phosphor [Materials] A substance that is capable of luminescence, *i.e.*, storing energy (particularly from ionizing radiation) and later releasing it in the form of light. If the energy is released after only a short delay (between 10^{-10} and 10^{-4} second) the substance is called a 'scintillator'.

phosphor bronze [Metallurgy] An alloy of copper (80%-95%), tin (5%-15%), and phosphorus (0.25%-2.5%). It is hard, tough, and elastic.

phosphorescence [Physics] A form of luminescence in which a substance emits light of one wavelength after having absorbed electromagnetic radiation of a shorter wavelength. Unlike fluorescence, phosphorescence may continue for a considerable time after excitation.

phosphorogen [Chemistry] A substance that promotes phosphorescence in another substance, as manganese does in zinc sulphide.

phosphorolysis [Chemistry] A reaction by which elements of phosphoric acid are incorporated into the molecule of a compound.

phosphorus [Chemistry] P. Element. At. No. 15. At.Wt. 30.9738, placed in group VA of the periodic table. It occurs in several allotropic forms, white phosphorus, red phosphorus being the commonest. The former is a waxy white, very inflammable and poisonous solid. Red phosphorus is a non-poisonous, dark red powder, that is not very inflammable.

phot [Optics] A unit of illumination; an illumination of one lumen per square centimetre; equal to 10^4 lux.

photocathode [Electronics] A cathode that emits electrons when it is illuminated, *i.e.*, as a result of the photoelectric effect.

photocell [Electronics] A solid-state photosensitive electron device whose voltage-current characteristic is a function of incident radiation. Also known as photoelectric cell; electric eye.

photochemical reaction [Chemistry] A chemical reaction initiated, assisted or accelerated by exposure to light, *e.g.*, hydrogen and chlorine combine explosively on exposure to sunlight but very slowly in the dark.

photochemistry [Chemistry] The branch of physical chemistry concerned with the effects of radiation on chemical reactions.

photochromism [Optics] A change of colour occurring in certain substances when exposed to light. Photochromic materials are used in sunglasses that darken in bright sunlight.

photoconduction [Physics] An increase in electrical conduction resulting from absorption of electromagnetic radiation.

photoconductive effect [Physics] A photoelectric effect in which the electrical conductivity of certain substances, notably selenium, increases with the intensity of the light to which the substance is exposed.

photodiode [Electronics] A semiconductor diode in which the reverse current varies with illumination.

photodisintegration [Nucl.Sc.] The break up of an atomic nucleus into two or more fragments as a result of bombardment by gamma radiation.

photodissociation [Chemistry] The dissociation of a chemical compound as the result of the absorption of radiant energy.

photoelasticity [Physics] An effect in which certain materials exhibit double refraction when subjected to stress. It is used in a technique for detecting strains in transparent materials.

photoelectric cell [Electricity] *See* photocell.

photoelectric effect The liberation of electrons from a metal surface exposed to electromagnetic radiation. The number of electrons emitted depends on the intensity of the radiation. The kinetic energy of the electrons emitted depends on the frequency of the radiation. The effect is a quantum process in which the radiation is regarded as a stream of photons, each having an energy hf, where h is the Planck constant and f is the frequency of the radiation. A photon can only eject an electron if the photon energy exceeds the work function, ϕ, of the solid, *i.e.* if $hf_o = \phi$ an electron will be ejected; f_o is the minimum frequency (or *threshold frequency*) at which ejection will occur. Apart from the liberation of electrons from metals other phenomena are also referred to as photoelectric effects. These are the *photoconductive effect* and the *photovoltaic effect*.

photoelectron [Electronics] An electron emitted from a surface as a result of photoelectric effect or by photoionization.

photoemissive cell [Electronics] A device which detects or measures radiant energy by measurement of the resulting emission of electrons from the surface of a photocathode.

photoemissivity [Electronics] The property of a substance that emits electrons when struck by light.

photography [Optics] The process of forming visible image directly or indirectly by the action of light or other form of radiation on radiation sensitive surfaces.

photoionization [Chemistry] The removal of one or more electrons of an atom or molecule as the result of exposure to radiation. If the frequency of the radiation is f, each photon will have an energy hf, where h is Planck's constant. Photons with energies in excess of the ionization potential of the atoms struck will cause ionization to occur.

photology [Optics] The scientific study of light.

photoluminescence [Physics] Luminescence caused by electromagnetic radiation. The emitted light always has a lower frequency than the radiation absorbed.

photolysis [Chemistry] The decomposition of a chemical compound as the result of irradiation by visible light or ultraviolet radiations. This process is important in photosynthesis in providing hydrogen donors by splitting of water as follows :

$$4H_2O \dashrightarrow 4[H] + 4[OH]$$

$$4[OH] \dashrightarrow 2H_2O + O_2$$

$$4[H] + CO_2 \dashrightarrow CH_2O + H_2O$$

photomeson [Physics] A meson produced by the interaction between a photon and an atomic nucleus.

photometer [Optics] An instrument for comparing the luminous intensity of sources of light. In astronomy, photoelectric photometers are used to measure the intensity of light from distant stars.

photomultiplier [Electronics] A photoelectric cell of high sensitivity used for detecting very small quantities of light radiation.

photon [Physics] A particle with zero rest mass consisting of a quantum of electromagnetic radiation. The photon may also be regarded as a unit of energy equal to hf, where h is the Planck constant and f is the frequency of the radiation in hertz. Photons travel at the speed of light. They are required to explain the photoelectric effect and other phenomena that require light to have particle character.

photoneutron [Nucl.Sc.] A neutron released from a nucleus in a photonuclear reaction.

photonuclear reaction [Nucl.Sc.] A nuclear reaction resulting from the collision of a photon with a nucleus.

photoperiodism [Biology] The response of an organism to

changes in day length. In plants, leaf fall and flowering are common responses to seasonal changes in day length, as are migration, reproduction, and winter-coat development in animals.

photophoresis [Physics] Production of unidirectional motion in a collection of very small particles suspended in a gas or falling in a vacuum by a strong light-beam.

photopic vision [Optics] Vision in which the cones in the eye are the principal receptors. It occurs under normal lighting conditions and colours can be distinguished.

photopolymer [Materials] A polymer of plastic that undergoes a change on exposure to light. The effect of light may be to cause further polymerization on crosslinking, or may cause degradation.

photoproton [Nucl.Sc.] A proton released form a nucleus in a photonuclear reaction.

photoreduction [Chemistry] Light-induction of an electron acceptor in photosynthetic cells.

photosensitive substance [Physics] Any substance that when exposed to electromagnetic radiation produces a photoconductive, photoelectric, or photovoltaic effect. [Chemistry] Any substance, such as the emulsion of a photographic film, in which electromagnetic radiation produces a chemical change.

photosphere [Astronomy] The visible, intensely luminous portion of the sun, which has an estimated temperature of about 6000 K.

photosynthesis [Biochemistry] Synthesis of chemical compounds in light by plants as well as bacteria to convert two inorganic substances (water and carbon dioxide) into carbohydrates with simultaneous liberation of oxygen. When light falls upon green plants the major part of the energy is absorbed by small particles called chloroplasts, which contain a variety of pigments, amongst them compounds called chlorophylls. The chlorophylls transform the energy of the light into chemical energy by a process that is not fully understood, but it is known to involve the photolysis of water and the activation of adenosine triphosphate (ATP). (*See also* photolysis). The energy-rich ATP subsequently energizes the fixation of the CO_2, after a series of reactions, so that sugar molecules are formed.

photosynthetic bacteria [Biology] A group of bacteria able to photosynthesize through possession of a green pigment, bacteriochlorophyll, slightly different to the chlorophyll of the plants. They do not use water as the hydrogen source, as do plants, and thus do not produce oxygen as a product of photosynthesis, but rather some oxidised by product. Photosynthetic bacteria include the green sulphur bacteria, purple sulphur bacteria, and purple nonsulphur bacteria.

photosynthetic pigments [Biochemistry] The plant pigments responsible for the capture of light energy during the light reactions of photosynthesis. The green pigment chlorophyll is the principal light receptor, absorbing blue and red light. However the carotenoids and various other pigments also absorb light energy and pass this on to the chlorophyll molecules.

phototrophism [Biology] A type of nutrition in which the source of energy for synthesis of organic compounds is light. Most phototrophic organisms are antotrophic; these include the green plants, blue-green algae, and some photosynthetic bacteria.

phototropism [Biology] A directional growth movement of part of a plant in response to light. The phenomenon is clearly shown by growth of shoots and coleoptiles towards light. [Physics] A reversible change in the structure of a solid exposed to light or other radiant energy, accompanied by a change in colour. *See also* photochromism.

phototropy *See* phototropism.

photovoltaic [Electronics] Capable of generating a voltage as a result of exposure to visible or other radiation.

photovoltaic cell [Electronics] A device that measures or detects electromagnetic radiation by generating a potential at a junction (barrier layer) between two types of material, upon absorption of radiant energy. Also known as barrier-layer cell; photronicphotocell.

photovoltaic effect [Electronics] A photoelectric effect in which light falling on a specially prepared boundary between certain pairs of substances (*e.g.* copper and cuprous oxide) produces a potential difference across the boundary.

photox cell [Electronics] Type of photovoltaic cell in which voltage is generated between a copper base and a film of cuprous oxide during exposure to visible light or other radiation.

photronic cell [Electronics] Type of photovoltaic cell in which voltage is generated in a layer of selenium during exposure to visible light or other radiation.

pH scale [Chemistry] A logarithmic scale for expressing the acidity or alkalinity of a solution. To a first approximation, the pH of a solution can be defined as $M \log_{10} c$, where c is the concentration of hydrogen ions in moles per cubic decimetre. A neutral solution at 298K has a hydrogen-ion concentration of 10^{-7} mol dm^{-3}, so the pH is 7. A pH below 7 indicates an acid solution; one above 7 indicates an alkaline solution. More accurately, the pH depends not on the concentration of hydrogen ions but on their activity, which cannot be measured experimentally. For practical purposes, the pH scale is defined by using a hydrogen electrode in the solution of interest as one half of a cell, with a reference electrode (*e.g.* a calomel electrode) as the other half cell. The pH is then given by $(E - E_R)F/2.303RT$, where E is the e.m.f. of the cell and E_R the standard electrode potential of the reference electrode. In practice, a glass electrode is more convenient than a hydrogen electrode.

physical cache [Computers] A high-speed cache memory that operates with physical addresses rather than the virtual addresses issued by the processor. Before it can be accessed, the virtual address needs to be translated into a physical address.

physical change [Chemistry] A change in which no new substances are formed, the substance may change its physical state, or some of its physical properties may change.

physical constant [Physics] A physical quantity which has a fixed and unchanging numerical value.

physcial property [Chemistry] 1. Property of a compound that can change without involving a change in chemical composition, *e.g.*, melting point and boiling point. 2. A property which does not depend on the effect of other materials or substances; *e.g.*, shape, density.

physics [Scientific Technology] The study of those aspects of matter and energy which can be understood in a fundamental way in terms of elementary laws and principles.

physiological saline [Physics] An isotonic solution of salts in distilled water used for preserving cells. Such solutions contain no food for the cells and their survival in them is therefore restricted.

physiology [Biology] The study of the functioning of the various organs of living beings by using physical and chemical methods.

physisorption *See* adsorption.

phytamins *See* auxins.

photo- [Scientific Technology] A prefix denoting plant.

phytotoxin [Biochem] 1. A substance toxic to plants. 2. A toxin produced by plants.

pickling [Metallurgy] Preferential removal of oxide from the surface of a metal by dipping usually in an acidic or alkaline solution.

pickoff [Electronics] A device used to convert mechanical motion into a proportional electric signal.

pico- [Scientific Technology] A prefix meaning 10^{-12}.

pictet's liquid [Materials] Liquid mixture of carbon dioxide and sulphur dioxide; used to produce low temperatures.

piezochemistry [Chemistry] The study of chemical reactions taking place under high pressures.

piezoeletric effect [Electronics] The generation of opposite electric charges in certain dielectric crystals as a result of the application of mechanical stress, conversely, the property of expansion along one axis and contraction along another when subjected to an electric field.

pig iron [Metallurgy] A crude form of iron having high carbon content obtained from iron ores by the blast furnace process.

pigment [Biochem] Any colouring matter in animal or plant cells. [Materials] Any substance used to impart colour to other materials; pigments reflect the light of certain wavelengths while absorbing light of other wavelengths, without producing appreciable luminescence.

pi meson [Physics] Collective name for three semistable mesons which have charges of +1, 0, −1 times the proton charge and form a charge multiplet, with approximate mass of 138 MeV, spin 0, negative parity negative G parity, and positive charge parity. Also known as pion.

pinchbeck [Metallurgy] An alloy of copper and zinc used as an imitation gold.

Pinch effect [Electronics] The constriction of a liquid conductor of electricity (*e.g.* mercury or molten metal) that occurs when a substantial current is passed through it. [Nucl] The constriction of a plasma due to the magnetic field

of a high current within the plasma.

ping [Electronics] A sonic or ultrasonic pulse sent out by an echoranging sonar.

pinking *See* knocking.

pion [Physics] A pi-meson. A type of meson. *See* pi meson.

pipelining [Computers] a computer design technique in which an operation is broken down into subsidiary tasks, each carried out in overlapped fashion.

pitch [Materials] Hard, dark substances that melt to viscous tarry liquids; they may be the residue from the destructive distillation of wood, coal-tar, asphalt, or various bitumens, etc. [Physics] The psychological property of sound characterized by loudness depending mainly upon frequency of the sound stimulus, sound pressure and wave form. [Science Technology] The inclination or degree of slope of an object or structure.

pitchblende [Mineral] U_3O_8. A natural ore consisting mainly of uranium oxide. It occurs in Saxony, Bohemia, East Africa, and Colorado. Pitchblende contains small amounts of radium, of which it is the principal source.

Pitot tube [Engineering] A device for measuring the velocity of a fluid, consisting of a tube with two openings, one facing the moving fluid and the other facing away from it. The difference in pressure created in the tube between the two openings, as measured by a manometer, allows the velocity of the fluid to be determined.

pK [Chemistry] A measure of the strength of an acid, defined as log $1/K$, where K is the equilibrium constant of the dissociation of the acid. The higher the value of pK, the weaker is the acid. In a solution of a weak acid, if the concentration of undissociated acid is equal to the concentration of the anion of the pK will be equal to the pH. 2. A measure of the completeness of a reversible reaction.

planck [Physics] A unit of action equal to the product of an energy 1 joule and time equal to 1 second.

Planck's constant h [Physics] A fundamental physical constant the elementary quantum of action; the radio of the energy of a photon to its frequency; equal to 6.6262×10^{-34} joules-second.

Planck's distribution law [Physics] A relation for the intensity of radiation emitted by a black body with a narrow band of frequencies, as a function of frequency, and of the body's temperature. Also known as Planck's radiation formula; Planck's law.

Planck's law of radiation [Physics] The energy of electromagnetic radiation (including light) is composed of discrete quanta, the

magnitude of which is given by the product of Planck's constant h and the frequency of the radiation v.

plane [Mathematics] A flat surface; mathematically defined as a surface containing all the straight lines passing through a fixed point and also intersecting a straight line in space.

plane of polarization [Physics] Plane containing the electric vector and the direction of propagation of eletromagnetic wave.

plane polarization [Physics] A type of polarization of electromagnetic radiation in which the vibrations take place entirely in one plane. It can be produced by reflection, or by transmission through a Nicol prism, or through Polaroid. Also known as linear polarization.

planetarium [Astron] 1. A complex system of optical projectors for representing the movements of the planets and stars on a domed ceiling 2. The building that houses such a system.

planetoids See asteroids.

planets [Astron] Heavenly bodies revolving in definite orbit around the sun. In order of increasing distance from the sun they are : Mercury, Venus, Earth , Mars, Jupiter, Saturn, Uranus, Neptune, and Pluto.

plane wave [Physics] The wave in which the wavefront is a plane surface.

planimeter [Engineering] A mechanical integrating instrument for measuring plane areas, consisting of a movable tracing arm the movements of which are recorded on a dial.

plant fermentation [Biochemistry] A type of plant metabolism in which carbohydrates are partially degraded without consuming molecular oxygen.

plant hormone [Biochemistry] Any organic compound that is synthesized in small quantities by one part of a plant and translocated to another part, where it influences physiological process. Some of the commonly occurring plant hormones are auxins;, gibberellins, cytokions. Also known as phytohormone.

plasma [Physics] 1. The region in a discharge in gases in which the numbers of positive and negative ions are approximately equal. 2. The very hot ionized gas in which controlled thermonuclear reaction experiments are carried out. In such a plasma, which has been described as the fourth state of matter, the ionization is virtually complete. Again the numbers of positive ions and electrons are approximately equal and the plasma is therefore virtually electrically neutral and highly conducting.

plasma control [Physics] Confinement of plasma into a desired region with the help of magnetic field or other means.

plasma diode [Electronics] A diode used to convert heat directly into electricity.

plasmasol [Biology] The inner, isolated zone of protoplasm in a pseudopodium.

plasmid [Biochemistry] A strand or fragment of genetic material existing outside the chromosomes in certain type of bacteria. They can be transferred from animals to man, as well as to the other harmful bacteria which also become resistant to antibiotics.

plasmin [Biochemistry] A proteolytic enzyme in plasma which can digest many proteins through the process of hydrolysis.

plasminogen [Biochemistry] The inert precursor, or zymogen, of plasmin. Also known as profibrinolysin.

plasmoid [Physics] An isolated collection of electrons, ions, and neutral particles which holds together for a duration many times as long as the collision times between particles.

plasmolysis [Biochemistry] An effect of osmosis on cells of living organisms. A cell placed in a solution of a greater molecular concentration than (*i.e.*, is hypertonic to) the contents of the cell becomes plasmolyzed; the water in the cell flows out through the cell wall and the cell contents contract.

plasmon [Physics] A quantum of collective longitudinal wave in the electron gas of a solid.

plasmosome [Chemistry] *See* nucleolus.

plaster of paris [Chemistry] $CaSO_4.1/2-H_2O$. Powdered calcium sulphate, obtained by heating gypsum to 390-400K. On mixing with water, it sets and hardens.

plastics [Materials] Materials that can be shaped by applying heat or pressure. Most plastics are made from polymeric synthetic resins, although a few are based on natural substances (*e.g.* cellulose derivatives or shellac). They fall into two main classes. *Thermoplastic materials* can be repeatedly softened by heating and hardened again on cooling. *Thermosetting materials* are initially soft, but change irreversibly to a hard rigid form on heating.

plasticity [Mechanics] The property of solids where by they undergo a permanent change in shape and size when subjected to a stress exceeding a particular value, called the yield value.

plasticizer [Materials] 1. A nonvolatile liquid added to paints and varnishes to prevent brittleness of the dried film. 2. A liquid or solid substance added to synthetic or natural resins to modify their flow properties.

plastic surgery [Medicine] Surgical repair, replacement, or alteration of lost, injured, or deformed parts of the human body by transfer of tissue.

platelet [Physics] *See* blood platelet.

platinoid [Metallurgy] An alloy of 60% copper, 24% zinc, 14% nickel, and about 2% tungsten.

platinum [Chemistry] Pt. Transition element. At. No. .78. At. Wt. 195.09, placed in group VIII of the periodic table. A hard silvery-white ductile and malleable metal, m.p. 2045K; b.p., 4070K: that is very resistant to both heat and acids. Its coefficient of expansion is very nearly equal to that of glass, which makes it useful in certain types of scientific equipment.

platinum black [Materials] Black finely divided platinum metal produced by vacuum evaporation and used as an absorbent and a catayst.

platinum metals [Chemistry] A group of six transition elements with similar metallic properties. They are : ruthenium, rhodium, palladium, osmium, iridium, and platinum.

platonic year [Physics] *See* great year.

Pleochroic [Chemistry] Denoting certain crystals that have different colours, depending on the direction from which they are observed.

pliotron [Electronics] General term for any hot cathode vacuum tube having one or more grids.

ploidy [Biology] Number of compete chromosome sets in a nucleus.

plumbago [Materials] Black-lead,

graphite. A natural allotropic form of carbon.

pluto [Astronomy] A planet with its orbit outside that of neptune. Its mean distance from the sun, is about 5.6×10^9 kilometres. Sidereal period ('year') 248.5 years. Mass approximately one tenth that of the earth, diameter approximately 5900 kilometres. Pluto's surface temperature is probably below $-200°C$.

plutonic [Geology] Pertaining to rocks formed at great depth.

plutonium [Chemistry] Pu. At. No. 94; m.p 914K, b.p. 3505K. Transuranic element. Different isotopes of plutonium can be produced by suitable nuclear reactions. The isotope ^{239}Pu is produced in nuclear reactors and is of considerable importance since it undergoes nuclear fission when bombarded by slow neutron. This isotope, which has a half-life of 24400 years, is also used in nuclear weapons, one kilogram having an energy equivalent of about 10^{14} joules.

pneumatic [Engineering] Operated by, or filled with, compressed air.

pnicogens [Chemistry] A collective term sometimes used (but not recommended) for the elements nitrogen, phosphorus, arsenic, antimony, and bismuth.

pn junction [Electronics] The interface between two regions in a semi-conductor crystal which have been treated so that one is a p-type semiconductor, and the other is n-type semiconductor and contains a permanent dipole charge layer.

pn pn diode [Electronics] A semiconductor device having four alternate layers of p-type and n-type semiconductor material, with terminal connections to the two outer layers. Also known as npnp diode.

pnp transistor [Electronics] A junction transistor having an n-type base between a p-type emitter and a p-type collector.

poise [Mechanics] A unit of viscosity in c.g.s. units defined as the tangential force per unit area (dynes per sq cm) required to maintain unit diference in velocity (cm per second) between two parallel planes separated by one centimetre of fluid. 1 poice = 10 N m^{-2}.

Poiseuille's equation [Mechanics] The volume V of a liquid flowing through a cylindrical tube per second is given by the equation $V = \pi p r^4/8\eta l$, where p is the pressure difference between two points on the axis of the tube at a distance apart, η is the coefficient of viscosity and r is the radius of the tube. The result assumes uniform stream-line flow, and also that the liquid in contact with the walls of the tube is at rest.

poison [Nucl] A substance that absorbs neutrons in a nuclear reactor. Piosons may be deliberately added to reduce the reactivity, or they may be fission products, such as xenon, which have to be periodically removed. [Chemistry] A substance that exerts inhibitive effects on catalysts, even if present in small quantities. [Electronics] A material that reduces the emission of electrons from the surface of a cathode [Medicine] A substance that is injurious to the health of a living organism.

poisson's ratio [Mechanics] The ratio of the lateral strain to the longitudinal strain in a stretched wire. Given by the ratio of d/D to $\Delta L/L$, where D = original diameter, L = original length, d = decrease in diameter, and ΔL = increase in length.

polar coordinates [Mathematics] A point in the plane may be represented by coordinates (r, θ), where θ is the angle between the positive x-axis and the ray from the origin to the point, and r the length of that ray.

polar covalent bond [Chemistry] A bond in which a pair of electrons is shared between two atoms of different electronegativity, so that the atom is held , more closely by the more electronegative atoms.

polarimeter [Chemistry] An instrument used to measure the rotation of the plane of vibration of polarized light by optically active substances. *See* polarization of light and optical activity. Also known as polariscope.

polariscope [Physics] *See* polarimeter.

polarization [Optics] 1. The process of confining the vibrations of the vector constituting a transverse wave to one direction. In unpolarized radiation the vector oscillates in all directions perpendicular to the direction of propagation. *See* polarization of light. [Electeric] 1. The formation of products of the chemical reaction in a voltaic cell in the vicinity of the electrodes resulting in increased resistance to current flow and, frequently, to a reduction in the e.m.f. of the cell. *See also* depolarization. 2. The partial separation of electric charges in an insulator subjected to an electric field. [Chemistry]3. The separation of charge in a polar chemical bond.

polarization of light : The process of confining the vibrations of the electric vector of light waves to one direction. In unpolarized light the electric field vibrates in all directions perpendicular to the direction of propagation. After reflection or transmission through certain substances (*see* Polariod) the electric field is confined to one direction and the radiation is said to be *plane-polarized· light*. The plane of plane-polarized light can be rotated when it passes through certain substances (*see* optical activity.)

In *circularly polarized light,* the tip of the electric vector describes a circular helix about the direction of propagation with a frequency equal to the frequency of the light. The magnitude of the vector remains constant. In *elliptically polarized light*, the vector also rotates about the direction of propagation but the amplitude changes; a projection of the vector on a plane at right angles to the direction of propagation describes an ellipse. Circularly and elliptically polarized light are produced using a retardation plate.

polar molecule [Chemistry] A molecule, the configuration of electric charge in which constitutes a permanent electric dipole.

polarography [Chemistry] A method of chemical analysis based on recording characteristic polarograms (curves representing variations of current strength with the applied voltage) for substances in solution. Compositions of solutions can be deduced from the form (characteristic "waves") of their polarograms.

Polaroid [Optics] A doubly refracting material that plane-polarizes unpolarized light passed through it. It consists of a plastic sheet strained in a manner that makes it birefringent by aligning its molecules.

polaron [Electronics] An excitation in a solid consisting of polar molecules resulting from the interaction between an electron and its strain field. The presence of a polaron can be detected by irregularities in the shape of the conduction band.

poling [Metallurgy] A technique used in refining of copper in which green-wood poles are dipped into molten metal to generate the reducing gases that react with the oxides in the liquid metal.

polling [Electronics] A process that involves interrogating in succession every terminal on a shared communications line to determine which of the terminals are in need of servicing.

pollution [sci] Destruction or impairment of the purity of environment, *i.e.*, introduction into any environment of substances that are not normally present therein and that are objectionable.

polonium [Chemistry] Po. A rare radioactive metallic element of group VIA of the periodic table; At. No.84; r.d. 9.32; m.p. 527K; b.p. 1235K. The principal isotope has a mass number of 209 and a half-life of 103 years.

polyamide [Chemistry] A polymer in which the units are linked by amide or thio-amide groupings. *See* nylon.

polybasic acid [Chemistry] An acid containing more than one atom of acidic hydrogen in molecule.

polycarbonates [Chemistry] Thermoplastic resins in which the structural units are linked through carbonate radicals. They usually consist of polyesters of carbonic acids and dihydric phenols. Their good dimensional stability and impact strength over a wide temperature range make them useful for electrical and other small components.

polychromatic [Optics] Denoting electromagnetic radiation that consists of a mixture of wavelengths.

polycyclic [Chemistry] Having more than one ring in a molecule; can be aromatic (such as naphthalene), aliphatic (bianthryl), or mixed (dicarbazyme).

polyene [Chemistry] Any organic compound containing more than two double bonds.

polygon [Mathematics] A plane figure bounded by straight lines.

polyhedron [Mathematics] A solid figure having polygons for its faces. A regular polyhedron has all its faces equal in all respects; the five possible types of regular polyhedra are : (I) tetrahedron, 4 triangular faces; (II) cube, 6 square faces; (III) octahedron, 8 traingular faces; (IV) dodecahedron , 12 five-sided faces; (V) isocahedron, 20 triangular faces.

polyhydric [Chemistry] Containing more than one hydroxyl group in the molecule; *e.g.* ethylene glycol (ethanediol) and glycerol (1,2,3-propanetriol) are polyhydric alcohols. Also known as polyols.

polymer [Chemistry] A product of polymerization.

polymerase [Biochem] An enzyme that catalyzes a biological polymerization reaction.

polymerization [Chemistry] A chemical reaction in which molecules join together to form a polymer. If the reaction is an addition reaction, the process is *addition polymerization;* condensation reactions cause *condensation polymerization,* in which a small molecule is eliminated during the reaction. Polymers consisting of a single monomer are *homopolymers;* those formed from two different monomers are *copolymers.*

polymorphism [Chemistry] The existence of the same substance in more than two different crystalline forms, such as diamond and graphite.

polynomial [Mathematics] An expression consisting of three or more terms.

polynucleotide [Biochem] A chain of nucleotides linked together as in a nucleic acid. Ribonucleic acid consists of a single chain, while deoxyribonucleic acid usually consists of a double helix comprising two polynucleotide chains.

polypeptide [Biochem] A chain of three or more amino acids each of which is joined to its neighbours by the peptide linkage. Polypeptide chains may consist of up to several hundred amino acid units. Proteins consist of polypeptide chains cross-linked together in a variety of ways.

polyploidy [Biochem] Having more than twice the normal haploid number of chromosomes in a cell. Artificial polyploidy can be induced (*e.g.* by colchicine), and is used to produce fertile hybrids with desired characteristics.

polysaccharides [Biochem] A large class of natural carbohydrates. The molecules are derived from the condensation of several molecules of simple sugars (monosaccharides). The class includes cellulose and starch.

polysomy [Biology] The occurrence in a nucleus of one or more individual chromosomes in a number higher than that of the remainder.

polyvalent [Chemistry] 1. Having more than a one valence. 2.Having a valence of more than one. [Biology] Containing more than one type of antibody and therfore effective against more microorganism.

polywater [Materials] A reported form of water differing in properties (such as density, viscosity) from normal water. These properties are due to the presence of colloidal particles derived from impurities rather than to any differences in the molecular structure of the water itself. Also known as anomalous water.

porcelain [Materials] A hard white material made by the firing of a mixture of pure kaolin (china clay) with felspar and quartz, or with other materials containing silica; used in pottery and as refractory material.

porphyrins [Materials] A class of naturally occurring complex nitrogenous compounds with a cyclic tetrapyrrolic structure in which the four pyrrole rings are joined through their α-carbon, atoms by four methane bridges $(\alpha\text{-})$ some of the important porphyrin derivatives are chlorophyll and hemoglobin.

position circle [Scientific Technology] A circle with its centre at an observed point and its radius such that the circumference passes through the place of observation. The portion of the circumference near the place of observations apporximates to a position line if the radius is large.

position line [Scientific Technology] A line of position on which the observer is situated at a given time. The intersection of two position lines, determined at the same time,fixes the position of the observer.

positive [Scientific Technology] In any convention of signs, regarded as being counted in the plus, or positive direction, as opposed to negative.

positive column [Electronics] A luminous region in a discharge of gases near to the positive electrode.

positive feedback *See* feedback.

positive rays [Electronics] Streams of ions bearing positive electric charges. They are produced by means of an electric discharge in a rarefied gas.

positron [Physics] An antiparticle whose mass and spin are the same as those of an electron, but whose electric charge is positive.

positronium [Physics] The bound state of an electron and a positron.

potash [Materials] An old name for potassium carbonate, potassium hydroxide (caustic potash), or any potassium salt.

potassium [Chemistry] Kalium. K. Element At. No. 19 At. Wt. 39.102, m.p. 336.8K; b.p. 1047K; placed in group IA of the periodic table. A silvery-white soft highly reactive alkali metal. It is essential to life and is found in all living matter. Its salts are used as fertilizers.

potassium-argon dating [Geology] A method of dating geological specimens based on the decay of the radioisotope potassium-40 to argon-40. The half-life of potassium-40 is about 1.3×10^{10} years and an estimate of the ratio of the two isotopes in a specimen give an indication of its age.

potential [Electricity] *See* electric potential.

Potential barrier [Physics] A region containing a maximum of potential that prevents a particle on one side of it from passing to the other side. According to classical theory a particle must possess energy in excess of the height of the potential barrier to pass it. However, in quantum theory there is a finite probabil-

ity that a particle with less energy will pass through the barrier (*see* tunnel effect).

potential difference [Electricity] If two points have a different electric potential there is said to be a potential difference (p.d.) between them; if the points are joined by an electric conductor, an electric current will flow between from the point of higher potential to the point of lower potential. Potential difference is defined as the work performed when a unit positive electric charge is moved from one of the points to the other. *See also* electromotive force, E.M.F. The practical unit of potential difference and E.M.F is volt.

potential energy [Mechanics] The energy that a body possesses by virtue of its position. It is measured by the amount of work the body performs in passing from that position to a standard position in which the potential energy is considered to be zero. The potential energy of mass, *m*, raised through, a height, *h* is *mgh* where *g* is the acceleration of free fall.

potential series *See* electromotive series.

potentiometric titration [Chemistry] A titration in which the end point is found by measuring the potential on an electrode immersed in the reaction mixture.

pour point [Mechanics] Lowest temperature at which a liquid starts flowing. [Metallurgy] The temperature at which a molten metal or alloy is cast.

powder metallurgy [Metallurgy] The production of small metal articles by sintering powdered metals under heat and pressure.

power [Mathematics] A quantity successively multiplied by itself is said to be raised to a power, the magnitude of the power being the number of times that the quantity occurs in the multiplication. Thus $3 \times 3 \times 3 \times 3$, is 3 raised to the fourth power, denoted as 3^4, 4 being the index or exponent. [Physics] Rate of doing work. Measured in units of work per unit time. The derived SI unit of power is the watt. *See also* horsepower.

power alcohol [Materials] Industrial ethanol used as a fuel.

power factor [Electericity] In an electrical circuit, the ratio of the power dissipated, p, to the product of the electromotive force, E, and the current, I. In single-phase and three-phase circuit the power factor is given by $\cos\phi$, where ϕ is the phase angle between the E.M.F. and the current, *i.e.*, $P = EI \cos \phi$.

powerline [Electericity] Two or more wires conducting electric power from one place to another. Also known as electric power line.

powerset [Mathematics] The set consisting of all subsets of a given set.

pp junction [Electronics] A region of transition between two regions having different properties in p-type semiconducting material.

practical units [Scientific Technology] The units of metre-kilogram-second-ampere system.

Prandtl number [Physics] A dimensionless number used in the study of forced and free convection; equal to the dynamic viscosity times the specific heat at constant pressure divided by the thermal coductivity. The lower the number, the higher is the convection capacity of the substance.

praseodymium [Chemistry] Pr. Element. At. No. 59. At. Wt. 140.907; m.p. 1207K; b.p. 3785K; placed in group III B of the periodic table ; one of the rare earth elements of the lanthanide group.

precessional motion [Mechanics] A rotating body is said to precess when, as a result of an applied couple, the axis of which is at right angles to the rotation axis, the body turns about the third mutually perpendicular axis.

precipitant [Chemistry] A substance that causes a precipitate to form when added to solution.

precipitate [Chemistry] An insoluble substance formed in a solution as the result of a chemical reaction.

precursor [Chemistry] An intermediate substance from which another is formed in the course of a chemical process.

predation [Biology] The killing and eating of an individual of one species by individual of another species.

presbyopia [Medicine] A defect of vision normally occurring in elderly people. The subject is able to see distant objects clearly, but is unable to accommodate the eye to see near objects distinctly. It is corrected by the use of convex spectacle lenses. Also known as longsightedness.

pressure [Mechanics] The force per unit area acting on a surface. 'Absolute pressure' is the pressure measured with respect to zero pressure. 'Gauge pressure' is the pressure measured by a gauge in excess of the pressure of the atmosphere. The SI unit of pressure is pascal (Nm^{-2}). The c.g.s. system uses the dyne per square centimetre (1 Pa = 10 dynes cm^{-2}). Other units are : bar (=10^5 Pa), the atmosphere (101325 Pa), the mmHg (=133.322 Pa).

pressurized water reactor [Nucl] PWR. A nuclear reactor in which water is the coolant and the moderator, but in which the water is maintained at a high pressure in order to prevent it boiling. The pressurized water is passed through a heat exchanger to generate steam for producing electric power in a conventional turbo-generator.

primary cell [Electricity] A cell usually irreversible, for producing an electromotive force and delivering an electric current as the result of a chemical reaction so that the cell cannot be recharged efficiently; *e.g.*,Daniell cell.

primary coil [Electricity] The input coil of a transformer or induction coil.

primary colours [Physics] Red, green, and bluish-violet. Any colour may be obtained by suitably combining light producing these (*see* colour vision). [Materials] The pigment colours red, yellow, and blue, which cannot be imitated by mixing any other pigment colours.

primary flow [Electronics] The current flow that is responsible for the major properties of a semiconductor device.

primary radiation [Physics] Radiations coming directly from its source without interaction with matter.

primary wave [Physics] A radiowave travelling by a direct path, as contrasted with skips.

prime number [Mathematics] A number possessing no divisors *i.e.*, divisible by no whole number, other than itself and one; *e.g.*, 3,5,7, and 11 are prime numbers.

primer [Materials] A prefinishing coat applied to a surface that is to be painted or otherwise finished.

principal axis [Optics] The line joining the centers of curvature of the faces of a lens, or the line normal to a reflector at the pole.

principal focus [Optics] Point of convergence of light coming from a source at infinite distance. '

principal plane [Optics] The plane

that is perpendicular to the optical axis of a lens and that passes through the optical centre. A thick lens has two principal planes, each passing through a principal point.

principal point [Optics] 1. Two points on the optical axis of a thick lens of combination lens system, such that if the object distance is measured from one and the image distance from the other, the equations obtained, relating object-image distance etc., are similar to those obtained for a thin lens. 2. The intersection of a principal plane with the optical axis.

principal quantum number [Physics] A quantum number for orbital electrons, which along with the orbital angular momentum and spin quantum numbers, lebels the electron wave function; the energy level and the average distance of an electron from the nucleus depend mainly upon principal quantum number.

principal section [Optics] A principal section of a crystal exhibiting double refraction is a plane passing through the optic axis and at right angles to one of the crystal surfaces.

principal series [Physics] A series in the line spectra of many atoms and ions with one, two, or three electrons in the valence shell, in which the total orbital angular momentum quantum number changes from 1 to 0.

principal of superposition [Physics]

See Huygen's principle of superposition.

principle of superposition [Physics] The resultant displacements at any point in a region through which two waves of the same type pass is the algebraic sum of the displacements that the two would separately produce at that point. Both waves leave the region of superposition unaltered.

printed circuit [Electronics] An electronic circuit in which the wiring between components, and certain fixed components themselves, are printed on to an insulating board. The board is coated with copper and the portion of the metal that represents the wiring or components is photographically covered with a protective film, the rest of the metal being etched away in an acid bath.

print out [Computers] A printed output of a data processing machine or system.

print through [Computers] Transfer of signals from one recorded layer of magnetic tape to the next on reel.

prism [Mathematics] A solid figure having two identically equal faces(bases) consisting of polygons in parallel planes; the other faces being parallelograms equal in number to the number of sides of one of the bases. [Optics] A triangular prism made of material transparent to the light being used ; *e.g.*, glass for visible light, quartz for ultraviolet and near infrared radiation.

prismatic error [Optics] The error due to lack of parallelism of the two faces of an optical element, such as mirror or a shade glass.

prismatic optical instrument [Optics] Instruments (field-glasses, etc.) in which a right-angled prism is used to invert the inverted image produced by the objective.

prism diopter [Optics] A unit used in measuring the deviating power of prism; this power in prism diopters is 100 times the tangent of the angle of deviation of a ray of light.

probability [Mathematics] A mathematical expression of the chance that a specified event will occur. If the event is certain to occur the probability is 1; if it is certain not to occur the probability is 0. Between these two extremes the probability of an event occurring is expressed as a number between 0 and 1. For example, if an event can happen in x ways and fail in y ways, and, except for the numerical difference between x and y, is as likely to happen as to fail, the mathematical probability of its happening is $x/(x+y)$ and its failing is $y/(x+y)$.

probability density [Physics] The square of the absolute value of the Schrodinger wave function for a particle at a given point; gives the probability per unit volume of finding the particle at that point.

probability of finding electrons [Physics] The probability that an electron within an atom will be at a certain point in space at a given time ; it is determined by the magnitude of the square of the wave function.

probe [Space] An instrumented vehicle moving through the atmosphere or space or land upon another celestial body to obtain information about the environment. [Physics] A small device which can be brought into contact with or inserted into a system in order to make measurments on the system.

procedural knowledge representation [Computers] A representation of knowledge about the world by a set of procedures-small programs that know how to do specific things (how to proceed in well-specified situations).

process control [Engineering] The control of complex industrial of chemical processes by electronic means.

process lens [Optics] A highly corrected, apochromatic lens used for precise colour-separation job.

processor [Computers] 1. A device that performs one or many functions, usually a central processing unit. 2. A program that transforms some input into some output such as an assembler, compiler, or linkage editor.

prochiral molecule [Chemistry] A symmetrical molecule that may react asymmetricaly with an enzyme having an asymmetric active site.

producer gas [Materials] A fuel gas produced by the partial combustion of coke or coal in a restricted supply of air, to which steam may have been added. The principal constituents of the gas are, carbon monoxide (25%-30%), nitrogen (50%-65%), hydrogen (10%-15%). Sometimes hydrocarbons and carbon dioxide may also be present.

product [Mathematics] The product of two or more quantities is the result of multiplying them together. [Engineer] An item or goods manufactured by an industry. [Chemistry] Substance that has passed through a processing operation (with or without a chemical reaction).

production reactor [Nucl] A nuclear reactor designed primarily for large-scale production of transmutation products, such as plutonium.

progesterone [Biochemistry] A steroid hormone secreted by the corpusluteum in the ovary after ovulation. It initiates the preparation of the uterus for implantation of the ovum, the development of the placenta, and development of the mammary gland in preparation for lactation.

progestogen [Biochemistry] Any hormone whose effects resemble those of progesterone.

program [Computers] The sequaence of instructions fed into a computer in order to enable it to carry out a process.

program generator [Computers] A program that permits a computer to write other programs automatically.

program level [Engineering] The level of the program signal in an audio system, expressed in volume units.

programming environment [Computers] The total programming setup that includes the interface, the languages, the editors, and other programming tools.

progressive wave [Physics] A wave which transfers energy from one part of the medium to another in contrast to a standing wave.

projectile [Physics] A body that is thrown or projected upward. If the projectile is discharged with a velocity $\overrightarrow{v}$ at an angle θ to the horizontal, the following formulae hold true if the resistance of the air is neglected (g being the acceleration of free fall) :

Time to reach highest-

Point of flight $= (v\sin\theta)/g$

Total time of flight $= (2v\sin\theta)/g$

Maximum height $= (v^2\sin^2\theta)/2g$

Horizontal range $= (v^2\sin 2\theta)/g$

projection microscope [Physics] An X-ray microscope which magnifies by image projection, either in contact microradiography or in projection microradiography.

prokaryotes [Biology] Simple unicelular organisms (*e.g.* bacteria and blue-green algae) with a single chromosome, no nuclear membrance-bound organelles.

prolapse [Medicine] The falling or sinking down of a part or an organ.

promethium [Chemistry] Pm. A radioactive element of the lanthanide series. At. No. 61. m.p. 1308K, b.p. 3270K. It occurs as a fission product of uranium in nuclear reactors.The most stable isotope, ^{145}Pm has a half-life of about eighteen years.

prominence [Astronomy] A volume of luminous, predominantly hydrogen gas which appears on the sun above the chromosphere occurs only in the region of horizontal magnetic fields because these fields support the prominences against gravity.

promoter [Chemistry] A substance which itself is a feeble catalyst, but greatly increases the catalytic activity of a given catalyst. [Biochem] A DNA sequence at which RNA polymerase may bind, leading to initiation.

prompt critical [Nucl] Capable of sustaining a nuclear fission chain reaction on the prompt neutrons alone, without contribution from delayed neutrons.

prompt neutron [Nucl] A neutron resulting from nuclear fission (either during the fission process or from freshly formed fission fragments) that are emitted without measurable delay, *i.e.*, in less than a millionth of a second.

proof spirit [Chemistry] Ethanol containing 49.28% alcohol by weight, or 57.10% by volume. Formerly defined as the weakest solution of alcohol that would fire gunpowder when brought into contact with it and ignited.

proof [Chemistry] A measure of the amount of alcohol (ethanol) in drinks. *Proof spirit* contains 42.28% ethanol by weight (about 57% by volume). Degrees of proof express the percentage of proof spirit present, so 70^0 proof spirit contains 0.7×57% alcohol Spirit contains $57.1 \times 70/100 = 39.97$% alcohol by volume.

propellant [Materials] 1. The explosive substance used to fill cartridges, shell cases, and solid fuel rockets. The term is also used to include the fuel and oxidant of rockets when these are separate. 2. A compressed gas used in aerosol preparations to expel the liquid contents through an atomizer.

propeller [Engineering] A bladed device that rotates on a shaft to produce a useful thrust in the direction of shaft axis.

proper fraction [Mathematics] A fraction in which the numerator is less than the denominator, *e.g.*, 5/6. In an 'improper fraction' the numerator is greater than the denominator, *e.g.*, 6/5.

proper motion [Astronomy] The component of a star's motion in space relative to the sun that is perpendicular to the line of sight; resulting in the change of star's apparent position relative to that of other stars; expressed in angular units.

prophase [Biology] The initial stage of mitotic or meiotic cell divison in which chromosomes are condensed from the nuclear material and split longitudinally to form pairs.

propositionallogic [Computers] An elementary logic that uses argument forms to deduce the truth or faleshood of a new proposition from known propositions.

propulsion [Mechanics] The process of causing a body to move by exerting a force against it.

Propulsion reactor *See* nuclear reactor.

prostaglandins [Biochemistry] A class of lipid-soluble, hormone-like regulatory molecules derived from arachidonic and other poly-unsaturated fatty acids.

prosthetic group [Biochemistry] A non-protein group (may be a metal ion or any organic group) combined to a protein, and serves as its active group, *e.g.*, the haem group in haemoglobin or the nucleic acid in nucleoprotein.

protactinium [Chemistry] Pa. Radioactive element. At. No.91. The most abundant natural isotope has a mass number 211 and a half-life of 12,480 years.

protargol [Materials] A powder containing finely divided silver and protein ; with water, it forms a colloidal solution silver.

protamine [Biochemistry] Any of a group of proteins of relatively low molecular weight found in association with the chromosomal DNA of vertebrate male germ cells. They contain a single polypeptide chain comprising about 67% arginine. Protamines are thought to protect and support the chromosomes.

protease [Biochemistry] A group or enzymes capable of breaking up proteins into amino acids, or building up amino acids into proteins, and of substituting one amino acid by another in protein molecules. Occurring in all living tissues; they conduct the processes of protein metabolism in the living organism. Also known as proteinase.

Protein [Biochemistry] Any class of complex nitrogenous organic compounds of high molecular wieght (18000-10000000) ; consist of hundreds of thousands of amino acids joined together by the peptide linkage into one or more inter-linked polypeptide chains, which may be folded in a variety of different ways. It is the sequence of the different amino acids that produces proteins.

proteinase *See* protease.

proteolysis [Biochemistry] Fragmentation of a protein molecule into amino acids by addition of water to the peptide linkage.

Proteolytic [Biochemistry] Having the power of decomposing or hydrolyzing proteins. Also known as proteoclastic.

protium [Nucl] The hydrogen isotope with mass number of one and consisting of a single proton and electron. Also known as light hydrogen.

Protolysis [Chemistry] A reaction involving the transfer of protons (hydrogen ions).

proton [Physics] A stable elementary particle with electric charge equal in magnitude to that of the electron but of opposite sign, and with mass 1836.12 times greater than that of the electron (1.672614×10^{-27} kilogram). The proton is a hydrogen atomic nucleus and is a constituent of all other atomic nuclei. The number of protons in a nucleus is its atomic number.

protonic acid [Chemistry] An acid that forms positive hydrogen ions (or, strictly, oxronium ions) in aqueous solution. The term is used to distinguish 'traditional' acids from Lewis acids or from Lowry-Bronsted acids in non-aqueous solvents.

proton number *See* atomic number.

protoplasm [Biology] The colloidal matter of which biological cells consist generally protein that composes the living material of a cell.

Protoplast [Biology] The living portion of a cell considered as a unit ; includes the cytoplasm, the nucleus, and the plasma membrane.

protostar [Astronomy] A developing star consisting of condensing interstellar gas and dust.

prototype [Engineering] A model suitable for use in complete evaluation of form, design, and working.

provitamin [Biochemistry] A substance from which a vitamin is formed; a vitamin precursor.

pseudo-aromatic [Chemistry] A ring compound containing conjugated double bonds in the manner of an aromatic compound, although its properties are different from those of an aromatic compound.

pseudohalogens [Chemistry] A group of compounds, including cyanogen $(CN)_2$ and thiocyanogen $(SCN)_2$, that have some resemblance to the halogens. Thus, they form hydrogen acids (HCN and HSCN) and ionic salts containing such ions as CN^- and SCN^-.

pseudo order [Chemistry] An order of a chemical reaction that appears to be less than the true order because of the experimental conditions used. Pseudo orders occur when one reactant is present in large excess. For example, a reaction of substance A undergoing hydrolysis may appear to be proportional only to [A] because the amount of water present is so large.

pseudo-scalar [Mathematics] A scalar quantity that changes in the transition from a right-handed to a left handed system of coordinates.

pseudo-vector [Mathematics] A vector quantity that changes sign in the transition from a right-handed to a left-handed system of coordinates.

P shell [Physics] The sixth layer of electrons about the nucleus of an atom, having electrons whose principal quantum number is 6.

psychrometry [Physics] The measurement of the humidity of the atmosphere.

ptomaines [Biochem] A class of extremely poisonous organic compounds formed during the purtrefaction of proteins of animal origin. Food poisoning, frequently misnamed ptomaine poisoning, is almost invariably due to cause other than the ptomaines.

ptyalin [Biochem] An enzyme that digests carbohydrates (see amylase). It is present in mammalian saliva and is responsible for the initial stages of starch digestion.

p-type conductivity [Electronics] The conductivity associated with holes in a semiconductor, which are equivalent to positive charges.

p-type semiconductor [Electronics] An extrinsic semiconductor in which the hole density exceeds the conduction electron density.

p^+ type semiconductor [Electronics]

A p-type semiconductor in which excess mobile hole concentration is very large.

puckering [Metallurgy] Corrugations in metals parts resulting from pressing or drawing.

puddling [Metallurgy] The preparation of nearly pure wrought iron from cast iron that contains a high percentage of carbon. The cast iron is heated with haematite, Fe_2O_3, the oxygen in which oxidizes the carbon, silicon, phosphorus, and manganese to their corresponding oxides.

pulsars [Astronomy] Stars that emit radio frequency electromagnetic radiation in brief pulses at extremely regular intervals. Many such objects have been located by radio telescope, a few of them have also been observed to emit pulses of light. The period of known pulsars range between 33 milliseconds and 3.75 seconds, and pulse duration range from 2 to about 150 milliseconds. It has been suggested that pulsars are neutron stars, emitting pulses of radiation as they rotate.

pulse [Physics] A brief increase in the magnitude of a quantity whose value is usually constant (e.g., current or voltage).

pulsed laser [Optics] A laser which generates light in a pulse lasting a few hundred microseconds or less, with a peak power of tens of kilowatts or more.

pulsed light [Optics] A beam of light whose intensity is modulated in some prescribed manner; analogous to a radar pulse.

pulse generator [Electronics] A generator that produces repetitive pulses or signal-initiated pulses.

pulse modulation [Electronics] A system of modulation in which the amplitude, duration, position, or the presence of discrete pulses may be so controlled as to represent the message to be communicated.

pulverizer [Engineering] Device for breaking down of solid lumps into fine powder especially by cleavage along crystal faces.

punch card [Computers] A card by means of which data are fed into a computer in the form of rectangular holes punched in card.

purging [Mechanics] The act or process of cleaning and purifying.

pushing [Electricity] A change in the resonant frequency of circuit due to changes in applied voltages.

putrefaction [Biochemistry] Chemical decomposition, by the action of bacteria, of the bodies of dead animals and plants; especially the decomposition of proteins with the production of offensive smelling compounds.

putty [Materials] A material composed of powdered chalk mixed with linseed oil, used in fastening glass and sealing crevices in wood work.

pyknometer [Electricity] A device for determining the density and coefficient of expansion of a liquid. It consits of a glass vessel graduated to hold a definite volume of liquid at a given temperature. By weighing it full of liquid at different temperatures, the variations in density, and therefore the apparent expansion, may be calculated.

pyramid [Mathematics] A solid figure having a polygon for one of its faces (the base), the other face being triangles with a common vertex. The volume of a pyramid is one-third of the product of the area of the base and the vertical height.

pyrazolone dye [Chemistry] An acid dye containing both -N=N- and =C=C= chromophore groups, such as tartrazine.

pyrex [Materials] A type of glass that is resistant to heat and chemical attack; it is widely used in laboratory glassware.

pyrgeometer [Engineering] An instrument for measuring radiation from the surface of the earth into space.

pyrites [Materials] Natural sulphides of certain metals, *e.g.*, iron pyrites (fools gold), FeS_2; copper pyrites, $CuFeS_2$.

pyro- [Chemistry] Prefix denoting an oxo acid that could be obtained from a lower acid by dehydration of two molecules. For example, pyrosulphuric acid is $H_2S_2O_7$ (*i.e.* $2H_2SO_4$ minus H_2O).

pyroelectricity [Electricity] The property of certain crystals, *e.g.*, tourmaline, of acquiring electric charges in opposite faces when the crystals are subjected to a change on temperature.

pyrogens [Biology] Substances produced by living bacteria (not yeasts or moulds) that cause a rise in body temperature on injection. Thus any material that has been infected may, despite subsequent sterilization contain pyrogens and be unsuitable for injection. Pyrogens are not destroyed by heat; sometimes water supplies can be pyrogenic.

pyrolusite [Mineral] MnO_2. Natural manganese dioxide. A black crystalline solid material; the principal ore of manganese.

pyrolysis [Chemistry] Chemical decomposition of complex molecules into simpler and lighter molecules by heating.

pyrometry [Engineering] The measurement of high temperatures (beyond the range of thermometers) using a *pyrometer*. Modern *narrow-band* or *spectral pyrometers* use infrared-sensitive photoelectric cells behind filters that exclude visible light. In the *optical pyrometer* (or disappearing filament pyrometer) the image of the incandescent source is focused in the plane of a tungsten filament that is heated electrically.

pyrophoric [Scientific Technology] Igniting spontaneously in air. *Pyrophoric alloys* are alloys that give sparks when struck. *See* misch metal.

pyroxene [Mineral] A group of minerals consisting principally of silicates of magnesium, iron, and calcium.

Pythagoras theorem [Mathematics] In a right-angled triangle, the square of the hypotenuse is equal to the sum of the squares of the other two sides.

Pythagorean numbers [Mathematics] Positive integers x, y, z, which satisfy the equation $x^2 + y^2 = z^2$.

Pythagorean scale [Physics] A musical scale such that the frequency intervals are represented by the ratios of integral powers of the numbers 2 and 3.

pyuria [Medicine] The presence of pus in the urine.

Q [Physics] A measure of the ability of a system with periodic behaviour to store energy equal to 2π times the averge energy stored in the system divided by the energy dissipated per cycle; *i.e.*, $\Delta E/E = 2\pi/\theta$. Also known as Q factor; quality factor; storage factor.

Q band [Electronics] A radio frequecny band of 36 to 46 gigahertz.

Q factor *See* Q.

Q point [Engineering] Data point

describing the position and movement of radar target, based on two or more radar observations.

quad [Electricity] A series of four separately insulated conductors, generally twisted together in pairs. [Physics] A unit of energy equal to 10^{15} Btu, which is the energy equivalent to 182 million barrels of oil, or 42 million tons of coal, 293 billion kwh of electricity.

quadrant [Mathematics] A sector of a circle bounded by an arc and two radii at right angles. Also known as quarter-circle.

quadratic equation [Mathematics] An equation involving the square or second power of the unknown quantity; satisfied by two values (known as roots) of the unknown quantity. Any quadratic equation may be written in the form: $ax^2+bx+c=0$ the roots of this equation are given by the expression

$$x=[-b\pm\surd(b^2-4ac)]/2a.$$

The sum of the root is $-b/a$ and their product is c/a.

quadrature [Astronomy] The position of the moon or any outer planet such that a line between it and the earth makes a right angle with a line joining the earth to the sun.

quadrilateral [Mathematics] A plane figure bounded by four straight lines.

quadrivalent [Chemistry] Having a valence of four. Also known as tetravalent.

qualitative [Scientific Technology] Dealing only with the nature, and not the amounts, of the substances under consideration.

qualitative analysis [Chemistry] The determination of the chemical nature of substance; identification of the elements, radicals, or the compounds present in the mixture.

quality control [Engineering] The application to the theory of mathematical probability of sampling the output of an industrial process, with the object of detecting and controlling any variations in quality.

quantitative [Scientific Technology] Dealing with quantities as well as the nature of the substance under consideration.

quantitative analysis [Chemistry] The determination of the amounts of substances present, by chemical means.

quantity of electricity [Electricity] The amount of electricity flowing through a circuit; the product of the current and the time for which it flows. The SI unit of the quantity of electricity is the coulomb.

quantization [Physics] The restriction of an observable quantity such as energy or angular momentum to discrete set of values.

quantized [Physics] A quantity is

said to be quantized if it is in accordance with quantum mechanics, it can only have certain discrete values (each of which is called quantum). Such a quantity cannot vary continuously, differences in value being separated by 'jumps'.

quantum [Physics] The minimum amount by which certain properties, such as energy or angular momentum, of a system can change. Such properties do not, therefore, vary continuously, but in integral multiples of the relevant quantum. This concept forms the basis of the quantum theory. In waves and fields the quantum can be regarded as an excitation, giving a particle-like interpretation to the wave or field. Thus, the quantum of the electromagnetic field is the photon and the graviton is the quantum of the gravitational field. *See* quantum mechanics.

quantum chemistry [Chemistry] A branch of physical chemistry concerned with the explanation of the chemical phenomena by means of laws of quantum mechanics.

quantum mechanics [Physics] A system of mechanics that was developed from quantum theory and is used to explain the properties of atoms and molecules. Using the energy quantum as a starting point it incorporates Heisenberg's uncertainty principle and the de Broglie wavelength to establish the wave-particle duality on which Schrodin-

ger's equation is based. This form of quantum mechanics is called *wave mechanics*. An alternative but equivalent formalism, *matrix mechanics*, is based on mathematical operators.

quantum numbers [Physics] Integral or half-integral numbers that specify the state of a system or its components in quantum mechanics. An electron within an atom, for example, is specified by four quantum numbers :

(1) The *principal quantum number* n gives the main energy level and has values, 1, 2, 3, etc. (the higher the number, the further the electron from the nucleus). Traditionally, these levels, or the orbits corresponding to them, are referred to as *shells* and given letters K, L, M, etc. The K-shell is the one nearest the nucleus.

(2) The *orbital quantum number l*, which governs the angular momentum of the electron. The possible values of l are $(n-1)$, $(n-2)$, ... 1, 0. Thus, in the first shell ($n=1$) the electrons can only have angular momentum zero ($l=0$). In the second shell ($n=2$), the values of l can be 1 or 0, giving rise to two *subshells* of slightly different energy. In the third shell ($n=3$) there are three subshells, with $l = 2$, 1, or 0. The subshells are denoted by letters $s(l=0)$, p $(l=1)$, $d(l=2)$, $f(l=3)$. The orbital quantum number is sometimes called the *azimuthal quantum number*.

(3) The *magnetic quantum number m*, which governs the energies of

electrons in an external magnetic field. This can take values of $+l$, $+(l-1)$, and 1, 0, -1, ... $-(l-1)$, $-l$. In an s-subshell (*i.e.* $l = 0$) the value of $m = 0$. In a p-subshell ($l = 1$), m can have values +1, 0, and -1; *i.e.* there are three p-orbitals in the p-subshell, usually designated p_x, p_y, and p_z. Under normal circumstances, these all have the same energy level.

(4) The *spin quantum number m_s*, which gives the spin of the individual electrons and can have the values +1/2 or -1/2. According to the Pauli exclusion principle, no two electrons in the atom can have the same set of quantum numbers. The numbers define the *quantum state* of the electron, and explain how the electronic structures of atoms occur.

quantum state [Physics] The condition of a physical system as described by a wave function, the function may be simultaneously an eigenfunction of one or more quantum-mechanical operator, the eignvalues are then the quantum numbers that lebel the state. For instance, the state of a hydrogen atom is described by the four quantum numbers n, l, m, m_s. In the ground state they have values 1, 0, 1, and 1/2 respectively.

quantum statistics [Physics] The statistical description of particles or systems of particles whose behaviour must be described by quantum mechanics rather than by classical mechanics.

quantum theory [Physics] The theory that grew up around Planck's introduction into physics of the concept of the discontinuity of energy. The system of quantum mechanics evolved from this theory during the first half of the twentieth century.

quantum yield [Chemistry] For a photochemical reaction, the number of moles of a reactant disappearing, or the number of moles of a product formed, per einstein of light of the stated wavelength absorbed.

quark [Physics] One of the hypothetical basic particles, having charges whose magnitudes are 1/3 or 2/3 of the electron charge, from which many of the elementary particles may, in theory, be built up; for example nucleons may be formed from three quarks and mesons from quark-anitquark combination. The experimental evidence for the actual existence of free quarks has not been available so far.

quart [Mechanics] Unit of capacity equal to one quarter of a gallon.

quarter wave [Physics] Having an electrical length of one quarter-wavelength.

quarter-wave plate [Optics] A thin sheet of mica or other doubly refracting material (*See* double refraction) cut parallel to the optic axis of the crystal, and of such a thickness that a phase difference of p/2 or 90° is introduced between the ordinary and extraordinary rays for light of a particular wavelength. Plane-polarized light incident normally upon such a plate, with its plane of vibration making an angle of 45° with the optic axis, emerges from the plate circularly polarized. A quarter-wave plate is often used in the analysis of polarized light.

quartic equation [Mathematic] Any fourth degree polynomial equation.

quartz [Mineral] SiO_2. Natural crystalline silica, which sometimes occurs in clear, colourless crystals (rock crystal); more frequently it occurs as a white, opaque mass. Quartz crystals exhibit the piezoelectric effect to a marked extent.

quartz clock [Engineering] A clock regulated by a quartz crystal, which vibrates with a definite constant frequency under the effect of an alternating electric field tuned to this resonance frequency of the crystal. It is used for astronomical and other very precise work.

quartz, fused [Chemistry] Pure silica that has been melted to yield a glass-like material on cooling; used for apparatus and equipment where its high melting point, ability to withstand large and rapid temperature changes, inertness, and transparency are required.

quartz lamp [Electricity] A mercury vopour lamp having a transparent envelope made of quartz instead of glass. It is capable of passing out even ultraviolet rays that are absorbed by ordinary glass.

quasars [Astronomy] A class of astronomical objects that appear on optical photographs as star-like but have large redshifts quite unlike those of stars. The redshifts are characteristic of galaxies flying outwards from the centre of the universe at enormous speeds as a result of the expansion of the universe. This *cosmological redshift* is the explanation of the high observed redshifts of quasars favoured by most astronomers. The origin of quasars is unknown but what we observe now are the emissions made by these objects 10^{10} years ago, when the galaxies are thought to have been forming. It has been suggested that the quasars may have been connected in some way with the birth of the galaxies. Also known as quasic stellar objects (QSO).

quasi-particle [Physics] An entity used in the description of a system of many interacting particles which has particle like properties such as mass, energy, and momentum, but which does not exist as a free particle, *e.g.* phonons.

quaternary ammonium compounds [Chemistry] Compounds of the general formula $[NR_4]^+OH^-$; they are theoretically derived from ammonium hydroxide, NH_4OH, by replacement of the hydrogen atoms by organic radicals.

quaternary structure [Biochemistry] The three-dimensional structure of an oligomeric protein; particularly the manner in which the sub-unit chains fit together.

queen's metal [Metallurgy] An alloy of tin containing antimony, zinc, lead, and/or copper.

quench aging [Metallurgy] Aging of metal induced by rapid cooling from solution heat-treatment temperatures.

quench annealing [Metallurgy] Annealing an austenitic ferrous alloy by heating followed by quenching from solution temperature.

quench bath [Metallurgy] A liquid medium, such as oil, fused salt, or water, into which a material is plunged for heat-treatment purposes.

quenching [Electricity] The process of terminating the discharge in a Geiger counter by preventing reignition. [Metallurgy] Rapid cooling by immersion into water or oil, to harden the steel. [Physics] The phenomenon in, which a strong electric field, such as crystal field, causes the orbit of an electron in an atom to precess rapidly so that the average magnetic moment associated with its orbital angular momentum is reduced to zero.

quick lime [Materials] CaO. Calcium oxide.

quick malleable iron [Metallurgy] Malleable iron containing 2.2% carbon, 1.5% silicon, 0.3% to 0.6% manganese, and 0.75% to 1.00% copper.

quick silver [Geology] A term applied to mercury where it occurs as a native mineral or has been mined but not yet used.

quiescent [Electronics] Condition of a circuit element which has no input signal, so that it does not perform its active function.

quiesent period [Electronics] Resting period, or the period between pulse transmissions.

quiet sun [Astronomy] The sun's condition when no sunspots solar flares or solar prominences are taking place. Radio frequency emission (*see* radio astronomy) from the sun, which has to be observed during the rare periods of the quiet sun, has enabled temperature measurements of the various layers of the solar atmosphere to be made.

quinitic equation [Mathematics] A fifth-degree polynomial equation.

quinhydrone electrode [Electricity] A half-cell consisting of a platinum electrode in an equimolar solution of quinone (cyclohexadiene-1, 4-dione) and hydroquinone (benzene-1, 4-diol). It depends on the oxidation-reduction reaction

$$C_6H_4(OH)_2 \rightarrow C_6H_4O_2 + 2H^+ + 2e^-$$

quinones [Chemistry] A series of aromatic compounds in whose molecules two hydrogen atoms in the same benzene nucleus are replaced by oxygen atom, forming carbonyl groups. The quinones are therefore diketones. The simplest member of the series is *p*-quinone (*p*-benzoquinone) $O=C_6H_4=O$; a yellow crystalline solid, used as an oxidizing agent, in dye manufacture, and in photography.

quotient [Mathematics] The result of dividing one quantity by another.

Q unit [Physics] A unit of energy used in measuring the heat energy of fuels reserves; equal to 10^{18} British thermal units, or approximately 1.055×10^{21} joules.

Q value [Nucl] The net amount of energy released or absorbed in a nuclear reaction; usually expressed in million electronvolts, MeV, per individual reaction. Also known as nuclear energy change; nuclear heat of reaction; disintegration energy.

Q wave [Physics] A type of surface wave having a horizontal motion that is shear or transverse to the direction of propagation. Its velocity depends only on density and rigidity modulus, and not on bulk modulus. Also known as Love wave.

racemase [Biochemistry] Any of a group of enzymes that catalyze racemization reactions.

racemic mixture [Chemistry] A mixture of equal quantities of the *d*- and *l*-forms of an optically active compound. Racemic mixtures are denoted by the prefix *dl* (*e.g. dl*-lactic acid). A racemic mixture shows no optical activity. Also known as racemate.

racemization [Chemistry] Conversion of an optically active compound into an optically inactive form by heat or by chemical means. In this process, half of the optically active substance become its mirror image (enantiomer). The change results in a mixture of equal quantities of dextro-and levorotatory isomers, as a result of which the compound does not show optical rotation.

rad [Nucl] The standard unit of absorbed dose of ionizing radiation. One rad is equal to the energy absorption of 100 ergs per gram (0.01 J kg^{-1}) of irradiated material.

radar [Engineering] An abbreviaiton of the words Radio Detection And Ranging. A method of detecting the presence, position, and direction of motion of distant objects (such as ships and aircraft) by means of their ability to reflect a beam of electromagnetic radiation of centimetric wavelengths. It consists of a transmitter producing radio-frequency radiation, often pulsed, which is fed to a movable aerial from which it is transmitted as a beam. If the beam is interrupted by a solid object, a part of the energy of the radiation is reflected back to the aerial. Signals received by the aerial are passed to the receiver, where they are amplified and detected. An echo from a reflection of a solid object is indicated by a sudden rise in the detector output. The time taken for a pulse to reach the object and be reflected back (*t*) enables the distance away (*d*) of the target to be calculated from the equation $d = ct/2$, where *c* is the speed of light. In some systems the speed of the object can be measured using the Doppler effect. The output of the detector is usually displayed on a cathode-ray tube.

radarscope [Electronics] Cathode-ray tube, serving as oscilloscope, the face of which is the radar viewing screen.

radiac [Nucl] Detection, identification, and measurement of the intensity of nuclear radiation in an area.

radial [Scientific Technology] Pertaining to directed or diverging from a centre.

radial motion [Mechanics] Motion in which a body moves along a line connecting it with an observer or reference point, *e.g.*, motion of stars which move away or toward the earth without a change in apparent position.

radian [Mathematics] The supplementary SI unit of plane angle defined as the angle subtended at the centre of a circle by an arc equal in length to the radius of the circle. 21 radians= 360°, 1 radian= 57.296°.

radiant emittance [Physics] The radiant flux per unit area that emerges from a surface. Also known as radiancy; radiant exitance.

radiant energy [Physics] Energy that is transmitted in the form of radiation, particularly electromagnetic radiation. Radiant energy is the only form in which energy can exist in the absence of matter.

radiant flux [Physics] The total power emitted or received by a body in the form of radiation per unit time. It is measured in watts. Also known as radiant power.

radiant heating [Engineering] Any method of space heating in which the heat-producing device is a surface that emits heat to the surroundings by radiation rather by conduction or convection.

radiant intensity [Physics] Ic. The radiant flux per unit solid angle emitted by a point source; measured in watts per steradian.

radiant power *See* radiant flux.

radiation [Physics] In general, the emission of any rays, wave motion, or particles (*e.g.*, alpha particles, beta particles, neutrons) source.

radiation hazard [Physiology] The potential danger to health resulting from exposure to ionizing radiation or the consumption of radioactive substance such as production of cancers, radiation ulcers, sterilization etc.

radiation ionization [Chemistry] Ionization of molecules or atoms of a gas or vapour by the action of electromagnetic radiation.

radiation potential [Physics] The energy (expressed in electron volts) necessary to transfer an electron from its normal position in an atom to some other possible position; i.e., to an energy level of greater energy.

radiation temperature [Physics] The surface temperature of a celestial body as calculated by Stefan's law, assuming that the body behaves as a black body. The radiation temperature is usually measured over a narrow portion of the electromagnetic spectrum.

radiative collision [NuCl] A collision between charged particles in which part of the kinetic energy is converted into electro-magnetic radiation.

radical [Chemistry] A group of atoms, present in a series of compounds, that maintains its identity through chemical changes affecting the rest of the molecule, but that is usually incapable of independent existence; e.g., the ammonium radical, NH_4^-; [Mathematics] Relating to a root. The symbol $\sqrt{}$ is called the 'radical sign'.

radio [Physics] A means of transmitting information in which the transmission medium consists of electromagnetic radiation. Information is transmitted by means of the modulation of a carrier wave in a transmitter; the modulated carrier wave is fed to a transmitting aerial from which it is broadcast through the atmosphere or through space. A receiving aerial forms part of a resonant circuit, which can be tuned to the frequency of the carrier wave, enabling the receiver that it feeds selectively to amplify and then to demodulate the transmitted signal. A replica of the original information is thus produced by the receiver.

radioactive [NuCl] Possessing, or pertaining to, radioactivity. Sometimes only the prefix 'radio-' is used to describe rodioactive nuclides or the substances containing them, e.g., radiocarbon is an abbreviation for radioactive carbon.

radioactive age [NuCl] The age of a mineral, fossil, or wooden object as estimated from its content of radioisotopes. This method assumes that the content of radioisotopes has remained unchanged except for radioactive decay. See also dating; potassium-argon dating; rubidium-strontium dating; radiocarbon dating.

radioactive decay [NuCl] The spontaneous transformation of a nuclide into one or more nuclides, accompanied by either the emission of particles from the nucleus, nuclear capature, or ejection of orbital electrons, or fission. Also known as radioactive disintegration; radioactive transformation; rodioactivity.

radioactive disintegration *See* radioactive decay.

radioactive displacement law [NuCl] The statement of the changes in mass number that take place during various nuclear transformations, is summarized in the table shown below :

Radiation			Changes in nulceus	
Type	Mass No.	Charge	Mass No.	Atomic No.
α	4	+2	decreases by 4	decreases by 2
β−	0	-1	no change	increases by 1
β+	0	+1	no change	decreases by 1
Electron capture			no change	decreases by 1
γ	0		no change	no change

radioactive element [NuCl] An element all of whose isotopes spontaneously transform into one or more differnt nuclides, giving off various types of radiation; *e.g.*, radium, uranium and promethium.

radioactive emanation [NuCl] A radioactive gas given off by certain radioactive elements; all of these gases are isotopes of the element radon.

radioactive equilibrium [NuCl] A state ultimately reached when a radioactive substance of slow decay yields a radioactive product on disintegration. This product may also decay to give a further radioactive substance, and so on. The amount of any of the daughter radioactive products present after equilibrium has been reached remains constant, the loss due to decay being counter-balanced by gain from the decay of the immediate parent.

radioactive series [NuCl] A series of radioisotopes, each except the first being the decay product of the previous one. The final member of the series, usually an isotope of lead is stable.

radioactive standard [NuCl] A sample of radioactive material containing a radioisotope of precisely known rate of decay that is used for the calibration of instruments measuring radiation.

radioactive tracing [NuCl] A method of tracing the course of an element through a biological, chemical, or mechanical system. Any two isotopes of an element are chemically identical. Thus, by introducing a small amount of a radioisotope, called a tracer, the course taken by the stable isotope of the same element can be followed or traced by detecting the course of the accompanying radioisotope by suitable means.

radioactive waste [NuCl] Disposal of waste containing radioisotope of spent nuclear reactor fuel. It presents a serious problem for which there is yet no completely satisfactory procedure. Such wastes remain radioactive for thousands of years and constitute a long-term hazard.

radio astronomy [Astronomy] The study of celestial objects by measurement and analysis of their emitted electromagnetic radiation in the wavelength range from about 1 millimetre to 30 millimetres.

radio atmometer [Engineering] An instrument used to measure the effect of sunlight upon evaporation from plant foilage; it consists of a porous-clay atmometer whose surface is blackened to absorb radiant energy.

radiobeam [Physics] A concentrated beam of radiation of radio frequency energy as used in radio ranges, microwave relays, and radar.

radiobiology [Biology] The branch of biology dealing with principles, mechanisms, and the effects of radiation on living organisms and the behaviour of radioactive materials, or the use of radioactive tracing, in biological systems.

radiocarbon dating *See* carbon dating.

radiochemistry [Chemistry] The branch of chemistry concerned with radioactive compounds and with ionization. It includes the study of compounds of radioactive elements and the preparation and use of compounds containing radioactive atoms.

radio detection [NuCl] The detection astronomy of the presence of an object by' radiolocation without precise determination of its position.

radio element [NuCl] A radioactive isotope of an element, or a specimen consisting of one or more radioactive isotopes of an element.

radio emission [Physics] The emission of radio-frequency electromagnetic radiation by oscillating charges or current.

radio energy [Physics] The energy carried by radio frequency electromagnetic radiation.

radio frequency [Physics] The frequency of electromagnetic radiation within the range used in radio, *i.e.*, 10 kilohertz to 100 gigahertz.

radio-frequency current [Electricity] Alternating current having a frequency higher than 10,000 hertz.

radio frequency heating [Engineering] Industrial induction of dielectric heating, particularly when the frequency of the alternating field is above about 25 kilohetz.

radio-frequency reactor [Electronics] A reactor used in electronic circuits to pass direct current and offer high impedance at high frequencies.

radio galaxies [Astronomy] Galaxies that emit electromagnetic radiation of radio frequencies as observed by the techniques of radio astronomy. The exact source of this galactic radiation is not always understood, but radiation has been received from galaxies that have been observed to be optically in collision.

radiogenic [NuCl] Resulting from radioactive decay, *e.g.*, production of lead from uranium decay.

radiography [Physics] The formation of images on fluorescent screens or photographic material by short wavelength radiation, such as X-rays and gamma rays.

radioimmunoassay [Biochemistry] Sensitive quantitative determination of trace amounts of hormone (or some other biomolecule) by its capacity to displace the radioactive form of combination with its specific antibody.

radio interferometer [Engineering] A type of radio telescope or radiometer that consists of two or more separate aerials, each receiving electromagnetic radiation of radio frequencies from the same source, and each joined to the same receiver. The instrument works on the same principle as the optical interferometer, but as the wavelengths of the incident radiation are much greater, the distance between aerials has to be correspondingly increased.

radioisotope [NuCl] An isotope of an element that is radioactive; used in research as source of radiation and as tracer in studies of chemical and bio-chemical reactions. Artificial radioisotopes are made by neutron bombardment in a nuclear reactor.

radiolocation [Electronics] The location of distant objects, such as ships or aircraft, by radar.

radiological agent [NuCl] Any substance capable of producing casualties of emitting radiation.

radiology [Medicine] The study of X-rays and radioactivity, including radiodiagnosis and radiotherapy.

radiolucent [Physics] Almost transparent to radiation, especially X-rays and radio waves.

radioluminescence [Physics] Lumenscence resulting form X-rays, γ-rays, or β-particles emitted in radioactive decay.

radiolysis [Chemistry] The chemical decomposition of substances as a result of irradiation.

radiometry [Physics] The detection and measurement of radiant electromagnetic energy especially that associated with infrared radiation.

radiomicrometer [Engineering] An extremely sensitive instrument for measuring heat radiations. It consists of a thermocouple connected directly into a single copper loop forming the coil of a sensitive galvanometer.

radiomimetic substances [Chemistry] Chemicals which cause biological effects similar to those caused by ionizing radiations.

radionuclide [NuCl] A nuclide of an atom that is radioactive.

radio opaque [Physics] Not appreciably penetrable by X-rays or other forms of radiation.

radiosonde [Engineering] A small balloon used to carry meteorological instruments into the earth's atomosphere. Measurements of temperature, pressure, etc. are transmitted by these instruments back to earth by radio.

radio source [Astronomy] A discrete source of electromagnetic radiation of radio frequencies outside the solar system. Such sources have been discovered by the techniques of radio astronomy, both within the Galaxy and outside it, but only a small number have been identified with stars that can be located with optical telescopes. Other sources are supernovae explosions and remnants, colliding galaxies and gas clouds, quasars and pulsars; some sources, however, remain unexplained.

radio spectrum [Physics] The entire range of frequencies in the radio range to about 300,000 megahertz. Also known as radio-frequency spectrum.

radio star *See* radio source.

radio telegraphy [Electronics] The transmission of coded messages (*e.g.* in Morse code) by radio.

radio telephony [Electronics] The use of radio, rather than wires or cables, for all or part of a telephone system.

radio telescope [Astronomy] An astronomical instrument used to detect and analyse the radio-frequency electromagnetic radiations of extraterrestrial sources. The two principal types of radio telescope are : (1) parabolic reflectors, which are usually steerable so that çan be pointed at any part of the sky, and which reflect the incoming radiation on to a small aerial at the focus of the paraboloid; and (2) fixed radio interferometers. The latter have greater position-finding accuracy and greater ability to distinguish a small source against an intense background; while the former are more versatile owing to their mobility.

radiotherapy [Medicine] The treatmet of disease by means of radiation, particularly X-rays and techniques involving radio-activity.

radio wave [Physics] An electromagnetic wave produced by reversal of current in a conductor at a frequency in the range from about 10 kilohertz to about 300,000 megahertz.

radio-wave propagation [Physics] The transfer of energy through space by electromagnetic radiation at radio frequencies.

radiotransparent [Physics] Transparent to radiation, especially to X-rays and gamma-rays.

radio window [Physics] A region of the electromagnetic spectrum in the radio-frequency band within which radio waves can be transmitted through the earth's atmosphere without significant reflection or attenuation by constituents of the atmosphere. It extends from about 10 megahertz to 100 gigahertz and enables radiation in this range from celestial radio sources to be picked up by radio telescopes on the earth's surface. Below 100 MHz incoming radio waves are reflected by the ionosphere and those above 100 GHz are increasingly affected by molecular absorption.

radium [Chemistry] Ra. Naturally occurring radioactive element. At. No. 88, placed in group II A of the periodic table. The most stable isotope, ^{226}Ra, has a half-life of 1620 yeas. A very rare metal, chemically resembling barium; m.p. 973k, b.p. 1413K.

radium emanation *See* radon.

radius of gyration [Math] The moment of inertia I, of mass m about a given axis can be expressed in the form $I=mk^2$, k being the radius of gyration about the axis.

radius ratio [Chemistry] The ratio of the radius of a cation to the radius of anion relative ionic radii pertinent to crystal lattice structure, particularly the determination of coordination number.

radius vector [Astronomy] A line drawn from a central body (the focus) to a planet in any position in its orbit. [Math] The position of any point P in space with respect to a given origin O may be completely defined by the direction and length of the line OP. This line is called the radius vector of the point P.

radix [Mathematics] A number that forms the base of a system of numbers, logarithms, etc., *e.g.* the radix of the binary notation is 2.

radon [Chemistry] Rn. Element. At. No. 86, m.p. 202K, b.p. 211.3K; placed in zero group of the periodic table. The most stable isotope,^{222}Rn, has a half-life 3.825 dyas. A naturally occurring radioactive gas, produced as the immediate decay product of radium. Sometimes it is referred as radium emanation.

raffinate [Chemistry] A refined, liquid especially an oil after its soluble components have been removed by solvent extraction.

rafting [Geology] Transportation of rocks or soil by means of attachement to ice, plants, or other floating material.

rainbow [Electronics] Technique which applies pulse-to-pulse frequency changes to identifying and discriminating against decoys and chaff. [Optics] A colour effect produced by the refraction and internal reflection of sunlight in minute droplets of water present in the air; the effect is visible only when the observer has his back towards the sun.

ram [Engineering] The forward motion of an air scoop or air inlet through the air. [Mechanics] A plunger, weight, or other guided structure for exerting pressure or drawing something by impact.

Raman effect [Physics] A type of inelastic scattering of electromagnetic radiation in which light suffers a change in frequency and a change in phase as it passes through a material medium. The intensity of Raman scattering is about one-thousandth of that in Rayleigh scattering in liquids.

In *Raman spectroscopy* light from a laser is passed through a substance and the scattering is analysed spectroscopically. The new frequencies in the *Raman spectrum* of monochromatic light scattered by a substance are characteristic of the substance. This enables Raman spectroscopy to be used as a means of determining molecular structure and as a tool in chemical analysis.

ram effect [Mehanics] The increased air pressure in a jet engine or in the manifold of a piston engine due to ram.

ramming [Engineering] Packing a powder metal or sand into a compact mass.

Ramsen eye-piece [Optics] An eye-piece consisting of two plane-convex lenses of equal focal length f, with their plane sides facing outward and separated by a distance of $2f/3$. The eye-piece has low spherical aberration, is fairly achromatic and is very useful when crosswires or a scale are desired in the eye-piece.

Raney nickel [Metallurgy] A powder of nickel, used as a catalyst, especially in the hydrogenation of fats and oils. It is made by dissolving aluminium in a nickel-aluminium alloy placed in sodium hydroxide.

Ranger program [Engineering] A series of nine spacecrafts, launched during 1961-65 designed to transmit photographs back to earth while on a collision course with the moon. Ranger 7, 8 & 9 transmitted high-resolution television pictures of the lunar surface upto the instant of impact.

Rankine temperature [Physics] °R. The absolute Fahrenheit scale. Zero degrees Rankine is –459.67°F. and therefore °F+459.67 = °R.

Raoult's law [Chemistry] The partial vapour pressure of a solvent is proportional to its mole fraction. If p is the vapour pressure of the solvent (with a substance dissolved in it) and x the mole fraction of solvent (number of moles of solvent divided by total number of moles) then $p = p_o x$, where p_o is the vapour pressure of the pure solvent. A solution that obeys Raoult's law is said to be an *ideal solution*. In general the law holds only for dilute solutions, although some mixtures of liquids obey it over a whole range of concentrations.

rarefaction [Physics] A reduction in pressure; the opposite of compression.

rarefied gas [Mechanics] A gas whose pressure is much less than the atmospheric pressure.

rare gas *See* noble gas.

rare metal [Chemistry] Any metal that is difficult to extract from ore and is rare and expensive commercially. Some of the commonly called rare metals are, beryllium, bismuth, cadmium, cobalt, gallium, germanium, hafnium, indium, lithium, platinum, selenium, tungsten, uranium, vanadium, zironium.

Raschig process [Chemistry] An industrial process for the production of phenol from benzene. It involves the chlorination of benzene followed by the treatment with aqueous sodium hydroxide solution at elevated temperatures and high pressure.

raster [Electronics] The pattern of lines that scan the fluorescent screen of a cathode ray tube in a television receiver.

rationalized units [Scitific Technology] A system of units in which the defining equations have been made to conform to the geometry of the system in a logical way. Thus equations that involve circular symmetry contain the factor $2p$, while those involving spherical symmetry contain the factor $4p$. SI units are rationalized; c.g.s. units are unrationalized.

rational number [Mathematics] A whole number, or a number that can be expressed as the ratio of two whole numbers.

ray [Optics] The rectilinear path along which any radiation, *e.g.*, light, travels in any direction from a point in the source of the radiation. Loosely used to denote radiation of any kind.

rayleigh [Optics] A unit of brightness, used to measure the brightness of the night sky and aurorae, equal to $10^{10}/4p$ quanta per square metre per second per steradian.

Rayleigh line [Physics] Spectrum line in the scattered radiation which has the same frequency as the corresponding incident radiation.

Ralyeigh wave [Physics] A type of surface wave having a retrograde, ellipitical motion at the free surface.

rayon [Materials] Formerly called 'artificial silk', the term is now restricted to two types of manmade cellulose fibres : (1) viscoserayon, made by forcing a solution of viscose through fine holes into a solution that decomposes the viscose to give thread of cellulose, and (2) cellulose acetate rayon, made by forcing a solution of cellulose acetate through fine holes into warm air and allowing the solvent to evaporate, thus leaving threads of cellulose acetate.

raysistor [Electronics] A device which contains a photosensitive semiconductor and a light source; light source can be used to control the conductivity of the semiconductor.

reactance [Electeric] X. A property of alternating current circuits that together with the resistance, R, makes up the impedance Z, according to the relation, $Z=(R^2+X^2)^{1/2}$. If the circuit comprise the resistance, an inductance L, and capacitance C all in series, the reactance is given by;

$$X = \omega\ L-1/\omega c$$

where ε is the angular frequency equal to $2\pi f$, f being the frequency of the alternating current.

reactant [Chemistry] A substance that takes part in a chemical reaction.

reaction, chemical *See* chemical reaction.

reactive [Chemistry] Readily entering into chemical reactions; chemically active.

reactive dyes [Materials] Dyes that react chemically with the substances being dyed, to form chemical compounds.

reactor [Chemistry] Any vessel in which a chemical reaction or any

desired change (especially industrial) is conducted., [Physics] A device for introducing reactance into an electrical circuit (*e.g.* a capacitor).

read [Computers] To acquire information, usually from some form of storage in a computer.

read-in program [Computers] Computer program that can be put into a computer in a simple binary form and allows other programs to be read into the computer in more complex forms.

read-out [Computers] The presentation of output information by means of lights, printed, or punched cards, or tape or other methods.

reagent [Chemistry] A chemical substance used to produce a chemical reaction.

real address [Computers] The actual hardware address of an information element in memory.

real gas [Chemistry] A gas which deviates from the ideal gas law due to interactions of gas molecules. Also known as imperfect gas.

real image [Optics] An optical image such that all the light from a point on an object that passes through an optical system actually passes close to or through a point on the image.

real-time [Computers] A method of operating a computer as part of a larger system, in which information from the computer output is available at the time it is required by the rest of the systems.

Reaumur scale [Physics] A temperature scale in which the melting point of ice is taken as $0°R$ and the boiling point of water as $80°R$.

reciprocal ohm *See* mho.

reciprocal proportions, law of *See* chemical combination, laws of.

recoil electron [Physics] An electron that has been set into motion by a collision.

recombinant DNA [Biochem] DNA formed by the joining of genes into new combinations.

recombination energy [Chemistry] The energy released when two oppositely charged portions of an atom or molecule rejoin to form a neutral atom or molecule.

recombination radiation [Physics] The radiation emitted in semiconductors when electrons in conduction band recombine with holes in the valence band.

rectangle [Mathematics] A quadrilateral with right angles between all the four sides.

rectification [Chemistry] The purification of a liquid by distillation. [Math] The process of determining the length of a curve. [Physics] The conversion of an alternating current into a direct current.

rectified spirit [Materials] Ethanol, usually obtained by fermentation on an industrial scale, and purified by fractional distillation.

rectifier [Physics] A device for transforming an alternating current into a direct current; it consists of an arrangement that presents a much higher resistance to an electric current flowing in one direction than in the other.

rectifying valve [Electronics] The thermionic valve commonly used for rectification is the diode. The valve will pass current only when the anode is at a positive potential with respect to the cathode. Hence if an alternating potential is applied to a circuit containing such a valve, a direct current will flow through the circuit.

rectilinear [Mathematics] In a straight line; consisting of straight lines.

rectilinear motion [Mechanics] A continuous change of position of a body so that every particle of the body follows a straight line path. Also known as linear motion.

red giant [Astronomy] A giant star thought to be in the later stages of stellar evolution. It has a surface temperature in the range 2000-3000 K and a diameter 10-100 times that of the sun.

red oil [Materials] A commercial grade of oleic acid having 70% oleic acid, 15% linoleic acid and 15% stearic acid.

redox [Chemistry] Short form of the term oxidation-reduction, as in redox reactions.

redox exchanger [Materials] A substance, usually a polymer, that can "exchange" (*i.e.*, transfer) electrons, thereby effecting redox reactions, when in contact with reacting ions or molecules. Redox exchangers may also act as ion exchangers. *See* ion exchange.

redox potential [Chemistry] Voltage difference at an inert electrode immersed in a reversible oxidation-reduction system; measure of the state of oxidation of the system.

redox reaction [Chemistry] A chemical reaction in which an oxidizing agent is reduced and a reducing agent is oxidized, thus involving the transfer of electrons from one atom, ion, or molecule to another.

red shift [Physics] A systematic displacement towards longer wavelengths of lines in the spectra of distant galaxies and also of the continuous portion of the spectrum; increases with the distance from the observer. Also known as Hubble effect.

reduced equation [Physics] An equation of state of a gas in which temperature, pressure, and volume are replaced by their reduced values. *See* reduced temperature, pressure, and volume.

reduced temperature, pressure, and volume [Physics] Ratios of the temperature, the pressure, and the volume to the critical temperature, critical pressure, and critical volume respectively.

reducing agent [Chemistry] A substance that removes oxygen from, or adds hydrogen to, another substance; in the more general sense, one that donates electrons that is decreases the positiveness of its valence.

reductase [Biochemistry] An enzyme that promotes a reduction reaction.

reduction [Chemistry] The removal of oxygen from a substance, or the addition of hydrogen to it. The term is also used more generally to include any reaction in which an atom gains electrons.

re-entry [Astronomy] The position, time, or act of re-entering the earth's atmosphere after a journey into space. The 'angle of re-entry' is critical because of the enormous quantity of heat generated by a spacecraft as it enters the atomsphere. This heat is generated by friction between the atoms and molecules of atmosphere and the great speed of the moving spacecraft; it is normally absorbed by the heat shield. Too sharp an angle of re-entry would cause the spacecraft to burn up while too oblique an angle would cause the spacecraft to bounce off the atmosphere.

reference point [Science Tech] A place in the network where one or more interfaces for specific functions may occur.

refine [Chemistry] Purification or to remove the impurities from a substance.

refinery gas [Materials] A mixture of hydrocarbon gases, produced in large scale cracking and refining of petroleum. The usual components are hydrogen, methane, ethane, propane, butanes, pentanes, ethylene, propylene, butenes, pentenes, and small amounts of other components.

reflectance [Optics] A measure of the extent to which a surface is capable of reflecting radiation, defined as the ratio of the intensity of the reflected radiation to the intensity of the incident radiation.

reflection [Physics] The process in which the radiation meeting the boundary between two media bounces back to stay in the first medium. Any kind of radiation-wave or stream of particles can be reflected. Reflection from a source makes the radiation appear to come from some where else, *i.e.*, the image of the source. For reflection by a plane surface :

(i) The image is formed at the same distance behind the surface as the object is in front.

(ii) The line joining each object point with the image is perpendicular to the surface.

(iii) The size of the image is equal to the size of the object.

(iv) The image is the same way up as the object but laterally inverted.

(v) The image is virtual.

For reflection by a curved surface, the relations between object and image depend upon how far the object is from the surface compared with the focal distance (*f*) of the surface the image distance (*v*) and the object distance (u) as:

$$1/f = 1/u + 1/v$$

reflection, angle of [Optics] The angle between a ray of light reflected from a surface, and the normal to the surface at that point.

reflection of light, laws of [Optics] 1. The incident ray, the reflected ray, and the normal to the reflecting surface at the point of incidence lie in the same plane.

2. The angle between the incident ray and the normal (*i.e.*, the angle of incidence) is equal to the angle between the reflected ray and the normal (*i.e.*, the angle of reflection).

reflection, total internal *See* total internal reflection.

reflector [Optics] Any surface that reflects radiation, particularly electromagnetic radiation. *See also* parabolic reflector).

reflex angle [Mathematics] An angle greater than 180° and less than 360°.

reflux condenser [Chemistry] A condenser in which in vapour over a boiling liquid is condensed to a liquid, which flows back into the vessel, so preventing its contents from boiling dry.

refraction [Physics] The change of direction of propagation of any wave, such as light or sound, when it passes from one medium to another in which the wave velocity is different.

refraction, angle of [Optics] The angle between the refracted ray and the normal to the surface at the point of refraction.

refraction correction [Optics] The small correction that has to be made to the observed altitude of a heavenly body due to the refraction of the light it emits or reflects by the earth's atmosphere. All bodies appear to be slightly higher than they actually are.

refraction, laws of [Optics] 1. The incident ray, the refracted ray, and the normal to the surface of separation of the two media at the point of incidence lie in the same plane.

2. The ratio of the sine of the angle of incidence to the side of the angle of refraction is a constant for any pair of media. *See* refractive index. Also known as Snell's law.

refraction loss [Optics] The portion of the transmission loss that is due to refraction resulting from non-uniformity of the medium.

refractive index [Optics] *n*. The *absolute refractive index* of a medium is the ratio of the speed of electromagnetic radiation in free space to the speed of the radiation in that medium. As the refractive index varies with wavelength, the wavelength should be specified. It is usually given for yellow light (sodium D-lines; wavelength 589.3 nm). The *relative refractive index* is the ratio of the speed of light in one medium to that in an adjacent medium. Also known as refractive constant.

refractivity [Optics] If the refractive index of a medium is μ, its refractivity is defined as μ-1. The 'specific refractivity' is given by $(\mu$-1$)/d$ where d is the density of the medium; the 'molecular refractivity; is defined as the specific refractivity multiplied by the molecular weight.

refractometer [Optics] An apparatus for the measurement of the refractive index of a substance.

refractor [Optics] Any surface that refracts radiation.

refractory [Chemistry] Any earthy material of low thermal conductivity which is not damaged by heating to high temperatures. Such materials are made into bricks and use for lining furnaces, etc.

refractory metals [Metallurgy] A metal alloy that is heat-resistant, and having high melting point.

refrigerant [Materials] A fluid used in the refrigerating cycle of a refrigerator, usually consisting of a liquid that will vapourize at a low temperature (*e.g.*, freon or ammonia).

refrigerating cycle [Physics] The cycle of operations that takes place in a refrigerator. The refrigerant absorbs heat from the cold chamber and its contents, which causes it to vaporize; it is then pumped to a compressor where it gives up heat and condenses back to a liquid; it again passes to the cold chamber, thus constituting a continuous cycle.

regelation [Physics] Phenomenon in which ice melts under high pressure and freezes again when the pressure is removed. The melting point of ice is lowered by increased pressure; therefore ice near its melting point is melted by sufficient pressure, and solidification or regelation takes place again when the pressure is removed.

register file [Computers] A small area of memory in which several data elements, or registers, can be accessed simultaneously, rather than one by one.

regulatory enzyme [Biochemistry] An enzyme having a regulatory function through its capacity to undergo a change in catalytic activity by non-covalent binding of a specific modulating metabolite.

Reimer-Tiemann reaction [Chemistry] Reaction for the preparation of phenolic aldehydes by heating a phenol with chloroform in the presence of alkali.

relative atomic mass *See* atomic weight.

relative denstiy [Physics] The ratio of the density of a solid or liquid at a specified temperature (often 20°C) to the density of water at the temperature of its maximum density (4°C). It is a pure number, but is numerically equal to the density in grams per cubic centimetre. The density in SI units ($kg\ m^{-3}$) is 1000 times greater than the relative density. If the relative density of a substance is less than 1 it will float on water, if it is greater than 1 it will sink. The relative density of gases is usually expressed with reference to air, both gases being at S.T.P. Also knwon as specific gravity.

relative humidity [Physics] The ratio of the actual vapour pressure of the air to the saturation vapour pressure. It is a dimensionless quantity. Abreviated RH.

relative moecular mass. *See* molecular weight.

relative momentum [Mechanics] The momentum of a body in a reference frame in which another specified body is fixed.

relative motion [Mechanics] The continuous change of position of a body with respect to another body.

relative refractive index. *See* refractive index.

relativistic mass [Physics] The mass of a body that is travelling at a speed comparable to the velocity of light. The relativistic mass, m, of a body travelling at a velocity, v, is given by:

$$m = m_o \sqrt{1-v^2/c^2}$$

where m_o is the rest mass and c is the velocity of light.

relativistic particle [Physics] A particle that has a speed comparable to the velocity of light; *i.e.*, a particle with a relativistic mass substantially in excess of its rest mass.

relativistic velocity [Physics] A velocity, approaching the velocity of light, at which the effect of the theory of relativity is significant.

relativity [Physics] Theory of physics which recognizes the universal character of the propagation speed of light and the consequent dependence of space, time, and other mechanical measurements on the motion of observer performing the measurements. This theory recognizes the impossibility of determining absolute motion and leads to the concept of a four-dimensional space-time continuum. The special theory, which is limited to the description of events as they appear to observers in a state of uniform motion relative to one another, is developed from two axioms : (*a*) the laws of natural phenomena are the same for all observers, (*b*) the velocity of light is the same for all observers irrespective of their own velocity. The

more important consequences of this theory are : (*i*) the mass of a body is a function of its velocity, (*ii*) the mass-energy equation for the interconversion of mass and energy; (*iii*) the Fitzgerald-Lorentz contraction appears as a natural consequence of the theory; (*iv*) time has no absolute value. The general theory, applicable to observers not in uniform relative motion, leads to a novel concept of the theory of gravitation. In this theory the presence of matter in space causes space to 'curve' in such a manner that the gravitational field is set up. Thus gravitation becomes a property of space itself.

relay [Electricity] A device by which the electric current flowing in one circuit can open or close a second circuit and thus control the switching on and off of a current in the second circuit. [Electronics] A microwave or other radio system used for passing on a signal from one radio communication link to another.

reluctance [Physics] The radio of the magnetomotive force acting in a magnetic circuit to the magnetic flux. Also known as magnetic reluctance.

reluctivity [Physics] The reciprocal of magnetic permeability.

rem [Nucl] A unit of ionizing radiation, equal to the amount that produces the same damage to human beings as 1 roentgen of high voltage X-rays.

remanence [Physics] The residual magnetization of a ferromagnetic substance subjected to a hysteresis cycle when the magnetizing field is reduced to zero.

remote control [Electronics] Control of a quantity which is separated by an appreciable distance from the controlling quantity.

renal [Biology] Pertaining to the kidney.

rep [NuCl] A unit of ionizing radiation, equal to the amount that causes absorption of 93 ergs per gram of soft tissue. Also known as parker.

replication [Engineering] Making a reverse image of a surface by means of an impression on or in a receptive material, usually applied to microscope techniques for obtaining plastic replicas of observed objects. [Biochem] 1. Refers to reproduction of the DNA molecule, which is composed of two inter-locking chains of nucleotides, the double helix structure. It reproduces itself by forming two identical daughter molecules each of which receives one of the two chains of the original molecule, the other in each case being synthesised from nucleic acids by enzyme. 2. Multiplication of phage in a bacterial cell.

repressible enzyme [Biochemistry] An enzyme whose synthesis is in-hibited when its reaction product is readily available to a bacterial cell.

repressor [Biochemistry] The protein that binds to the regulatory sequence or operator for a gene and blocks its transcription.

residual charge [Electricity] The charge remaining on the plates of a capacitor after initial discharge.

residual current [Electronics] Current flowing through a thermionic diode when there is no anode voltage, due to the velocity of the electrons emitted by the heated cathode.

resistance [Electricity] The opposition that a device or material offers to the flow of direct current, equal to the ratio of the potential difference between the ends of a conductor to the electrical current flowing in the conductor. The extent to which a conductor resists the flow of a given current depends upon its physical dimensions, the nature of the material of which it is made, its temperature, and in some cases the extent to which it is illuminated. The derived SI unit of resistance is the ohm.

resistance wire [Medicine] Wire made from a metal or alloy having high resistance per unit length; used in wire-would resistors and heating element.

resistivity [Electricity] A constant for any material equal to the reciprocal of its conductivity. The resistivity is defined as the resistance offered by a cube of the material at 273K. Thus the resistivity, r equals RA/l where R is the resistance of a uniform conductor of length l and cross-sectional area A. It is usually expressed in ohm metres. Also known as specific resistance.

resistor [Electricity] A device used in electronic circuits primarily for its resistance. The most common types are either 'wire-wound', or made of finely ground carbon particles mixed with a ceramic binder.

resnatron [Electronics] A microwavebeam tetrode containing cavity resonators, mainly used for generating large amounts of continuous power at high frequencies.

resolution of forces [Physics] The division of forces into components that act in specified directions.

resolving power [Optics] A measure of the ability of an optical instrument to form separable images of close objects or to separate close wavelengths of radiation. The *chromatic resolving power* for any spectroscopic instrument is equal to l/dl, where dl is the difference in wavelength of two equally strong spectral lines that can barely be separated by the instrument and l is the average wavelength of these two lines.

resonance [Chemistry] The description of the structure of a molecule in terms of definite valence states of its atoms, and integral numbers of valence bonds between the atoms, gives an over-simplified picture of the actual state of the molecule, whose characteristics, *e.g.*, electron-density distribution, may be inconsistent with any classical formula. The resonance or valence-bond method of describing approximately the actual structure of a compound uses a number of classical structures ("resonance forms"), in terms of which the actual structure (the "resonance hybrid" is described. Also known as quantum-chemical resonance; mesomerism. [Physics] If, to a system capable of oscillation, a small periodic force is applied, the system is in general set into forced oscillations of small amplitude. As the frequency, f, of the exciting force approaches the natural frequency of the system, f_o, the amplitude of the oscillations builds up, becoming a maximum when $f=f_o$. The system is then said to be in resonance with the exciting force, or simply in resonance. [NuCl] Resonance is said to occur in nuclear reactions if the energy of an incident particle or photon is equal, or near to, the value of an appropriate energy level of the compound nucleus. Thus a resonance neutron is one whose energy corresponds to a particular energy level of a nucleus that will readily absorb it.

resonance frequency [Physics] A frequency at which some measure of the response of a physical system to an external periodic driving force is a maximum.

resonance hybrid [Chemistry] A molecule that may be considered an intermediate between two or more valence bond structures.

resonance spectrum [Physics] An emission spectrum resulting from illumination of a substance (usually a molecular gas) by radiation of a definite frequency or definite frequencies.

resonance vibration [Mechanics] Forced vibration in which the frequency of the disturbing force is very close to the natural frequency of the system, so that the amplitude of vibration is very large.

resonant cavity [Physics] A space enclosed by electrically conducting surfaces, in which electromagnetic energy may be stored or excited. The frequency of the oscillations within a resonant cavity will depend upon its physical dimensions.

resonant circuit [Electricity] An electric circuit that contains inductance, capacitance, and resistance of such values as to give resonance at an operating frequency.

resonant coupling [Electricity] Coupling between two circuits that reaches a sharp peak at a certain frequency.

resonate [Physics] To bring to resonance, as by tuning.

resorption [Geology] The process by which a magma redissolves previously crystallized minerals. [Physics] Absorption of material by a body from which the material was previously released.

respiration [Physiology] The process by which tissues and organisms exchange gases with their environment. Aerobic respiration is the process by which living organisms, or their components take oxygen from the atmosphere to oxidize their food to obtain energy. Anaerobic respiration is the process by which organisms or their components, obtain energy from chemically combined oxygen when they do not have access to free oxygen. Certain bacteria depend entirely on anaerobic respiration.

respiratory pigment [Biochemistry] Any of various conjugated proteins such as haemoglobin, present in blood cells or blood plasma that is capable of combining loosely and reversibly with oxygen.

respiratory quotient [Physiology] RQ. The ratio of the volume of carbon dioxide expired by an organism or tissue to the volume of oxygen consumed by it during the same period.

rest density [Physics] The density of a small portion of a fluid in a Lorentz in which that portion of the fluid is at rest.

rest energy [Physics] The rest mass of a body expressed in energy terms according to the relationship

$E_o = m_o c^2$, where m_o is the rest mass of the body and c is the speed of light.

rest frame [Physics] The Lorentz frame in which the total momentum of a system equals zero; for an accelerated system, the rest frame varies from instant to instant.

restitution coefficient [Physics] e. A measure of the elasticity of colliding bodies. For two spheres moving in the same straight line, $e = (v_2 - v_1)/(u_1 - u_2)$, where u_1 and u_2 are the velocities of bodies 1 and 2 before collision $(u_1 > u_2)$ and v_1 and v_2 are the velocities of 1 and 2 after impact $(v_2 > v_1)$. If the collision is perfectly elastic $e = 1$ and the kinetic energy is conserved; for an inelastic collision $e < 1$.

rest mass [Physics] The mass of a body when at rest relative to the observer. The mass of a body varies with its velocity.

resultant [Physics] A single force or velocity that produces the same effect as the two or more forces or velocities acting together.

ret [Biochemistry] The reduction or digestion of fibres with the help of enzymes.

retardation [Physics] The rate of decrease of velocity. Also known as deceleration; negative acceleration.

retort carbon *See* gas carbon.

retrograde orbit [Astronomy] Motion in an orbit opposite to the usual orbital direction of celestial bodies within a given system; specifically, of a satellite, motion in a direction opposite to the direction of rotation of the primary.

retroreflection [Optics] Reflection in which the reflected rays of radiation return along paths parallel to those of their corresponding incident rays.

retrorse [Biology] Bent backward or downward.

retrovirus [Biochemistry] RNA virus containing a reverse transcriptase, *i.e.*, an RNA-directed DNA polymerase.

reverberatory furnace [Metallurgy] A furnace designed for operations in which it is not desirable to mix the material with the fuel; the roof is heated by flames, and the heat is radiated down on to the material off the roof.

reversible process [Chemistry] Any process in which the variables that define the state of the system can be made to change in such a way that they pass through the same values in the reverse order when the process is reversed. It is also a condition of a reversible process that any exchanges of energy, work, or matter with the surroundings should be reversed in direction and order when the process is reversed. Any process that does not comply with these conditions when it is reversed is said to be an *irreversible process*.

Reynolds number [Mechanics] Re. A dimensionless quantity applied to a liquid flowing through a cylindrical tube, given by (Rc) = udl/η, where u=velocity of flow, d=density of the liquid, l=the diameter of the tube, and η=the coefficient of viscosity of the liquid. At low velocities, the flow of the liquid is streamline. At a certain value of (Re), corresponding to a critical velocity u_c, the flow becomes turbulent.

R_F value [Chemistry] (in chromatography) The distance travelled by the solvent front divided by the distance travelled by a given component. For a given system at a known temperature, it is a characteristic of the component and can be used to identify components.

rhe [Mechanics] The unit of fluidity; the reciprocal of the poise.

rhenium [Chemistry] Re. Element. At. No. 75. At.Wt. 186.20, placed in group VIIB of the periodic table. A hard, and heavy grey metal, m.p. 3453K, b.p. 5900K; used in thermocouples and as a catalyst.

rheology [Mechanics] The study of the deformation and flow of matter in terms of stress, strain and time.

rheopexy [Chemistry] The process by which certain thixotropic substances set more rapidly when they are stirred, shaken, or tapped. Gypsum in water is such a *rheopectic substance*.

rheostan [Metallurgy] An alloy of 52% copper, 25% nickel, 18% zinc, and 5% iron; used for electrical resistance wire.

rheostat [Electricty] A variable electrical resistor constructed so that its resistance may be changed without interrupting the circuit to which it is connected.

rhesus factor [Biochemistry] A group of antigens in the red blood cells of some humans (said to be *Rh* positive) but absent in some individuals (*Rh* negative). If a women is *Rh* negative and her husband *Rh* positive, the fetus may be *Rh* positive., having inherited the factor from its father. Blood from fetus may pass through some defect in the placenta into the maternal blood stream and stimulate the formation of antibodies to the *Rh* factor by the white cells of the mother. Then, when this women becomes pregnant second time, some of these antibodies may pass through the placenta into the child's bloodstream, and cause the clumping of its red cells. In extreme cases so many red cells are destroyed that the fetus dies before birth. Also known as *Rh* factor.

rhinology [Medicine] The scientific study of the anatomy, functions, and diseases of the nose.

rhodium [Chemistry] Rh. Element. At. No. 45. At.Wt. 102.905, placed in group VIII of the periodic table. A silvery-white hard metal, m.p. 2239K, b.p. 4000K. It occurs with and resembles platinum; used in alloys, catalysts, and thermocouples.

rhombus [Mathematics] A quadrilateral having all its sides equal.

rH scale [Chemistry] A scale of hydrogen pressures that gives a measure of the strength of a reducing agent. The rH value is defined as $\log_{10} 1/[H]$, where $[H]$ is the hydrogen pressure that would produce the same electrode potential as that of a given redox reaction at the same pH value.

ribonuclease [Biochemistry] An enzyme that catalyzes the hydrolysis of ribonucleic acid.

ribonucleic acid [Biochemistry] A long-chain, usually single stranded nucleic acid consisting of repeating nucleotide units containing four kinds of heterocyclic organic bases : adenine, cytosine, guanine, and uracil; they are conjugated to the pentose sugar ribose and held in sequence by phosphodiesterbonds. RNA is the chief constituent, together with protein, of many types of virus, and it appears to be responsible for the self-replication of the virus. 'Messenger' RNA transmits the coded information contained by the chromosomes of the nucleus of a cell to the protein-making ribosomes of the cytoplasm. 'Transfer' RNA or t-RNA transfers the activitated amino acids on to the messenger RNA.

ribosomes [Biochemistry] Small granules (about 10^{-8} metre in diameter) that occur in the cytoplasm of cells and appear to be the sites of protein synthesis.

Richter scale [Geology] A scale of numerical value of earthquake magnitude ranging from 1 to 9.

rig [Engineering] A drill machine complete with auxilary and accessory equipment needed to drill.

rigidity [Mechanics] The property or state of resisting change in form.

ring compound [Chemistry] A chemical compound in the molecule of which some or all of the atoms are linked in a closed ring. *See* carbocyclic compounds; heterocyclic compounds.

RNA *See* ribonucleic acid.

roast [Metallurgy] To heat ore to effect some chemical change that will facilitate smelting.

Rochon prism [Optics] A prism used for obtaining plane-polarized light (*see* polarization of light) and in other related problems. Such a prism, made of quartz, may be used for work with ultraviolet radiation.

rock crystal [Mineral] SiO_2. A pure natural crystalline form of silica.

rock salt [Mineral] NaCl. Natural crystalline sodium chloride.

roentgen [Physics] The amount of X-ray or gamma-radiation that will produce ions carrying 2.58×10^{-4} coulomb of electricity of either sign in 1 cm^3 of dry air.

Roentgen rays *See* X-ray.

rolled gold [Metallurgy] Same as gold-filled except that the proportion of gold alloy to total weight of the article may be less than 1/20; fineness of the gold alloy may not be less than 10 carat.

root [Mathematics] 1. One of the equal factors of a number of quantity. The square root, $^2\sqrt{}$ or $\sqrt{}$, is one of two equal factors; *e.g.*, $9=3\times3$ or $-3 \; x \; -3$; hence $^2\sqrt{9}=\pm3$. Similarly the cube or third root is denoted by $\sqrt{3}$ etc. It may also be denoted by a fractional index; thus $^2\sqrt{x}=^{1/2}$. 2. The root of an equation is a value of the unknown quantity that satisfies the equation.

root mean square value of alternating quantity [Scientific Technology] If x is a periodic function of t, of period T, the root mean square (RMS) value of x is the square root of the mean of the square of x taken over a period. [Electeric] The RMS value I of an alternating current is important since it determines the heat generated (RI^2) in a resistance R. All ordinary AC measuring instruments gives RMS values of current, etc. If the alternating quantity can be represented by a pure sine wave, the RMS value of the quantity A is related to the maximum value a of the quantity (*i.e.* amplitude) by the expression $A=a\sqrt{2}$. The RMS value of a current is also known as the 'effective value of the current'. Similarly, the RMS value of an alternating *e.m.f.* is known as the 'effective *e.m.f,*.

root mean square value of variable R.M.S. [Scientific Technology] The square root of the average of the squares of a number of values, given by :

RMS=

$$\sqrt{\frac{\text{(Sum of square of the individual values of the variable}}{\text{(total number of values)}}}$$

Rose's metal [Metallurgy] An alloy of 50% bismuth, 25% lead, and 25% tin; m.p.367K.

rotary converter [Electricity] An alternating current electric motor mechanically coupled to a direct current generator; used for converting an AC supply into DC.

rotary dispersion *See* optical activity.

rotational constant [Physics] The constant inversely proportional to moment of inertia of a linear molecule; used in calculations of microwave spectroscopy quantums.

rotational motion [Mechanics] The laws relating to the rotation of a body about an axis are analogous to those describing linear motion. The *angular displacement* (θ) of a body is the angle in radians through which a point or line has been rotated in a specified sense about a specified axis. The *angular velocity* (ω) is the angular displacement divided by the time, *i.e.* $\omega = d\omega/dt$, and the *angular acceleration* (α) is the rate of increase of angular velocity, *i.e.* $\alpha = = d\omega/dt = d^2\theta/dt^2$.

The equations of linear motion have analogous rotational equivalents, *e.g.* :

$$\omega_2 = \omega_1 + \alpha t$$
$$\theta = \omega_1 t + \alpha t^2/2$$
$$\omega_2^2 = \omega_1^2 + 2\theta$$

The counterpart of Newton's second law of motion is $T = Ia$, where T is the torque causing the angular acceleration and I is the moment of inertia of the rotating body.

rotatory power [Optics] The capability of a substance to rotate the plane of polarization of polarized electromagnetic radiation. [Chemistry] The product of the specific rotation of a substance and its atomic/molecular weight.

rotor [Engineering] The rotating part of a turbine, electric motor, or generator.

routine [Computers] A set of digital computer instructions designed and constructed so as to accomplish a specific function.

rubber [Materials] An elastic solid obtained from the latex of the Hevea brasiliensis tree. Raw natural rubber consists mainly of the cis-form of polyisoprene, $[CH_2=CH\text{-}C(CH_3)=CH_2]_n$, a hydrocarbon polymer, with molecular weight of about 300 000. Nearly all rubber articles are made by 'compounding' raw rubber, *i.e.*, mixing it with other ingredients and then vulcanizing

it in moulds by heating with sulphur and accelerators.

rubber, synthetic [Materials] A class of synthetic elastomers made from polymers or copolymers (*see* polymerization) of simple molecules. *See* butyl rubber; neoprene; nitrile rubber; styrene-butadiene rubber (SBR); silicone rubber; stereoregular rubber.

rubidium [Chemistry] Rb. Element. At. No. 37. At.Wt. 85.47; m.p. 312K; b.p. 961K; placed in group IA of the periodic table. A soft, extremely reactive, white metal resembling sodium.

rubidium-strontium dating [Physics] A method of dating some rocks, used for specimens over 10^9 years old. It is based on the decay of rubidium-87 (half-life 5×10^{11} years) to yield strontium-87. An estimate of the sample's age is given by the ratio of the two isotopes.

ruby [Mineral] A red form of corundum, Al_2O_3, that owes its colour to traces of chromium; used in lasers and as a gemstone.

run [Computers] A single complete execution of a computer program, or one continuous segment of computer processing, used to complete one or more tasks for a single computer or application.

Russell-Saunders coupling [Physics] A method for building many-electron single-particle eigen-functions of orbital angular momentum and spin; the orbital functions are combined to make an eigenfunction of the total orbital angular momentum, the spin functions are combined to make an eigenfuction of the total spin angular momentum, and then the results are combined into eigenfunctions of the total angular momentum of the system. Also known as LS coupling.

rust [Metallurgy] The iron oxides formed on corroded ferrous metals 'and alloys due to electrochemical interaction between iron and atmospheric oxygen and moisture. The reaction is very fast in the presence of moist or humid air.

rust prevention [Engineering] Surface protection of ferrous structures and equipments to prevent formation of oxides of iron; can be done by coatings, surface treatment, plating, cathodic arrangement, or other means.

ruthenium [Chemistry] Ru. Transition element. At. No. 44. At.Wt. 101.07; m.p. 2583K, b.p. 4173K; placed in group VIII of the periodic table. A hard brittle metal; occurs together with platinum; used in alloys and as a catalyst.

rutherford [NuCl] A unit used to express the decay rate of radioactive material; equal to 10^6 disintegrating atoms per second.

rutile [Mineral] TiO_2. A reddish brown to black mineral composed of titanium oxide; may contain upto 10% iron.

rydberg [Physics] A unit of energy used in atomic physics, equal to the square of the charge of the electron divided by twice the Bohr's radius; equal to 13.60583 $\pm$ 0.00004 electron volts.

Rydberg constant [Physics] R. A constant that occurs in the formulae for atomic spectra and is related to the binding energy between an electron and an atomic nucleus. It is connected to other constants by the relationship $R = \mu_o^2 m e^4 c^3 / 8h^3$, where μ_o = magnetic constant m and e = mass and charge of an electron, c = speed of light, and h =. Planck constant. It has the value $1.097 \times 10^7 \, m^{-1}$.

rydberg formula [Physics] A formula similar to that of Balmer, for expressing the wave numbers (λ) of the lines in a spectral series.

$$\lambda = R \left[\frac{1}{(n+a)^2} - \frac{1}{(m+b)^2} \right]$$

where n and m are integers and $m > n$, a and b are constants for a particular series, and R is the Rydberg constant.

S *see* sulphur.

sabin [Physics] A unit of sound

absorption for a surface, equivalent to 1 square foot (0.09290304 square metre) of perfectly absorbing surface.

Sabin vaccine [Medicine] Living attenuated poliovirus which can be given orally; produces active immunity against poliomyelitis.

sac [Medicine] A small pouch or cystlike cavity.

saccharase [Biochemistry] An enzyme that catalyses the hydrolysis of disaccharide to monosaccharides, specifically of sucrose to dextrose and levulose. Also known as invertase; sucrase.

saccharide [Biochemistry] A simple sugar; a monosaccharide.

saccharimeter [Engineering] An instrument for measuring the concentration of a sugar solution by measuring the angle of rotation of the plane of vibration of polarized light passing through a tube containing the solution.

saccharometer [Engineering] A type of hydrometer used for finding the concentration of sugar solutions by determining their density or the gases produced by fermentation; usually graduated to directly read the percentage of sugar.

saccharose *See* sucrose.

Saha ionization [Physics] The ionization of a gas which exist when the gas is in thermal equilibrium at a given temperature, in the absence of external influences; it increases with the rise in temperature. Also known as thermal ionization.

sal ammoniac [Chemistry] NH_4Cl. A white crystalline mineral, mainly composed of ammonium chloride.

saline [Chemistry] Containing salt, especially the salts of alkaline metals and magnesium. A 'saline solution' is a solution of salts in water. The saline which is isotonic with body fluids, is called physiological saline.

salinometer [Engineering] A type of hydrometer used for determining the concentration of salt solutions by measuring their density.

sal soda [Chemistry] Na_2CO_3. $10H_2O$. Sodium carbonate decahydrate. Also known as washing soda.

salt [Chemistry] A chemical compound formed when the hydrogen of an acid has been replaced by a metal. A salt is produced, together with water, when an acid reacts with a base.

salt bridge [Chemistry] An electrical connection made between two half cells. It usually consists of a glass U-tube filled with agar jelly containing a salt, such as potassium chloride. A strip of filter paper soaked in the salt solution can also be used.

salt cake [Chemistry] Na_2SO_4. Impure sodium sulphate (90-99%); used in making paper-pulp, detergents and soaps, ceramic glazes, and dyes.

salt effect *See* salting-out.

salting-out [Chemistry] Reduction in the solubility of a substance by addition of another (usually a salt) that lowers its solubility; *e.g.*, soaps can be salted-out by common salt (sodium chloride) from solutions in water.

samarium Sm. Element, At. No. 62. At.Wt. 150.35; m.p. 1345K, b.p. 2064K; placed in group IIIB of the periodic table; a rare-earth of the lanthanide group.

sand [Mineral] Hard, granular powder, generally composed of granules of impure silica, SiO_2 the shape of the grains vary from almost spherical to angular, with a diameter range from 1/16 to 2 millimetres.

sandstone [Mineral] Rock formed from sand and/or quartz particles cemented together with clay, calcium carbonate, and iron oxide.

sandwitch compound [Chemistry] A complex in which an atom of a transition element, is sandwitched between parallel benzene or other aromatic rings, *e.g.*, ferrocene, $(CH_2)_5Fe(CH_2)_5$.

sanitizer [Materials] A special class of disinfectant prepared for use on food-processing equipment and glassware in restaurants.

saponification [Chemistry] The hydrolysis of an ester; the term is often confined to the hydrolysis of an ester using an alkali, thus forming a salt (a soap in the case of the higher fatty acids) and free alcohol.

saponification number [Chemistry] One of the characteristics of a fat or oil; the number of milligrams of potassium hydroxide required for the complete saponification of one gram of the fat or oil.

saponins [Biochemistry] Glucosides, derived from plants, that form a lather with water; used as foaming agents and detergents.

sapphire [Mineral] A natural crystalline form of blue, transparent corundum (alumina, Al_2O_3); the colour being due to traces of cobalt or other metals.

satellite [Astronomy] Any of a solid body that rotate in orbits round other bodies of greater mass under the influence of the mutual gravitational field. Particularly heavenly bodies such as moons, that rotate around planets, *e.g.*, the moon is a satellite of the earth. *See also* satellite, artificial.

satellite, artificial [Astronomy] Any vehicle designed to orbit the earth or other heavenly body. The first artificial satellite, Sputnik I, was launched by Russia on 4th October 1957. Communication satellites are artificial earth satellites used for relaying radio and television signals around the curved surface of the earth.

satellite DNA [Biochemistry] Highly repeated, nontranslated segments of DNA in eukaryotic chromosomes.

saturated activity [NuCl] The maximum activity obtainable by activation in a definite flux in a nuclear reactor.

saturated colour [Optics] A pure colour which cannot be contaminated by white.

saturated compound [Chemistry] An organic compound with all carbon bonds satisfied by univalent radicals or groups. It does not form addition compounds; a compound the molecule of which contains no double or multiple valence bonds between the atoms.

saturated solution [Chemistry] A solution that can exist in equilibrium with excess of solute. The saturation concentration is a function of the temperature.

saturated vapour [Chemistry] 1. A vapour that can exist in equilibrium with its liquid. A vapour whose temperature equals the temperature of boiling at the pressure existing on it.

saturation [Physics] The condition in which a further increase in some case produces no further increase in the resultant effect.

saturation current [Electronics] 1. The maximum current which can be obtained under certain conditons. 2. In a semiconductor, the maximum current which just precedes a change in conduction mode.

saturn [Astronomy] The second largest planet in the solar system with ten small satellites, and surrounded by characteristic rings (*see* saturn's rings). Its orbit lies between those of jupiter and uranus. Mean distance from the sun, 1427.01 million kilometers. Sidereal period ('year'), 29.46 years. Mass, approximately 95.14 times that of the earth, diameter 1193000 kilometers, surface temperature, about-423K

saturn's rings [Astronomy] Three concentric rings, probably composed of the remains of a broken-up satellite, which are seen round the planet saturn.

sawtooth waveform [Electronics] A waveform in which the shape resembles the teeth of a saw. The voltage builds slowly and linearly up to a maximum value and then falls perpendicularly to zero in each cycle.

scalar meson [Physics] A meson which has spin 0 and positive parity, may be described by a scalar field.

scalar operation [Mathematics] Mathematical operations performed on random data elements rather than on sequential data elements.

scalar processing [Mathematics] Calculations performed one at a time.

scalar quantity [Mathematics] Any quantity that is sufficiently defined when the magnitude is given in appropriate units.

scaler [Electronics] An electronic device or circuit that produces an output pulse when a prescribed number of input pulses has been received. If the prescribed number is two (or ten) the circuit is referred to as a binary (or decade) scaling circuit or scaler.

scalp [Metallurgy] To remove surface layers, and thereby defects from ingots, billets or slabs machining.

scandium [Chemistry] Sc. Element. At. No. 21. At.Wt. 44.965; m.p. 1813K, b.p. 3123K; placed in group IIIB of periodic table.

scanner [Engineering] Any device that examines an area or region point by point in a continuous systematic manner, repeatedly sweeping across until the entire area or region is covered.

scanning [Engineering] The repeated and controlled traversing of; (a) a mosaic in a television camera, or a screen in a cathode-ray tube, with an electron beam; (b) an airspace with a radar aerial; or more generally (c) any area or volume with a moving detector in order to measure some quantity or detect some object.

scanning electron microsocoe [Electronics] SEM. A type of electron microsope in which a beam of electrons, a few hundred angstroms in diameter, systematically sweeps over the specimen; the intensity of secondary electrons generated at the point of impact of the beam on the sample is measured, and the resulting signal is fed into a cathode ray tube display which is scanned in synchronism with the scanning of the sample.

scattering [Physics] The deflection of any radiation as a result of its interaction with matter, e.g., the change in direction of a particle or photon on interacting with a nucleus or electron.

scattering of light [Optics] When a beam of light traverses a material medium, scattering of the beam takes place. Two types of scattering occur; (a) by random reflection; i.e., small particles suspended in the medium act as tiny mirrors and being randomly orientated with respect to the beam, produce random reflections. This type of scattering occurs when the size of the particles is large in comparison with the wavelength of the light; (b) by diffraction, this occurs when particles that are small compared with the wavelength of the light are present in the medium. Owing to diffraction phenomena, the particles act as centres of radiation and each particle scatters the light in all directions. In this type, the degree of scattering is proportional to the inverse fourth power of the wavelength of the light. Thus, blue light is scattered to a greater extent than red. The blue colour of the sky is due to scattering by the actual molecules of the atmosphere.

scavenger [Chemistry] Any substance added to a mixture or system to consume or inactivate traces of impurities. [Metallurgy] A reactive metal added to a molten metal to combine with and remove dissolved gases.

Schiff's base [Chemistry] *RR'C=NR"* Any of a class of derivatives of condensation of aldehydes or ketones with primary amines; weakly basic; hydrolysed by water and strong acids to form carbonyl compounds and amines.

Schiff's reagent [Chemistry] A reagent used to test for aldehydes. It consists of rosaniline which has been decolorized with sulphur dioxide or sulphurous acid. Aldehydes oxidize the reduced form of the dye back to its original colour.

Schottky defect [Physics] 1. A defect in an ionic crystal in which a single ion is removed from its interior lattice site and relocated in a lattice site at the surface of the crystal. 2. A defect in an ionic crystal consisting of the smallest number of positive-ion vacancies and negative ion vacancies to leave the crystal electrically neutral.

Schottky effect [Electronics] The enhancement of the thermionic emission of a conductor resulting from an electric field at the conductor surface.

Schrodinger wave equation [Physics] A partial differential equation governing the Schrodinger wave function ψ of a system of one or more nonrelativistic particles. It is based on de Broglie's concept that every moving particle is associated with a wave of wavelength h/mv (where h is Planck's constant and m and v are the mass and velocity of the

particle). In three dimensional system the equation has the form:

$$\nabla^2\psi + (8\pi^2 m/h^2)(E-V)\varphi = 0$$

where ∇^2 is the Laplace operator, c is the wave function, E is the total energy and U is the potential energy of the particle. *See also* eigen function.

Schwartzchild radius [Astronomy] A critical radius of a body of given mass that must be exceeded if light is to escape from that body. It is equal to $2GM/c^2$, where G is the gravitational constant, c is the speed of light, and M is the mass of the body. If the body collapses to such an extent that its radius is less than the Schwartzchild radius the escape velocity becomes equal to the speed of light and the object becomes a black hole.

scintillation [NuCl] A flash of light produced in a phosphor by an ionizing particle or photon. [Optics] Rapid changes of brightness of stars or other distant, celestial bodies caused by variation in density of air through which the light passes.

scintillation counter [NuCl] A device in which light flashes, produced by a scintillator when exposed to ionizing radiation, are converted into electrical pulse by a photomultiplier, thus enabling the number of ionizing events to be counted.

scintillation spectrometer [Engineering] A device for determining the energy distribution of a given radiation. It consists of a scintillation counter that incorporates a pulse height analyser.

scintillator *See* phosphor.

sclerometer [Engineering] An instrument for measuring the hardness of a material, usually by measuring the pressure needed to scratch it.

scleroscope [Engineering] An instrument used to determine the hardness of a material by measuring the height to which a standard ball rebounds from its surface when dropped from a standard height.

scleroprotein [Biochemistry] A class of complex, insoluble, fibrous proteins, (*e.g.*, keratin, collagen, elastin) that occur in the surface coatings of animals and form the framework of binding cells together in animal tissues.

-scope [Scientific Technology] Suffix applied to names of instruments for observing or watching, usually as distinct from measuring, *e.g.*, telescope.

scopometer [Optics] An instrument used to measure the absorption or scattering of light in a solution containing solid particles by measuring the contrast between an illuminated line placed behind the solution and a light field of constant brightness.

scotophor [Materials] A material such as an alkali halide, that can be used on the screen of a cathode ray tube instead of the usual phosphor when day light viewing and long persistant are needed. The material darkens due to electron bombardment to produce black-and-white picture which can be erased on heating.

scotopic vision [Medicine] Vision in which the rods in the eye are the principal receptors. This type of vision occur when the level of light is low and colours cannot be distinguished.

scotoscope [Electronics] A telescope which empolys an image intensifier to see in the dark.

Scott connection [Electronics] A type of transformer which transmits from two-phase to three-phase systems or vice-versa.

scouring [Engineering] 1. Physical or chemical attack on process equipment surfaces. 2. The cleaning of fabric before dyeing. 3. Removal of dirt and grease from wool.

scram [Nucleonics] A sudden shutting down of a nuclear reactor, usually by dropping safety rods, when a predetermined neutron flux or other dangerous situation occurs.

screen grid [Electronics] A grid placed between the anode and control grid of a thermionic valve, usually held at a fixed positive potential.

screening [Physics] The reduction of the electric field about a nucleus by the space charge of the surrounding electrons.

scrubber [Engineering] A device

for the removal, or washing out of entrained liquid droplets or dust. Also known as wet collector.

scum [Materials] 1.A film of impurities or any material that rises to or is formed on the surface of a liquid. 2. A slimy film formed on the surface of a solid object.

secant [Mathematics] A straight line cutting a circle or other curve.

second [Scientific Techniques] The SI unit of time defined as the duration of 9192631770 periods of the radiation corresponding to the transition between two hyperfine levels of the ground state of the caesium-133 atom. [Mathmatic] A measure of angle; 1/60 of a minute.

secondary [Electricity] Low-voltage conductors of a power distributing system.

secondary alcohol [Chemistry] An organic alcohol with molecular structure R_1R_2CHOH, where R_1 and R_2 may be alkyl or aryl, identical or different groups.

secondary amine [Chemistry] An organic compound of the formula R_1R_2NH, where R_1 and R_2 may be alkyl of aryl, identical or different groups.

secondary cell *See* accumulator.

secondary colour [Optics] A colour *e.g.* green or orange, obtained by mixing two primary colours.

secondary emission [Electronics] When a primary beam of rapidly moving electrons strikes a metal surface secondary electrons are emitted from the surface. The effect is of importance in the thermionic valve, the photomultiplier, etc. In the thermionic valve, the emission occurs when the electrons strike the anode, and may be suppressed or controlled in multi-electrode tubes (tetrode, pentode) by various grids called the suppressor and screen grid.

secondary radiation [Physics] Particles of photons produced by action primary radiation on matter, such as Compton recoil electrons, delta rays, secondary cosmic rays, and secondary electrons.

secondary voltage [Electricity] The voltage across the secondary winding of a transformer.

secondary wave [Optics] One of the waves that radiate from each point on a wavefront, according to Huygen's principle.

second law of thermodynamics *See* laws of thermodynamics.

second order reaction [Chemistry] A reaction whose rate of reaction is determined by the concentration of two chemical species, [*e.g.*, A+B→C+D]

secretion [Physiology] The act or the process of producing a substance which is specialized to perform a certain function within the organism or is secreted from the body.

secular movements [Geology]

Systematic, persistent movements of the earth's crust, either upward or downward, that take place slowly and imperceptibly over long periods of geologic time.

secular parallax [Astronomy] An apparent angular displacement of a star, resulting from the sun's motion.

secular variation [Astronomy] A relatively larger, slow change in the part of the earth's magnetic field caused by the internal state of the planet and having a form roughly to be expected from a simple but not quite uniformly polarized sphere.

sedative [Medicine] A drug that reduces nervousness and excitement by inducing relaxation and varying degrees of depression of the central nervous system.

sedimentation [Chemistry] The process of separating an insoluble solid from a liquid in which it is suspended by allowing it to fall or settle to the bottom of the containing vessel, with or without agitation or centrifuging.

*See*beck effect [Electronics] If two wires of different metals are joined at their ends to form a circuit and the two junctions are maintained at different temperatures, an electric current flows round the circuit.

seeding [Chemistry] The addition of fine particles to a solution to induce crystallization. Each particle (often a tiny crystal of the solute) acts as a nucleus upon which the new crystal grows. Also known as impfing.

Seger cone [Engineering] A device for estimating the approximate temperature of a furnace; the cones are made of material softening at a definite temperature.

seismic [Geology] Pertaining to an earthquake or earth vibration, including those that are artificially induced.

seismic-electric effect [Geology] A phenomenon in which a periodic change in current is caused to flow between two electrodes inserted in the ground when a seismic wave passes through the region between them.

seismic wave [Geology] A general term for all elastic waves produced by earthquakes or generated artificially by explosion. It includes both body waves and surface waves.

seismogram [Geology] The record obtained from a seismograph.

seismograph [Engineering] An instrument for recording earthquake shocks.

seismology [Geology] The scientific study of earthquakes and the phenomena associated with them.

selenides [Chemistry] Binary compounds of selenium with other more electropositive elements. Selenides of nonmetals are covalent.

selenuim [Chemistry] Se. Element. At No. 34. At. Wt. 78.96;

m.p.490 K, b.p.958 K; placed in group VI A of the periodic table. It is a non-metal resembling sulphur in its chemical properties; exits in several allotropic forms. The so-called 'metallic' selenium, a silvery-grey crystalline solid, its electrical resistance varies on exposure to light and is used in photoelectric cells.

selenium cell [Electronics] Either of two types of photoelectric cell; one type relies on the photoconductive effect, the other on the photovoltaic effect (see photoelectric effect). In the photo conductive selenium cell an external e.m.f. must be applied; as the selenium changes its resistance on exposure to light the current produced is a measure of the light energy falling on the selenium. In the photovoltaic selenium cell, the e.m.f. is generated within the cell.

selenium rectifier [Electronics] A rectifier that consists of alternate layers of iron and selenium in contact.

selenodesy [Astronomy] The branch of applied ˉ mathematics that determines, by observation and measurement, the exact positions of points on the surface of the moon, as well as the shape and size of the moon.

selenology [Astronomy] The scientific study of the moon, its natural, origin and movements. Now that samples of the moon's surface are available for study on earth, selenology has become a branch of chemistry as well as astronomy.

self-absorption [Nucleonics] The decrease in the radiation from a raioactive material caused by the absorption of a part of the radiation by the material itself.

self-diffussion [Physics] The spontaneous movement of an atom to a new site in crystal of its own species.

self-energy [Physics] The contribution to the energy of a particle due to virtual emission and absorption of other particles, in particular, photons and mesons.

self-exciting [Electricity] 1. Having magnets that are excited by current drawn from the output of the generator. 2. Operating without an external source of alternating current power.

self-induction [Electricity] The production of a voltage or e.m.f. in a circuit by a varying current in the same circuit. The e.m.f. is sometimes known as the back e.m.f. as its direction opposes the current change in accordance with Lenz's law. The relationship between the e.m.f. (E) and the rate of change of current (dI/dt) where L is a constant for the coil known as the self-inductance. The SI unit of inductance is the henry(H).

semen [Biology] A fluid containing spermatozoa, produced by male animals. The testes produce the spermatozoa, and the other constituents of the semen are produced by the prostate gland and seminal vesicles.

semicarbazone [Chemistry] $R_2C=N-NH-CONH_2$. A condensation product of an aldehyde or ketone with semicarbazide.

semiconductor [Physics] A solid crystalline material whose electrical conductivity is intermediate between that of a metal and an insulator, ranging from about 10^{-5} mhos to 10^{-7} mho per metre, and is usually strongly temperature-dependent. Semiconductors may be elements or compounds, for example , germanium, silicon, selenium, and lead telluride. In general, semiconductors consist of covalent crystals, 'ideal' examples of which at the absolute zero of temperature would pass no electric current as all the valence electrons would be held by the covalent bonds. At normal temperatures, however, some of the electrons have sufficient thermal energy to break free from the bonds leaving holes. Electrons liberated in this way will have random thermal motions, but in an imposed electric field there will be a net drift against the field resulting in so called n-type conductivity. The behaviour of the holes is more complex , but they may be regarded as positive charge free to move about the crystal giving rise to p-type conductivity. The total current passed by such an intrinsic semiconductor is therefore the sum of the electron current and the hole current in the direction of the field. A rise in temperature will create more carriers due to more bonds being broken by thermal energy, and thus lower resistance.

semiconductor device [Electronics] Electronic device in which the characteristic distinguishing electronic conduction takes place within a semiconductor.

semiconductor diode. [Electronics] A semiconductor device, either based on a semiconductor junction or on point contact, with two electrodes. It is used for rectification.

semiconductor junction [Electronics] A plane that separates two layers of a semiconductor each of which have different electrical characteristics. For example, a p-n junction separates the p-region (in which holes are the majority carrier) from the n-region (in which electrons are the majority carrier)

semicustom IC [Electronics] A chip made under a process where the chip maker provides basic circuit elements for the customer to develop the IC.

semimicro-analysis [Chemistry] A chemical analysis procedure in which the weight of the sample to be analysed lies between 10-100 milligrams.

semipermeable membrane [Physics] A membrane allowing the passage of some substances and not of others; a partition that permits the passage of pure solvent molecules more readily than those of the dissolved substance, e.g., copper ferrocyanide, $Cu_2Fe(CN)_6$. is permeable to water, but only very slightly permeable to dissolved substances.

semipolar bond [Chemistry] A valence bond in which two electrons are donated by one atom(donor) to another atom (acceptor).

semistandard IC [Electronics] A chip made where the chip develops large-scale integrated and very large-scale integrated functions that can be modified to the chip's purpose. This is equivalent to one electrovalent bond and one covalent bond, and is therefore called a semipolar bond.

separation energy [Nucleonics] The energy required to remove a particle such as a proton or a neutron from a atomic nucleus.

sequencer [Computers] The component of a processor that controls the program flow by implementing branches for subroutine processing and handling interrupts.

sequestering agent [Chemistry] A substance that restricts a metal ion in a solution by forming a complex ion that does not show the chemical reaction of the ion that is complexed; can be a chelating or complexing agent.

sequestration [Chemistry] The process of forming coordination complexes of an ion in solution. Sequestration often involves the formation of chelate complexes, and is used to prevent the chemical effect of an ion without removing it from the solution(e.g. the sequestration of Ca^{2+} ions in water softening)

series [Electricity] Elements in an electric circuit are in series if connected so that each carries the same current in turn. For resistors in series, the resulting resistance is:

$$R=R_1+R_2+R_3+\ldots\ldots$$

For cells in series, the resulting e.m.f. is :

$$E=E_1+E_2+E_3+\ldots\ldots$$

For capacitors in series, the resulting capacitance is:

$$1/C=1/C_1+1/C_2+1/C_3+\ldots\ldots$$

[Mathematics] A sequence of numbers or mathematical expressions such that the nth term may be written down in general form, and any particular term (say, the rth) may be obtained by substituting r for n; e.g., x^n is the general term of the series $1, x, x^2, x^3 \ldots x^n$.

series-parallel circuit [Electricity] A circuit in which some of the components or elements are connected in parallel, and one or more of these parallel combinations are in series with other components of the circuit.

serology [Biology] The *in vitro* study of reactions between antigens and antibodies in the blood serum. Various serological tests involving specific types of reaction enable indentification of blood groups, pathogens, diseases, etc.

Serpek process [Chemistry] A process for the fixation of atmospheric nitrogen. Aluminium is made to react with nitrogen to form aluminium nitride, which is then decomposed by steam to give ammonia.

serum [Physiology] The liquid that remains after the clotting and removal of blood cells and fibrin from the blood; any similar body liquid. It differs from plasma by the absence of fibrinogen.

servomechanism [Engineering] A mechanism that converts a small low powered mechanicl motion into a mechanical motion reqiring considerably greater power.The output power is always proportional to the input power, and the system may include a negative feedback device.

sesqui-[Chemistry] A prefix meaning one and a half. Often used for salts in which proportions of metals oxide to acidic anhydride are 2:3, or vice-versa.

sewage [Engineering] The fluid discharge from domestic, industrial, and medical sanitary appliances.

sewer gas [Materials] The gas evolved from the decomposistion of sewage, having a high methane and hydrogen sulphide content; can be used as a fuel gas.

sex chromosomes [Biology] The chromosomes that determine sex in most animals. They are of two types : X chromosome and Y chromosome. In heterogamretic sex(XY) they can usually be distinguished from the other chromosomes, because the Y chromosome is much shorter than the X chromosome with which it is paired.

sextant [Engineering] An instrument for determining the angle between two objects (*e.g.* horizon and star); commonly empoyed for determining the radius of a position circle.

shadow [Optics] A dark patch formed by a body that obstructs rays of right. A shadow cast by an object in front of a point source of light in a sharply defined area; a source of light of appreciable size produces two distinit regions, the umbra or full-shadow, and the penumbra or half-shadow.

shadow bands [Astronomy] A series of wavy shadow bands that fall across the earth just before and after totality in a solar eclipse. It is due to differences in density of the atmosphere.

shear [Mechanics] A stress applied to a body in the plane of one of its faces.

sheath [Electricity] A protective covering on a cable. [Electronics] A space charge formed by ions near an electrode in a gas tube.

sheave [Engineering] A grooved wheel or pulley.

shell [Physics] A set of orbital electorn states that have the same principal quantum number and, therefore, have approximately the same energy level and average distance from the nucleus. The shells are designated by the letters K to P (equivalant to values of principal quantum number n from 1 to 6) in order of increasing distance from the nucleus. The number of electrons in each shell is restricted but each shell is capable of containing $2n^2$ electrons. Within each shell, elctrons are further classified into sub-shell (or energy sub-levels) according to their orbital angular momentum, which is represented by their azimuthal quantum number, l. The separate subshells are distinguished by the letters s, p, d, and f (corresponding to valves of l that is 0, 1, 2, and 3).

shellac [Materials] A yellowish natural resin secreted by the lac insect; it consists of several polyhydroxy organic acid ($C_{15}H_{20}O_6$) together with 3%-5% of wax. Shellac produces smooth, durable films from alcoholic solutions and alkaline dispersions, which adhere to a variety of surfaces; used in varnishes, polishes, leather dressings, and sealing wax.

sherardizing [Metallurgy] A method of plating iron or steel with zinc, to form a corrosion resistant coating. The iron or steel is heated in contact with zinc powder to a temperature slightly below the melting point of zinc.

At this temperature the two metals amalgamate forming internal layer of zinc-iron alloy and an external layer of pure zinc.

shielded wire [Electricity] Insulated wire covered with a metal shield usually of tinned braided copper wire.

shielding [Metallurgy] Placing a nonconducting object in an electrolytic bath during plating to alter the current distribution. [Nucleonics] Reducing the ionzing radiation reaching one region of space from another region by using any suitable device.

shim rod [Nucleonics] A control rod used for making some coarse adjustments in the reactivity of a nuclear reactor.

shock [Mechanics] A pluse or transient motion or force lasting for extemely small time period which is capable of exciting mechanical resonances; *e.g.*, blast produced by explosives.

shock wave [Physics] 1. A very narrow region of high pressure and tempertature in which air flow changes from subsonic to supersonic. 2. A fully developed compression wave of large amplitude, across which density, pressure, and particle velocity change drastically.

shooting star *See* meteor.

short circuit [Electricity] A low resistance connection across a voltage source or between both sides of a circuit line, usually accidental and resulting in excessive current flow that may cause damage.

short sight *See* myopia.

shortwave propagation [Electronics] The propagation of radio waves that has a wavelength in the range 10 to 100 metres at frequencies in the range from about 1600 to 30,000 kilohertz.

shower [Astronomy] The production by one high-energy particle, originating from cosmic rays or accelerators, of several fast particles. 'Cascade'; showers (or soft showers) consist of electrons, positrons, or photons formed by successive pair productions or radiative collisions. 'Penetrating' showers contain nucleons and muons capable of penetrating upto about 20 cm of lead. 'Auger' showers (or extensive showers) extend over areas of upto 1000 square metres.

shunt [Electricity] A device for reducing the amount of electric current flowing through a piece of apparatus, such as a galvanometer. It consists of a conductor connected in parallel with the apparatus.

shunt regulator [Electricity] A regulator that maintains a constant output voltage by controlling the current through a dropping resistance in series with the load.

sial [Geology] A petrologic name for the upper layer of the earth's crust, composed of rocks that are rich in silica, SiO_2, and alumina, Al_2O_3.

sideband [Physics] The band of frequencies laying on either side of a modulated carrier wave; the width of each sideband is equal to the highest modulating frequency.

side chain [Chemistry] An aliphatic group or radical attached to a straight chain or to a benzene ring or other cyclic group in the molecule of an organic compound; *e.g.*, in toluene, C_6H_6-CH_3, the methyl group, CH_3-, is a side chain attached to a benzene ring.

sidereal [Astronomy] Referring to a quantity, such as time, to indicate that it is measured in relation to the apparent motion or position of the stars.

sidereal day [Astronomy] The period of a complete rotation of earth upon its axis, with respect to the fixed stars. It is 4.09 minutes shorter than a mean solar day.

sidereal period [Astronomy] The length of time required for one revolution of a celestial body about its primary, with respect to the stars.

sidereal year [Astronomy] The time period relative to the stars of one revolution of the earth around the sun, it is about 365.2564 mean solar days.

siemens [Electricity] The unit of electrical conductance susceptance, and admittance. An element possesses a condutance of one siemens if it has electrical resistance of one ohm. The seimens has replaced the 'mho'.

sievert [Nucleonics] The SI unit of dose equivalent.

sigma bond [Chemistry] A covalent bond directed along the line joining the centres of two atoms.

sigma particle [Physics] S-particle. An elementary particle classified as a hyperon. It exists in three charged states positive, negative, and neutral.

sigma pile [Nucleonics] An assembly consisting of a neutron source and a moderator, without any fissile material, which is used to study the properties of moderators.

sign digit [Automatic data processing] A digit containing one to four binary bits, associated with a data item and used to devote an algebraic sign.

silanes [Chemistry] A class of silicon hydrides of the general formula Si_nH_{2n+2}, forming a homologous series analogous to the alkanes.

silent mutation [Biochemistry] A mutation in a gene that causes no detectable change in the biological characteristic of the gene product.

silica [Minerals] A hard, insoluble white or colourless solid having chemical composition SiO_2, and very high melting point. It is very abundant in nature in the forms of quartz, rock-crystal, flint, and as silicates in rocks; used in the form of a white powder in the manufacture of glass, ceramics, and abrasives.

silica gel [Materials] A form of silica, SiO_2, with a slighlty porous structure capable of adsorbing (*see* adsorption) 40% of its weight of water from a saturated vapour; used in gas drying and as a catalyst support.

silicate [Chemistry] A compound whose crystal structure contains SiO_4 tetrahedra, either isolated or joined through one or more of the oxygen atoms to form groups, chains, sheets or three dimensional strucutures with metallic elements. Silicates are classified according to crystal structure: neosilicate, sorosilicate, cyclosilicate, inosilicate, phyllosilicate and tectosilicate.

silicol process [Chemistry] The manufacture of hydrogen by the (action of sodium hydroxide caustic soda, NaOH) solution on silicon.

silicon [Chemistry] Si. Element. At No. 14. At Wt. 28.086; m.p. 1683 k, b.p. 2628 k; a nonmetal placed in group IV A of the periodic table. It occurs in two allotropic forms; a brown amorphous powder and as dark grey crystals. It is the second most abundant element in the earth's crust; occurring in sand and rocks as silica and as silicates. The pure element is used in semiconductors; it is also used in alloys and in the form of silicates in glass.

silicone rubbers [Materials] Rubber-like polymers of various organo silicon compounds, such as siloxanes (in particular, dimethylsioxane, $(CH_3)_2Sio$, having valuable characteristics, such as high stability over wide ranges of temperature, outstanding water repellence, high resistance to chemical action, good electrical properties, etc.

silicones [Chemistry] The term originally applied to compounds of general formula R_2SiO,. where R stands for hydrocarbon radicals. They are now defined as polymeric organic siloxanes of the general type $(R_2Sio)n$; used as lubricants, for water-repellent finishes high temperature resisting resins, and lacquers.

siliconizing [Metallurgy] Diffusing silicon into solid metal at an elevated temperature.

silicon-steel [Metallurgy] A variety of steel that contains 0.5-4.5% silicon; used in electric transformer coils.

siloxanes [Chemistry] A group of compounds containing silicon atoms bound to oxygen atoms, with organic groups linked to the silicon atoms, e.g. $R_3SiOSiR_3$, where R is an organic group. Silicones are polymers of siloxanes.

silver [Chemistry] Ag. Element. At. No.47. At.Wt. 107.87; m.p.1235k, b.p. 2485k; placed in group IB of periodic table. A white, soft, extremely malleable metal; occurs as the metal, and as argentite or silver glance. Ag_2S; horn silver, AgCl; used in coinage and jewellery; compounds are used in photography.

silver plating [Metallurgy] The process of depositing a layer of silver on the surface of metal articles, usually by eletctrolytic methods.

silvichemical [Chemistry] A chemical derived from wood, e.g., lignins, vanillin, yeast.

silvicide [Materials] A nonselective herbicide used to kill or defoliate bashes and small trees, e.g., ammonium sulphamate.

sima [Geology] The rocks that form the earth's oceanic crust and underlie the upper crust. These are basaltic rock types rich in silica (SiO_2) and *magnesium* (Mg), hence the name. The sima is denser and more plastic than the sial that forms the continental crust.

simple harmonic current [Electricity] Alternating current, the instantaneous value of which is equal to the product of a constant, and the cosine of an angle varying linearly with time. Also known as sinusoidal current.

simple harmonic motion [Mechanics] S.H.M. A point is said to move in simple harmonic motion when it oscillates along a line about a central point, O, so that its acceletiation towards O is always proportrioral to its distance from O. Thus, if a point P moves in a circle, centre O and radius *r*, with a constant angular velocity *w*, the projection of *P* on any diameter will move in S.H.M. If the distance from *O* of the projection of *P* on a vertical diameter is *y*, at time *t*, then a graph of *y* against *t* will give a 'sine wave' of amplitude *r* and equation *y* = *r* sin ωt. This equation may be rewritten in the more general form:

$$y = r\sin 2\pi(t/T - x/\lambda)$$

where *T* is the period of the wave, λ its wavelength and *x* the distance it has travelled from *O* in time *t*.

simple protien [Biochemistry] A protein that yields only amino acids on hydrolysis.

simultaneous equation [Mathematics] A set of equations in which the values of the variables will satisfy all the equations; if the equations contain *n* variables, then to obtain a solution there must be at least *n* equations.

simulation [Computers] A process in which a software program takes a circuit netlist, models, and input stimulus and returns outputs that ideally are the same as those that would be returned by the physical circuit.

sine wave [Physics] A wave that has an equation in which one varible is proportional to the sine of the other. Also known as sinusoidal wave.

sintering [Metallurgy] Compressing metal particles into a coherent solid body. The process is carried out under heat, but at a temperature below the melting point of the metal. Certain nonmetals, such as ceramics and glass, may also be sintered.

sinusoidal [Physics] Having the characteristics of a sine wave.

sinusoidal oscillator *See* harmonic oscillator.

sinusoidal wave *See* sine wave.

SI units [Scientific Techniques] Systems International d' Unites. An internationally agreed coherent system of units, derived from the m.k.s. system now in use for all scientific purpose and there by replacing the c.g.s. system and the f.p.s. system. The seven basic units are ; the metre (symbol m), kilogram (kg), second (s), ampere (A), kelvin (K), mole (mol), and candela (cd). The radian (rad) and steradian (sr) are suppementary units. Derived units are obtained by combining, by multiplication and or division, two or more base units. Derived units are obtained by combining, by multiplication and or division, two or more base units. Thus the coulomb, which is the derived unit of charge, is formed from a combination of charge, is formed from a combination of one ampere times one second. Some of the derived units having special names and symbols are the hertz (*Hz*), netwton (N), joule (J) , watt (W), coulomb (C), volt (V), farad (F), ohm (V), weber (Wb), tesala (T), henry (H), lumen (lm), lux (lx), pascal (Pa), siemens (S), becquerel (Bq), and gray (Gy). *See also* appendices.

skelp [Metallurgy] A strip or sheet of steel that can be rolled and welded to form a tube.

skewness [Scilentific Techniques] The condition of being disordered or lacking symmetry.

skiascope [Optics] An instrument used to study optical refraction within the eye.

skip distance [Physics] The minimum distance to that radio waves can be transmitted between two points on the earth by reflection from the ionosphere, at a specified time and frequency.

skip zone [Physics] A region in the air surrounding a source of sound in which no sound is heard, although the sound becomes audible at greater distances. Also known as zone of silence.

sky wave [Physics] A radio wave may travel from transmitting aerial to receiving aerial by one of the two paths; either directly along the ground (*See* ground wave), or by reflection from the ionosphere. In the latter case it is called a sky wave or ionospheric wave.

slag [Metallurgy] Non-metallic material obtained during the smelting of metallic ores; it is generally formed as a molten mass floating on the molten metal. It is now utilized especially in construction and for making fertilizers.

slaking [Chemistry] The Treatment of lime with water to give hydrated (slaked) lime. [Geology] The crumbling and disintegration of earth materials upon exposure to air and/or moisture, especially the breaking-up of dry clay or soil when saturated with or immersed in water.

slate [Minerals] A natural form of aluminium silicate formed from clay hardened by pressure.

slewing [Engineering] Moving a radar antenna or a sonar transducer rapidly in a horizontal or vertical direction, or both.

slide rule [Engineering] A mathematical instrument used for rapid calculations; it consists of a grooved ruler with a scale, with another similarly marked ruler sliding inside the groove. Multiplication and division are carried out by adding or subtracting lengths on the two rulers, the divisions of which are in a logarithmic scale.

slow neutron [Nucleonics] A neutron whose kinetic energy does not exceed about 10-20 electronvolts.

slow reactor [Nucleonics] A nuclear reactor in which fission is induced primarily by slow neutrons, as in a thermal reactor.

slow wave [Physics] A wave having a phase velocity less than the velocity of light.

sludge [Materials] Residue left after acid treatment of petroleum oils.

slug [Physics] A heavy copper ring placed on the core of a relay to delay operation of the relay. [Nucleonics] A short fuel rod inserted in a hole or channel in the active lattice of a nuclear reactor.

slurry [Materials] A thin paste consisting of a suspension of a solid in a liquid.

slushing agent [Materials] A non-drying oil, grease, or similar material used to coat metals to afford temporary protection against corrosion.

smelting [Metallurgy] The extraction of a metal from its ores by a process involving heat. Generally the process is one of chemical reduction of the oxide of the metal with carbon in a suitable furnace.

smetic crystal [Chemistry] A liquid crystal in which the molecules are arranged in layers with their axes parallel and perpendicular to the plane of the layers.

smog [Chemistry] A dark, dust and soot-laden, sulphurous fog that under certain meteorological conditions, pollutes the atmosphere of many industrial cities and affects the lungs of their inhabitants.

smoke [Chemistry] A suspension of fine particles of a solid in a gas; smoke from coal consists mainly of fine particles of carbon.

Snell's law *See* refraction, laws of.

soap [Chemistry] A mixture of the sodium salts of stearic acid, $C_{17}H_{35}COOH$, palmitic acid, $C_{15}H_{31}COOH$, and oleic acid, $C_{17}H_{33}COOH$; or of the potassium salts of these acids (soft soap). Soaps are made by the action of sodium or potassium hydroxide on fats, the process of hydrolysis or saponification giving the soap, with glycerol as a by-product.

soapstone [Minerals] A massive metamorphic rock composed essentially of talc, with varying amounts of micas, chlorite, and other minerals.

soda ash [Chemistry] Na_2CO_3. Commercial name of sodium carbonate.

soda-lime [Chemistry] A solid mixture of sodium hydroxide NaOH, and calcium hydroxide, $Ca(OH)_2$; made by slaking, quicklime (*see* calcium oxide) with a solution of sodium hydroxide and drying by heat.

sodalite [Minerals] $Na_4Al_3 Si_2O_{12}Cl$. A blue mineral of the feldspathoid group; occurs in various sodium-rich igneous rocks.

soda water [Materials] Water containing carbon dioxide, CO_2 under pressure; releasing the pressure lowers the solubility of the gas, and thus causes effervesence.

sodium [Chemistry] Na (Natrium). Element. At. No.11. At. Wt. 22.9898; m.p. 370.6K, b.p.1165K; placed in group IA of the periodic table. A soft silvery-white metal. It is very reactive, tarnishing rapidly in air. The metal is used in the perparation of organic compounds and as a coolant in some types of nuclear reactor.

sodium-vapour lamp [Electronics] A luminous discharge obtained by passing an electric current between two electrodes in a tube containing sodium vapour at low pressure; used for outdoor illumination as the characteristic yellow light is less absorbed by fog and mist than white light.

soft iron [Metallurgy] Iron containing little carbon, as distinct from steel; iron that does not retain magnetism permanently, but loses most of it when the magnetizing field is removed.

softner [Chemistry] A substance that reduces the hardness of water by removing or sequestering calcium and magnesium ions. Also known as softening agent.

soft radiation [Physics] The radiation of relatively long wavelength and low penetrating power. The particles or photons of such radiation have low energy.

soft soap *See* soap.

software [Automatic data processing] The programs used in a computer, especially the general programs supplied by the computer manufacturer. The 'hardware ' is the actual equipment of the computer itself.

soft water [Chemistry] Water that forms an immediate lather with ordinary soap and is free from magnesium and calcium salts; water containing not more than 60 mg per litre of hardness-forming constituents expresed as $CaCO_3$ equivalent.

soil [Materials] A mixture of inorganic matter derived from weathered rocks and organic components (such as humus) resulting from decay of prior vegetation.

soil conditioner [Materials] Any material added to topsoil to reduce acidity (such as lime) and promote growth (such as bone meal).

sol [Chemistry] A colloidal slution consisting of a suitable dispersion medium, which may be gas, liquid, or solid, and the colloidal substance, the disperse phase, which is distributted throughout the dispersion medium.

solar activity [Astronomy] Disturbances on the surface of the sun, *e.g.*, solar flares, sunspots, and prominences.

solar battery [Electronics] An array of solar cells, usually connected in parallel and series.

solar burst [Astronomy] A sudden increase in the radio frequency energy radiated by the sun, generally associated with visible solar flares.

solar cell [Electronics] An electric cell that converts the sun's electromagentic energy into a usable electric current. Most solar cells consist of a single-crystal silicon *p-n* junction. When photons of light energy from the sun fall on or near the semiconductor junction the electron-hole pair created are forced by the electric field at the junction to separate so that the holes pass to the p-region and the electrons pass to the n-region. This displacement of free charge creates an electric current when a load is connected across the terminals of the device.

solar constant [Physics] The rate at which solar energy is received per unit area at the outer limit of the earth's atmosphere at the mean distance between the earth and the sun. Its value is 1.353 kwm^{-2}

solar cycle [Astronomy] The periodic change in the number of sunspots, the maxima and minima of which takes place after about 11.1 years.

solar day See mean solar day.

solar eclipse [Astronomy] An eclipse that takes place when the new moon passes between the earth and the sun and the shadow formed reaches the earth. See also eclipse.

solar energy [Physics] The energy transmitted from the sun in the form of electromagnetic radiation.

solar faculae [Astronomy] Bright streaks or regions on the surface of the sun, especially near solar sunspots.

solar flares [Astronomy] Short high temperature outbursts seen as bright areas in the chromo-

sphere of the sun. Jets of particles, known as the solar wind and strong radio frequency electromagnetic radiations are emitted during solar flares; associated with sunspots; cause magnetic and radio disturbance on earth.

solar pond [Engineering] An experimental means of storing solar energy in either fresh or salt water. The 'pond' is a large, suitably lined open container of varying dimensions.

solar prominence [Astronomy] The large eruptions of luminous gas from the sun's surface; they appear dark against the sun's disc but bright against the dark sky, and occur only in regions of horizontal magnetic fields.

solar radio noise [Astronomy] Radio noise originating at the sun. Its intensity increases during sunspots and solar flares. It is heard as a hissing noise on short wave radio receivers. Also known as solar noise.

solar still [Engineering] A device for converting sea water into fresh water by evaporation. The sea water is collected in one or more pools and a transparent cover made of glass or plastic is placed over it. Heat from the sun evaporates the water, leaving behind a residue of salts; the vapours from the evaporated water condenses on the surface of the receiving tank.

solar system [Astronomy] The sun and the celestial bodies moving about it; includes nine planets: mercury, venus, the earth mars, jupiter, saturn, uranus, neptune and pluto; and of the belt of asteroids revolving in elliptical orbits round the sun, along with satellites of the planets, comets, and meteor swarms. *See also* appendices.

solar wind [Astronomy] The supersonic flow of gas containing streams of electrically charged particles such as ionized hydrogen, electron and helium, which continuously flows from the sun out through the solar system with velocities of 300 to 1000 kilometres per second emitted by the sun, predominantly during solar flares and sunspot activity.

solation [Chemistry] The change of a substance from a gel to sol.

solder [Metallurgy] 1. To join by means of solder. 2. A low melting such as of zinc and copper, or of tin and lead, used to join metallic surfaces.

sole [Electronics] Electrode used in magnetrons and backward-wave oscillators to carry a current that generates a magnetic field in the desired direction.

solenoid [Electricity] A coil of wire wound uniformly on a cylindrical former, having a length that is large compared with its radius. When a current I is passed through the solenoid, a uniform magnetic field 4 is produced inside the coil parallel to its axis. If I is in amperes and n is the number of turns per metre. $H=nI$ amperes per metre.

solid [Mathematics] A three-di-mensional figure having length, breadth, and thickness; a figure occupying space or having a measurable volume. [Physics] A substance that has a definite mass, volume and shape, and resists forces that tend to alter its volume and shape.

solidifying point [Physics] The constant temperature at which a liquid solidifies under a given pressure, usually the standard atmosphere.

solid solution [Chemistry] A crystalline material that is a mixture of two or more components, with ions, atoms, or molecules of one component replacing some of the ions, atoms, or molecules of the other component in its normal crystal lattice. Solid solutions are found in certain alloys. For example, gold and copper form solid solutions. Mixed crystals of double salts (such as alums) are also examples of solid solutions. Compounds can form solid solutions if they are isomorphous (*See* isomorphism).

solid state [Physics] The physical state of matter in which the constituent molecules, atoms, or ions have no translatory motion although they vibrate about the fixed positions that they occupy in a crystal lattice. A solid is said to possess cohesion, retaining the same shape, unless changed by external forces.

solid state chemistry [Chemistry] Study of exact arrangements in solids, especially crystals, with particular emphasis on imperfections and irregularities in the electronic and atomic patterns in a crystal, and the effects of these on electrical and chemical properties.

solid-state circuit [Electronics] Complete electric or electronic circuit formed from a single block of semiconductor material.

solid state device [Electronics] A device, other than a conductor, which uses magnetic, electrical and other properties of solid materials, as opposed to vacuum or gaseous devices.

solion [Electricity] An electrochemical device in which amplification is obtained by controlling and monitoring a reversible electrochemical reaction.

soliquid [Chemistry] A system in which solid particles are dispersed in a liquid.

Solomon R unit [Nucleonics] A unit of radiation dose rate of X-rays, equal to 2100 roentgens hour. Also known as R unit.

solstice [Astronomy] The time (or, more accurately, the point) at which the sun reaches its greatest declination north or south. The points are situated upon the ecliptic half-way between the equinoxes; the times are approximately 21 June and 22 December.

solubility [Chemistry] The quantity of solute that dissolves in a given quantity of solvent to form a saturated solution. Solubility is measured in kilograms per metre cubed, moles per kilogram of solvent, etc. The solubility of a substance in a given solvent depends on the temperature. Generally, for a solid in a liquid, solubility increases with temperature; for a gas, solubility decreases.

solubility product [Chemistry] K_s. The product of the concentrations of ions in a saturated solution. For instance, if a compound $A_x B_y$ is in equilibrium with its solution

$$A_x B_y(s) \rightleftharpoons x A^+(aq) + y B^-(aq)$$

the equilibrium constant is

$$K_c = [A^+]^x [B^-]^y / [A_x B_y]$$

Since the concentration of the undissolved solid can be put equal to 1, the solubility product is given by

$$K_s = [A^+]^x [B^-]^y$$

The expression is only true for sparingly soluble salts. If the product of ionic concentrations in a solution exceeds the solubility product, then precipitation occurs.

solute [Chemistry] A substance that is dissolved in a solvent to form a solution.

solution [Chemistry] A homogeneous molecular mixture of two or more substances of dissimilar molecular sturcture; the word is usually applied to solutions of solids in liquids. Other types of solutions include gases in liquids, the solubility of gases decreasing with rise in temperature; gases in solids; liquids in liquids; and solids in solids *e.g.*, some alloys.

solution poison [Nucleonics] A soluble nuclear poison, such as boric acid, added to the coolant of a nuclear reactor for controlling the nuclear activity; generally used during shutdown periods.

solutrope [Chemistry] A ternary system with two liquid phases and a third component distributed between the phases, or selectively dissolved in one or the other phases; analogous to azeotrope.

solvation [Chemistry] 1. The combination of solvent molecules with molecules or ions of the solute. The compound so formed is called a 'solvate'. 2. The adsorption of water or other solvent on individual dispersed particles of a solution or dispersion.

Solvay process [Chemistry] An inudstrial preparation of sodium carbonate or washing-soda, $Na_2CO_3 . 10H_2O$, from common salt, NaCl, and calcium carbonate, $CaCO_3$. By the action of ammonia, NH_3, and carbon dioxide (obtained by heating $CaCO_3$) on common salt solution, the less soluble sodium hydrogen carbonate, $NaHCO_3$, is precipitated. The action of heat on this compound gives the required sodium carbonate, while the ammonia is recovered from solution by the action of the lime (calcium oxide).

solvent [Chemistry] A substance (usually liquid) having the power of dissolving other substances in it that component of a solution which has the same physical state as the solution itself, *e.g.*, in a solution of salt in water, water is the solvent, while salt is the solute.

solvent extraction [Chemistry] The process of separating one constituent from a mixture by dissolving it in a solvent in which it is soluble but in which the other constituents of the mixture are not. The process is usually carried out in the liquid phase, in which case it is also known as *liquid-liquid extraction*.

solvolysis [Chemistry] A reaction in which a solvent reacts with the solute to form a new substance.

somatic [Biology] Pertaining to the body. Somatic cells are the cells of which the body of an organism is constructed, as opposed to the reproductive or germ cells. *See also* mutation. In another sense the word refers to the body as opposed to the mind; *e.g.*, psychosomstic medicine is the study of the influence of psychological factors upon physiological illness.

somatic cell [Biology] Any cell of the body of an organism except the gametes and the cells from which they develop.

sonar [Engineering] Sound Navigation Ranging. An apparatus for locating submerged objects by transmitting a high frequency sound wave and collecting the reflected wave. The time for the wave to travel to the object and return gives an indication of the depth.

sonic boom [Physics] The loud noise created by the shock wave set up by an aircarft or missile travelling at supersonic speeds, *i.e.*, at speeds faster than that of sound.

sonolysis [Chemistry] The breaking up of molecules by ultrasonic radiation, for example, sonolysis of pure water produces hydrogen atoms, hydroxyl radical, molecular hydrogen, oxygen and hydrogen peroxide.

soot [Materials] Impure black carbon with oily compounds; formed as a result of incomplete combustion of resinous materials, oils, coal, or wood.

sorbent [Chemistry] Any agent that can provide a sorption function, such as adsorption, absorption, or desorption.

sorosilicate [Chemistry] A class of structural type of silicate characterized by the linkage of two SiO_4, tetrahedra by sharing of one oxygen, with a Si:O ratio of 2:7; *e.g.*, hemimorphite, $Zn_4Si_2O_7(OH)_2.H_2O$.

sorption [Chemistry] Adsorption (a surface process) or absorption (a volume process). The term is often used when the mechanism of a particular process is not known or is not specified. This term can also be used for ion-exchange, ion-exclusion, ion-retardation, chemisorption, and dialysis.

sound [Physics] Elastic waves in which the direction of particle motion is longitudinal, *i.e.*, parallel with the direction of propagation. The term is sometimes restricted to such waves in air and water, but is also applied to wave motion in solids.

source [Electricity] The electrode in a field-effect transistor from which electrons or holes enter the inter-electrode space. [Nucleonics] A radioactive material packaged so as to produce radiation for experimental or industrial use.

space [Physics] That part of the boundless four-dimensional continuum in which matter can be physically (rather than temporally) extended. [Astronomy] The part of the universe that lies beyond the earth's atmosphere, in which the density of matter is very low.

space capsule [Engineering] A container generally attached to a space vehicle manned or unmanned, used for carrying out an experiment or operation in space.

spacecraft [Engineering] A vehicle capable of travelling in space.

space flight [Engineering] Travel beyond the earth's atmosphere; it may be an orbital flight about the earth or it may be a more extended flight beyond the earth into space.

space lattice [Crystallography] The

regular and repeated three-dimensional arrangement of atoms or ions in a crystal. Also known as Bravias lattice; crystal lattice.

space probe [Engineering] A rocket-propelled missile that has sufficient velocity to escape from the earth's atmosphere. Space probes are used for making measurements of conditions within the solar system that cannot be made by terrestrial observation. The measurements are made by miniaturized electronic equipments within the probe, the results of which are signalled back to earth by radio.

spallation [Nucleonics] A nuclear reaction in which a high energy incident particle, or photon, causes several particles or fragments to be emitted by the target nucleus.The mass number and atomic number of the target nucleus may thus be reduced by several units.

spar [Geology] A term loosely applied to any transparent or translucent light-coloured mineral, usually readily cleavable and somewhat lustrous, especially one occurring as a valuable non-metallic mineral; e.g., fluro-spar, barite.

spark coil See induction coil.

spark plug [Engineering] A device for providing an electric spark for exploding the mixture of air and petrol vapour in the cylinder of the internal-combustion engine. Also known as sparking plug.

spatial [Physics] Of or pertaining to space; occupying space, occurring in, or conditioned by, space; considered with relation to space.

spatter [Metallurgy] Particles of metal expelled during arc or gas welding.

specific [Physics] When the adjective 'specific' is used before the name of an extensive physical quantity, it generally implies 'divided by mass', e.g., specific heat capacity is heat capacity per unit mass. When the extensive quantity is denoted by a capital letter (e.g., V for volume), the specific quantity is usually denoted by the corresponding small letter ($v=V/m$ for specific volume). In some older physical quantites the world had other meanings e.g, specific resistance), but such uses are now deprecated.

specific activity [Nucleonics] The activity per unit mass of a pure radioisotope; or the activity of a radioisotope in a material per unit mass of that material. It is expressed in curies per gm or disintegrations per second per kg.

specific charge [Electricity] The electric charge to mass ratio of an elementary particle or other charged body.

specific gravity [Mechanics] The former term for the ratio of the density of a substance to that of water. As the word specific now has a different usage the term relative density is now used for this concept.

specific heat capacity [Physics] Heat capacity divided by mass. The quantity of heat required to raise the temperature of unit mass of a substance by one degree. It is expressed in joules per kg per kelvin (SI units), calories per gram per °C. (c.g.s. units), or British thermal units per lb per °F (F.P.S. units). The two most important specific heat capacities of a gas are *(i)*. that measured at constant pressure Cp, and *(ii)* that measured at constant volume., Cv. Cp is greater than Cv because when a gas is heated at constant pressure it has to do work against the surroundings in expanding. The ratio Cp/Cv, usually denoted by γ (gamma) varies from 1.66 for monatomic gases to just over 1 for more complex molecules. The value of γ (gamma) therefore gives some indication of the number of atoms in the molecules of a gas.

specific impulse [Engineering] A measure of the thrust available from a rocket propellant. It is the ratio of the thrust produced to the fuel consumption.

specific ionization [Nucleonics] The number of ion pairs formed per unit distance along the track of an ion passsing through matter.

specific power [Nucleonics] The power produced per unit mass of fuel present in a nuclear reactor.

specific resistance *See* resistivity.

specific rotation [Optics] The angle of rotation plane-polarized light,

CD-line of sodium passing through a substance at 298K' at a specified concentration and light path.

specific surface [Chemistry] The total surface area per unit mass of a given substance, *e.g.*, a powder or a porous material. It is usually expressed in m^2 kg^{-1} or square centimetres per gram. It represents the actual surface area available for processes, such as adsorption, and may be very large for fine powders and highly porous substances.

specific volume [Physics] The volume at a specified temperature and pressure, occupied by unit mass (usually 1 kg) of a substance; the reciprocal of the density.

specific weight [Physics] The weight per unit volume of a substance.

spectral classification [Astronomy] A form of classification used for stars, based on their spectra. It is based on the seven star types known as O, B, A, F, G, K, M; O hottest blue stars; ionized helium lines dominant B hot blue stars; neutral helium lines dominant, no ionized helium A blue blue-white stars; hydrgen lines dominant, F white stars; metallic lines strengthen, hydrogen lines weaken G yellow stars; ionized calcium lines dominant K orange-red stars; neutral metallic lines dominant, some molecular bands M coolest red stars; molecular bands dominant characteristics revealed by study of their spectra ; the six classes B, A, F, G, K, and M include 99% of all known stars.

spectral colour [Optics] A colour corresponding to light of a pure frequency; the basic spectral colours are violet, blue, green, yellow, orange, and red.

spectral line [Physics] A particular wavelength of light emitted or absorbed by an atom, ion, or molecule.

spectral series [Physics] The emission spectrum of any substance may be analysed into one or more groups of frequencies (or wavelength) the frequencies in each group forming a series. For example, the spectrum of the hydrogen atom possesses series given by the expression

$$\nu = k(1/n_o{}^2 - 1/n^2),$$

where ν is the frequency of the spectral lines and k is a constant. For the different series, n_o takes the values 1,2,3,4, etc. For any one value of n_o, n may have all integral values from $n_o + 1$ upwards, the expression then giving the frequencies of all the lines in that particular series.

spectrograph [Engineering] 1. An instrument by which spectra may be photographed. 2. A photograph taken by means of such an insturment. *See* spectrographic analysis.

spetrographic analysis [Chemistry] An investigation of the chemical nature of a substance by the exmination of its spectrum, using the fact that the position of emission and absorption lines and bands in the spectrum of a substance is characteristic of it.

spectroheliograph [Engineering] An instrument used to photograph the sun with light of a particular wavelength.

spectrometer [Engineering] 1. A type of spectroscope so calibrated that it is suitable for the precise measurements of refractive indices. 2. An instrument for measuring the energy distribution of a particular type of radiation, e.g., a scintillation spectrometer.

spectrophotometer [Engineering] A photometer for comparing two light radiations wavelength by wavelength.

spectroscope [Engineering] An instrument for spectrograhic analysis of spectra. The simplest type is the prism spectroscope. This consists of a collimator, which collects the light from the source and throws it on to the face of a glass prism. The specturm so formed, after refraction by the prism, is viewed through a telescope. The angle between the collimator and the telescope can be varied.

spectroscopic binary [Astronomy] A binary star system that cannot be seen as two stars by a telescope, but which show a Doppler effect in their line spectrum as these stars revolve about each other.

spectroscopy [Physics] The study of

matter and energy that deals with the production, measurement, and interpretation of electromagnetic spectra arising from either emission or absorption of radiant energy by various substances.

spectrum [Physics] 1. A range of electromagnetic radiation emitted or absorbed by a substance under specific conditions. 2. The set of frequencies, wavelengths, or related quantities, involved in some process; for example, each element has a characteristic discrete spectrum for absorption and emission of light. A continuous spectrum is one in which all wavelengths, between certain limits, are present. A line spectrum is one in which only certain wavelengths or 'lines' appear. The emission and absorption spectra of a substance are fundamental characteristics of it and are often used as a means of identification. Such spectra arise as a result of transitions between different stationary states of the atoms or molecules of the substance, electromagnetic waves being emitted or absorbed simultaneously with the transition.

specular reflection [Optics] Perfect or regular reflection of electromagnetic rediation, *e.g.*, light. It occurs whenever the reflecting surface is flat to approximately 1/8 of a wavelength of the radiation incident upon it.

speculum [Optics] A reflecting mirror, especially a metallic mirror (*see* speculum metal) used in a reflecting telescope.

speculum metal [Metallurgy] A hard and brittle alloy of 2/3 copper and 1/3 tin; used for mirrors and reflectors.

speed [Physics] The ratio of the distance covered to the time taken by a moving body. Speed in a specified direction is velocity.

speleology [Geology] The exploration and scientific study of caves, including their genesis, morphology, and mineralogy.

spelter [Metallurgy] Commercial zinc, about 97% pure, conatining lead and other impurities; used in galvanizing.

spermatocyte [Biology] A male gametocyte that undergoes meiosis to form spermatids, which change into spermatozoa.

spermatozoon [Biology] A male gamete, four of which are derived by meiosis from a single spermatocyte. Also known as sperm.

sphenoid [Crystallography] An open crystal form having two nonparallel faces that are symmetrical to an axis of two fold symmetry. It occurs in monoclinic crystals of the sphenoidal class.

sphere [Mathematics] A solid figure generated by the revolution of a semicircle about a diameter as axis. The flat surface of a section cut by a plane through the centre is a great circle; the surface of a section cut off by any other plane is a small circle. The solid cut off by a plane of a great circle is a hemisphere; that cut off by a small circle is a segment. The volume of a sphere having radius r is $4\pi r^3/3$; surface area $= 4\pi r^2$.

spherical aberration *See* aberration, spherical.

spherical coordinates [Mathematics] Three-dimentional polar coordinates. A point in space defined by the length of its radius vector and the angle this vector makes with two perpendicular planes.

spherical mirror [Optics] A mirror, either concave or convex, whose surface forms the part of a sphere.

spherical triangle [Mathematics] A triangle drawn on a spherical surface, bounded by the arcs of three great circles. The properties of such triangles differ from those of plane triangles ; calculation relating to them form the purpose of spherical trigonometry.

spheroid [Mathematics] A solid figure generated by an ellipse rotating about its minor axis (oblate spheroid, a 'flattened sphere), or about its major axis (prolate spheroid, an 'elongated sphere').

spherometer [Engineering] An instrument for the accurate measurement of small thickness, or curvature of spherical surfaces.

spiegel [Metallurgy] A form of pig iron containing 15-30% of manganese and 4-5% of carbon. It is added to steel in a Bessemer converter as a deoxidizing agent and to raise the manganese content of steel.

spill [Nucleonics] The accidental release of radioactive material.

spin [Mechanics] The rotation of a body about its own axis. [Physics] The intrinsic angular momentum of an elementary particle or nucleus, which exists even when the particle is at rest, as distinguished from orbital angular momentum. Quantum considerations limit the magnitude of the spin angular momentum of orbital electrons to two values, given by $Jh/2p$ or Jh where J is the spin quantum number, which can have the values $\pm \frac{1}{2}$. The plus and minus signs indicate that the spin can be clockwise or anti-clockwise. For all baryons and leptons J is half integral ($\frac{1}{2}$, $1\frac{1}{2}$), but for mesons and photons it is integral (0,1,2).

spinel [Minerals] A group of minerals having the general composition $MO.R_2O_3$, M being a bivalent metal (magnesium, ferrous iron, manganese, zinc) and R a tetravalent metal such as aluminium, chormium, ferric iron).

Spin quantum number *See* spin.

spiral galaxies [Astronomy] Galaxies in which the stars, dust, and gas clouds are concentrated in the arms of a spiral. Spiral galaxies are believed to have evolved from 'elliptical' galaxies. The Galaxy to which the solar system belongs is also spiral in form.

spirits [Chemistry] An absolescent and ambiguous term usually meaning the distilled essence of a substance. [Medicine] In pharmacy the term refers to alcoholic solution of a volatile substance.

splash down [Engineering] The landing of a spacecraft or missile on water.

splice [Electricity] A joint used to connect two lengths of conductors with good mechanical strength and good conductivity.

sponge iron [Metallurgy] Iron in porous or powder form made without fusion by heating iron ore in a reducing gas or with charcoal.

sponge metal [Metallurgy] Any porous metal made by decomposition or reduction of a compound with melting.

spontaneous [Physics] Occurring without application of an external energy, because of the inherent properties of an object or the system.

spontaneous combustion [Chemistry] The cumbustion of a substance of low ignition point, which results from the heat produced within the substance by slow oxidation.

spurious radiation [Physics] Any emission from a radio transmitter at frequencies outside its frequency band.

spore [Biology] An asexual reproductive element, usually unicellular, of an organism, such as protozoan or cryptogamic plant, which can directly develop into an adult.

sporozoa [Biology] Class of protozoa; have no special method of locomotion and are parasitic; one kind is the human parasite causing malaria.

sputtering [Electronics] The ejection of atoms or group of atoms from the surface of the cathode of a vacuum tube as a result of heavy-ion impact. [Metallurgy] A process for depositing a thin uniform film of a metal on to a surface.

square [Mathematics] 1. A quadrilateral having all its sides equal and all its angles right angles. 2. The square of a quantity is that quantity raised to the second power, *i.e.*, multiplied by itself.

square root [Mathematics] A square root of a real of complex number n is a number a for which $a^2=n$.

square wave [Electricity] A wave motion that alternates between two fixed values for equal lengths of time, the time of transition between the two values being negligible compared to the duration of each fixed value.

SSI [Electronics] Small-scale integration; fewer than 100 transistors in a circuit.

stability constant [Chemistry] Refers to equilibrium reaction of a metal ion and a ligand to form a complex; the absolute stability constant is expressed by the product of the concentration of products divided by the product of the concentration of the reactants.

stabilization [Chemistry] The prevention of the chemical occurence of a reaction such as chemical decomposition of a substance, by the addition of a 'stabilizer' or 'negative catalyst'.

stable equilibrium [Mechanics] A body at rest is in stable equilibrium. If when slightly displaced, it tends to return to its original position of equilibrium. If the displacement tends to increase, the body is said to be in unstable equilibrium. Positions of stable equilibrium are position of minimum potential energy, those of unstable equilibrium are of higher potential energy.

stainless steel [Metallurgy] A class of corrosion-resistant iron chromium steels containing 70%-90% iron, 12%-20% chromium, 0.1%-0.7% carbon. Sometimes containing other elements such as nickel, silicon, tungsten, and niobium.

stalgamometry [Physics] The measurement of surface tension by determining the mass (or volume) of a drop of the liquid hanging from the end of a tube.

standard cell [Electricity] A primary cell, *e.g.*, the Weston cell, characterized by a high constancy of e.m.f. over long periods of time. The e.m.f. is a function of the temperature, and in the Weston cell it decreases by about 10^{-5} per 1° rise in temperature.

Standard chip [Electronics] A predesigned SSI or MSI circuit that can be combined with others on the same chip.

standard electrode *See* hydrogen electrode ; calomel electrode.

standard electrode potential [Chemistry] The reversible or equilibrium potential of an electrode in an environment where reactants and products are at unit activity.

standard potential [Chemistry] The potential of an electrode composed of a substance in its standard state, in equilibrium with ions in their standard compared to a hydrogen electrode.

standard state [Scientific Techniques] A state of a system used as a reference value in thermodynamic measurements. Standard states involve a reference value of pressure (usually 1 atmosphere, 101.325 kPa) or concentration (usually 1 M). Thermodynamic functions are designated as 'standard' when they refer to changes in which reactants and products are all in their standard and their normal physical state.

standard volume [Chemistry] The volume of 1 mole of a gas at pressure of 1 atmosphere and 273K. Also known as normal volume.

standing wave [Physics] A wave produced by the simultaneous transmission of two similar wave motions in oppsite directions. In acoustics, standing waves are caused by interference between waves of the same frequency in such a way that the combined intensity varies between maxima and minima over the region of interference. Also known as stationary wave.

stand oil [Materials] A drying oil that has been thickened by heating in an inert atmosphere (without the addition of driers). The thickening is due to polymerization of some of the constituents.

star [Astronomy] A celestial body consisting of large, self luminous mass of hot gas or gases held together by its own gravity, and that produces it energy by thermonuclear reactions, e.g., the sun.

star cloud [Astronomy] An aggregation of millions of stars spread over hundreds of light years.

star cluster [Astronomy] A group of stars held together by gravitational attraction; the two main types are open clusters and globular clusters.

stat-[Electricity] Prefix attached to the name of electrical units to indicate the corresponding electrostatic units, e.g., statcoulomb.

statcoulomb [Physics] The electrostatic unit of electric charge in the c.g.s. system. It is equal to 3.3356×10^{-10} coulomb.

static charge [Electricity] An electric charge accumulated on an object.

static electricity [Electricity] 1. Electricity characteristic of charges at rest, as opposed to current electricity. 2. The study of the effects of macroscopic charges, including the transfer of a static charge from one object to another by actual contact or by means of spark that bridges an air gap between objects.

statics [Mechanics] The branch of mechanics; the mathematical and physical study of the behaviour of matter under the action of forces, dealing with cases where no motion is produced, *i.e.*, of forces which act on bodies in equilibrium.

stationary orbit [Engineering] A circular, equitorial orbit in which the satellite revolves about the primary body at the angular rate at which the primary body roates about its axis, the satellite thus appears to be stationary over a point above the primary body; a stationary orbit must be synchronous, but the reverse need not to be true.

stationary state [Physics] A term used in quantum mechanics[e]; if only certain energy values or energy levels for the total energy of a system are permissible, the energy is said to be quantized. These levels are characteristic of the state of the system; such states are called stationary states. A transition from one stationary state to another state can occur only with the emission or absorption of energy in the form of photons; *i.e.*, electromagnetic radiation is emitted or absorbed.

stationary wave *See* standing wave.

statistical mechanics [Physics] The branch of physics which deals with the study of the mechanical properties of large assemblies of particles or components in terms of statistics, *e.g.*, the kinetic theory of gases treats the molecules of a gas in terms of statistical mechanics.

statistics [Mathematics] The collection and study of numerical facts or data and their interpretation in mathematical terms, with special reference to the theory of probability.

stator [Electricity] The fixed part of any electric motor or generator that contains the stationary magnetic ciruits and their associated windings.

steady state [Physics] The condition of a body or system in which the conditions at each point do not change with time, that is after initial transients or fluctuations have disappeared.

steady-state conduction [Physics] Heat conduction in which the temperature and heat flow at each point does not change with time.

steady-state current [Electricity] An electric current that does not change with time.

steady-state wave motion [Physics] Wave motion in which the wave quantities at each point in the region through which the wave is passing repeat themselves periodically.

steam bronze [Metallurgy] An alloy containing 88% copper, 6% tin, 4.5% zinc, 1.5% lead; used for making steam valve bodies, bearings, and gears.

steam distillation [Chemistry] A type of distillation in which vapourization of volatile substances takes place at a lower temperatures by introducing steam directly into the mixture; used for removal of essential oils from plant components and purifying organic substances that decompose near their boiling points; also known as hydrodistillation.

steam point [Physics] The temperature at which the maximum vapour pressure of water is equal to the standard atmospheric pressure (101325 Pa). On the Celsius scale it has the value 100ºC.

steel [Metallurgy] Iron containing from about 0.2% to 1.5% carbon in the form of cementite (iron carbide, Fe_3C). The properties of different steels vary according to the percentage of carbon and of metals other than iron present, and also according to the method of preparation. Steel is prepared by the open-hearth and Bessemer processes, and in electric-arc furnaces. *See also* stainless steel.

steel bronze [Metallurgy] A hardened bronze containing 92% copper and 8% tin; used as a substitute for ordinary steel in guns.

steelyard [Engineering] A weighing machine with a counterbalanced arm supporting the load to be weighed on the smaller end.

Stefan-Boltzmann law [Physics] The total energy radiated from a blackbody is proportional to the fourth power of the absolute temperature of the body. Also known as Stefan's law; Stefan's law of radiation.

stellar [Astronomy] Relating to or consisting of stars.

stellar evolution [Astronomy] The changes in the spectrum and luminosity that take place in the life of a star. It is thought that stars are born from condensation of gas (mostly hydrogen), which is compressed as a result of the gravitational field between the constituents. The compression is so high in the interior of the gas that thermonuclear reactions occur during which hydrogen is converted into helium (and possibly heavier elements) with the evolution of energy. On a Hertzprung-Russell diagram the stars remain on the 'main sequence' until they have consumed about 10% of their hydrogen. They then become red gaints and consume their hydrogen at increased rates so that eventually they contract and become white dwarfs.

stellar flare [Astronomy] Ejection of material from a star in an eruption that may last from a few minutes to an hour or more.

stellite [Metallurgy] A family of the hard, wear-resistant, and corrosion-reistant alloys of cobalt (20%-65%), chromium (11%-32%), tungsten (2%-5%), molybdenum (0%-40%), and iron (0%-5%); used for making surgical instruments and jet engine turbines.

step-up transformer [Electricity] A transformer used to convert low voltage current into current of higher voltage.

steradian [Physics] The supplementary SI unit of solid angle; the solid angle that encloses a surface on the sphere equal to the square of the radius.

stere [Mechanics] A metric unit of volume; 1 cubic metre.

stereo- [Physics] A perfix used to designate a three-dimensional characteristic.

stereo amplifier [Electricity] An audio frequency amplifier having two or more channels, as required for use in a stereo sound system.

stereochemistry [Chemistry] Chemistry involving consideration of the arrangement in space of the atoms in a molecule. If a molecule is considered as a three-dimensional entity in space, possibilities of stereoisomerism or space isomerism arise; thus, a molecule consisting of four different radicals or atoms attached to a central-carbon atom can exist in two distinct space arrangements, one being a mirror image of the other. Such isomerism is associated with optical activity.

stereo effect [Physics] Reproduction of sound in such a manner that the listner receives the sensation that as if individual sounds are coming from different sources, just as did the original sound reaching the stereo microphone system.

stereoisomerism [Chemistry] Isomerism caused by possibilities of different arrangement in three-dimensional space of the atoms within a molecule, resulting in two isomers that are mirror images of each other.

stereoregular [Chemistry] Having a regular arrangement in space of the atoms and groups within a molecule. *See* stereoregular rubber.

stereoregular rubber [Materials] Any of the group of synthetic rubbers manufactured by the solution polymerization process using special catalysts that control the stereoisomeric (*see* stereoisomerism) regularity of the products. These materials can therefore be made to resemble closely the structure of natural rubber. In *cis*-1,4-polyisoprene, the structure of natural rubber is substantially duplicated, and this elastomer can be used for many of the purposes that were the exclusive preserve of natural rubber. A similar product is *cis*-1, 4-polybutadiene, which is also used in place of natural rubber. *See also* ethylene-propylene rubber.

stereoscope [Optics] An optical device by which two-dimensional pictures are given the appearance of depth and solidity.

stereospecific [Chemistry] Having a particular arrangement in space of the atoms and groups within a molecule. *See also* tactic polymer.

steric hindrance [Chemistry] An effect in which a chemical reaction is slowed down or prevented because large groups on a reactant molecule hinder the approach of another reactant molecule.

sterlization [Biology] 1. The process of destroying all bacteria and other infectious organisms on and/or in an object. 2. Rendering a living being incapable of reproduction by radiation or chemical treatment or minor surgery.

steroid [Biochemistry] Any of a group of lipids derived from a saturated compound called cyclopentanoperhydrophenanthrene, which has a nucleus of four rings. Some of the most important steroid derivatives are the steroid alcohols, or sterols. Other steroids include the bile acids, which aid digestion of fats in the intestine; the sex hormones (androgens and oestrogens); and the corticosteroid hormones, produced by the adrenal cortex.

sterol [Biochemistry] Any of a group of steroid-based alcohols having a hydrocarbon side-chain of 8-10 carbon atoms. Sterols exist either as free sterols or as esters of fatty acids. Animal sterols *(zoosterols)* include cholesterol and lanosterol. The major plant sterol *(phytosterol)* is beta-sitosterol, while fungal sterols *(mycosterols)* include ergosterol.

stigma [Biology] A light sensitive pigmented spot found in the cells of some primitive animals.

stigmatism [Physiology] A condition of the refractive media of the eye in which rays of light from a point are accurately brought to a focus on the retina.

still [Engineering] A metal or glass apparatus used for the distillation of liquids.

stimulated emission *See* maser and laser.

stoichiometric [Chemistry] A compound is said to be stoichiometric when its component elements are present in the exact proportions represented by its chemical formula.

stochiometric mixture [Chemistry] A stoichiometric mixture is one that will yield on reaction a stoichiometric compound (*e.g.* two molecules of hydrogen and one molecule of oxygen consti-

tute a stoichiometric mixture because they yield exactly two molecules of water on combustion).

stoichiometry [Chemistry] Chemical study dealing with the composition of substances; more particulary with the determination of combining proportions or chemical equivalents.

stokes [Physics] St. A c.g.s. unit of kinematic viscosity equal to the ratio of the viscosity of a fluid in poises to its density in grams per cubic centimetre. $1 \, St = 10^{-4} \, m^2 s^{-1}$.

Stokes's law [Physics] A law that predicts the frictional force F on a spherical ball falling through a viscous medium under gravity. According to this law $F = 6\pi r \eta v$, where r is the radius of the ball, v is its velocity, and h is the viscosity of the medium.

storage battery *See* accumulator.

storage factor *See* Q.

storage ring [Engineering] A large evacuated toroidal ring forming part of some large particle accelerators. Particles from the accelerator are injected into the ring, around which they can be made to circulate for many months. In some devices two beams of particles circulate in opposite directions. At the intersections of these two beams very high collision energies occur, enabling interactions to be studied.

store [Computers] A part of the hardware of a computer in which information is stored.

S.T.P. [Scientific Techniques] Standard temperature and pressure, formerly known as *N.T.P.* (normal temperature and pressure). The standard conditions used as a basis for calculations involving quantities that vary with temperature and pressure. These conditions are used when comparing the properties of gases. They are 273.15K (or 0°C) and 101 325 Pa (or 760.0 mmHg).

straight chain molecule [Chemistry] A hydrocarbon molecule in which the carbon atoms are linked together in one long straight chain with no side chains attached.

strain [Mechanics] The resistance to deform, that is, the tendency to resume its original shape, of a material subjected to a static or dynamic force; it increases as a function of stress. When a body is deformed by an applied stress, the strain is the ratio of the dimensional change to the original or unstrained dimension. The strain may be a ratio of lengths, areas, or volumes.

strain hardening [Metallurgy] An increase in the hardness and tensile strength of a metal due to cold working, that cause a permanent alteration like (distortion of its crystalline structure. Also known as work hardening.

strand [Engineering] A number of steel wires twisted together to form a wire rope or cable or electrical conductor.

stratopause [Geology] The boundary between the stratosphere and the mesosphere.

stratosphere [Geology] A layer of the atmosphere beginning approximately 11 kilometres above the surface of the earth.

streamline [Mechanics] A streamline is a line in a fluid such that the tangent to it at every point is in the direction of the velocity of the fluid particle at that point, at the instant under consideration. When the motion of the fluid is such that, at any instant, continuous streamlines can be drawn through the whole length of its course, the fluid is said to be in streamline flow.

stress [Mechanics] The deformation undergone by a material when subjected to a definite force, and is equal to a force per unit area. When a stress is applied to a body (within its elastic limit) a corresponding strain is produced, and the ratio of stress to strain is a characteristic constant for that body.

stroboscope [Engineering] An instrument with the aid of which it is possible to view objects that are moving rapidly with a periodic motion (*see* period) and to *see* them as if they were at rest.

strong acid [Chemistry] An acid, such as nitric acid, that is completely dissociated into ions in solution, *i.e.*, whose degree of ionization is almost one.

strong base [Chemistry] A base with a high degree of dissociation in solution, such as sodium hydroxide, NaOH, *i.e.*, whose degree of ionization is almost one.

strong interaction [Physics] An interaction that occurs between hadrons. It occurs only at very short range (about 10^{-15} metre and is the force that holds the nucleons together in an atomic nucleus. The strong interaction is some 100 times stronger than the electromagnetic interaction at this short range. The force between hadrons (sometimes called an exchange force) can be visualized as the exchange of virtual mesons between the particles.

strontium [Chemistry] Sr. Element. At. No. 38. At.Wt. 87.62 m.p. 1042K; b.p. 1657K; a reactive metal, placed in group IIA of periodic table. Its compounds impart crimson colour to flame and it is used in fireworks. The radioisotope strontium-90 is present in the fall-out from nuclear explosions. It poses a health hazard as it has a relatively long half-life of 28 years and, owing to its chemical similarity to calcium, can become incorporated into bone.

strontium unit [Nucleonics] SU. A measure of the concentration of strontium-90 in an organic medium (*e.g.* milk, bone, soil, etc.) relative to the concentration of calcium in the same medium. 1 $SU = 10^{-12}$ curie of strontium-90 per gram of calcium.

structural formula [Chemistry] A chemical formula that in addition to showing the atoms present in a molecule, also gives an indication of its strucutre, *e.g.*, the structural formula of benzene (C_6H_6) is represented by the hexagonal benzene ring.

sub- [Scientific Techniques] Prefix denoting under, below. In chemistry it is used to indicate either that the element mentioned is present in a lower proportion than usual *e.g.*, sub-oxide, or that the compound is basic, *e.g.*, subacetate.

subatomic [Physics] Consisting of particles smaller than, or forming a part of the atom.

subcarrier [Electronics] A carrier wave that is used to modulate a second different carrier.

subcritical [Nucleonics] Said of a nuclear reactor in which the effective multiplication constant is less than unity, and in which the nuclear chain reaction is therefore not self-sustaining.

subgiant star [Astronomy] A member of the family of stars whose luminosity is intermediate between giant and the main sequence in the Hertzsprung-Russell diagram; spectral classes G and K mostly belong to this class.

sublate [Chemistry] The product collected by ion flotation.

sublimate [Chemistry] A solid obtained by the direct condensation of a vapourized solid without passing through the liquid state.

sublimation [Chemistry] The conversion of a solid direct into vapour, and subsequent condensation, without melting.

sublimation cooling [Physics] Cooling caused by the extraction of energy to produce sublimation.

sublimation energy [Chemistry] The increase in internal energy when 1 mole of a solid is converted into a gas at constant pressure and temperature.

subroutine [Computers] 1. A statement in FORTRAN used to define the beginning of a closed subroutine. 2. A body of computer instruction designed to be used by other routines to accomplish some specific job.

subshell [Physics] Electrons of an atom within the same energy level (shell) having the same azimuthal quantum numbers; designated by the letters s, p, d, and f. Also known as sublevel.

subsonic [Physics] Moving at, or relating to, a speed that is less than the acoustic velocity.

subsonic frequency [Physics] A frequency of value less than the audio-frequency range.

substance [Materials] Any chemical element or compound, occurring in macroscopic amounts.

substitution product [Chemistry] A compound obtained by replacing an atom or group by another atom or group in a molecule. The new atom or group is known as the 'substituent'.

substitution reaction [Chemistry] The replacement of one element or radical by another as a result of chemical reaction, $e.g.$, chlorination of benzene to produce chlorobenzene involves the replacement of one hydrogen atom by chlorine atom in the benzene molecule.

substrate [Chemistry] A substance whose reactivity is increased by a specific enzyme. [Metallurgy] Any solid surface on which a coating or layer of different material is deposited.

subtend [Mathematics] Two points, A and B, and said to subtend the angle ACB at the point C.

subtractive process [Optics] The process of producing colours by mixing three different dyes or pigments together. The final colour is produced by the absorption of different wavelengths of light.

sucrase See saccharase.

sucroclastic [Biochemistry] Sugar-splitting; applied to enzymes that have the power of hydrolyzing complex carbohydrates, $e.g.$, invertase.

sugar [Materials] In general, any sweet, water soluble, monosaccharide or disaccharide. The word is commonly applied to sucrose.

sulpha drug [Medicine] Any of a family of drugs of sulphonamide type having bacteriostatic properties.

sulphating [Electricity] The formation of an insoluble layer of lead sulphate on the electrodes of a lead accumulator, when it is not in use and is left discharged for any length of time.

sulphation [Chemistry] 1. The conversion of a compound into a sulphate by the oxidation of sulphide, e.g., oxidation of Na_2S, to sodium sulphate, Na_2SO_4. 2. The addition of sulphate group into a compound.

sulphofication [Biochemistry] Oxidation of sulphur and its compounds into sulphates in the soil by the bacterial action.

sulphonation [Chemistry] Substitution of-SO_2OH group or groups for hydrogen atoms in organic compounds; e.g., conversion of benzene, C_6H_6, into benzensulphonic acid, C_6H_5-SO_2OH.

sulphur [Chemistry] S. Element. At. No. 16. At.Wt. 32.064. A nonmetallic element, placed in group VI A of the periodic table; occurring in serveral allotropic forms. The stable form under ordinary condition is rhombic or alphasulphur, a pale-yellow brittle crystalline solid, r.d. 20.7. Sulphur occurs as the element in many volcanic regions and as sulphides of many metals; used in the manufacture of sulphuric acid, carbon disulphide, for vulcanizing rubber, in the manufacture of dyes, medicines, and various chemicals, used for killing moulds and insects.

sulphuration [Chemistry] Chemi-cal combination of an element or compound with sulphur.

sulhpur bacteria [Biology] Anaerobic bacteria that obtain the energy needed in metabolism by oxidation of sulphides or elemental sulphur, and build up carbohydrates from carbon dioxide.

sulphur point [Chemistry] The temperature of equilibrium between liquid sulphur and its vapour at a pressure of one standard atmosphere; equals 717.6K

sun [Astronomy] The incandescent, approximately spherical heavenly body around which the earth and other planets rotate in elliptical orbits. Mean distance from the earth is approximately 149.6×10^6 kilometres, and the distance to nearest star is approximately 40×10^{12} km. The diameter of the sun is about 1392000 km, its mass is approximately 2×10^{30} kilograms, and its average density is 1.4 grams per cm^3. The sun is composed of about 90% hydrogen, 8% helium, and only 2% of the heavier elements. The sun is held together by its own gravity; thermonuclear reactions take place in the deep interior of the sun converting hydrogen into helium, thus releasing a large amount of energy.

sunspot [Astronomy] A large patch which appears black by contrast with their surroundings, visible upon the surface of the sun.

Owing to the rotation of the sun, they appear to move across its surface; sunspots represent lower temperature and consisting of a dark central umbra surrounded by a penumbra which is intermediate in brightness between the umbra and the surrounding surface of the photosphere.

sunspot cycle [Astronomy] The variation of the size and number of sunspots in an 11-year cycle which is shared by all other forms of solar activity.

superalloy [Metallurgy] A thermally resistant iron-base, cobalt-base, or nickel-base alloy; it is oxidation resistant, and create thermal resistance to an unusual degree. Superalloys can withstand high stresses and chemical reaction with oxygen upto 2200°C. They are used in jet engine parts, turbo superchargers, and extreme high temperature applications.

superconductivity [Physics] The absence of measurable electrical resistance in certain substances at temperatures close to 0K. The theoretical explanation of the phenomenon was given by the *BCS theory*. According to this theory an electron moving through an elastic crystal lattice creates a slight distortion of the lattice as a result of Coulomb forces between the positively charged lattice and the negatively charged electron. If this distortion persists for a finite time it can affect a second passing electron. Effect of this phenomenon is for the current to be carried in superconductors not by individual electrons but by bound pairs of electrons, called *Cooper pairs*. The *BCS theory* is based on a wave function in which all the electrons are paired. Because the total momentum of a Cooper pair is unchanged by the interaction between one of its electrons and the lattice, the flow of electrons continues indefinitely. This phenomenon has been used to produce large magnetic fields without the expenditure of appreciable quantities of electrical energy.

supercooling [Physics] The metastable state of a liquid cooled below its freezing point. A supercooled liquid will usually freeze on the addition of a small particle of the solid substance, and often on the addition of any solid particle or even on shaking; the temperature then rises to the freezing point.

supercurrent [Electricity] In the two fluid models of superconductivity, the current resulting due to motion of super conducting electrons, in contrast to the normal current.

superexchange [Physics] A phenomenon in which two electrons from a di-negative anion (such as O^{-2}) in a solid go to different cations and couple with their spins, giving rise to a strong antiferromagnetic coupling between the cations, which are too far apart to have a direct exchange interaction.

superfluid [Physics] A fluid that flows without friction and has an abnormally high thermal conductivity. The particles of a superfluid are in the lowest energy state allowed by quantum mechanics, having zero entropy and zero resistance to motion, *e.g.*, helium below 2.186 K.

supergiant star [Astronomy] A star of exceptionally high luminosity, low density, and a diameter some hundreds of times greater than that of the sun.

superheating [Physics] Heating a liquid above its boiling point, when the liquid is in a metastable state. *See* supercooling.

superheterodyne [Electronics] Abbreviation of 'supersonic heterodyne'. A method of radio reception in which the frequency of the carrier wave is changed in the receiver to a 'supersonic' intermediate frequency (*i.e.* a, frequency above the audible limit for sound) by a heterodyne process. Also known as superhet.

super high frequency [Physics] S.H.F. Radio frequencies in the range 3000 to 30000 megahertz, corresponding to wavelengths from 1 to 10 centimetres.

superior conjunction [Astronomy] A conjunction when a celestial body is opposite the earth on the other side of the sun.

superior planet [Astronomy] Any of the planets that are farther than the earth from the sun, *i.e.*, Mars, Jupiter, Saturn, Uranus, Neptune, and Pluto.

supermalloy [Metallurgy] A soft magnetic alloy having very low hysteresis loss; consists of 79% nickel, 16% iron, and 5% molybdenum.

supermassive star [Astronomy] A star with a mass exceeding about 50 times that of the sun.

supernatant liquor [Chemistry] The liquid above settled solids.

supernova [Astronomy] The star that suddenly bursts into very great brilliance as a result of its blowing up. The explosions are believed to be caused when a star runs out of hydrogen and contracts under its own gravitational field. The contraction causes a sufficiently high temperature in the interior for thermonuclear reactions to occur, which produce heavy elements. The formation of heavy elements, with atomic number in excess of about 40, absorbs energy and the star collapses inwards, increasing its speed of rotation and ultimately flinging a large portion of its matter into space.

superoxide [Chemistry] A compound characterized by the presence of O_2^- ion in its structure, *e.g.*, sodium superoxide, NaO_2.

superphosphate [Materials] A commercial phosphate mixture consisting mainly of monocalcium phosphate. Single-superphosphate is made by treating phosphate rock with sulphuric acid; the product contains 16-20% 'available' P_2O_5:

$$Ca_{10}(PO_4)_6F_2 + 7H_2SO_4 \rightarrow$$
$$3Ca(H_2PO_4)_2 + 7CaSO_4 + 2HF.$$

Triple-superphosphate is made by using phosphoric (V) acid in place of sulphuric acid; the product contains 45-50% 'available' P_2O_5:

$$Ca_{10}(PO_4)_6F_2 + 14H_3PO_4 \rightarrow$$
$$10Ca(H_2PO_4)_2 + 2HF$$

supersaturation [Chemistry] The metastable state of a solution holding more dissolved solute than is required to saturate the solution.

supersonic [Physics] Moving at, or relating to the speeds greater than the speed of sound.

supplementary angles [Mathematics] Angles together totalling 180°, or two right angles.

suppressor grid [Electronics] A grid, placed between the screen grid and the anode of a thermionic valve, to reduce the secondary emission of electrons between them.

surd [Mathematics] An irrational quantity; a root that cannot be expressed as an exact number or fraction; e.g. 2^{-1}

surface-active agent [Chemistry] Any substance which when introduced into a liquid reduces surface tension in order to affect (usually to increase) its spreading, wetting, and similar properties. Many detergents fall into this class. Also known as surfactant.

surface barrier [Electronics] A potential barrier formed at a surface of a semiconductor by the trapping of carriers at the surface.

surface colour [Physics] Certain reflecting surfaces, e.g. metal surfaces, exhibit selective reflection of light waves; e.g., they reflect some wavelengths more readily than others. When illuminated by white light, such surfaces reflect light deficient in certain wavelengths, and the body appears coloured. The body is then said to show surface colour, as opposed to pigment colour. Bodies showing surface colour when viewed by transmitted light appear to be of the complementary colour to that observed when viewed by reflected light. Substances that show pigment colour appear of the same colour whether viewed by reflected or transmitted light.

surface tension [Mechanics] g. The property of a liquid that makes it behave as if its surface is enclosed in an elastic skin. The property results from intermolecular forces : a molecule in the interior of a liquid experiences a force of attraction from other molecules equally from all sides, whereas a molecule at the surface is only attracted by molecules below it in the liquid. The surface tension is defined as the force acting over the surface per unit length of surface perpendicular to the force. It is measured in newtons per metre. It can equally be defined as the energy required to increase the surface area by one square metre, *i.e.* it can be measured in joules per metre squared (which is equivalent to Nm^{-1}).

surface wave [Physics] A wave that can travel along an interface between two different mediums without radiation; the interface must be essentially straight in the direction of propagation; the commonest interface used is that between air and the surface of a circular wire.

surfactant *See* surface-active agent.

surge [Electricity] A momentary large increase in the current or voltage in an electric system.

susceptance [Electricity] B. The imaginary part of the admittance, Y, of a circuit which is given by $Y=C+iB$, where C is the conductance. For a circuit containing both resistance, R, and reactance X, the susceptance is given by $B=-X$ (R^2+X^2). It is the reciprocal of the reactance and is measured in siemens.

susceptibility, magnetic [Physics] X_m. The dimensionless quantity describing the contribution made by a substance when subjected to a magnetic field to the total magnetic flux density present. If is equal to $\mu_r - 1$, where μ_r is the relative permeability of the material. Diamagnetic materials have a low negative susceptibil-

ity, paramagnetic materials have a low positive susceptibility, and ferromagnetic materials have a high positive value. (susceptibility electric) X_e. The dimensionless quantity referring to a dielectric equal to P/e_oE, where P is the electric polarization, E is the electric intensity producing it, and e_o is the electric constant. The electric susceptibility is also equal to e_r-1, where e_r is the relative permittivity of the dielectric.

suspension [Chemistry] A two-phase system (*see* phase) consisting of very small solid particles distributed in a liquid dispersion medium.

swaging [Metallurgy] Tapering a rod or tube or reducing its diameter by any of methods such as forging, hammering, or squeezing.

sweep [Electronics] The steady movement of the electron beam across the screen of a cathode ray tube, producing a steady bright line when no signal is present.

sweet gas [Materials] A petroleum natural gas containing no corrosive substances like hydrogen sulphide and mercaptans.

sylvine [Material] KCl. Natural potassium chloride, usually containing 57% sodium chloride as an impurity; an important source of potassium compounds.

symbiosis [Biology] A relationship between two different types of organisms that live together for their mutual benefit, *e.g.*, the relationship between cellulose digesting bacteria and the herbivores whose alimentary tract they inhabit.

symbol [Chemistry] A letter or a combination of letters representing an atom of an element; *e.g.* O represents one atom of oxygen. Often loosely taken to mean the element in general, *e.g.*, Na represents an atom of sodium.

symmetry [Mathematics] The correspondence of parts of a figure with reference to a plane, line or point of symmetry. Thus, a circle is symmertical about any diameter; a sphere is symmertical about a plane of any great circle.

symmetry element [Physics] Some combination of rotations and reflections and translations which bring a crystal into a position that cannot be distinguished from its original condition. Also known as symmetry operation; symmetry transformation.

synapse [Biology] A junction between neurones by which nerve impulses are transferred within the nervous systems of animals. A synapse is usually formed between the axon of one neurone and the cell body or denrite of another.

synchro- [Scientific Techniques] Occurring or made to occur at the same time.

synchrocyclotron [Nucleonics] A type of cyclotron that enables relativistic velocities to be achieved by modulating the frequency of the accelerating electric field.

synchronism [Physics] Condition of two periodic quantities which have the same frequency, and whose phase difference is either constant or varies around a constant average value.

synchronization [Engineering] The maintianance of one operation in step with another, as in keeping the electron beam of a television picture tube in step with the electron beam of the television camera tube at the transmitter.

synchronous motor [Electricity] An alternating current electric motor whose speed of rotation is proportional to the frequency of its power supply.

synchronous orbit [Engineering] The orbit of an artificial earth satellite that has a period of 24 hours. The altitude corresponding to such an orbit is about 35 700 km; a satellite in a circular orbit parallel to the equator at this altitude would appear to be stationary in the sky. Communication satellites in synchronous orbits are used for relaying radio signals between widely separated points on the earth's surface.

synchrotron [Nucleonics] A device for accelerating electrons or protons in which the **magnetic** field is maintained at a **constant** frequency so as to keep **the orbit** radius constant.

synchrotron radiation [Physics] High energy electrons with in a synchrotron emit light as a consequence of their acceleration in a strong magnetic field; this emission is known as synchrotron radiation. The term is also used to describe the emission of radio frequency electromagnetic radiations from interstellar gas clouds in radio galaxies as this emission is believed to be an analogous phenomenon.

synecology [Biology] The study of the relationships between communities and their environments.

synersis [Chemistry] The contraction of a gel on standing, with exudation of liquid, *e.g.*, separation of blood serum from a blood clot.

synergism [Chemistry] A chemical phenomenon in which the effect of two active components of a mixture is more than the sum of their individual effects, *i.e.*, is greater than the equivalent volume or reactivity of either component alone.

synfuel [Materials] A term generally applied to fuels derived from coal gasification *e.g.*, shale oil.

syngamy [Biology] Sexual reproduction involving union of gametes in fertilization.

syngenetic [Geology] Said of a mineral deposit formed at the same proceses, as the enclosing rocks.

synodic month [Astronomy] A month based on the phases of moon.

synodic period [Astronomy] The time period between two successive astronomical conjunctions of the same celestial body.

syntax [Computers] The set of rules needed to construct valid expressions or sentences in a computer language.

synthesis [Chemistry] The formation of substance by the union of simpler elements or compounds by means of one or more chemical reactions. [Nucleonics] The formation of elements by a nuclear change. Synthesis of elements has occurred since 1940 and elements of atomic number from 93 to 106 as well as a vast variety of radioisotopes of natural elements have been synthesized by nuclear bombardment of various types.

synthesis gas [Chemistry] Any of the serveral gaseous mixtures used for synthesizing a wide range of compounds; for example, carbon monoxide and hydrogen to make hydrocarbons or other organic compounds, or hydrogen and nitrogen to make ammonia.

syntony [Electricity] Condition in which two oscillating circuits have the same resonating frequency.

syrup [Materials] Commercial name for an aqueous solution of cane or beet sugar.

systematics [Biology] The study that deals with the classification and diversity of organisms and their relationships.

systole [Physiology] The contraction phase of the heart cycle; the interval between the first and second heart sounds during which blood is forced into the aorta and pulmonary arteries.

syzgy [Astronomy] 1. One of the two points in a celestial object's orbit where it is in conjunction with or opposition to the sun. 2. Those points in the moon's orbit where the moon, earth, and sun are in a straight line.

szechtman cell [Chemistry] An electrolytic process for the manufacture of chlorine that is a variation of both the mercury cell and the molten salt cell.

T

TAB [Medicine] A vaccine used to produce active artificial immunity in man against typhoid and paratyphoid fever.

table sugar See sucrose.

tabulate [Computers] To order a set of data into a table form, or to print a set of data into a table.

tachometer [Engineering] An instrument for measuring the rate of revolution per minute or the angular speed of a revolving shaft.

tachymeter [Engineering] A surveying instrument designed for use in rapid determination of distance, direction, and differ-

ence of elevation from a single observation, using a short base, which may be an integral part of the instrument. Range-finders with self-contained bases belong to this class.

tachyon [Physics] A hypothetical particle that travels faster than the velocity of light. To satisfy the special theory of relativity such a particle would have imaginary energy and momentum if it had a real rest mass, or imaginary rest mass if the energy was real.

tacnode [Mathematics] A point at which two branches of a curve touch each other and have a common tangent.

tactic polymer [Materials] A polymer in which the groups attached to the polymer chain are regularly arranged, giving a sterospecific and a stereoregular structure.

tacticity [Chemistry] The regularity or symmetry in the molecular arrangement or structure of a polymer molecule.

tagged atom [Chemistry] A radioactive isotope used in tracing the behaviour of a substance in biochemical and engineering research.

talc [Minerals] $Mg_3Si_4O_{10}(OH)_2$. An extermely soft, light green or light grey monoclinic mineral. It has characteristic soapy touch and can be easily cut with knife; used as a filler, coating, and dusting agent in rubber, plastics, lubricant, in talcum powder and ceramics.

tall oil [Materials] A mixture of rosin acids, fatty acids, and other materials obtained by acid treatment of the alkaline liquors from the pulping of pine wood; used in soaps and paints.

tallow [Materials] Animal fat with carbon chains containing 16-18 carbon atoms, derived from solid fat or 'suet' of cattle, sheep, or horses by dry or wet rendering. It consists of various glycerides; used in soaps, leather dressings, candles, food, and greases.

tamp [Engineering] To tightly pack a drilled hole with clay or other stemming material after the charge has been placed.

tandem generator [Electricity] An accelerator of the electrostatic generator type. The name is derived from the fact that it consists essentially of two Van de Graff generators in series, thus enabling twice as much energy to be obtained for a given accelerating potential as could be obtained from a single machine. Negative ions are accelerated from ground potential, the electrons are then 'stripped' off and the positive particle accelerated back to ground potential.

tangent [Mathematics] A line is tangent to a curve at a fixed point P if it is the limiting position of a line passing through P and a variable point K on the curve, as K approaches α.

tangent galvanometer [Physics] A galvanometer consisting of a coil of wire (n turns of radius r) held in a vertical plane parallel to the earth's magnetic field, H, with a small magnetic needle pivoted at the centre of the coil that is free to rotate in a horizontal plane. A direct electric current, I, flowing through the coil produces a magnetic field at right angles to that of the earth. The needle takes up the direction of the resultant of these two fields; if θ is the angle of deflection of the needle from its equilibrium position parallel to the earth's field, then the current will be given by :

$$I = Hr \, \tan\theta / 2\pi n.$$

tannin [Biochemistry] One of a group of complex organic chemicals commonly found in leaves, unripe fruits, and the bark of trees. Their function is uncertain though the unpleasant taste may discourage grazing animals. Some tannins have commercial uses, notably in the production of leather and ink.

tanning [Engineering] A process of preserving animal hides by the conversion of raw animal hide into leather by the action of substances containing tannin, tannic acid, or other agents. This process makes the leather immune to bacterial attack; raises the shrinkage temperature; and prevents the collagen fibres from sticking together on dyeing, so that the material remains soft, porous, and flexible.

tantalum [Chemistry] Ta. Element At. No. 73. At.Wt. 180.948; m.p. 3269K; b.p. 5700K; placed in group V B of the periodic table. A greyish-white metal that is very ductile and malleable. It occurs together with niobium in a few rare minerals and is extracted by reduction of the oxide with carbon in an electric furnace; used for electric lamp filaments, in alloys, in cemented carbides for very hard tools, and in electrolytic rectifiers.

tantiron [Metallurgy] A ferrous alloy containing 84.87% iron, 13.5% silica, 1% carbon, 0.4% manganese, 0.15% phosphorus, and 0.5% sulphur. It is used for making chemical equipments because of its acid-resistant property.

tar [Materials] A viscous material of high molecular weight compounds; obtained from the distillation of petroleum or destructive distillation of coal or wood.

tar base [Chemistry] A basic nitrogen compound derived from coal tar, such as pyridine, picoline, and quinoline.

tare [Chemistry] The weight of a container or wrapper which is deducted in determining the net weight of a substance.

tarnish [Metallurgy] Discolouration of a metallic surface due to the formation of a thin film of oxide, sulphide, or some other corrosion product. It is easily removable with a cleaning compound and is not a true form of corrosion.

tau particle [Physics] A heavy lepton that reacts by the weak interaction. It has a very short lifetime (about 5×10^{-12} second) and a mass of approximately 1800 MeV (*i.e.*, about 3500 times heavier than an electron).

tautomerism [Chemistry] A type of isomerism in which a compound exists as a mixture of two isomers in equilibrium. The two forms are convertible one into the other, and removal of one of the forms from the mixture results in the conversion of part of the other to restore the equilibrium; but each of the two forms may give rise to a stable series of derivatives.

toxon [Biology] A named group of organisms of any rank, such as particular species, family, or class; also the name applied to that unit. A taxon may be designated by a formal Latin name or by a letter, number or other symbol.

taxonomy [Biology] The theory and practice classifying plants and animals. The term toxonomy and systematics are usually distinguished, the latter having broader connotation, but they may also be used more or less synonymously.

technetium [Chemistry] Tc. Element. At. No. 43; m.p. 2445K; b.p. 5150K; placed in group VII B of periodic table. The most stable, isotope, ^{97}Tc, has a half-life of 2.6×10^6 years. It is not found in nature but is formed as fission product of uranium; used to absorb slow neutrons in reactor technology.

tectonics [Geology] A branch of geology dealing with the broad architecture of the outer part of the earth, that is, the major structural or deformational features and their relations, origin, and historical evolution. It is closely related to structural geology which generally deals with larger features.

tectosilicate [Chemistry] A class of structural type of silicates characterized by the sharing of all four oxygen atoms of the SiO_4 tetrahedra with neighbouring tetrahedra, and with a Si:O ratio of 1:2; e.g., SiO_2, quartz.

teeming [Metallurgy] Pouring molten metal, usually a ferrous metal, into an ingot moild from a furnace.

telecommunications [Engineering] The study and application of means of transmitting information, either by wires or by electromagnetic radiation.

telemeter [Engineering] Any apparatus for recording a physical event at a distance, e.g., an instrument in an artificial satellite that tansmits measurements made in space back to earth by radio.

telephoto lens [Optics] A combination of a convex and a concave lens, used to replace the ordinary lens of a camera in order to magnify the normal image. The telephoto lens system increases the effective focal length without the necessity of increasing the distance between the film and the lens.

telescope [Optics] An optical instrument used for viewing magnified images of distant objects. In the refracting telescope the objective is a large convex lens that produces a small, bright real image; this is viewed through the eyepiece, which is another convex lens, serving to magnify the image. In the reflecting telescope a large concave mirror is used instead of the objective lens to produce real image which is then magnified by the eye-piece.

television [Electronics] The transmission of visible, moving images by electrical means. In 'closed circuit' television the transmission is ·by line; in 'broadcast' television it is by radio waves. In either case light waves are converted into electrical impulses by a television camera and reconverted into a picture on the screen of a cathode-ray tube in the receiver.

television satellite [Engineering] An orbiting satellite that relays television signals between ground stations.

telluric [Astronomy] Pertaining to the earth (as a planet), or the earth soil especially depths of the earth, e.g., as applied to natural electric fields or currents. [Chemistry] Referring to compounds of tellurium in VI oxidation state.

tellurium [Chemistry] Te. Element. At. No. 52. At. Wt. 127.6; m.p. 722.6K, b.p. 1263K; placed in group VI A of periodic table. A silvery-white brittle nonmetal, resembling sulphur in its chemical properties. It exists in several allotropic forms; used in alloys, for colouring glass, and in semiconductors.

temper [Metallurgy] 1. The normal carbon content of steel. 2. To annneal or toughen glass. 3. An alloy added to pure tin to make the finest powder. 4. To soften hard steel or cast iron by reheating to a temperature below the eutectiod temperature and quinching in oil or water.

temperate phage [Biology] A phage whose DNA may be incorporated into the host-cell genoms without being expressed. In contrast, a virulent phage destroys the host cell.

temperature [Physics] A property of an object which determines the thermal state and the direction of heat flow when the object is placed in thermal contact with another object; heat flows from higher temperature to lower temperature. The temperature of a body is a measure of its 'hotness', which can be defined as a property determining the rate at which heat will be transferred to or from it. Temperature is thus a measure of the kinetic energy of the molecules, atoms or ions of which matter is composed. The basic physical quantity, the thermodynamic temperature, is expressed in kelvins.

template [Biology] A macromolecular mold or pattern for the synthesis of an informational macromolecule.

temporal algebra [Mathamatics] An extension of Boolean algebra that allows time to be explicitly represented and manipulated.

temporary hardness [Chemistry] The portion of total hardness of water that can be removed by boiling whereby the soluble calcium and magnesium bicarbonates are precipitated as insoluble carbonates.

tenacity [Mechanics] Strength per unit weight of a filament, fibre, or wire, expressed as grams per denier or newtons per square metre.

tenderization [Biochemistry] Treatment of meats with ultraviolet radiation or with certain enzyme preparations to accelerate softening of the collagen fibres and thus reduce the time necessary to hang the meat.

tenorite [Mineral] A naturally occurring oxide of copper that consists of small black scales. It is found in volcanic regions and in copper veins.

tensile strength [Physics] The tensile (pulling) stress that has to be applied to a material to break it. It is measured as a force per unit area; *e.g.* newtons per square metre; dynes per square centimetre; pounds or tons per square inch.

tensimeter [Engineering] An instrument for measuring vapour pressure.

tensiometer [Engineering] 1. An apparatus for measuring the surface tension of a liquid. 2. An apparatus for measuring the tension in a wire, fibre, or beam. 3. An apparatus for measuring the moisture content of soil.

tensor [Mathematics] A magnitude or set of functions by which the components of a system are transformed from one system of coordinates to another; a quantity expressing the ratio in which the length of a vector is increased.

tephra [Geology] A collective term for all clastic materials ejected from a volcano and transported through air. It includes volcanic dust, ash ciners, scoria, pumice, boms and blocks.

tera- [Scientific Techniques] T. A prefix used in the metric system to denote one million million times. For example 10^{12} volts = 1 teravolt (TV).

terbium [Chemistry] Tb. Element. At. Wt. 158.924. At. No. 65; m.p. 1633K; b.p. 3314K; placed in group IIIB of periodic table; a rare-earth element of yttrium subgroup.

terminal [Electricity] The point at which an electrical connection is made; the point, or the connecting device, at which current enters or leaves a piece of electric equipment. [Computers] An input or output device connected to a computer; it may be a line printer, a punched-card reader, teletype, or a visual-display unit.

terminal velocity [Mechanics] If a body free to move in a resisting medium is acted upon by a constant force such as a body falling under the force of gravity through the atmosphere, the body accelerates until a certain terminal velocity is reached, after which the velocity remains constant.

terminal voltage [Electricity] The voltage at the terminals connected to the source of electricity or an electric machine.

terminating [Electricity] Closing of a circuit at either end of a link or transducer by connecting some device thereto.

termination codons [Biochemistry] The three codons UAA, UAG, and UAG, which signal termination of a polypeptide chain.

termination factors [Biochemistry] Protein factors of the cytosol; required in releasing a completed polypeptide chain from the ribosome.

termination sequeuce [Biochemistry] A DNA sequence that appears at the end of a transcriptional unit and signals the end of transcription.

terminator [Astronomy] The line on the surface of the moon, or a planet, that separates the dark and light hemispheres.

termolecular reaction [Chemistry] A chemical reaction in which there are three reactant molecules, e.g., $2H_2 + O_2 \rightarrow 2H_2O$.

ternary compound [Chemistry] A chemical compound consisting of three elements, e.g., phosphoric acid, H_3PO_4.

ternary fission [Nucleonics] A very rare form of nuclear fission as a result of which a heavy nucleus breaks up into three fragments of comparable mass. The term is also used for the more frequent case in which one of the three fragments (e.g. an alphaparticle) is much lighter than the others.

terpene [Chemistry] Any of the class of hydrocarbons composed of recurring isoprene units and occurring in many fragrant essential oils of plants. They are colourless liquids, generally with a pleasant smell. Terpenes include pinene, $C_{10}H_{16}$, the chief ingredient of turpentine; and limonene, $C_{10}H_{16}$, found in the essential oils of oranges and lemons.

terpolymer [Chemistry] A polymer made from three monomers.

terrestrial [Scientific Techniques] Of or pertaining to the earth.

terrestrial gravitation [Physics] The effect of gravitational attraction of the earth.

terrestrial guidance [Engineering] A method of missile or rocket guidance in which the missile steers itself with reference to the strength and direction of the earth's gravitational or magnetic field (magnetic guidance).

terrestrial telescope [Optics] A telescope for use on land or sea, as opposed to an astronomical telescope.

tertiary alcohol [Chemistry] A trisubstituted alcohol in which the hydroxyl group is attached to a carbon atom that is joined to three other carbon atoms, e.g., tert-butyl alcohol, $(CH_3)_3C\text{-}OH$.

tertiary amine [Chemistry] A trisubstituted amine in which three alkyl or aryl groups are attached to nitrogen, e.g., trimethyl miline, $(CH_3)_3N$.

tertiary colour [Optics] A colour obtained by mixing two secondary colours, e.g., brown and grey.

tertiary structure of a protein [Biochemistry] The three dimensional confirmation of the polypetide chain of a globular protein in its native folded state.

tervalent [Chemistry] Trivalent, having a valence of three.

tesla [Physics] T. The SI unit of magnetic flux density equal to one weber of magnetic flux per square metre, i.e. $1\ T = 1\ Wbm^{-2}$.

Tesala coil [Electricity] A transformer for producing high voltages at high frequencies, consisting of a coil the primary circuit of which has a small number of turns includes a spark gap and a fixed capacitor. The secondary winding has a large number of turns and the secondary circuit is tuned by means of a variable capacitor, to resonate with the primary circuit.

testosterone [Biochemistry] An androgen secreted by the testes under the influence of the luteinizing hormone. Its secretion during adult life is responsible for the development, function and maintainence of secondary male sexual characteristics, male sex organs, and spermatogenesis. It is also secreted from the adrenal cortex and the ovaries.

test vector [Electronics] The specification of a set of stimuli of a set of stimulie to the input pins of a chip and the expected response at the output pins.

tetramer [Chemistry] An oligomer whose molecule is composed of four molecules of the same chemical compositions.

tetrode [Electronics] A thermionic valve containing four electrodes; a cathode, an anode or plate, a control grid, and (between the two latter) a screen grid.

thallium [Chemistry] Tl. Element. At. No. 81. At.Wt. 204.37; m.p. 576.5K; b.p. 1733K; a white malleable metal placed in group III A of periodic table; used in alloys; its salts are used in insecticides and as rat poison.

theodolite [Engineering] An optical instrument used for the measurement of angles in surveying. It consists essentially of a telescope moving along a circular scale graduated in degrees.

theorem [Scientific Techniques] A statement or proposition that is proved by logical reasoning from given facts and justifiable assumptions.

theory [Scientific Techniques] An attempt to explain a certain class of phenomena by deducing them as necessary consequences of other phenomena regarded as more primitive and less in need of explanation.

theory of games [Mathematics] A mathematical treatment of competitive games with special reference to the strategic and tractical decisions that have to be made in situations involving conflicting interests in the light of specific odds and probabilites. The theory is extended for use in military and commercial situations.

therm [Physics] A particle unit of quantity of heat; equal to 100000 British thermal units, 25200000 calories, or 1.05506×10^8 joules.

thermal [Physics] Pertaining to or caused by heat.

thermal analysis [Chemistry] Any analysis of physical or thermodynamical properties of substances in which heat is directly involved; for example, boiling, freezing, solidification point determinations.

thermal barrier [Physics] The limit to the speed with which an aircraft or rocket can travel in the earth's atmosphere due to overheating caused by friction with the atmospheric molecules.

thermal battery [Electricity] A voltage source consisting of a number of bimetallic junctions connected to produce a voltage when heated in a flame.

thermal black [Materials] Carbon made from natural gas by the thermatomic process. Production by pyrolysis of bituminous coal is an alternative method.

thermal capacity [Physics] The quantity of heat required to raise the temperature of unit mass of a body through 1°C; is equal to the product of mass, temperature change, and specific heat.

thermal conductivity [Physics] The heat flow across a surface per unit time, divided by the negative of the rate of change of temperature with the distance in a direction perpendicular to the surface.

thermal converter [Electricity] A device that converts heat energy directly into electrical energy using Seebeck effect. Also known as thermoelement.

thermal cracking [Chemistry] A process involving decomposition, rearrangement, or sometimes recombination of hydrocarbons by the application of heat in absence of any catalyst.

thermal cross-section [Nucleonics] A nuclear cross-section as measured with thermal neutrons.

thermal diffusion [Chemistry] If a temperature gradient is maintained over certain volume of gas containing molecules of different masses, the heavier molecules tend to diffuse down the temperature gradient, and the lighter molecules in the opposite direction. This forms the basis of a method of separating the different isotopes of an element in certain cases.

thermal energy [Nucleonics] Energy which is characteristic for thermal neutrons at room temperature, about 0.025 electronvolt. *See also* thermal neutrons.

thermal equilibrium [Physics] The state of a system in which there is no net flow of heat between its components.

thermal gradient [Physics] The rate of change of temperature with distance.

thermal inductance [Physics] The product of temperature difference and time divided by entropy flow.

thermal instability [Mechanics] The instability resulting in free convection in a fluid heated at a boundary.

thermalize [Nucleonics] To bring neutrons into thermal equilibrium with their surroundings; to reduce the energy of neutrons with a moderator; to produce thermal neutrons.

thermal neutrons [Nucleonics]

Neutrons of very slow speed and consequently of low energy. Their energy is of the same order as the thermal energy of the atoms or molecules of the substance through which they are passing, *i.e.*, about 0.025 electron-volt, which is equivalent to an average velocity of about 2200 metres per second. Thermal neutrons are responsible for numerous types of nuclear reactions, including nuclear fission. Also known as slow neutron.

thermal pollution [Biology] Heat introduced into rivers or estuaries by power plants or other industrial cooling waters and chemical wastes, which has adverse effect on estuarine ecology.

thermal reactor [Nucleonics] A nuclear reactor in which most of the nuclear fissions are caused by thermal neutrons. [Chemistry] A device or vessel in which chemical reactions take place because of heat; for example, thermal cracking, thermal reforming.

thermal reforming [Chemistry] A petroleum refining process using heat (but no catalyst) to effect molecular rearrangement of a low-octane naphtha to form high-octane gasoline.

thermal relay [Electricity] A relay operated by the heat produced by the current flow.

thermal resistivity [Physics] The reciprocal of thermal conductivity.

thermal spike [Nucleonics] The zone of high temperature briefly produced in a substance along the path of high energy particle or nuclear fission fragment.

thermal wave [Physics] A sound wave in a solid which has a short wavelength.

thermatomic process [Chemistry] The production of amorphous carbon and hydrogen by cracking methane or natural gas over hot bricks at a temperature of about 1143K.

thermie [Physics] A unit of heat energy equal to the heat needed to raise the temperature of 1 tonne of water from 14.5°C to 15.5°C, at a constant pressure of 1 atmosphere; equal to $(4.1855 \pm 0.0005) \times 10^6$ joules.

thermion [Electronics] An ion emitted by a hot body; as by the hot cathode of thermionic tube.

thermionic emission [Electronics] The emission of electrons from a heated metal, especially in thermionic valves.

thermistor [Electronics] A semiconductor, the electrical resistance of which decreases rapidly with increase of temperature; *e.g.*, the resistance may be of the order of 10^5 ohms at 20°C and only 10 ohms at 100°C; used as a sensitive temperature-measuring device.

thermite [Chemistry] A stoichiometric powdered mixture of iron(III) oxide and aluminium for the reaction :

$$2Al + Fe_2O_3 \rightarrow Al_2O_3 + 2Fe$$

The reaction is highly exothermic and the increase in temperature is sufficient to melt the iron produced. It has been used for localized welding of steel objects (*e.g.* railway lines) in the *Thermit process*.

thermochemistry [Chemistry] The branch of physical chemistry dealing with the quantities of heat absorbed or evolved during chemical reactions.

thermocouple [Engineering] A device for the measurement of temperature; consists of two wires of different metals joined at each end. One junction is at the point where the temperature is to be measured and the other is kept at a lower fixed temperature. Owing to this difference of temperature of the junctions, a thermoeletric e.m.f. is generated, causing an electric current to flow in the circuit. This current can be measured by means of a galvanometer in the circuit, or the thermoelectric e.m.f. can be measured using a potentiometer.

thermodynamic process [Physics] The change of any property of a collection of matter and energy accompanied by thermal effects.

thermodynamic property [Physics] A quantity which is either an attribute of an entire system or is a function of position which is continuous and does not vary rapdily over microscopic dis-

tances, except possibly for abrupt changes at boundaries between phases of the system, *e.g.*, concentration, pressure, temperature, volume, surface tension, and viscosity. Also known as macroscopic property.

thermodynamics [Physics] The study of the laws that govern the conversion of energy from one form to another, the direction in which heat will flow, and the availability of energy to do work.

thermodynamics, laws of *See* laws of thermodynamics.

thermodynamic temperature [Physics] The thermodynamic temperature is a basic physical quantity that depends on the concept of temperature as a measure of the thermal energy of random motion of the particles of a system in thermal equilibrium. Originally, thermodynamic temperature was defined in terms of the ice point and steam point of water using a gas thermometer. However, this was replaced by a definition using only one fixed point, the triple point of water, which was fixed as 273.15 Kelvins exactly. The magnitude of the unit of thermodynamic temperature, the kelvin, is the same as the degree on the International Practical Scale of Temperature.

thermoelectric effect *See* Seebeck effect.

thermoelectricity [Physics] Electricity produced by the direct conversion of heat energy into electrical energy.

thermofor [Materials] A heat transfer medium.

thermoforming [Engineering] Forming or shaping a thermoplastic sheet by heating above its melting point and then pulling it down onto a mold surface to shape it.

thermograph [Engineering] A self-registering thermometer; an apparatus that records temperature variations during a period of time on a graph.

thermographic analysis [Chemistry] Any of the various methods of chemical analysis based on recording changes of mass (thermogravimetric analysis) due to decomposition, or of temperature ("heating curves") due to endothermic or exothermic processes, when substances that undergo chemical changes on heating are heated at a definite rate.

thermoluminescen [Physics] Luminescence produced in a solid when its temperature is raised. It arises when free electrons and holes, trapped in a solid as a result of exposure to ionizing radiation, unite and emit photons of light.

thermometric analysis [Chemistry] A method of determination of the transformations of a substance undergoes while being cooled or heated at an essentially constant rate.

thermo-milliammeter [Electricity] An instrument for measuring small alternaing electric currents. The current passes through a wire made of constantan or platinum, which is in contact with or very close to a thermocouple. The thermocouple is connected to a sensitive milliammeter, the heat of the constantan wire producing a thermoelectric current in the thermocouple; this current is recorded by the milliammeter.

thermonuclear [Nucleonics] Refering to any process in which a very high temperature is used to bring about the fusion of light nuclei, with the liberation of energy.

thermonuclear device [Nucleo A device such as fusion bo used for peaceful purpos experiments, or tests.

thermonuclear reaction [Nucleonics] A nuclear fusion reaction in which the interacting particles or nuclei possess sufficient kinetic energy, as a result of their thermal agitation, to initiate and sustain the process.

thermophilic [Biology] Describing micro organisms that require high temperatures (around 60°C) for growth. It is exhibited by certain bacteria that grow in compost and manure.

thermopile [Engineering] An array of thermocouples connected either in series to give higher voltage output or in parallel to give higher current output; used for measuring temperature or radiant energy, or for converting radiant energy into electric energy.

thermoplastic [Chemistry] A high polymer that becomes plastic on being heated; a plastic material that can be repeatedly melted or softened by heat without change of properties.

thermoset [Chemistry] A high polymer that solidifies or 'sets' irreversibly when heated. This property is usually associated with a cross-linking reaction of the molecular constituents induced by heat or radiation, as with proteins, and in the baking of doughs.

thermosphere [Geology] The region of the upper atmosphere in which the temperature increases with altitude with height starting at 70 or 80 kilometres.

thermostat [Engineering] An instrument for maintaining a constant temperature by the use of a device that cuts off the supply of heat when the required temperature is exceeded and automatically restores the supply when the temperture falls below that required. It usually consists of a bimetallic strip so arranged that when it is heated (or cooled) the power supply contacts are opened (or closed).

thia [Chemistry] Prefix indicating the presence of sulphur in a heterocyclic ring.

thickening agent [Materials] Any

of a variety of hydrophilic substances used to increase the viscosity of liquid mixtures and solutions to aid in maintaining stability by their emulsifying properties, *e.g.*, starch, gums, silicates; etc.

thin film [Electronics] A film few molecules thick deposited on a glass, ceramic, or semiconductor material to form a capacitor, resistor, coil, cryotron, or any other circuit component.

thin-layer chromatography [Chemistry] A technique for the analysis of liquid mixtures using chromatography. The stationary phase is a thin layer of an absorbing solid (*e.g.* alumina) prepared by spreading a slurry of the solid on a plate (usually glass) and drying it in an oven. A spot of the mixture to be analysed is placed near one edge and the plate is placed upright in a solvent. The solvent rises through the layer by capillary action carrying the components up the plate at different rates (depending on the extent to which they are adbsorbed by the solid). After a given time, the plate is dried and the location of spots noted. It is possible to identify constituents of the mixture by the distance moved in a given time. The technique needs careful control of the thickness of the layer and of the temperature.

thio- [Chemistry] Prefix denoting sulphur, in the naming of chemical compounds.

thioester [Chemistry] An ester of a carboxylic acid with a thiol or mercaptan.

thio ethers [Chemistry] A group of compounds with the general formula RSR', where R and R' are hydrocarbon radicals.

thiokol [Chemistry] Trademark for rubber-like polymer materials of the general formula $(RSx)n$, where R is an organic bivalent radical and x is usually between 2 and 4. They are very resistant to the swelling action of oils, and undergo a form of vulcanization on being heated with certain metallic oxides.

thiolate [Chemistry] Metallic salts of thiols, formerly known as "mercaptides"; sulphur analogues of alcoholates.

thiol [Chemistry] Any of the class of organic compounds of the general formula RSH, with sulphur attached directly to carbon; they are the sulphur analogues of alcohols, containing SH instead of OH groups. Formerly called mercaptans.

third harmonic [Physics] A sine-wave component having three times the fundamental frequency of a complex signal.

third law of thermodynamics *See* laws of thermodynamics.

third order reaction [Chemistry] A chemical reaction in which the rate of reaction is determined by the concentration of three reactants, *e.g.*, A+B+C → D+E.

thixotropy [Chemistry] The property of certain colloidal substances, to weaken or change from a gel to a sol when disturbed but to increase in strength upon standing. The rate of change of viscosity with time, certain liquids, *e.g.* some paints, possess the property of increasing in viscosity with the passage of time when the liquid is left undisturbed. On shaking, the viscosity returns to its original value.

Thomson effect [Physics] A thermo-electric effect in which heat flows into or out of a homogeneous conductor when an electric current flows between two points in the conductor at different temperatures; the direction of the heat flow depends upon whether the current flows from colder to warmer conductor or vice-versa.

Thomson voltage [Physics] The voltage that exists between two points that are at different temperatures in a conductor.

thoride [Nucleonics] Any of the natural radioisotopes that occurs in the radioactive series containing thorium.

thorite [Minerals] $ThSiO_4$. A mineral consisting of thorium silicate; used as a source of thorium.

thorium [Chemistry] Th. Element. At. No. 90. At.Wt. 232.038; m.p. 2023K; b.p. 5063K; a dark grey radioactive and metallic element of the actinide series, placed in group III B of periodic table. The most stable isotope, ^{232}Th, has a half-life of 1.4×10^{10} years. Its compounds occur in monazite and thorite; used in alloys and as a source of nuclear energy.

thorium decay series [Nucleonics] The series of radioactive elements produced as successive intermediate products when thorium undergoes its spontaneous natural radioactive disintegration into lead. Many of these are severe radioactive poisons when ingested or inhaled in the form of thorium dust particles .

thoron [Chemistry] A gaseous radio-isotope of radon, ^{86}Rn, produced by the disintegration of thorium; half-life 51.5 seconds.

three-phase circuit [Electricity] A circuit energized by alternating-current voltages that differ in phase by one third of a cycle or 120°.

three-phase current [Electricity] Current delivered through three wires, with each wire serving as the return for the other two and with the three current components differing in phase successively by one-third cycle, or 120 electrical degrees.

threshold [Physics] The lowest value of any stimulus, signal, or agency that will produce a specified effect, *e.g.*, threshold frequency.

threshold frequency [Physics] Light incident on a metal surface will give rise to the emission of electrons only if the frequency of the light is greater than a certain threshold value, which is characteristic of the metal used.

threshold voltage [Electronics] The gate-to-source voltage of an FET that causes current to stop (or start) flowing from the drain to the source.

thrombin [Biochemistry] An enzyme formed in the blood of vertebrates that acts upon fibrogen to form fibrin; it is therefore essential to the process of blood clotting. Thrombin is formed from a blood protein, prothrombin.

thrombocytes [Biochemistry] One of the minute protoplasmic discs found in vertebrate blood. Also known as blood platelet; platelet.

thrust [Mechanics] The propulsive force produced by a reaction propulsion motor *i.e.*, the force exerted in any direction by a fluid engine or a rocket engine.

thulium [Chemistry] Tm. Element. At. No. 69. At.Wt. 168.934; m.p. 1818K; b.p. 1998K; a rare-earth element of the lanthanide group, placed in group III B of periodic table.

thyratron [Electronics] A gasfilled thermionic valve (usually a troide) in which a voltage applied to the control grid initiates, but does not limit the anode current; used as an electronic switch.

tides [Astronomy] The regular rise and fall of the water level in the earth's oceans as a result of the gravitational forces between the earth, moon, and sun. The forces involved are complex, but the moon is approximately twice as effective as the sun in causing tides.

time [Physics] The dimension of the physical universe which, at a given place, orders the sequence of events.

time delay [Physics] The time required for a signal to travel between two points in a circuit or for a wave to travel between two points in space.

time delay circuit [Electronics] A circuit in which the output signal is delayed by a specified time interval with respect to the input signal. Also known as delay circuit.

timer [Electronics] A circuit used in electronic navigation systems to start pulse transmission and synchronize it with other actions such as, the start of a cathode ray sweep. [Engineering] A device for automatically starting or stopping a machine or other device.

time-sharing [Computers] The simultaneous utilization of computer system from multiple terminals.

time standard [Physics] A recurring phenomenon, used as a reference for estabilishing a unit of time; the presently accepted standard is the second, defined as 9,192,631,770 transitions between two specified hyperfine levels of the ceasium-133 atom.

timing diagram [Electronics] A diagram that shows the time relationships between the various input and output signals of a circuit.

tin [Chemistry] Sn. (Stannum.) Element At. No. 50. At.Wt. 118.69; m.p. 505K, b.p. 2543K; a silvery-white metallic element placed in group IVA of periodic table; is soft, malleable, and ductile and is unaffected by air or water at ordinary temperatures; occurs in two allotropic forms, white tin, the normal form of the metal, which below 286.3K passes into the powdery form known as grey tin. This causes tin plaque but can be prevented by the addition of small amounts of antimony or bismuth; used for tin-plating and in many alloys.

tincal [Materials] An impure form of sodium tetraborate (borax).

tincture [Materials] An alcoholic extract or solution of a drug or chemical; less volatile than spirits.

tinning [Metallurgy] Covering or protecting a substance with a thin layer of tin.

tin pest [Metallurgy] Transformation of tin to a brittle, grey variety occurring spontaneously at temperatures below 273.15K.

tin plaque [Metallurgy] An allotropic change (*see* allotropy) in which white tin changes into a grey powdery form at 286.3K.

tin plate [Materials] Iron coated with a thin layer of tin, by dipping it into the molten metal.

tin sweat [Metallurgy] Exudation of tin rich low melting material from a tin-bronze surface as a result of inverse segregation in bronze casting, or overheating of the alloy.

tintometer [Optics] An instrument for comparing the colour of solutions with a series of standard solutions on stained glass slides.

tints [Optics] Colours that have the same hue but different saturation.

tissue [Biology] A collection of similar cells and intercellular material, which forms the structural material of a plant or animal.

tissue culture [Biology] The process of growing of fragments of the tissues or cells of organisms for biochemical examination in vitro. Tissue cultures are usually maintained in correctly balanced physiological medium.

titanium [Chemistry] Ti. Element At. No.22 At.Wt. 47.90; m.p. 1933K,b.p. 3560K; malleable and ductile metal of 3d-transition element series; placed in group IVB of periodic table. Titanium is widely used where strong and light alloys are required, as in aircraft, missiles, etc.

titre [Chemistry] 1. The least amount or volume needed to give a desired result in titration 2. The concentration in solution of a dissolved substance as shown by titration.

titrant [Chemistry] A standard solution of known concentration and composition used in analytical titrations.

titration [Chemistry] Any of a number of methods for analyzing the composition of solution by adding known amounts of a standardized solution until the desired reaction has reached the desired stage.

tokamak [Nucleonics] A Russian term for an assembly for producing nuclear fusion. It consists of a doughne t-shaped evacuated chamber called a torus, through which the plasma moves. The torous is surrounded by a powerful electromagnetic field to confine the energized plasma sufficiently to achieve the required density of 10^{14} particles per cubic centimetre per second and a temperature above 44 million K.

tomography [Medicine] The use of X-rays to photograph a selected plane of a human body with other planes eliminated. The CAT (computerzied axial tomography) scanner is a ring-shaped X-ray machine that rotates through 180° around the horizontal patient, making numerous X-ray measurements every few degrees. The vast amount of information acquired is built into a three-dimensional image of the tissues under examination by the scanner's own computer.

tone [Physcis] A sound oscillation capable of exciting an auditory sensation having pitch.

toner [Chemistry] An organic pigment which does not contain inorganic pigment or inorganic carrying base.

tonometer [Engineering] 1. An instrument for measuring the pitch of a sound, usually consisting of a set of calibrated tuning forks. 2. An instrument for measuring vapour pressure. 3. An instrument for measuring blood pressure, or the pressure within an eye-ball.

topaz [Minerals] A crystalline mineral, consisting of aluminium fluosilicate.

topoisomerazes [Biochemistry] Enzymes capable of positive or negative supercoiling of duplex DNA circles.

topology [Mathematics] A branch of geometry concerned with the way in which figures are 'connected', rather than with their shape or size. Topology is thus concerned with the geometerical factors that remain unchanged when an object undergoes a continuous deformation (*e.g.* by bending, stretching, or twisting) without tearing or breaking.

torque [Mechanics] A force, moment of a force, or system of forces that tends to produce rotation.

tor. [Mechanics] A unit of pressure used in the field of high vacuum; equivalent to 1 mm of mercury or equal to 133.322 pascals

Torricellain vacuum [Physics] The space, containing mercury vapour, that is produced at the top of a column of mercury when a long tube sealeᴜ at one end is filled with mercury. The mercury sinks inverted in a trough in the tube until it is balanced by the atmospheric pressure.

torsion [Mechanics] A twisting deformation about an axis, produced by the action of two opposing couples acting in parallel planes.

torus [Mathematics] A 'doughnut' or anchor-ring shaped solid of circular or elliptical cross-section. If the cross-section is a circle of radius a, and the ring has a radius b, the volume of the torus is $2\pi^2 a^2 b$.

total internal reflection [Optics] A phenomenon in which light passes from one medium to another that is optically less dense, *e.g.*,. from glass to air, the ray is bent away from the normal. If the incident ray meets surface at such an angle that the refracted ray must be bent away at an angle of more than 90°, the light cannot emerge at all, and is totally internally reflected.

totality [Astronomy] The period in

a total eclipse of the sun, during which the bright surface of the sun is totally obscured from view on earth by the moon.

tourmaline [Minerals] A class of natural crystalline minerals, consisting of silicates of various metals and containing boron. The crystals show some interesting pyroelectric, piezoelectric and optical effects.

toxicity [Biochemistry] The ability of a substance to cause damage to living tissues, impairment of central nervous system, severe illness, or in some cases, death, when indigested, inhaled, or absorbed by the skin.

toxicology [Medicine] The study of poisonous substances including their nature, effects, and detection, and methods of treatment.

toxin [Biochemistry] The name is generally confined to intensely poisonous substances produced by certain bacteria, which cause dangerous effects when they attack food or the human body; includes phytotoxine and zootoxins.

trace analysis [Chemistry] Analysis of very small quantity of a substance by sensitive techniques such polarography or spectroscopy.

trace element [Biochemistry] An element required in very small quantities by an organism. Such elements often form essential constituents of enzymes, vitamins, or hormones. Also known as guest element; microelement.

tracer [Chemistry] A radioactive substance, that is mixed with or attached to a given substance so the distribution or location of the latter can be determined. Also known as tracer element.

tracer element *See* tracer.

trajectory [Mechanics] The curve described by an object moving through space, as of a meteor through the atmosphere, a projectile fired from a gun.

tranquillizer [Medicine] A drug used to reduce tension and anxiety, without impairing alertness or causing drowsiness.

transaction [Computers] General description of updating data relevent to any item.

transamination [Chemistry] A biochemical reaction in amino acid metabolism in which an amine group is transferred from an amino acid to a keto acid to form a new amino acid and keto acid. The coenzyme required for this reaction is pyridoxal phosphate.

transcendental [Mathematics] 1.(Of a number or quantity) Not capable of being expressed as the root of an algebraic equation with rational coefficients, *e.g.* x or e. 2. (Of a function). Not capable of being expressed by a finite number of algebraic operations, *e.g.*, $\sin x$, e^x. (*See* exponential.)

transconductance [Electronics] The mutual conductance between the control grid of a thermionic valve and its anode; it is usually expressed in siemens.

transocibe [Computers] To copy with or without translating from one external computer storage medium to another.

transducer [Electricity] A device for converting a nonelectrical signal, such as sound, light, heat, etc, into an electrical signal, or vice versa. Thus microphones and loudspeakers are electroacoustic transducers. An active transducer · is one that can itself introduce a power gain and has its own power source. A passive transducer has no power source other than the actuating signal and cannot introduce gain.

transduction [Biology] Transfer of genetic material from one cell to another cell by means of a viral factor.

transference number [Chemistry] The portion of the total electrical current carried by any ionic species in a fluid state electrolyte. Also known an transport number.

transfer orbit [Astronomy] In interplanetary travel, an elliptical trajectory tangent to the orbits of both the departure planet and target planet.

transfer RNA [Biochemistry] A class of RNA molecules (Mol. Wt. 25,000-30,000, each of which combines covalently with a specific amino acid as the first step in protein synthesis . Also known as *t*RNA.

transfer time [Computers] The time required to move an information element from memory to processor.

trans-form *See cis-trans* - Isomerism

transformation [Nucleonics & Nuclear physics] The change of one nuclide into another due to natural radioactivity or by artificial means.

transformation constant *See* disintegration constant.

transformer [Electronics] A device by which an alternating current of one voltage is changed to another voltage, without alteration in frequency. A step-up transformer, which increases the voltage and diminishes the current.

transformer oil [Materials] A liquid having the property of insulating the coils of transformers, both electrically and thermally.

transient [Mathematics] A function whose value tends to zero as the independent variable tends to infinity. [Physics] A pulse, damped oscillation, or other short-level phenomenon in a system caused by a sudden change of voltage, current, or load.

transient motion [Physics] An oscillatory or other irregular motion occurring while a quantity is changing to a new steady-state value.

transistor [Electronics] A semiconductor device capable of amplification in a similar manner to thermionic valves. It consists of two *p-n* semiconductor junctions back to back forming either a *p-n-p* or *n-p-n* structure. In a *p-n-p* transistor the thin central n-region is called the base, one *p*-region is called the emitter, the other the collector. In an *n-p-n* transistor the *p*-region is the base. In order to obtain amplification an *n-p-n* transistor is included in a circuit that supplies a positive voltage to the collector (*n*-region) and a negative voltage to the emitter (the other *n*-region). The collector in this type of transistor therefore corresponds to the anode of a thermionic value while the emitter corresponds to the cathode. The base (*p*-region) is also positively biased and is analogous to the control grid. With this arrangement the large number of electrons in the emitter region is attracted to the *p*-layer, which, if it is sufficiently thin, will allow the electrons to pass through it and be attracted into the positive collector.

transit [Astronomy] 1. The passage of a smaller celestial body across the larger one. 2. A celestial body's movement across the meridian of a place.

transition [Nucleonics & Nuclearn physics] A change in the configuration of an atomic nucleus. It may involve a transformation (*e.g.* by alpha-or beta-particle emission) or a change in energy level by the emission of gamma ray.

transition elements [Chemistry] The elements in the periodic table in which filling of electrons in an inner *d*-or *f*-level occurs. With increasing proton number, electrons fill atomic levels upto argon, which has the electronic configuration, $1s^2 \, 2s^2 \, 2p^6 \, 3s^2 \, 3p^6$. In this shell, there are five-*d*-orbitals, which can each contain 2 electrons. However, at this point the subshell of lowest energy is not the 3*d* but the 4*s*. The next two elements, potassium and calcium, have the configurations $[Ar]4s^1$ and $[Ar]4s^2$ respectively. For the next element, scandium, the 3*d* level is of lower energy than the 4*p* level, and scandium has the configuration $[Ar]3d^14s^2$. This filling of the inner *d*-level continues up to zinc $[Ar]3d^{10}4s^2$, giving the first transition series. There is a further series of this type in the next period of the table: between yttrium ($[Kr]4d^1s^2$) and cadmium ($[Kr]4d^{10}5s^2$). This is the second transition series. In the next period of the table the situation is rather more complicated. Lanthanum has the configuration $[Xe]5d^16s^2$. The level of lowest energy then becomes the 4*f* level and the next element, cerium, has the configuration $[Xe]4f^15d^16s^2$. There are 7 of these *f*-orbitals, each of which can contain 2 electrons, and filling of the *f*-levels continues up to lutetium ($[Xe]4f^{14}5d^16s^2$). Then the filling of the 5*d* levels continues from hafnium to mercury. The series of 14 elements from cerium to lutetium is a 'series within a series', sometimes called an *inner transition series*. This one is the lanthanide series. In the next period there is similar inner transition series, the actinide series, from thorium to lawrencium. Then filling of the d- level continues from element 104 onwards.

transition point [Crystallography] The temperature at which one crystalline form of a substance changes to another form.[Physics] 1. The temperature at which a substance changes phase. 2. The temperature at which a substance becomes superconducting. 3. The temperature at which some other change, such as a change of magnetic properties, takes place. Also known as transition temperature.

transition state [Chemistry] An activated form of a molecule (or atom) in which it is capable of undergoing a chemical change.

transition zone [Geology] A region within the upper mentle of the earth bordering the lower mentle at a depth of 410-1000 Km, characterized by a rapid increase in density of about 20%, and an increase in siesmic-wave velocities.

translation [Biochemistry] The process in which the genetic information present in an *m*RNA molecule directs the sequence of amino-acids during protein synthesis.

translatory motion [Mechanics] A motion that involves a nonreciprocating movement of matter from one place to another.

translocation [Botany] Movement of water, mineral salts, and organic substances from one part of a plant to another.

translucent [Optics] Permitting the passage of light in such a way that an object cannot be seen clearly through the substance; *e.g.*, frosted glass.

transmission coefficient [Optics] T. When a beam of light (or other electromagnetic radiation) passes through a medium the radiation is absorbed to a greater or lesser extent (depending upon the medium and the wavelength of the radiation) and the intensity of the beam decreases. The ratio of the intensity after passing through unit distance of the medium to the original intensity is called the transmission coefficient. Also known as transmittance.

transmission electron microscope [Engineering] An electron-optical microscope that utilizes an assembly of magnetic lenses and a beam of high energy electrons that are transmitted through thin specimen. The main advantage of this microscope is its high resolution which results from the very small wavelengths of electrons.

transmittance *See* transmission coefficient.

transmitter [Electronics] The equipment required to broadcast electromagnetic radiation of radio frequencies. The transmitter consists of devices for producing the carrier wave, modulating it , and feeding it to the aerial system.

transmutation [Nucleonics & Nuclearn physics] A nuclear process in which one chemical element changes into another. Artificial transmutation by suitable nuclear reactions forms the basis or experimental nuclear physics.

transparent [optics] Permitting the passage of light in such a way that objects can be seen clearly through the substance.

transpiration [Biology] The passage of water from the roots of the plants to the atmosphere via the vascular system and leaves.

transplantation [Biology] The artificial removal of a part of an organism and its replacement in the body of the same or of a different individual. [Biology] To remove a plant from one location and replant it in another place.

transponder [Electronics] Electronic equipment designed to receive a specific signal and automatically transmit a reply.

transport number *See* Transference number.

transport overhead bits [Computers] The part of a digital transmission that is appended to the actual data for support function such as parity checking and **frame** identification.

transposition [Biochemistry] The movement of a gene or set of genes from one site in the genome to another.

transectification [Electricity] Rectification that occurs in one circuit when an alternating voltage is applied to another circuit.

transuranic elements [Chemistry] Elements beyond uranium in the periokic table; *i.e.*, elements of atomic number greater than 92. Such elements do not occur in nature, but may be obtained by suitable nuclear reactions ; they are all radioactive and members of the actinide group.

transverse [Mathematics] Crosswire ; in a direction at right angles to length of the body under consideration.

transverse wave [Physics] A wave in which the vibration or displacement takes places in a plane at right angles to the direction of propagation of the wave; *e.g.*, electromagnetic radiation.

travelling wave [Physics] A wave in which energy is transported from one part of the medium to another, in contrast with a standing wave.

triangle of forces [Mechanics] If three forces acting at the same point can be represented in magnitude and direction by the sides of a triangle taken in order, they will be in equilibrium.

triangle of velocities [Mechanics] If

a body has three component velocities that can be represented in magnitude and direction by the sides of a triangle taken in order, the body will remain at rest.

triangular pulse [Electronics] An electrical pulse in which the voltage rises linearly to some value, and immediately reduces linearly to the original value.

triatomic [Chemistry] Having three atoms in the molecule, *e.g.*, ozone, O_3; water, H_2O.

tribasic acid [Chemistry] An acid having three ionizable hydrogen atoms of acidic nature in the molecule, thus giving rise to three possible series of salts; *e.g.*, orthosphoric acid, H_3PO_4 can give rise to trisodium orthophosphate, Na_3PO_4, disodium hydrogen orthophosphate, Na_2HPO_4, and soidum dihydrogen orthophosphate, NaH_2PO_4.

tribo- [Physics] A prefix meaning to or resulting from friction.

triboelectric series [Electricity] A list of materials which produces an electrostatic charge when rubbed together; arranged in an order such that a material has a positive charge when rubbed with a material below it in the series, and has a negative charge when rubbed with a material above it in the series.

tribology [Mechanics] The study of friction and lubrication.

triboluminescence [Physics] The emission of light when certain crystals (*e.g.*, cane-sugar) are crushed.

trigonometrical ratios [Mathematics] If a perpendicular *AB* is drawn from any point on arm *OA* of an angle *AOB* to the other arm, the following ratios are constant for the particular angle : *AB/OB*, sin (sin *AOB*); *OB/AO*, cosine (cos *AOB*); *AB/OB*, tangent (tan *AOB*); *AO/AB*, cosecant (cos *AOB*); *AO/OB*, secant (sec *AOB*); and *OB/AB*, cotangent (cot *AOB*).

trigonometry [Mathematics] A branch of mathematics using the fact that numerous problems may be solved by the calculation of unknown parts (*i.e.* sides and angles) of a triangle when three parts are known. The solution of such problems is greatly assisted by the use of the trigonometrical ratios.

trihydric [Chemistry] Containing three hydroxyl groups in the molecule, *e.g.*, glycerol.

trillion [Scientific Technology] 10^{18}, a million million million (British); 10^{12}, a million million (American).

trimer [Chemistry] A substance composed of molecules that are formed from three molecules of a monomer, *e.g.*, trioxane; tripropylene.

triode [Electronics] A thermionic valve containing three electrodes; an anode or plate, a cathode, and a control grid.

triol [Chemistry] Any of the trihydric alcohols derived from aliphatic hydrocarbons by the substitution of hydroxyl groups for three of the hydrogen atoms in the molecule, *e.g.*, glycerol.

triose [Biochemistry] A sugar containing three carbon atoms in the molecule.

triple bond [Chemistry] Three covalent bonds linking two atoms in a chemical compound, *e.g.*, ethyne, $HC \equiv CH$

triple point [Physics] The temperature and pressure at which the vapour, liquid, and solid phases of a substance are in equilibrium. For water the triple point occurs at 273.16K and 611.2 Pa. This value forms the basis of the definition of the kelvin and the thermodynamic temperature scale.

triplet state [Physics] Electronic state of an atom or molecule whose total spin angular momentum quantum number is equal to one.

tris [Chemistry] A prefix indicating that a certain chemical grouping occurs three times in a molecule.

trisaccharide [Biochemistry] Any of a group of sugars the molecules of which consist of three monosaccharides.

trisistor [Electronics] Fast-switching semiconductor consiting of an alloyed junction *p-n-p* device in which the collector is capable of electron injection into the base; characteristics resemble those of a thyratron electron tube, and

switching time is in nånosecond range.

tritium [Nucleonics & Nuclear physics] T ^3_1H. A radioactive isotope of hydrogen with mass number 3 and atomic mass 3.016. The abundance of tritium in natural hydrogen is only one atom in 10^{17}, and its half life is 12.5 years.

triton [Nucleonics & Nuclear physics] The nucleus of a tritium atom; consisting of a proton and two neutrons.

trivalent [Chemistry] Tervalent, having a valence of three.

tRNA *See* transfer RNA.

trochoid [Mathmatic] A curve formed by a point on the radius of a circle as the circle rolls along a straight line. If the point is on the circumference of the circle the curve is a cycloid.

trochotron [Electronics] A multielectrode thermionic valve used as a scaler.

tropical month [Astonomy] The average period of revolution of the moon about the earth with respect to vernal equinox, a period of about 27 days 7 hours 43 minutes 4.7 seconds.

tropical year [Astronomy] The average period of one revolution of the earth about the sun measured between successive vernal equinoxet, it is equal to 365.2422 mean solar days.

tropic hormone [Biochemistry] A peptide hormone that stimulates its target gland to secrete its hormone, *e.g.*, thyrotropin of the pituitary stimulates secretion of thyroxine by the thyroid. Also known as tropin.

tropic of cancer [Astronomy] A parallel circle on the earth, latitude 23° 45' north of equator.

tropin *See* tropic hormone.

tropism [Biology] Orientation movement of a sessile organism in response to a stimulus.

tropopause [Geology] The boundary between the troposphere and the stratosphere.

troposphere [Geology] The lower part of the earht's atmosphere, *i.e.*, the portion of the atmosphere next to earth's surface in which temperature decreases with height, except for local areas of temperature inversion.

Trouton's rule [Chemistry] The ratio of the molar latent heat of vapourization to the boiling point in kelvin is a constant for all substances. The rule is only an approximation.

trypsin [Biochemistry] An enzyme produced by the pancreas. In the process of digestion it breaks up proteins into amino acids.

tumor [Medicine] Any abnormal mass of cells resulting from excessive cellular multiplication.

tungsten [Chemistry] W. Wolfram. Element. At. No. 74. At.Wt. 183.85; m.p. 3683K; b.p. 5933K; A transition element placed in **groups VIB** of periodic table. A

grey hard ductile malleable metal that is resistant to corrosion, used in alloys, in cemented carbides for hard tools and for electric lamp filaments.

tuning [Computers] Making changes in a program to improve its performance without altering its results.

tuning fork [Engineering] A two-pronged metal fork that when struck, produces a pure tone of constant specified pitch; used in acoustics and for tuning musical instruments.

tunnel diode [Electronics] A heavily doped junction diode that has negative resistance over a part of its operating range. It consists of a *p-n* semiconductor junction in which both the *p-*and n-regions contain very large numbers of impurity atoms, thus producing a high potential barrier at the junction. If a small voltage is applied to the device, positive at the p-region, an electron current will flow (despite the high potential barrier) as a result of the tunnel effect. After a certain voltage has been reached this effect is reduced and the current declines with increasing voltage, thus exhibiting the negative resistance characteristic. At higher voltages the normal majority carrier current flows and the current again increases with voltage; used in switching circuits and where low noise amplification is required up to frequencies of about 1000 megahertz.

tunnel effect [Physcis] An effect in which electrons are able to tunnel through a narrow potential barrier that would constitute a forbidden region if the electrons were treated as classical particles. That there is a finite probability of an electron tunnelling from one classically allowed region to another arises as a consequence of quantum mechanics.

turbine [Engineering] Any motor in which a shaft is steadily rotated by the impact or reaction of a current of steam, air, water, or other fluid upon blades of a wheel.

turbogenerator [Engineering] A steam turbine coupled to an electric generator for the production of electric power. It is the usual arrangement in a 'conventional' power station.

turbulent flow [Mechanics] The type of fluid flow in which the motion at any point varies rapidly in direction and magnitude.

Turkey-red oil [Materials] A mixture of sulphate esters obtained by treatment of castor oil with sulphuric acid; used in dyeing and printing.

turnover number [Biochemistry] The number of times an enzyme molecule transforms a substrate molecule per mintue under conditions giving maximal activity.

tuyere [Metallurgy] An opening through which a stream of hot air is introduced into a blast furnace or cupola to support combustion.

twinning [Crystallography] A rational intergrowth of two or more single crystals of the same material in a mathematically describable manner so that some lattices are parallel whereas others are in reversed position. The symmetry of the two parts may be reflected about a common plane.

twinning axis [Crystallography] The crystal axis about which one individual of a twin crystal may be rotated, usually by 180°, to bring it into coincidence with the other individual. It cannot be coincident with the axes of twofold, fourfold, or sixfold symmetry.

two-phase current [Electricity] The current passing through two conducting wires at a phase difference of one-quarter cycle (90°) between the current in two wires.

Tyndall effect [Optics] The scattering of light by particles of matter in the path of the light, thus making a visible 'beam', such as is caused by a ray of light illuminating particles of dust floating in the air of a room. The effect causes the appearance of a visible cone of light through the suspended particles. This principle is utilized in the ultramicroscope.

type metal [Metallurgy] An alloy of 75-95% lead, 3-18% antimony, and a little tin and sometimes copper. Owing to the presence of antimony it expands on solidifying and thus gives a sharp cast.

tyvelose [Biochemistry] A dideoxy sugar found in bacterial lipopolysaccharides.

U *See* uranium.

U format [Computers] A record format to which the input control system treats as completely unknown and unpredictable.

ulmin [Materials] A class of amorphous and alkali soluble compounds formed as a result of the decomposition of cellulose and lignite tissue of plants.

ultimate lines [Astronomy] Special spectral lines that can be used to indicate the existence of an element in the sun or other star.

ultracentrifuge [Engineering] A high speed centrifuge; used for the determination of the molecular weights of larger molecules in high polymers and proteins and to settle colloidal particles of larger size.

ultrafiltration [Chemistry] Separation of colloidal or very fine solid particles by filtration through microporous or semipermeable mediums.

ultraforming [Chemistry] A catalytic reforming process used to

increase the octane ratings of petroleum nephtha.

ultra-highfrequency [Physics] U.H.F. Radio frequencies in the 300 to a megahertz,corresponding to wavelengths of 10 cm to 1 meters.

ultramarine [Materials] An artificial form of lapis lazuli, made by heating together clay, sodium sulphate, carbon, and sulphur.

ultramicrobalance [Engineering] A balance for weighing accurately upto 10^{-8} gram.

ultramicroscope [Optics] An instrument, making use of the Tyndall effect for showing the presence of particles that are too small to be seen with the ordinary microscope. A powerful beam of light is brought to a focus in the liquid that is being examined; suspended particles appear às bright specks by scattering the light.

ultraphotowaves [Physics] Rays outside the visible part of the spectrum, includes infrared and ultraviolet rays.

ultrashort waves [Physics] Radio waves shorter than 10 metres in wavelength; corresponding to frequencies above 30 megahertz.

ultrasonic generator [Engineering] A device for the production of pressure waves of ultrasonic frequency.

ultrasonics [Physics] The study of pressure waves that are of the same nature as sound waves, but that have frequenc·

audible limit. Also known as supersonics.

ultrasonic wave [Physics] A sound wave that has frequency more than 20,000 hertz.

ultraviolet lamp [Electronics] A lamp capable of providing a high proportion of ultraviolet radiation, such as various forms of mercury-vapour lamp.

ultraviolet microscope [Optics] A microscope in which the object is illuminated by ultraviolet radiation. Quartz lenses are used and the image is recorded photographically. As ultraviolet radiation is of shorter wavelength than visible light, greater magnification can be obtained than with an optical microscope.

ultraviolet radiation [Physics] Electromagnetic radiation in the wavelength range of approximately 4×10^{-7} to 5×10^{-9} metre; *i.e.*, between visible light waves and X-rays. The longest ultraviolet waves have wavelengths just shorter than those of violet light, the shortest perceptible by the human eye.

umbra [Optics] A region of complete shadow.

uncertainty principle [Physics] It is impossible to determine with accuracy both the position and the momentum of a particle (*e.g.*, an electron) simultaneously. If the position is known with accuracy, the less accurately can be momentum be determined. If the uncertainty in

position is Δx, and the uncertainty in momentum is Δp, then

$$\Delta p . \Delta x \geqslant h/2\pi$$

where h is Planck's constant. The principle arises from the dual, i.e., particle as well as wave nature of matter. Also known as Heisenberg's uncertainty principle; interdeterminancy principle.

under cooling [Metallurgy] Cooling a metal below its transformation temperature without obtaining the transformation.

uniaxial crystal [Optics] A double refracting crystal possessing only one optic axis, *e.g.*, a tetragonal or hexagonal crystal.

unicellular [Biology] Consisting of only one cell (*e.g.*, bacteria, protozoa, etc.).

unified field theory [Physics] A theory that attempts to describe the electromagnetic and gravitational fields in one set of equations. No such satisfactory theory has yet been devised. To achieve complete unification the theory would also have to explain strong and weak interactions.

uniform corrosion [Metallurgy] Corrosion that attacks uniformly over the entire exposed surface.

unilateral conductivity [Electronics] Conductivity only in one direction, as in perfect rectifier.

unimolecular reaction [Chemistry] A chemical reaction involving only one molecular species as the reactant, *e.g.*, $H_2S \rightarrow H_2 + S$.

unit [Science Technology] A quantity or dimension adopted as a standard of measurement.

unitary symmetry [Physics] A method (SU_3) of classifying elementary particles according to their properties in a similar manner to the classification of atomic properties in the periodic table. SU_3 has successfully predicted the existence of particles that have subsequently been detected experimentally, *e.g.*, omegaminus.

unit cell [Crystallography] The group of particles (atoms, ions, or molecules) in a crystal that is repeated in three dimensions in the crystal lattice. *See also* crystal system.

Universal asynchronous receiver-transfer(UART)[Computers] circuit that receives data in an asynchronous serial bit stream and converts it into a byte format.

universal donor [Biology] An individual of 'O' blood group; can donate blood to persons of all blood groups.

universal recipient [Biology] An individual of 'AB' blood group; can receive blood of all types, *i.e.*, A, B, AB, or O.

universe [Astronomy] The total of all the matter, energy, and space that man is capable of experiencing, or whose existence he can deduce or has grounds for postulating. The universe is currently best described in terms of a four-dimensional curved space-time continuum.

unsaturation [Chemistry] A state of a chemical compound in which atomic bonds of an organic compound's chain or ring are not completely satisfied. In such compounds the extra bonds formed usually double or triple bonds are reactive than saturated compounds, *e.g.*, ethene, $H_2C=CH_2$; ethyne, $HC\equiv CH$.

unstable [Chemistry] Easily decomposed. [Science Technology] Capable of undergoing spontaneous change, as in a radioactive nuclide.

unstable equilibrium [Physics] An equilibrium state of a system in which any departure of the system from equilibrium gives rise to forces moving the system further away from the equilibrium.

upper atmosphere [Physics] The upper atmosphere of the earth is usually taken to include its gaseous envelope from 30 kilometres upwards (*i.e.* the part of the atmosphere that is inaccessible to direct observations by balloons). Up to about 100km the composition of the upper atmosphere is similar to that at ground level. Above this height the dissociation of oxygen into atoms is almost complete, and at above 150 km the nitrogen separates out owing to its greater mass so that monoatomic oxygen predominates. There is considerable ionization in the upper atmosphere as a result of solar ultra-violet radiation and x-rays.

upper mantle [Geology] That part of the earth's mantle which lies above a depth of about 1000 km and has a density of about 3.40 gm/cm3.

uranium [Chemistry] U. Naturally occurring radioactive element a member of actinide series, At. No. 92. At.Wt. 238.03; m.p. 1405K; b.p. 4091K; hard white metal. The natural element consists of 99.28% ^{238}U (half-life 4.51×10^9 years) and 0.71% ^{235}U (half-life 7.13×10^8 years). The latter isotope is capable of sustaining a nuclear chain reaction and is of greater importance in nuclear reactors and nuclear weapons.

uranium age [Geology] The age of a mineral as calculated from the number of uranium atoms present originally, now, and when equlibrium is reached with uranium.

uranium decay series [Nucleonics & Nuclear physics] The series of elements produced as successive intermediate products when the uranium undergoes spontaneous natural radioactive disintegration into lead; radium and radon are members of this series.

uranium enrichment [Nucleonics & Nuclear physics] A process in which the ratio of the abundance of the isotope uranium-235 to that of the isotope uranium-238 is increased above that found in natural uranium.

uranium-lead dating [Geology] A group of methods of dating certain rocks that depends on the decay of the radioisotopes uranium-238 to lead-206 (half-life 4.5×10^9 years) or the decay of uranium-235 to lead-207 (half-life 7.1×10^8 years).

uranogrophy [Astronomy] The science of mapping stars and groups of stars.

uranus [Astronomy] A planet, seventh in the order of distance from the sun, possessing five satellites, with its orbit laying between those of saturn and neptune. Mean distance from the sun, 2869.6 million kilometres. Sidereal period ('year') 84 years. Mass approximately 14.52 times that of the earth, diameter 47100 kilometres. The surface temperature, of uranus is about 93K.

urea-formaldehyde resin [Materials] Thermosetting resins with good oil resistant properties, produced by the condensation polymerization of urea and formaldehyde.

urease [Biochemistry] An enzyme capable of splitting urea into ammonia and carbon dioxide.

urethane resin [Materials] Any of a class of polymers chemically related to urethanes, generally made by condensation of isocyanates with polyhydric compounds. They form valuable materials for a number of purposes, including the manufacture of coatings and foam plastics. Also known as polyurethene.

urokinase [Biochemistry] An enzyme present in human urine, which catalyzes the conversion of plasminogen to plasmin.

user interface [Computers] The collection of screen formats, editing tools, commands, and software tools by which a user interacts with a computer.

uviol glass [Materials] A type of glass that is highly transparent to ultraviolet radiation.

V *See* vanadium.

vacancy [Physics] A defect that occurs in a crystal lattice when a site normally occupied by an atom or ion is unoccupied.

vaccination [Medicine] Inoculation of any antigenic material or viral or bacterial organisms to produce immunity in the living being.

vaccine [Medicine] A suspension containing viruses or other microorganisms (either killed or of attenuated virulence) that is introduced into the human system to stimulate the formation of antibodies. In this way immunity (partial or complete) to subsequent infection by this type of microorganism is conferred.

vacuole [Biology] A membrane bound cavity within a cell.

vacuum [Physics] A space in which there are no molecules or atoms

i.e., there is no matter. A perfect vacuum is unobtainable, since every material that surrounds a space has a definite vapour pressure, the term is generally taken to mean a space containing air or other gas at pressure about 10^{-6} mmHg. Ultra-high vacuum, *i.e.*, vacuum in which the pressure does not exceed 10^{-9} mmHg or 10^{-7}Pa) occur naturally at heights of more than 800 kilometres above the earth's surface. By using special techniques pressure of the order of 10^{-13} torr can be achieved in the laboratory.

vacuum distillation [Chemistry] The process of distillation carried out at reduced pressure. The reduction in pressure is accompanied by a depression in the boiling point of the substance to be distilled thus lower temperatures can be employed. This process therefore enables substances to be distilled, which while boiling at normal pressure would decompose.

vacuum deposition [Chemistry] A technique for covering solid surface with a thin layer of a substance which is heated in a vacuum, the atoms escaping from its surface being allowed to condense on the surface to be coated. The coatings obtained range in thickness from 0.01 to as much as 3 mils. A vacuum of the order of 10^{-6} atm. is necessary for this purpose. Also known as vacuum coating.

vacuum pump [Engineering] A device used to produce a low pressure in a closed system. The common type of rotary oil pump can produce pressures down to 10^{-3} Pa, below this pressure a condensation pump is required.

vacuum tube [Electronics] An electron tube evacuated to such an extent that its electrical characteristics are unaffected by the presence of residual gas or vapour.

valence [Chemistry] The combining power of an atom; the number of hydrogen atoms that an atom will combine with or replace.

valence band [Electronics] The range of highest energies in a semiconductor corresponding to states that can be occupied by the valence electrons binding the crystal together. Electrons missing from the valence band give rise to holes.

valence bond [Chemistry] The link holding atoms together in a molecule. In the case of two univalent atoms joined together, a single valence bond holds them together; it is possible for an atom to satisfy two or three valence bonds of another atom, giving rise to a double or triple bond.

valence electrons [Chemistry] Electrons that occupy the outermost energy level of an atom and generally take part in formation of a valence bond.

valence shell [Chemistry] The electrons that form the outermost shell of an atom.

valve [Electronics] An active device in which two or more electrodes are enclosed in an envelope, one of the electrodes acting as primary source of electrons. The electrons are provided by thermionic emission and the device may be either evacuated or gas filled.

vanadium [Chemistry] V. Transition element At. No. 23. At.Wt. 50.942; m.p. 2163K; b.p. 3653K; placed in group V B of periodic table. A very hard white metal; used in making vanadium steels; its compounds, especially V_2O_5 are used as catalyst.

Van Allen radiation belt [Physics] Two belts of charged particles trapped within the earth's magnetic field, which were discovered from the results of artificial satellite and space probe experiments. The inner belt, ranging from 2400 to 5600km above the earth's surface, is believed to consist of secondary charged particles emitted by the earth's atmosphere as a consequence of the impact of cosmic rays. The outer belt lies between 13000 and 19000km above the earth, and it is believed that the particles it contains originate from the sun.

Van de Graff generator [Electricity] A high-voltage electrostatic generator used for accelerating charged particles of atomic magnitudes, *e.g.*, protons, to high energies.

Van der Waal's adsorption [Chemistry] Adsorption in which the cohesion between gas and solid arises from Van der Waal's forces.

Van der Waal's attraction *See* Van der Waal's force.

Van der Waal's equation [Chemistry] An empirical equation of state which takes into account the finite size of the molecules and the attractive forces between them; $(P + a/V^2)$ $(V-b)=RT$ for a mole of a substance in the gaseous and liquid phases where P=pressure, V=volume, T=absolute temperature, R=the gas constant; a/v^2 is a correction for the mutual attraction of the molecules and b is a correction for the actual volume of the molecules themselves. The equation represents the behaviour of ordinary gases more correctly than the perfect gas equation $PV=RT$.

Van der Waal's force [Chemistry] The attractive force existing between atoms or nonpolar molecules of all substances. The force arises as a result of electrons in neighbouring atoms or molecules (*see* atom, structure of) moving in sympathy with one another. This force is responsible for the term a/V^2 in Van der Waals' equation of state. In many substances this force is small as compared with the other inter-atomic attractive and repulsive forces present.

Van't Hoff equation [Chemistry] An equation for the variation with temperature T of the equilibrium constant k of a gaseous reaction in terms of the heat of reaction at constant pressure, ΔH: $d(InK)/dT = \Delta H/RT^2$.

Van't Hoff factor [Chemistry] The ratio of the observed osmotic pressure of a solution to that predicted by Van't Hoff's law.

Van't Hoff's law [Chemistry] The osmotic pressure of a dilute solution is equal to the pressure that the solute would exert in the gaseous state, if it occupied a volume equal to the volume of the solution, at the same temperature.

vapour [Physics] A substance in the gaseous state that can be liquefied by increasing the pressure without altering the temperature. A gas below its critical temperature.

vapour density [Chemistry] A measure of the density of a gas or vapour; usually given relative to oxygen or hydrogen. The latter is the ratio of the mass of a certain volume of the gas to the mass of an equal volume of hydrogen, measured under the same conditions of temperature and pressure. Numerically this ratio is equal to half of the molecular weight of the gas.

vapour pressure [Physics] For a liquid or solid, the pressure of the vapour in equilibrium with the liquid or solid.

varactor [Electronics] A semiconductor device characterized by a voltage-sensitive capacitance that resides in the space charge region at the surface of a semiconductor bounded by an insulating layer. Also known as varactor diode; varicap.

variable [Mathematics] A symbol of term that assumes, or to which may be assigned, different numerical values. An 'independent variable' is a variable in a function that determines the value of other variables. A 'dependent variable' has its value determined by other varibles, e.g., in $y = 6x^2+4$, x is the independent variable and y is the dependent variable. [Computers] A data item in main memory, that can assume any of a set of values.

variance [Chemistry] The number of degrees of freedom that a system can have. [Mathmatic] The square of the mean deviation.

variate [Mathematics] A variable that can have any of a set of values according to specified probabilities.

variation [Mathematics] If a quantity y is some function of another quantity x, i.e., if $y=f(x)$, then, as x varies, y varies in a manner determined by the function. If $f(x)=x$ x a(where a is a constant), then y is said to vary directly as x, or to be directly proportional to x; $y=ax$. If $f(x)=a/x$, y is said to vary inversely as x, or to be inversely proportional to x; $y=a/x$.

variometer [Physics] A variable inductor that usually consist of two coils in series, arranged so that one coil can rotate within the other. It is also used as a means of measuring inductance.

varistor [Electronics] A two-electrode semiconductor device having a voltage-dependent non-linear resistance. It can be formed a p-n junction diode. Its resistance drops as the applied voltage is increased. Also known as voltage-dependent resistor.

vat dye [Materials] Any of a class of insoluble dyes that are applied by first reducing them to leucocompounds, which are soluble in alkalies. The solution is applied to the material, and the insoluble dye is regenerated in the fibres by oxidation. Indigo and many synthetic dyes belong to this class.

V band [Electronics] A band of radio-frequencies ranging from 46-56 gigahertz.

vector [Physics] Any physical quantity that requires a direction to be stated in order to define it completely, e.g., velocity, force and field strength are vector quantities.

vector processor [Computers] A processing unit whose architecture is optimized to perform mathematical operations on sequentially ordered data known as vectors.

vector processing [Computers] A method for carrying out many repetitive mathematical operations with a single computer instruction.

vegetable black [Materials] Carbon made by the destructive distillation or incomplete combustion of vegetable matter such as wood.

vegetable oil [Materials] Oil obtained from the leaves, fruits, or seeds of plants; they consist of esters of fatty acids and glycerol.

velocity [Mecanics] n. The rate of change of position of a body, equal to distance travelled divided by time; it is a vector quantity having direction as well as magnitude.

velocity modulation [Electronics] The modulation of the velocity of a stream of electrons by alternately accelerating and decelerating them.

velocity ratio [Meanics] The ratio of the distance through which the point of application of the applied force moves, to the distance through which the point of application of the resistance moves in the same time. For an 'ideal' machine, which requires no energy to move its component parts, the velocity ratio is equal to the mechanical advantage.

Venetain white [Materials] A mixture of white lead and barium sulphate, $BaSO_4$, in equal parts; used in paints.

Venturi tube [Engineering] A device for measuring the rate of flow of a fluid; it consists of an open ended tube flated at each end, so that the fluid velocity in the narrow central portion is higher than at the flated ends. The fluid velocity can be calculated from the difference in pressure between the centre and the ends.

venus [Astronomy] A planet second in distance from the sun, with its orbit between those of mercury and the earth. Mean distance from the sun, 108.21 million kilometres. Sidereal period ('year'), 224.701 days. Its mass is approximately 0.815 that of the earth and its diameter is 12300 kilometres.

verdigris [Chemistry] A blue-green deposit formed upon copper; it consists of basic copper carbonate of variable composition.

vermicide [Medicine] A substance used to kill intestinal worms.

vermiculite [Materials] A group of low-grade micas that expand and exfoliate on heating to light water-absorbent material; used in the exfoliated form as heat and sound insulating material, and in special (potting) soil.

vermifuge [Medicine] A substance used for expelling intestinal worms.

vermilion [Materials] HgS. A scarlet form of mercuric sulphide, used as a pigment.

vernal equinox [Astronomy] The sun's position on the celestial sphere about 21st March; at this time the sun's path on the ecliptic crosses the celestial equator.

vernier [Engineering] A short, auxilary scale which slides along the main instrument scale to permit accurate fractional readings of the least main division of the main scale.

vernier engine [Engineering] A small rocket motor used to correct the flight path or velocity of a missile or spacecraft. Also known as vernier rocket.

vertex [Mathematics] The point on a geometrical figure farthest from the base. [Astronomy] The point on the celestial sphere towards which, or from which, a star appears to move. [Optics] One of the points where the surface of a lens intersects the optical axis.

very high frequency [Physics] VHF. The band of radio frequencies in the range 30 to 300 megahertz corresponding to wavelengths of 1-10 metres.

very low frequency [Physics] VLF. The band of radio frequencies below 30 kilohertz, corresponding to wavelenghts of 10-100 kilometres.

vibration [Mechanics] A continuing periodic change in displacement with respect to a fixed reference.

vibrational energy [Chemistry] For a diatomic molecule, the difference between the energy of the molecule idealized by setting the rotational energy equal to zero, and that of a further idealized molecule which is obtained by gradually stopping the vibration of nuclei without placing any new constraint on the electronic motions.

vibrational spectrum [Physics] The molecular spectrum resulting from transitions between vibrational levels of a molecule which behaves like a quantum-mechanical harmonic oscillator.

vibration damping [Engineering] The techniques used for converting the mechanical vibrational energy of solids into heat energy.

vibrator [Electronics] A device that produces an alternating current by periodically interrupting or reversing a continuous steady current from a direct current source. [Mechine] An instrument which produces mechanical oscillations.

vibrotron [Electronics] A triode electron tube having an anode that can be moved or vibrated by an externally applied force.

vicinal (vic) [Chemistry] Designating a molecule in which two atoms or groups are linked to adjacent atoms. For example, 1, 2-dichloroethane (CH_2ClCH_2Cl) is a vic dihalide.

video [Electronics] Pertaining to picture signals or to the section of a TV system that carry these signals in modulated or unmodulated form.

video frequency [Electronics] The frequency of any component of the output signal of a TV camera; it may be of any value from 1-10 megahertz.

video signal [Electronics] In TV, the signal containing all of the visual information along with blanking and synchronizing pulses.

vinasse [Materials] The residual liquid obtained after fermentation and distillation of beetroot molasses; used as a source of potassium carbonate.

vinegar [Materials] A liquid containing 3%-6% acetic acid, obtained by the oxidation of ethanol by the action of bacteria on wine, beer, or fermented wort.

virial equation [Physics] A gas law that attempts to account for the behaviour of a real gas. It usually takes the form;

$$pV = RT + Bp + Cp^2 + Dp^3 ...$$

where B, C, D are empirical constants known as the virial coefficients.

virion [Biology] The extra cellular inert phase of a virus.

virology [Biology] The study of viruses and the disease they cause.

virtual cathode [Electronics] The surface located in a space-charge region between the electrodes of a thermionic-valve, at which the electric potential is a mathematical minimum and the potential gradient is zero. It may be considered to behave as if it were the source of electrons.

virtual image [Optics] An optical image from which rays of light only appear to diverge, without actually being focused there.

virtual state [Nucleonics & Nuclear physics] An unstable state of a compound nucleus which has a life time many times longer than the time it takes a nucleon, with the same energy as it has in the virtual state, to cross the nucleus.

virtual work [Mechanics] If a body, acted upon by a system of forces, is imagined to undergo a small displacement, then in general the forces will do work, termed the virtual work of the forces. If the body is in equilibrium, the total virtual work done is zero.

virus [Biology] A large group of infectious agents too small to be seen by an optical microscope but visible with an electron microscope. Viruses are only capable of multiplication within a living cell, each type of virus requiring a specific host cell. Viruses differ from organisms in that they lack metabolism, are unable to utilize oxygen, to synthesize macromolecules, or to grow. They have the ability to mutate; they are also antigenic and thus initiate formation of antibodies. The simplest viruses consist of a single helical strand or ribonucleic acid coated with protein molecules. The active principle of these viruses resides in the RNA as it is only this part of the particle that enters the cell. Viruses are associated with the formation of some tumours and a variety of diseases such as small-pox and common cold.

viscometer [Engineering] An instrument used for the measurement of viscosity of a fluid.

viscose [Materials] A thick brownish

fluid consisting mainly of a solution of cellulose xanthate in dilute sodium hydroxide. It is made from cellulose by the action of sodium hydroxide and carbon disulphide; used for the production of viscose rayon and of cellulose film.

viscosity [Physics] A measure of the resistance to flow that a fluid offers when it is subjected to shear stress. For a newtonian fluid, the force, F, needed to maintain a velocity gradient, dv/dx, between adjacent planes of a fluid of area A is given by: $F = \eta A(dv/dx)$, where η is a constant, called the coefficient of viscosity. In SI units it has the unit pascal second (in the c.g.s. system it was measured in poise). Non-Newtonian fluids, such as clays, do not conform to this simple model. *See also* kinematic viscosity.

viscous [Mecanics] Having high viscosity; a viscous liquid drags in a treacle-like manner.

visible spectrum [Physics] The range of electromagnetic radiations that are visible to human eye.

visual display unit [Automatic data processing] A computer display device whose output is a cathode-ray tube for displaying text or diagrams. It may have an input device consisting of a keyboard or it may be a light pen.

vitamin [Biochemistry] Any number of complex organic compounds, present in natural products or made synthetically, which are essentially required in small amounts in the diet of animals and man. The absence or shortage of vitamins leads to various deficiency diseases. Some vitamins (A, D, K) are fat-soluble, other are water-soluble (B-complex, C). Vitamins are essential for the normal processes of growth and maintainance of the body. Though vitamins do not furnish energy, but are essential for energy transformation and regulation of metabolism.

vitamin A [Biochemistry] The term includes both retinol (previously called preformed vitamin A) and carotene (previously called vitamin A precursor). It is essential for formation of glucoproteins of the tissue by acting as a carrier for the monosacharides involved; thus maintaining normal condition of the moist epithelial tissues lining mouth, respiratory and urinary tract. Vitamin A is essential for growth. It occurs in milk, butter, cheese, liver, cod-liver oil, green vegetables, carrorts, and palm oil. In the body, carotene is oxidised to retinol. Deficiency in vitamin A can result in reduced reistance to disease, particularly those which can enter through the skin, and in night blindness. Daily recommended intake of this vitamin is 750 μg for adult (2,500 I.U.). Vitamin A content of foods expressed as retinol equivalents : 1 μg retinol=6μg beta-carotene=12 μg other active carotenoides =3.3

I.U. retinol=10 I.U. beta-carotene.

vitamin A₂ [Biochemistry] Old name for dehydroretinol, the form found in livers of fresh water fish; has 40% of biological activity of retinol.

vitamin B_c [Biochemistry] Vitamin essential in the synthesis of purines and pyrimidines and certain amino acids. Its deficiency causes megaloblastic anaemia. It occurs in foods as a variety of derivatives of pteroylglutamic acid (PGA). It is found in fresh, dark green vegetables, kidney and liver. The recommended intake is 0.4 mg per day. Also known as folic acid; vitamin M.

vitamin B complex [Biochemistry] A group of more than ten water soluble vitamins, which tend to occur together. They may be obtained from whole grains of cereals, meat and liver, and yeast. Since B vitamins are present in most unprocessed food, deficiency diseases occur only in persons living on restricted diets. Most of B vitamins act as coenzymes, involved in the normal oxidation of carbohydrates during respiration.

The vitamins of B complex group include thiamin (vitamin B₁), riboflavin (vitamin B₂), nicotinic acid (niacin), pentathenic acid (vitamin B₃), pyridoxine (vitamin B₆), cynocobalamin (vitamin B₁₂), biotin, lipoic acid, and folic acid (PGA).

vitamin B_p [Biochemistry] Called the antiperosis factor for chicks, but can be replaced by manganese and choline.

vitamin B_T [Biochemistry] An essential dietary factor for the mealworm and certain related species.

vitamin B_w [Biochemistry] Probably identical with biotin.

viamin B_x [Biochemistry] Obsolete name for p-aminobenzoic acid.

vitamin B₁ [Biochemistry] A member of the vitamin B complex. The deficiency of this vitamin leads to impaired metabolism of carbohydrate. It occurs in cereal grains, yeast, meat, pulses, and egg. It is one of the more labile vitamins and is destroyed by heat under alkaline conditions, by sulphur dioxide, is lost by leaching into the cooking water, and during food processing. It is water-soluble vitamin and there is little storage in the body. The daily requirement of this vitamin is 0.6 mg per 1000 non-fat calories, or 0.4 mg per 1000 total calories. Also known as thiamin.

vitamin B₂ [Biochemistry] $C_{17}H_{20}N_4O_6$. A water-soluble, yellow-orange fluorescent pigment that is essential to human nutrition as a component of the coenzymes flavin mononucleotide and flavin adenine dinuelotide. In combination with a number of different proteins it forms a group of coenzymes called flavoproteins that are essential for the oxidation of carbohydrates. The deficiency of vitamin B$_2$ impairs cell oxidation and results is cracking of the skin at the corners of the mouth fissuring of the lips and tongue changes. This vitamin is lost during food processing and roasting. 50% of this vitamin present in the milk can be destroyed in 2 hours by exposure to bright sunlight. It is present in cereals, egg fruits, liver, cheese, milk, and meat. The recommended intake of vitamin B$_2$ is 0.55 mg per 1000 kcal or an average of 1.5 mg per day.

vitamin B$_3$ [Biochemistry] Name given to substance that was probably pantothenic acid.

vitamin B$_4$ [Biochemistry] Name given to a substance what was later identified as a mixture of arginine, glycine, and cystine.

vitamin B$_5$ [Biochemistry] Name given to a substance later presumed to be identical with vitamin B$_6$ or possibly nicotinic acid.

vitamin B$_6$ [Biochemistry] A vitamin that exists in three chemically related and water-soluble forms; 2-methylpyridine, pyridoxine (previously known as pyridoxol); the aldehyde pyridoxal; and the amine pyridoxamine; all equally active. It is found in green vegetables and liver; deficiency causes convulsions and skin disorder in rats, abnormal red cells in dairy cattle, anaemia in dogs and, nervousness and insomnia in humans. It functions as coenzyme for specific amino

acid decarboxylases and deaminases, trans-aminases, and trans-methylases. It is rarely deficient in human diets; its recommended intake is about 2 mg per day.

vitamin B$_7$ [Biochemistry] When a new factor was discovered that was claimed to be essential for chick growth and feathering the claimant stated that as nine factors were known the new factors should be named vitamin B$_{10}$ and B$_{11}$. In fact the B vitamins had been numbered only upto B$_6$, hence B$_7$, and B$_8$, and B$_9$ have never existed.

vitaman B$_8$ *See* vitamin B$_7$.

vitamin B$_9$ *See* vitamin B$_7$.

vitamin B$_{10}$ [Biochemistry] Name given to a substance which was later found to be a mixture of vitamin B$_{12}$ and folic acid.

vitamin B$_{11}$ *See* vitamin B$_7$.

vitamin B$_{12}$ [Biochemistry] A group of closely related polypyrole compounds containing trivalent Cobalt (Co III) *i.e.,* cynocobalamin, hydroxocobalamin, and nitritocobalamin. It is essential for hemopoiesis. Although a dietary essential, cases of dietary deficiency have been seen in very rare instances only in individuals solely living on fruits and vegetables, since it is found mainly in foods from animal origin. Its richest sources are meat, liver, and kidney; recommended dose in about 5 mg daily.

vitamin B$_{13}$ [Biochemistry] Name given to uracil-4-carboxylic acid;

an intermediate in the biosynthesis of pyrimidines; a growth factor for certain micro-ogranisms; not an established vitamin. Also known as orotic acid.

vitamin B$_{14}$ [Biochemistry] A substance found in human urine which increases the rate of cell-proliferation in bone-marrow culture; not an established vitamin.

vitamin B$_{15}$ [Biochemistry] Name given to di-isopropyl derivative of glucuronic acid; acts as a powerful methylating agent concerned with respiratory enzymes in cells. There is no evidence that it is dietary esential. Also known as pangamic acid.

vitamin C [Biochemistry] C$_6$H$_8$O$_6$. A white, crystalline, water-soluble vitamin found in many plant materials, especially citrus fruits. It controls production of intercellular cementing substances because it is essential for the hydroxylation of proline to hydroxyproline, a step in the synthesis of collagen. It is easily oxidised, especially in foods kept hot, and leached into cooking water. It is used as antioxidant and bread improver; recommended intake is 30-45 mg per day. Also known as ascorbic acid.

vitamin D [Biochemistry] A vitamin formed in the skin from its precursors under the action of ultraviolet light which converts 7-hydrocholesterol into vitamin D$_3$ or cholecalciferol; also synthesized as vitamin D$_2$ or ergocalceferol by irradiation of ergosterol. The term vitamin D$_1$ was given originally to an impure mixture and is not used now. The main sources of this vitamin are fish-liver oil, butter, milk, cheese, egg-yolk, and liver. Its main action is to increase the absorption of calcium and phosphorus from the intestine and calcium turnover in bones. The deficiency of this vitamin results in inadequate deposition of calcium in the bones, causing rickets in young children. The recommended intakes of this vitamin are 10μg for infants and children and 2.5 μg for adults.

vitamin E [Biochemistry] C$_{29}$H$_{50}$O$_2$. Any of the series of eight related compounds called tocopherols, alphotocopherols having the highest biological activity. These compounds are antioxidants with varying potencies and their natural occurances in vegetable oils protects the latter against rancidity ; occurs in wheat germ and other oils and is believed to be needed in certain human physiological processes.

vitamin F [Biochemistry] Collective name for two unsaturated fatty acids, linoleic and arachidonic acid, found in animal tissues. The term, vitamin F is no longer in use, now known as essential fatty acids.

vitamin G [Biochemistry] Obsolete name for vitamin B$_2$, riboflavin.

vitamin H [Biochemistry] $C_{10}H_{16}O_3^-$ N_2S. A crystalline substance, A vitamin of the B complex, widely distributed in nearly all living cells in very small quantities. It apperas to be of importance in the metabolism of carbohydrates, fats and proteins; also known as biotein; coenzyme R.

vitamin K [Biochemistry] Any of the three yellowish oils which are fat-soluble, nonsteroid, and non saponifiable; required to catalyze the synthesis of prothrombin, a blood clotting factor, in the liver. Intestinal micro-organisms are capable of synthesising considerable amount of this vitamin in the intestine and this, together with dietary supply, ensures that the deficiency is unlikely to occur in any but the new born.

vitamin L [Biochemistry] Vitamin L_1 and L_2 are factors in yeast, said to be essential for lactation; they have not become established.

vitamin M [Biochemistry] Name given to folic acid. Also known as vitamin B_C.

vitamin P [Biochemistry] Name formerly given to a group of plant flavanoid substances that affect the strength of the walls of blood capillaries. Now it is considered that the effect is pharmacological and that they are not dietary essentials.

vitamin P-P [Biochemistry] Name to nicotinic acid, vitamin of B complex. The amide, nicotinamide has the same biological function and both are known according to internationally agreed nomenclature as niacin. It functions as a coenzyme in the oxidation of carbohydrates as nicotinamide adenie dinucleotide.

vitamin T [Biochemistry] A factor found in insect cuticle, mould mycelia and yeast fermentation liquor; said to be a mixture of folic acid, vitamin B_{12}, and deoxyribosides and not a new factor.

vitrain [Geology] A brilliant black coal lithotype with vitreous lustre and cubical clevage. Also known as pure coal.

vitreous [Materials] Pertaining to, composed of, or resembling glass.

vitreous state [Physics] A solid state in which the atoms or molecules are not arranged in any regular order, as in a crystal, and which crystallizes after long time. Also known as glassy state.

vitrification [Geology] Formation of glassy or noncrystalline material.

vitrify [Chemistry] To convert into glass or a glassy substance by fusion.

vitriol [Chemistry] Concentrated sulphuric acid, H_2SO_4, oil of vitriol, copper sulphate, $CuSO_4$, $5H_2O$, blue vitriol, ferrous sulphate, $FeSO_4 \cdot 7H_2O$, green vitriol; zinc sulphate, $ZnSO_4 \cdot 7H_2O$, white vitriol.

void [Physics] Empty space of molecular dimensions occurring between closely packed solid particles.

void ratio [Physics] The ratio of the volume of void space to that of the solid material.

volatile [Chemistry] Passing readily into vapour; having a high vapour pressure.

volcanic [Geology] Pertaining to activities, structures, or rock types of a volcano.

volcanic gases [Geology] Volatile matter released during volcanic eruption, that has previously dissolved in magma. It consists of about 90% water vapours, carbon dioxide, sulphur dioxide, hydrogen sulphide, hydrogen chloride, and nitrogen.

volcanic glass [Geology] Natural glass formed by the cooling of molten lava, or one of its liquid fractions, too rapidly to allow crystallization.

volt [Electricity] n. The derived SI unit of electric potential defined as the difference of potential between two points on a conducting wire carrying a constant current of one ampere when the power dissipated between these points is one watt. Also the unit of potential difference and electromotive force. 1 volt=10^8 electromagnetic units. Symbol V (=W/A).

voltage [Electricity] The potential, potential difference, or electro motive force of a supply of electricity, measured in volts.

voltage divider [Electricity] A potential divider, potentiometer, tapped resistor, or adjustable resistor. A resistor or series of resistors connected across a source of voltage (V) and tapped at a point to give a fraction (v) of the total voltage.

voltage doubler [Electronics] A transformerless rectifier electronic circuit that delivers a direct current voltage approximately twice the peak alternating current voltage it feeds on. It usually consists of two rectifiers whose outputs are connected in series.

voltage drop [Electricity] The voltage developed across a component or conductor by the flow of current through the resistance or impedance of that component or conductor.

voltage multiplier [Electronics] A rectifier circuit that produces an output voltage amplitude that is an integral multiple of that of a single rectifier, *i.e.*, the peak value of applied alternating voltage.

voltage regulator [Electricity] A device that maintains the terminal voltage of a generator or other voltage source within required limits despite variations in input voltage.

voltage transformer [Electricity] A transformer whose primary winding is connected in parallel with a circuit in which the voltage is to be measured or controlled.

voltaic cell [Electricity] A primary cell consisting of two dissimilar metal electrodes in a solution that react chemically with one or both electrodes to produce voltage.

voltaic pile [Electricity] The earliest electric battery devised by Volta. It consists of a number of cells joined in series, each consisting of a sheet of zinc and copper separated by a piece of cloth moistened with dilute sulphuric acid.

voltameter [Electricity] An electrolytic cell in which a metal, generally silver or copper, is deposited by electrolysis of a salt of the metal upon the cathode. From the increase in mass of the cathode and a knowledge of the electrochemical equivalent of the metal, the quantity of electricity that has passed through the circuit may be calculated.

voltammeter [Electricity] An instrument which can be used either as a voltmeter or ammeter.

volt-ampere [Electricity] The SI unit of apparent power, defined as the product of the root-mean-square values of voltage and current in an alternating-current circuit.

volt-amper-hour [Electricity] A unit for expressing the integral of apparent power over time; equal to the product of 1 volt-ampere and 1 hour, or 3600 joules.

voltmeter [Electricity] An instrument for measuring the potential difference between two points. In principle, it consists of an arrangement similar to an ammeter with a high resistance in series incorporated in the instrument, the scale being calibrated in volts. When the instrument is connected in parallel between the points at which the potential difference is being measured, very little current flows through it, and a correct reading of the voltage is obtained.

volume control [Electricity] A potentiometer used to vary the loudness of a reproduced sound by varying the audiofrequency signal voltage at the input of the audio amplifier.

volumetric analysis [Chemistry] A group of methods of quantitative chemical analysis involving the measurement of volumes of the reacting substances. The amount of a substance present is determined by finding the volume of a solution of another substance, of known concentration, that is required to react with it. The added volume is measured by adding the reacting solution from a burette; the end point of the reaction is often shown by a suitable indicator.

vulcanite [Materials] A hard insulating material made by the action of rubber on sulphur.

vulcanization [Chemistry] A chemical change resulting from crosslinking of the unsaturated hydrocarbon chain of polyisoprene (rubber) with sulphur, by the application of heat. It increases strength and resilliency of the rubber.

vulcanized fibre [Materials] A laminated plastic obtained from chemically treating layers of 100% rag-content paper dried under high pressure after chemical treatment. It forms a hard and tough material having high mechanical strength.

vulcan power [Materials] A highly explosive mixture consisting of 30% nitroglycerin; 52.5% sodium nitrate, 10.5% charcoal, and 7% sulphur.

Vycor glass [Materials] An almost pure variety of silica glass formed from sodium borosilicate glass without the production problems of fused silica.

W *See* tungsten.

Waeker process [Chemistry] A process for the conversion of ethylene into acetaldehyde utilizing oxygen in presence of palladium chloride and cupric chloride.

wafer [Electronics] A thin semiconductor slice on which microcircuits have been fabricated or which can be cut into individual dice for fabricating single transistors and diodes.

warning odour [Chemistry] A distinctive odour imparted to fuel gases for safety purposes, as they have little or no odour of their own.

water equivalent *See* heat capacity.

water gas [Materials] A fuel gas obtained by the action of steam on glowing hot coke; the gas formed consits of carbon monoxide and hydrogen. The formation of water gas is accompanied by absorption of heat (an endothermic reaction); thus the coke is rapidly cooled and has to be reheated at intervals by blast of hot air, which causes partial combustion and makes the coke incandescent again.

water glass [Materials] A viscous colloidal solution of sodium silicates in water used to make silica gel and as a preservative.

water of crystallization [Chemistry] Water chemically combined in many crystallized substances; can be removed by heating, with the loss of crystalline properties.

water of hydration [Chemistry] The portion of water of crystallization that, in some hydrated salts, is retained more tenaciously than the rest. Thus, cupric, sulphate, $CuSO_4.5H_2O$, when heated to 100°C looses 4 molecules of water of crystallization and becomes $CuSO_4.H_2O$, but the last molecule is retained till the temperature reaches 250°C.

water paint [Materials] A paint in which binder or vehicle is dissolved in water.

water pollution [Biology] Contamination of water by materials such as sewage effluent, chemicals, detergents, and fertilizer runoff.

water repellent [Materials] Chemicals used for treating textiles, paper, wood, or leather to make them resistant to wetting by water; includes various types of resins, aluminium and zirconium acetates.

water softening [Chemistry] The removal of the causes of hardness of water (*see* hard water). It generally depends on the precipitation or removal from solution of the metals the salts of which cause hardness.

water vapour [Physics] Water in the gaseous or vapour state; it is present in the atmosphere in varying amounts.

watt [Electricity] W. The derived SI unit of power, equal to one joule per second. The energy expended per second by an unvarying electric current of 1 ampere flowing through a conductor the ends of which are maintained at a potential difference of 1 volt.

wattage [Electricity] Power measured in watts.

watt-hour [Electricity] A unit of energy used in electrical measurements; equal to energy consumed at a rate of 1 watt during a period of 1 hour, or 3600 joules.

wattmeter [Engineering] An instrument for the direct measurement of power in watts, of an electrical circuit.

watt-second [Electricity] A unit of work or energy equivalent to one joule.

wave [Physics] A periodic disturbance in a medium or in space that involves the elastic displacement of material particles or a periodic change in some physical quantity, such as temperature, pressure, electric potential, electromagnetic field strength, etc.

wave equation [Physics] Any of several equations which relate the spatial and time dependence of a function characterizing some physical entity which can propagate as a wave, including quantum-wave equations for particles. For example, the equation of wave mechanics that gives mathematical expression to wave motion is :

$$\nabla^2 \psi = 1/c^2 \, \delta\psi / \, dt^2$$

where ∇^2 is the Laplace operator, ψ is the wave function, c is the velocity of light, and t is the time at any instant. The physical significance of the wave function is that the square of its absolute value, $|\psi|^2$, at a point is proportional to the probability of finding the particle in a small element of volume, $dx.dy.dz,$ at that point. For an electron in an atom, this gives rise to the idea of atomic and molecular orbitals.

wave form [Physics] The shape of a wave, illustrated graphically by plotting the values of the periodic quantity against time.

wave front [Physics] The locus of adjacent points in the path of a wave motion that possess the same phase.

wave guide [Electronics] A hollow metal conductor through which microwaves may be propagated; used to guide ultrahigh frequency electromagnetic waves propagated along its length.

wave intensity [Physics] The average amount of energy transported by a wave in the direction of wave propagation, per unit area per unit time.

wavelength [Physics] λ. The distance between successive points of equal phase of a wave. The wavelength is equal to the velocity of the wave motion divided by its frequency. For electro-magnetic radiation, $\lambda = c/f$, where c is the velocity of light and f is frequency.

wave intensity [Physics] The average amount of energy transported by a wave in the direction of wave propagation, per unit area per unit time.

wave mechanics [Physics] The version of nonrelativistic quantum mechanics in which a system is characterized by a wave function which is a function of the coordinates of all the particles of the system and time, and follows a equation. Wave mechanics is based on Schrodinger's wave equation relating the energy of a system to its wave function, only certain values for which are allowed.

wavemeter [Engineering] A device for measuring the wavelength of a radio frequency electromagnetic radiation.

wave motion [Physics] The process by which a disturbance at one point is propagated to another point more remote from the wave-source, with no net transport of the material of the medium itself. The wave motion moves forward a distance equal to its wavelength in the time taken for the displacement at any point to undergo a complete cycle about its mean position.

wave number [Physics] The number of waves in unit length. It is the reciprocal of wavelength.

wave period [Phyics] The time between the attainment of successive maxima, at a fixed point, of a quantity characterizing a wave.

wave plate [Optics] A plate of material which is linearly birefringent.

wave propagation See wave motion.

wave theory of light [Optics] Theory which assumes that light is propagated as a wave motion formerly the existence of a medium, the ether, was postulated for the transmission of light waves. This hypothesis has been rejected as unnecessary, and the classical wave theory has been modified to include the dual particle (photon) wave concept, which is required to explain all the observed phenomena.

wave train [Physics] A series of waves, particularly a small group of waves of limited duration produced by the same disturbance.

wax [Materials] Any of a group of substances that are simple lipids consisting of esters of higher fatty acids than are found in fats and oils, with monohydric alcohols. The term is often loosely applied to solid, non-greasy, insoluble substances that soften or melt at fairly low temperatures, *e.g.*, paraffin wax.

weak acid [Chemistry] An acid, such as acetic acid or carbonic acid, that is only partly dissociated in aqueous solution.

weak interaction [Physics] An interaction between elementary particles that is some 10^{12} times weaker than strong interactions. Beta decay is a form of weak interaction. It is thought that such interactions are the result of an exchange of virtual particles (*see* virtual state) called intermediate vector bosons.

W band [Physics] A band of microwave frequencies ranging from 56 to 100 gigahertz.

weathering [Geology] The destructive processes by which rocks are changed on exposure to atmospheric agents at or near the earth's surface, with little or no transport of the loosened or altered material.

weber [Physics] Wb. The derived SI unit of magnetic flux defined as the flux that, linking a circuit of one turn, produces in it an E.M.F. of one volt as it reduces to zero at a uniform rate in one second.

weight [Mechanics] The force of attraction of the earth on a given mass is the weight of that mass. Being a force, weight is correctly measured in units of force, such as the newton. The weight of a mass m, being equal to mg, where g is the acceleration of free fall. Thus the weight of a body depends on its geographical position because of the variation in the value of g.

weightlessness [Mecanics] A condition in which no acceleration, whether of gravity or other force, can be detected by an observer within the system under consideration. Also known as zero gravity.

Weiss magneton [Physics] A unit of magnetic moment, equal to 1.853×10^{21} erg/oersted, about one-fifth of the Bohr magneton.

welding [Metallurgy] Joining of two metal surfaces by raising their temperature sufficiently to melt and fuse them, with or without filler metal.

Weston cell [Electricity] A primary cell used as a standard of E.M.F. It produces 1.018636 volts at 293K. It consists of a mercury anode covered with mercurous sulphate and a cadmium amalgam cathode coated with cadmium sulphate crystals. The electrolyte is a saturated solution of cadmium sulphate. Also known as cadmium cell.

wet cell [Electricity] A primary cell in which there is substantial amount of free electrolyte in liquid or solution form.

wet collector *See* scrubber.

wet contact [Electricity] Contact through which direct current flows.

wet gas [Geology] A natural gas containing liquid hydrocarbons.

wetting [Electronics] The coating of a contact surface with an adherent film of mercury.

wetting agent [Materials] A substance that lowers the surface tension of a liquid.

whale oil [Materials] Animal fat obtained from the fatty layer of blubber of true whales. After extraction it is divided into various fractions and used for soap manufacture and other purposes; on hydrogenation a hard tasteless edible fat is obtained.

Wheatstone bridge [Electricity] A four armed electrical circuit used for the measurement of resistances by comparing it with a known standard resistance.

white dwarf [Astronomy] A class of small, highly dense stars of low luminosity. They are the remnants of stars that have consumed nearly all their available hydrogen. Owing to their small size they have high surface temperatures and therefore appear white. *See* supernovae.

white light [Optics] Light can be resolved into a continuous spectrum of wavelengths (*i.e.*, colours); *e.g.* the light from an incandescent 'white-hot' solid.

white spirit [Materials] A mixture mainly of alkanes of boiling range 423-473 K; used as a solvent and in the paint and varnish industry.

wide-angle lens [Optics] A camera lens with a wide angle of view generally greater than 80° and a short focal length.

width control [Electronics] Control that adjusts the width of the pattern on the screen of a cathode-ray tube in a TV or oscilloscope.

Wiedemann effect [Electricity] The twist produced in a current-carrying wire when placed in a longitudinal magnetic field.

Wiedemann-Franz law [Electronics] The law that the ratio of the thermal conductivity of a metal to its electrical conductivity is a constant, independent of the metal, times the absolute temperature.

Wien's displacement law [Physics] For a black body, $\lambda m T$ =constant, where λm is the wavelength corresponding to the maximum radiation of energy and T is the thermodynamic temperature of the body. Thus as the temperature rises the maximum of the spectral energy distribution curve is displaced towards the short-wavelength end of the spectrum.

Wignger effect [Physics] The effect produced when the atoms in a crystal are displaced as a result of irradiation. If graphite, for example, is bombarded with neutrons, the shape of the crystal lattice is altered and the material suffers a change of physical dimensions.

Wigner energy [Physics] Energy stored within a crystalline substance as a result of the Winger effect. In a nuclear reactor in which graphite is used as the moderator, some of the energy lost by the neutrons is stored in the graphite; this is known as the Wigner energy.

Wigner force [Nucleonics] A short range nonexchange force between nucleons, postulated to explain various phenomena.

Wigner nuclides [Nucleonics] An important class of mirror nuclides, comprising pairs of odd-mass-number isobars for which the atomic number and the neutron number differ by 1, *e.g.*, $_1H^3$ and $_2He^3$.

Winchester formatter [Computers] A circuit that organizes data for storage on a hard disk.

window [Electronics] A wavelength band to which a particular medium is transparent. The atomosphere, for example, has a

radio window in the range 8mm-20 m.

wiping effect [Metallurgy] Activation of a metallic surface by mechanical rubbing or wiping to enhance the formation of a conversion coating.

wolframite [Minerals] (Fe, Mn) WO_4. A mineral, occurring in monoclinic crystals; a principal ore of tungsten.

wood's metal [Metallurgy] An alloy of 50% bismuth, 25% lead, 12.5% tin, 12.5% cadmium; m.p. 341K; used for automatic sprinkler plugs.

woofer [Engineering] A loudspearker designed to reproduce the lower audio-frequency sounds.

word [Computers] The smallest number of bits of information that a particular computer can conveniently process as a single unit; usually 12 to 64 bits.

work [Mechanics] The work done by a force f when it moves its point of application through a distance d is equal to $fd\cos$ u, where u is the angle between the line of action of the force and the displacement. The derived SI unit of work is the joule.

work function [Physics] The minimum energy required to remove an electron from the Fermi level of a metal to infinity; expressed in electron volts.

work hardening [Metallurgy] An increase in the hardness of metals as a result of working them cold.

It causes a permanent distortion of the crystal structure.

Workload [Computers] The mix of different types of program typically run at a given worksite; major characteristics include *I/O* requirements, and kinds of computation, and degree of vectorization possible.

wrought alloy [Metallurgy] An alloy that has been mechanically worked after easting.

wrought iron [Metallurgy] The commercial form of iron nearly free from carbon but containing iron silicate embedded in a ferrite matrix. It is very tough and fibrous and can be welded.

W stars [Astronomy] Stars of W spectral class; their spectra shows an abundance of highly ionized elements such as helium, carbon, nitrogen, and oxygen; they are very hot with surface temperatures of about 50,000 to 100,000 K.

wulfenite [Minerals] $PbMoO_4$. A tetragonal mineral, occurs in tabular crystals and in granular masses; an ore mineral of molybdenum.

Wurtz-Fitting reaction [Chemistry] A chemical reaction in which an aromatic halide reacts with an alkyl halide in the presence of sodium and ether to form alkylated aromatic hydrocarbons.

Wurtz reaction [Chemistry] A method of synthesizing hydrocarbons by treating an alkyl halide with sodium in ether, *e.g.*,

$$2CH_3I + 2Na \rightarrow CH_3\text{-}CH_3 + 2NaI$$

wye [Electronics] Polyphase circuit whose phase differences are 120° and which when drawn resembles the letter '*Y*'.

wye level [Engineering] A levelling instrument having a removal telescope, with attached spirit level, supported in Y-shaped rests, in which it may be rotated about its longitudinal axis, and from which it may be lifted and reversed, end for end, for testing and adjustment.

xanthene dye [Chemistry] A group of dyes whose molecular structure is related to the xanthene; having (C_6H_4) group as the chromosphore group.

xanthine oxidase [Biochemistry] An enzyme found in the tissue of animals; catalyzes the oxidation of certain purines.

xanthophyll [Biochemistry] A class of yellow orange pigments derived from carotene, the commonest being lutein; found in certain flowers, fruits and leaves.

X-band [Physics] A band of radio-frequencies extending from 5200 to 10,900 megahertz, corresponding to wavelengths of 5.77 to 2.77 centimetres.

X-chromosome [Biology] The larger of the two types of sex chromosome carrying many sex-linked genes; occurs in double amount in heterogametic sex.

xenoblast [Geology] A mineral that has grown in a rock during metamorphism without developing its characteristic crystal faces.

xenogamy [Botany] Cross-fertilization between flowers of different plants.

xenograft *See* graft.

xenolith [Geology] A foreign inclusion in an igneous rock.

xenon [Chemistry] Xe. Element. At. No. 54. At.Wt. 131.3; m.p. 153.2K; b.p. 166K; noble gas, placed in zero group of periodic table. It occurs in exceedingly minute amounts in the air (about 0.006 parts per million by volume), and used in filling certain types of thermoionic valves, fluorescent tubes, and light bulbs.

xerography [Engineering] A dry method of photography of photocopying in which an electrostatic image is formed on a surface coated with selenium when it is exposed to an optical image. A dark powder (consisting of graphite and a thermoplastic resin), oppositely charged to the electrostatic image, is dusted on to the surface after exposure so that particles adhere to the charged regions; the image thus formed is then transferred to a sheet of charged paper and fixed by heating. Coloured prints are possible by the use of suitable developing resins. Various materials other than paper can also be printed by this technique.

xeromorphic [Biology] Adapted to withstand dry conditions, *e.g.*, xerophytes.

xerophyte [Botany] A plant with very low water requirements; a desert plant.

X-radiation [Physics] The penetrating electromagnetic radiation of frequencies between that of ultraviolet radiation and gamma rays, and sometimes higher and wavelengths in the range 10^{-5}A 10^3A ; produced when a hard material is bombarded with sufficiently energetic electrons; can also be produced as a result of electronic transitions from higher to lower energy levels within an atom. X-radiation can be reflected, refracted, polarized, also exhibit diffraction and interference. They interact with matter to produce relatively high energy electrons. Also known as roentgen rays; X-rays.

X-ray astronomy [Astronomy] The study of X-ray sources by rockets and balloons in the earth's atmosphere and by satellites beyond it.

X-ray crystallography [Physics] The study of crystalline substances by observation of the diffraction patterns that occur when a beam of X-rays is passed through a crystal. It is principally as a result of the use of X-ray crystallography that the structure of certain proteins (*e.g.*, haemoglobin) and nucleic acids has been analysed.

X-ray diffraction *See* X-ray crystallography.

X-ray fluorescence [Physics] The emission of X-rays from excited atoms produced by the impact of high-energy electrons, other particles, or a primary beam of other X-rays.

X-ray spectrum [Physics] Each element, when bombarded by electrons, emits X-rays of a characteristic frequency, which depends upon the atomic number; a photograph of the line spectrum corresponding to various elements may thus be obtained from the X-rays emitted.

X-ray star [Astronomy] A star that emit X-rays; discovered by instruments carried outside the earth's atomsphere by space probes. The nature of the stars, or their mechanism of X-ray emission, is not fully understood.

X-ray tube [Electricity] A vacuum tube designed to produce X-rays by accelerating electrons to a high velocity by means of an electrostatic field, then suddenly stopping them by collision with a target. It consists of an electron gun and a heavy metal target forming part of a massive anode. The metal emits X-rays when it is bombarded by high-energy electrons. The spectrum of the radiation depends on the voltage between the cathode and the anode, the temperature of the cathode, and the metal of the target.

X unit [Scientific Technology] X.U. Unit of length, 10^{-11} cm; used mainly for expressing X-rays wavelengths.

xylan [Biochemistry] A complex polysaccharide that occurs closely associated with cellulose in plants.

xylem [Botany] The water-conducting tissue in vascular plants; composed of tracheids, vessel members, fibres, and parenchyma.

X-Y plotter [Engineering] A graphical instrument that produces a chart showing the relationship between two varying signals of two quantities. One of the signals causes the pen to move in the direction of x-axis and the other independently causes it to move in the direction of y-axis.

X-Y recorder [Engineering] A recorder that traces on a chart the relation of two variables, neither of which is time.

Y *See* yttrium.

Yagi aerial [Electronics] A sharply directional aerial array from which most aerials used for TV and radioastronomy have been developed. It consists of one or two dipoles, a parallel reflector, and a series of directors in front of the dipole, all so arranged that radiation is focussed on to the dipole.

yaw acceleration [Mechanics] The angular acceleration of an aircraft or missile about its normal or z-axis.

yaw axis [Mechanics] A vertical axis through an aircraft, rocket, or missile; about which the body yaws.

Yb See ytterbium.

Y-chromosome [Biology] The smaller of the two types of chromosome, found only in the heterogametic sex.

year [Astronomy] A unit of time based on the revolution of the earth about the sun. The civil year has an average value of 365.2425 mean solar days; 3 succesive years consisting of 365 days, the fourth or leap year of 366 days. Century years do not count as leap years unless divisible by 400.

yeasts [Botany] A group of unicellular fungi many of which belong to the Ascomycetes. Certain species of the genus *Saccharomyces* are used in the baking and brewing industries.

yellow [Optics] The hue evoked in average observer by radiation in the approximate range 577-597 nanometres, however the same sensation can also be produced in a variety of other ways.

yellow brass [Metallurgy] A brass containing 34-37% zinc, it has excellent fabrication properties and is corrosion resistant; used for structural and decorative purposes.

yield point [Mechine] If a wire or rod of a material, such as steel, is subjected to a slowly increasing tension, the elongation produced is at first proportional to the tension (Hooke's law). If the tension is increased beyond the elastic limit, a point is reached at which a sudden increase in elongation occurs with only a small increase in tension; this is known as yield point.

yield strength [Mechanics] The stress at which a material exhibits a specified deviation from proportionality of stress and strain.

yield stress [Mechanics] The lowest stress at which extension of the tensile test piece increases without increase in load.

yoke [Electronics] A piece of ferromagnetic material; used to connect permanently two or more magnetic cores and thus complete a magnetic circuit without surrounding it by a winding of any kind.

yolk [Biochemistry] The yellow mass present in central portion of an egg; consists of proteins and fats.

Young's modulus [Mechanics] Elastic modulus applied to a stretched wire or to a rod under tension or compression; the ratio of the stress on a cross-section of the wire or rod to the longitudinal strain.

ytterbium [Chemistry] Yb. A rare earth metal of yttrium subgroup. At. No. 70; At. Wt. 173.04; m.p. 1092K, b.p. 1469K; placed in group IIIB of periodic table; used in chemical research lasers, garnet doping, and X-ray tubes.

yttrium [Chemistry] Y. At. No. 39. At. Wt. 88.905; m.p. 1795K, b.p. 3611K; a rare-earth metal, placed in group IIIB of periodic table; used in alloys and nuclear technology and as a metal deoxidiser.

Yukawa force [Physics] The strong, short-range force between nucleons; calculated on the assumption that this force is due to the exchange of a particle of finite mass (Yukawa meson), just as electrostatic forces are interpreted in quantum electrodynamics as being due to the exchange of photons.

Z

Zeeman displacement [Physics] The separation in wave numbers, of adjacent spectral lines in the normal Zeeman effect in a unit magnetic fidd.

Zeeman effect [Physics] A splitting of spectral lines in the radiation emitted by atoms or molecules in a static magnetic field. From the separation of lines, information about the structure of atoms is deduced.

Zeeman energy [Physics] The energy of interaction between an atomic or molecular magnetic moment and an applied magnetic field.

zemotherapy [Medicine] The treatment by raw meat-juice; used for anaemia, meurasthenia, in convalescence.

Zener breakdown [Electronics] Nondestructive breakdown in a semi-conductor, occurring when the electric field across the barrier region becomes high enough to produce a form of field emission that suddenly increases the number of carriers in this region. Also known as Zener effect.

Zener current [Electronics] The current in a semiconductor, consisting of electrons that have escaped from the valence band into the conduction band under the influence of a strong electric field.

Zener diode [Electronics] A special type of p-n junction diode that acts like a rectifier until the applied voltage reaches a value known as Zener voltage or avalanche breakdown voltage; at this point the diode becomes conducting, with the voltage drop across the diode remaining constant independent of current. Also known as avalanche diode.

Zener effect See Zener breakdown.

zenith [Astronomy] The highest point on the celestial sphere directly overhead and directly opposite to nadir; the term also denotes the stretch of sky overhead.

zeolite [Minerals] A large class of white or colourless or sometimes yellow aluminosilicates, both natural synthetic, used as ion exchangers and as adsorbents.

zero bais [Electronics] The condition in which the control grid and cathode of an electron tube are at the same direct-current voltage.

zero gravity *See* weightlessness.

zero order reaction [Chemistry] A reaction for which reaction rate is independent of the concentration of the reactants, *e.g.*, photochemical reactions.

zero point energy [Physcis] The energy possessed by the atoms or molecules of a substance at the absolute zero of temperature.

zeta-potential [Physics] The electrical potential that exists across the interface of all solids and liquids. Also known as electrokinetic potential.

Ziegler catalyst [Chemistry] A catalyst capable of initiating the polymerization of ethylene and propylene at normal temperatures and pressures, *e.g.*, titanium trichloride and aluminium alkyl.

zinc [Chemistry] Zn. Element. At. No. 30. At.Wt. 65.37; m.p. 692.7K; b.p. 10,80K; hard, bluish-white metal; placed in group II B of periodic table; used in alloys, especially brass, and in galvanized iron.

zinc blende [Minerals] ZnS. Natural zinc sulphide; an important ore of zinc.

zinc-copper couple [Metallurgy] Metallic zinc coated with a thin film of copper by immersing zinc in copper sulphate solution. It evolves hydrogen when treated with hot water.

zincite [Minerals] Natural zinc oxide, (Zn, Mn)O. An important ore of zinc.

zircon [Chemistry] $ZrSiO_4$. Trade name of zirconium silicate. A colourless of yellowish insoluble substance, m.p. 2823K; used as a gemstone when transparent and a refractory when coloured.

zirconium [Chemistry] Zr. Element. At. No. 40. At.Wt. 91.22; m.p. 2126K; b.p. 4649K; a rare metal placed in group IV B of periodic table; used in alloys, abrasives, and flame proofing compounds.

zodiac [Astronomy] A band that passes round the celestial sphere, extending 9° on either side of the ecliptic. It includes the apparent paths of the sun, moon, and planets (except Pluto). The band is divided into the twelve *signs of the zodiac,* each 30° wide. These signs indicate the sun's position each month in the year.

zodiacal light [Astronomy] A faint luminous patch seen in the sky, on the western horizon after sunset or on the eastern horizon before sunrise, believed to be due to the scattering of sunlight by meteoric matter revolving round the sun.

zone axis [Crystallography] The line or crystallographic direction through the centre of a crystal which is parallel to the intersection edges of the crystal faces defining the crystal zone.

zone of sphere [Mathematics

portion of the surface of a sphere cut off by two parallel planes. Its area is given by $2prd$, where r is the radius of the sphere and d the distance between the two planes.

zone refining [Metallurgy] A technique to purify metals in which a narrow molten zone is moved slowly along the complete length of the specimen to bring about impurity segregation. It is based on the principle that the solubility of an impurity x in a main component c in the solid state may differ from the solubility of x in c in the liquid state. When a narrow molten zone is made to pass (*e.g.*, by movement of a heater outside a tube containing a long bar of the material) along a bar or impure c, the distribution of x between the solid and liquid material alters so that the impurity x tends to segregate towards one end of the bar, with pure material at the other end.

zone of silence *See* skip zone.

zoology [Biology] The scientific study that deals with the knowledge of animal life including their taxonomy, behaviour and morphology.

zoomar lens [Optics] Trademark for a type of zoom lens used in a TV camera in which the focal length is determined by the cameraman by means of a handle which he moves backward or forward.

zoom lens [Optics] A system of lenses used as cinematic or television camera lens in which two or more parts are moved with respect to each other and whose focal length can be adjusted continously to vary the magnification without loss of focus while the image is kept in the same image plane. Also known as varifocal lens.

zwitterion [Chemistry] An ion that has a positive and negative charge on the same group of atoms. Zwitterions can be formed from compounds that contain both acid groups and basic groups in their molecules. Also known as ampholyte ion; dipolar ion.

zygote [Biology] A fertilized ovum before cleavage; the product formed by the union of two gametes.

zymase [Biochemistry] A complex of enzymes present in yeast that catalyzes glycosis and acts on sugar with the formation of alcohol and carbon dioxide.

zymogen [Biochemistry] The inactive precursor of on enzyme; liberates an active enzyme on reaction with appropriate kinose. Also known as proenzyme.

zymology [Biochemistry] The study of fermentation reactions and reactions and reactions involving enzymes.

zymophore [Biochemistry] The active portion of an enzyme.

zymotachygraph [Engineering] An instrument that measures the gas produced in a fermenting dough and the amount escaping from the dough.